IET MATERIALS, CIRCUITS AND DEVICES SERIES 71

Radio Frequency and Microwave Power Amplifiers

Other volumes in this series:

Volume 2	**Analogue IC Design: The current-mode approach** C. Toumazou, F.J. Lidgey and D.G. Haigh (Editors)
Volume 3	**Analogue–Digital ASICs: Circuit techniques, design tools and applications** R.S. Soin, F. Maloberti and J. France (Editors)
Volume 4	**Algorithmic and Knowledge-Based CAD for VLSI** G.E. Taylor and G. Russell (Editors)
Volume 5	**Switched Currents: An analogue technique for digital technology** C. Toumazou, J.B.C. Hughes and N.C. Battersby (Editors)
Volume 6	**High-Frequency Circuit Engineering** F. Nibler *et al.*
Volume 8	**Low-Power High-Frequency Microelectronics: A unified approach** G. Machado (Editor)
Volume 9	**VLSI Testing: Digital and mixed analogue/digital techniques** S.L. Hurst
Volume 10	**Distributed Feedback Semiconductor Lasers** J.E. Carroll, J.E.A. Whiteaway and R.G.S. Plumb
Volume 11	**Selected Topics in Advanced Solid State and Fibre Optic Sensors** S.M. Vaezi-Nejad (Editor)
Volume 12	**Strained Silicon Heterostructures: Materials and devices** C.K. Maiti, N.B. Chakrabarti and S.K. Ray
Volume 13	**RFIC and MMIC Design and Technology** I.D. Robertson and S. Lucyzyn (Editors)
Volume 14	**Design of High Frequency Integrated Analogue Filters** Y. Sun (Editor)
Volume 15	**Foundations of Digital Signal Processing: Theory, algorithms and hardware design** P. Gaydecki
Volume 16	**Wireless Communications Circuits and Systems** Y. Sun (Editor)
Volume 17	**The Switching Function: Analysis of power electronic circuits** C. Marouchos
Volume 18	**System on Chip: Next generation electronics** B. Al-Hashimi (Editor)
Volume 19	**Test and Diagnosis of Analogue, Mixed-Signal and RF Integrated Circuits: The system on chip approach** Y. Sun (Editor)
Volume 20	**Low Power and Low Voltage Circuit Design with the FGMOS Transistor** E. Rodriguez-Villegas
Volume 21	**Technology Computer Aided Design for Si, SiGe and GaAs Integrated Circuits** C.K. Maiti and G.A. Armstrong
Volume 22	**Nanotechnologies** M. Wautelet *et al.*
Volume 23	**Understandable Electric Circuits** M. Wang
Volume 24	**Fundamentals of Electromagnetic Levitation: Engineering sustainability through efficiency** A.J. Sangster
Volume 25	**Optical MEMS for Chemical Analysis and Biomedicine** H. Jiang (Editor)
Volume 26	**High Speed Data Converters** A.M.A. Ali
Volume 27	**Nano-Scaled Semiconductor Devices** E.A. Gutiérrez-D (Editor)
Volume 29	**Nano-CMOS and Post-CMOS Electronics: Devices and modelling** Saraju P. Mohanty and Ashok Srivastava
Volume 30	**Nano-CMOS and Post-CMOS Electronics: Circuits and design** Saraju P. Mohanty and Ashok Srivastava
Volume 32	**Oscillator Circuits: Frontiers in design, analysis and applications** Y. Nishio (Editor)
Volume 33	**High Frequency MOSFET Gate Drivers** Z. Zhang and Y. Liu
Volume 34	**RF and Microwave Module Level Design and Integration** M. Almalkawi
Volume 35	**Design of Terahertz CMOS Integrated Circuits for High-Speed Wireless Communication** M. Fujishima and S. Amakawa
Volume 38	**System Design with Memristor Technologies** L. Guckert and E.E. Swartzlander Jr.
Volume 39	**Functionality-Enhanced Devices: An alternative to Moore's law** P.-E. Gaillardon (Editor)
Volume 40	**Digitally Enhanced Mixed Signal Systems** C. Jabbour, P. Desgreys and D. Dallett (Editors)
Volume 43	**Negative Group Delay Devices: From concepts to applications** B. Ravelo (Editor)
Volume 47	**Understandable Electric Circuits: Key concepts, 2nd Edition** M. Wang
Volume 58	**Magnetorheological Materials and Their Applications** S. Choi and W. Li (Editors)
Volume 60	**IP Core Protection and Hardware-Assisted Security for Consumer Electronics** A. Sengupta and S. Mohanty
Volume 62	**VLSI and Post-CMOS Devices, Circuits and Modelling** R. Dhiman and R. Chandel (Editors)
Volume 68	**High Quality Liquid Crystal Displays and Smart Devices, vol. 1 and vol. 2** S. Ishihara, S. Kobayashi and Y. Ukai (Editors)
Volume 69	**Fibre Bragg Gratings in Harsh and Space Environments: Principles and applications** B. Aïssa, E I. Haddad, R.V. Kruzelecky, W.R. Jamroz
Volume 70	**Self-Healing Materials: From fundamental concepts to advanced space and electronics applications, 2nd Edition** B. Aïssa, E I. Haddad, R.V. Kruzelecky, W.R. Jamroz

Radio Frequency and Microwave Power Amplifiers

Volume 1: Principles, Device Modeling and Matching Networks

Edited by
Andrei Grebennikov

The Institution of Engineering and Technology

Published by The Institution of Engineering and Technology, London, United Kingdom

The Institution of Engineering and Technology is registered as a Charity in England & Wales (no. 211014) and Scotland (no. SC038698).

First published 2019

The Institution of Engineering and Technology
Michael Faraday House
Six Hills Way, Stevenage
Herts, SG1 2AY, United Kingdom

www.theiet.org

British Library Cataloguing in Publication Data
A catalogue record for this product is available from the British Library

ISBN 978-1-83953-036-4 (Hardback Volume 1)
ISBN 978-1-83953-037-1 (PDF Volume 1)
ISBN 978-1-83953-038-8 (Hardback Volume 2)
ISBN 978-1-83953-039-5 (PDF Volume 2)
ISBN 978-1-83953-040-1 (Hardback Volumes 1 and 2)

Typeset in India by MPS Limited

Contents

Preface

The main objective of this two-volume edited book is to present by world-class technical experts all relevant information required for RF and microwave power amplifier design including well-known historical and recent novel schematic configurations, theoretical approaches, circuit simulation results, and practical implementation techniques. This comprehensive book can be very useful for lecturing to promote the systematic way of thinking with analytical calculations, circuit simulation, and practical verification, thus making a bridge between theory and practice of RF and microwave engineering. As it often happens, a new result is the well-forgotten old one. Therefore, the demonstration of not only new results based on new technologies or circuit schematics is given, but some sufficiently old ideas or approaches are also introduced and clearly explained that could be very useful in modern design practice or could contribute to appearance of new general architectural ideas and specific circuit and system design techniques. As a result, this unique two-volume comprehensive book is intended for and can be recommended to *university-level professors* as a comprehensive reference material to help in lecturing for graduate and postgraduate students, to *researchers and scientists* to combine the theoretical analysis with practical design and to provide a sufficient basis for innovative ideas and circuit and system design techniques, and to *practicing designers and engineers* as an anthology of many well-known and novel practical circuits, architectures, and theoretical approaches with detailed description of their operational principles and applications.

The book is divided into two volumes. Volume 1 comprises six chapters and Volume 2 comprises ten chapters. Volume 1 begins with introductory Chapter 1 explaining the basic principles of power amplifier design including basic classes of operation, load-line definition, power gain and stability, impedance matching concept and application aspects, push–pull and balanced structures, and transmission-line transformers and combiners. Chapter 2 covers basics of the empirical nonlinear device models implemented in CAD tools focusing on GaN HEMT including its physical phenomena like thermal effects, breakdown, dispersion, and self-heating. Harmonic load-pull tuners are important systems for characterizing power transistors and amplifiers and finding the impedances needed for gaining optimum performance levels. Chapter 3 includes history, techniques, progress, and challenges in power amplifier load-pull characterization using passive and active tuning.

Different matching network design techniques are described in Chapters 4–6 with many practical examples performed using MATLAB® programing software.

Chapter 4 is dedicated to automated Darlington synthesis to construct the lossless matching networks with lumped and distributed elements via correction techniques using low-pass, bandpass, and high-pass network functions. Chapter 5 covers basic "real-frequency" techniques to construct lossless matching networks by assessing the best performance and solving the generalized single and double-matching problems. Chapter 6 describes the design of broadband RF and microwave single-stage and multi-stage power amplifiers based on the "simplified real frequency" techniques using lumped elements, commensurate transmission lines, and mixed lumped and distributed elements.

Modern commercial and military communication systems require high-efficiency long-term operating conditions. In Volume 2, Chapter 1 describes in detail the possible load-network solutions to provide a high-efficiency power amplifier operation based on using Class-F, inverse Class-F, and different Class-E operation modes depending on the technical requirements. In Class-F power amplifiers analyzed in the frequency domain, the fundamental and harmonic load impedances are optimized by short-circuit termination and open-circuit peaking to control the voltage and current waveforms at the drain of the device to obtain maximum efficiency. In Class-E power amplifiers analyzed in the time domain, an efficiency improvement is achieved by realizing the on/off switching operation with special current and voltage waveforms so that high voltage and high current do not exist at the same time.

Chapter 2 describes the basic Doherty approach to the power amplifier design, operational principle, and modern trends in Doherty amplifier design techniques using asymmetric multi-way, multistage, inverted, and broadband architectures with examples of the integrated and monolithic Doherty amplifier implementations. Envelope tracking technology is used in actual smartphone to improve efficiency as well as linearity for RF and microwave power amplifiers for LTE and Wi-Fi communication signals. Chapter 3 presents the envelope-tracking fundamentals as well as the architecture implementation such as fast dc–dc, multilevel supply, and hybrid architectures. Outphasing architectures generate load modulation through phase control of multiple nonlinear PAs, offering the potential for linear amplification with high efficiency over a wide range of output powers. Chapter 4 describes an overview of outphasing history, fundamental principles, modern techniques, and implementation approaches that are making outphasing an attractive option for linear-efficient RF and microwave power amplifiers. Chapter 5 has focused on the importance of the combiner in the design of Doherty and outphasing power amplifiers that plays a detrimental role for the efficiency enhancement in both these architectures since it provides the desired mutual active load modulation between two amplifying branches. Several of the functions that traditionally are part of the combiner realization, such as impedance matching, offset lines, impedance inversion, transistor scaling are absorbed into the synthesized combiner network. This results in a continuum of new outphasing and Doherty solutions that were used to design power amplifiers with higher efficiency, better linearity, greater gain, and smaller size.

It is now well established that power amplifier designers need to control the internal mode of operation of transistors at the current-source reference planes to better optimize the efficiency of power amplifiers. The traditional approach has been to rely on multi-harmonic load and source pulling while monitoring the load lines at the current-source reference planes using a de-embedding model. However, given the tremendously huge search space for the load and source multi-harmonic terminations required to find the desired internal waveforms, it is greatly preferable to use a nonlinear embedding device model described in Chapter 6 to obtain a single simulation, the required multi-harmonic impedances at the package or extrinsic reference planes which implement the desired class of operation. Various examples of design techniques for high-efficiency single-ended power amplifiers, two-way and four-way Chireix and Doherty structures are presented.

Chapter 7 focuses on the basic circuit schematics of the CMOS power amplifiers for different RF and microwave applications including common-source, common-gate, cascode, differential pair, and stacked configuration techniques including power combining. CMOS performance issues such as low breakdown voltage, hot carrier degradation, effect of substrate and device parasitics, and practical integrated circuit implementation features are discussed, as well as efficiency-enhancement techniques for microwave and mm-wave CMOS power amplifiers.

Chapter 8 describes the basic principles of behavioral modeling and analog and digital linearization of power amplifiers used in radio frequency transmitters and presents the analog linearization structures such as feedforward compensation and analog predistortion. Measures and models of the power amplifier nonlinearity are reviewed. Most of the spectrum efficient techniques proposed in modern communication systems such as carrier aggregation require either wideband operation (in-contiguous carrier aggregation) or multiband operation (in the case of non-contiguous operation). Chapter 9 focuses on investigating the practical implementation of spectrum efficient techniques proposed for 4G/5G communication systems and provide software-defined solutions for the power efficient operation of transmitter/receiver system.

Finally, the basic principles of distributed amplification and circuit implementation of microwave GaAs FET distributed amplifiers are introduced and described in Chapter 10. Different architectures such as cascode and cascaded distributed power amplifiers and different techniques based on using tapered lines and extended resonant approach are given, with several examples of monolithic implementation of distributed power amplifiers based on pHEMT, GaN HEMT, and CMOS technologies.

Andrei Grebennikov

List of contributors

Mustafa Acar	NXP Semiconductors, Netherlands
Iltcho Angelov	Chalmers University of Technology, Sweden
Florinel Balteanu	Skyworks Solutions, USA
Taylor Barton	University of Colorado Boulder, USA
Neil Braithwaite	Consultant, USA
Christian Fager	Chalmers University of Technology, Sweden
Paolo de Falco	University of Colorado Boulder, USA
Andrei Grebennikov	Sumitomo Electric Europe Ltd., UK
William Hallberg	Chalmers University of Technology, Sweden
Narendra Kumar	University of Malaya, Malaysia
Mustafa Özen	Ericsson AB, Sweden
Karun Rawat	Indian Institute of Technology Roorkee, India
Meenakshi Rawat	Indian Institute of Technology Roorkee, India
Patrick Roblin	Ohio State University, USA
Mury Thian	Queens University Belfast, UK
Mattias Thorsell	Chalmers University of Technology, Sweden
Christos Tsironis	Focus Microwaves, Canada
Tudor Williams	Mesuro, UK
Siddik Yarman	Istanbul University-Cerrahpasa, Turkey

Chapter 1

Power amplifier design principles

Andrei Grebennikov[1]

This introductory chapter presents the basic principles for understanding the power amplifiers design procedure in principle. Based on the spectral-domain analysis, the concept of a conduction angle is introduced, by which the basic Classes A, AB, B, and C of the power amplifier operation are analyzed and illustrated in a simple and clear form. The frequency-domain analysis is less ambiguous because a relatively complex circuit often can be reduced to one or more sets of immittances at each harmonic component. Classes of operation based upon a finite number of harmonics are discussed and described. The mixed-mode Class-C is introduced and nonlinear effect of collector capacitance is shown and analyzed. The possibility of the maximum power gain for a stable power amplifier is discussed and analytically derived. The design and concept of push–pull and balanced power amplifiers are presented including transmission-line impedance transformers and combiners. In addition, the basics of the load-line concept and impedance matching are discussed and illustrated.

1.1 Basic classes of operation: A, AB, B, and C

As established yet in 1920s, power amplifiers can generally be classified in three classes according to their mode of operation: *linear mode* when its operation is confined to the substantially linear portion of the vacuum-tube characteristic curve; *critical mode* when the anode current ceases to flow, but operation extends beyond the linear portion up to the saturation and cutoff (or pinch-off) regions; and *nonlinear mode* when the anode current ceases to flow during a portion of each cycle, with a duration that depends on the grid bias [1]. When high efficiency is required, power amplifiers of the third class are employed since the presence of harmonics contributes to the attainment of high efficiencies. In order to suppress harmonics of the fundamental frequency to deliver a sinusoidal signal to the load, a parallel resonant circuit can be used in the load network, which bypasses harmonics through a low-impedance path and, by virtue of its resonance to the fundamental, receives energy at that frequency. At the very beginning of 1930s, power amplifiers operating

[1]Sumitomo Electric Europe Ltd., Elstree, Hertfordshire, UK

in first two classes with 100% duty ratio were called the Class-A power amplifiers, whereas the power amplifiers operating in third class with 50% duty ratio were assigned to Class-B power amplifiers [2].

The best way to understand the electrical behavior of a power amplifier and the fastest way to calculate its basic electrical characteristics such as output power, power gain, efficiency, stability, or harmonic suppression is to use a spectral-domain analysis. Generally, such an analysis is based on the determination of the output response of the nonlinear active device when applying the multiharmonic signal to its input port, which analytically can be written as

$$i(t) = f[v(t)] \tag{1.1}$$

where $i(t)$ is the time-varying output current, $v(t)$ is the time-varying input voltage, and $f(v)$ is the nonlinear transfer function of the device. Unlike the spectral-domain analysis, time-domain analysis establishes the relationships between voltage and current in each circuit element in the time domain when a system of equations is obtained applying Kirchhoff's law to the circuit to be analyzed. As a result, such a system will be composed of nonlinear integro-differential equations describing a nonlinear circuit. The solution to this system can be found by applying the numerical-integration methods.

The voltage $v(t)$ in the frequency domain generally represents the multiple-frequency signal at the device input which is written as

$$v(t) = V_0 + \sum_{k=1}^{N} V_k \cos(\omega_k t + \phi_k) \tag{1.2}$$

where V_0 is the constant voltage, V_k is the voltage amplitude, ϕ_k is the phase of the k-order harmonic component ω_k, $k = 1, 2, \ldots, N$, and N is the number of harmonics.

The spectral-domain analysis, based on substituting (1.2) into (1.1) for a particular nonlinear transfer function of the active device, determines the output spectrum as a sum of the fundamental-frequency and higher-order harmonic components, the amplitudes, and phases which will determine the output signal spectrum. Generally, it is a complicated procedure that requires a harmonic-balance technique to numerically calculate an accurate nonlinear circuit response. However, the solution can be found analytically in a simple way when it is necessary to only estimate the basic performance of a power amplifier in terms of the output power and efficiency. In this case, a technique based on a piecewise-linear approximation of the device transfer function can provide a clear insight into the basic behavior of a power amplifier and its operation modes. It can also serve as a good starting point for a final computer-aided design and optimization procedure.

The piecewise-linear approximation of the active device current–voltage transfer characteristic is a result of replacing the actual nonlinear dependence $i = f(v_{in})$, where v_{in} the voltage applied to the device input, by an approximated one that consists of the straight lines tangent to the actual dependence at the specified points. Such a piecewise-linear approximation for the case of two straight lines is shown in Figure 1.1(a).

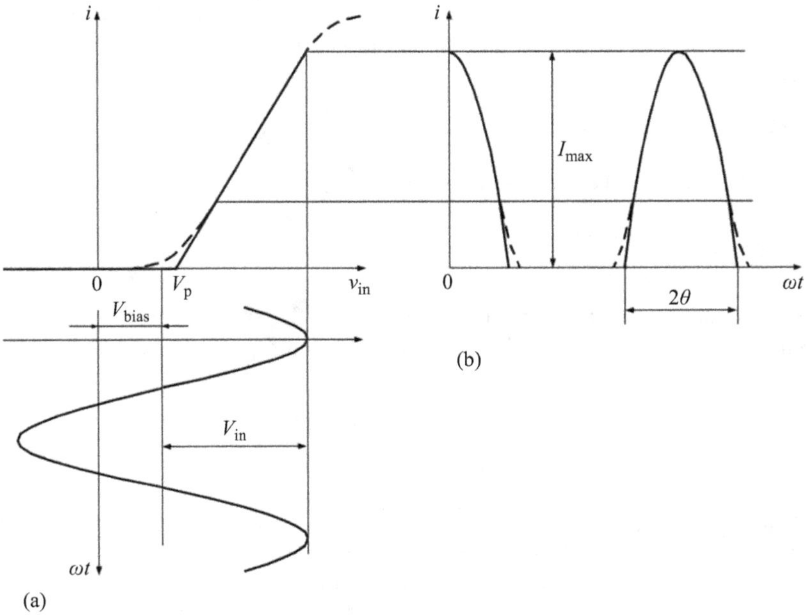

Figure 1.1 Piecewise-linear approximation technique

The output current waveforms for the actual current–voltage dependence (dashed curve) and its piecewise-linear approximation by two straight lines (solid curve) are plotted in Figure 1.1(b). Under large-signal operation mode, the waveforms corresponding to these two dependences are practically the same for the most part, with negligible deviation for small values of the output current close to the pinch-off region of the device operation and significant deviation close to the saturation region of the device operation. However, the latter case results in a significant nonlinear distortion and is used only for high-efficiency operation modes when the active period of the device operation is minimized. Hence, at least two first output current components (dc and fundamental) can be calculated through the Fourier-series expansion with a sufficient accuracy. Therefore, such a piecewise-linear approximation with two straight lines can be effective for a quick estimate of the output power and efficiency of the linear power amplifier.

The piecewise-linear active device current–voltage characteristic is defined as

$$i = \begin{cases} 0 & v_{in} \le V \\ g_m\left(v_{in} - V_p\right) & v_{in} \ge V_p \end{cases} \tag{1.3}$$

where g_m is the device transconductance and V_p is the pinch-off voltage.

Let us assume the input signal to be in a cosine form:

$$v_{in}(\omega t) = V_{bias} + V_{in}\cos\omega t \tag{1.4}$$

where V_{bias} the input dc bias voltage.

At the point on the plot when the voltage $v_{in}(\omega t)$ becomes equal to a pinch-off voltage V_p and where $\omega t = \theta$, the output current $i(\theta)$ takes a zero value. At this moment:

$$V_p = V_{bias} + V_{in} \cos\theta \tag{1.5}$$

and the phase angle θ can be calculated from:

$$\cos\theta = -\frac{V_{bias} - V_p}{V_{in}} \tag{1.6}$$

As a result, by substituting (1.4) into (1.3), the output current represents a periodic pulsed waveform described by the cosine pulses with maximum amplitude I_{max} and width 2θ as

$$i(\omega t) = \begin{cases} I_q + I\cos\omega t & -\theta \le \omega t < \theta \\ 0 & \theta \le \omega t < 2\pi - \theta \end{cases} \tag{1.7}$$

where $I_q = g_m\,(V_{bias} - V_p)$ is the quiescent current, $I = g_m V_{in}$ is the output current amplitude, and the conduction angle 2θ indicates the part of the RF current cycle, during which a device conduction occurs. When the output current $i(\omega t)$ takes a zero value:

$$I_q = -I\cos\theta \tag{1.8}$$

For a piecewise-linear approximation, (1.7) can be rewritten for $i > 0$ by

$$i(\omega t) = g_m V_{in}(\cos\omega t - \cos\theta) \tag{1.9}$$

When $\omega t = 0$, then $i = I_{max}$ and

$$I_{max} = I(1 - \cos\theta) \tag{1.10}$$

The Fourier-series expansion of the even function when $i(\omega t) = i(-\omega t)$ contains only even components of this function and can be written as

$$i(\omega t) = I_0 + I_1\cos\omega t + I_2\cos 2\omega t + \ldots + I_n\cos n\omega t \tag{1.11}$$

where the dc, fundamental-frequency, and nth-harmonic current amplitudes are obtained by

$$I_0 = \frac{1}{2\pi}\int_{-\theta}^{\theta} g_m V_{in}(\cos\omega t - \cos\theta)d\omega t = I\gamma_0(\theta) \tag{1.12}$$

$$I_1 = \frac{1}{\pi}\int_{-\theta}^{\theta} g_m V_{in}(\cos\omega t - \cos\theta)\cos\omega t\, d\omega t = I\gamma_1(\theta) \tag{1.13}$$

and

$$I_n = \frac{1}{\pi}\int_{-\theta}^{\theta} g_m V_{in}(\cos\omega t - \cos\theta)\cos n\omega t\, d\omega t = I\gamma_n(\theta) \tag{1.14}$$

where $\gamma_n(\theta)$ are called the coefficients of expansion of the output-current cosine waveform, or the current coefficients [3,4]. They can be analytically defined for the dc and fundamental-frequency components as

$$\gamma_0(\theta) = \frac{1}{\pi}(\sin\theta - \theta\cos\theta) \tag{1.15}$$

$$\gamma_1(\theta) = \frac{1}{\pi}(\theta - \sin\theta\cos\theta) \tag{1.16}$$

and for the second- and higher-order harmonic components as

$$\gamma_n(\theta) = \frac{1}{\pi}\left[\frac{\sin(n-1)\theta}{n(n-1)} - \frac{\sin(n+1)\theta}{n(n+1)}\right] \tag{1.17}$$

where $n = 2, 3, \ldots$.

The dependences of $\gamma_n(\theta)$ for the dc, fundamental-frequency, second-, and higher-order current components are shown in Figure 1.2. The maximum value of $\gamma_n(\theta)$ is achieved when $\theta = 180°/n$. Special case is $\theta = 90°$, when odd current coefficients are equal to zero, that is, $\gamma_3(\theta) = \gamma_5(\theta) = \ldots = 0$. The ratio between the fundamental-frequency and dc components $\gamma_1(\theta)/\gamma_0(\theta)$ varies from 1 to 2 for any values of the conduction angle, with a minimum value of 1 for $\theta = 180°$ and a maximum value of 2 for $\theta = 0°$, as shown in Figure 1.2(a). Besides, it is necessary to pay attention to the fact that the current coefficient $\gamma_3(\theta)$ becomes negative within the interval of $90° < \theta < 180°$, as shown in Figure 1.2(b). This implies the proper phase changes of the third current harmonic component when its values are negative. Consequently, if the harmonic components with $\gamma_n(\theta) > 0$ achieve positive maximum values at the time moments corresponding to the middle points of the current waveform, the harmonic components with $\gamma_n(\theta) < 0$ can achieve negative maximum values at these same time moments. As a result, the combination of different harmonic components with proper loading will result in flattening of the current or voltage waveforms, thus improving efficiency of the power amplifier.

To analytically determine the operation classes of the power amplifier, consider a simple resistive stage shown in Figure 1.3, where L_{ch} is the ideal RF choke inductor with zero series resistance and infinite reactance at the operating frequency, C_b is the dc-blocking capacitor with infinite value having zero reactance at the operating frequency, and R_L is the load resistor. The dc-supply voltage V_{cc} is applied to both plates of the dc-blocking capacitor, being constant during the entire signal period. The active device behaves as an ideal voltage- or current-controlled current source having zero saturation resistance.

For an input cosine voltage given by (1.4), the operating point must be fixed at the middle point of the linear part of the device transfer characteristic with $V_{in} \leq V_{bias} - V_p$. Normally, to simplify an analysis of the power amplifier operation, the device transfer characteristic is represented by a piecewise-linear approximation. As a result, the output current is cosinusoidal:

$$i = I_q + I\cos\omega t \tag{1.18}$$

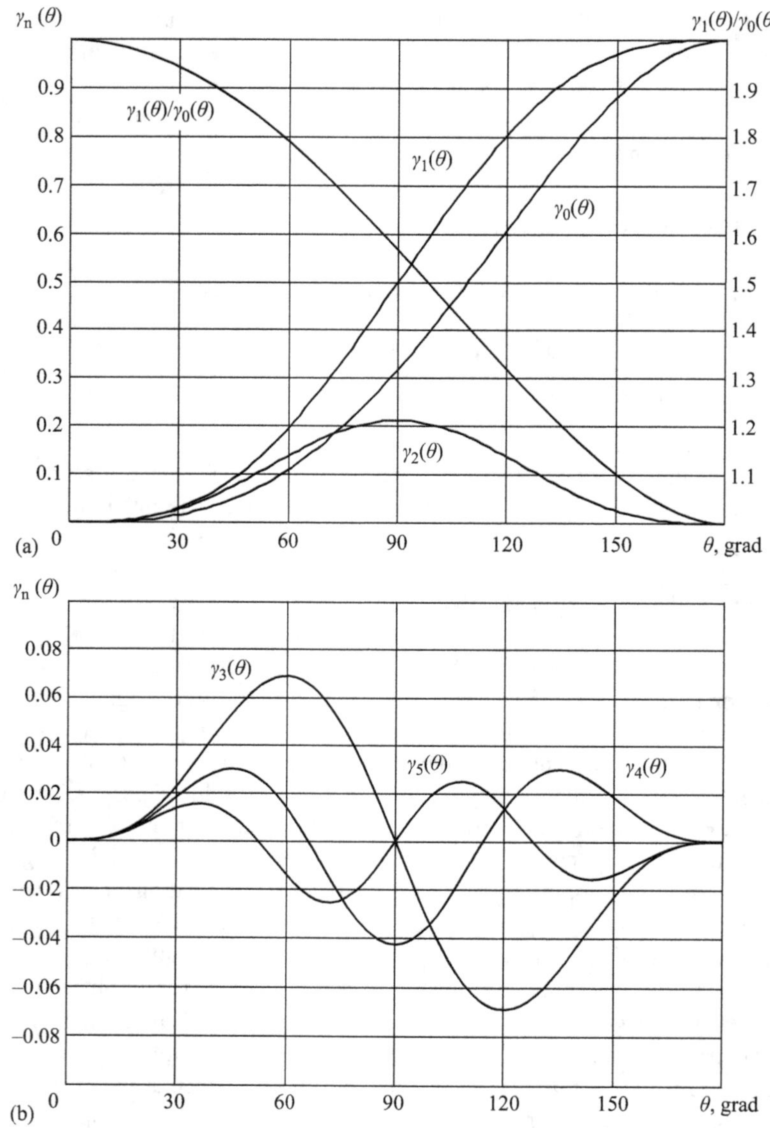

Figure 1.2 Dependences of γ_n (θ) for dc, fundamental, and higher-order current components

with the quiescent current I_q greater or equal to the collector current amplitude I. In this case, the output collector current contains only two components—dc and cosine—and the averaged current amplitude is equal to a quiescent current I_q.

The output voltage v across the device collector represents a sum of the dc supply voltage V_{cc} and cosine voltage v_R across the load resistor R_L. Consequently, the greater output current i, the greater voltage v_R across the load resistor R_L and the

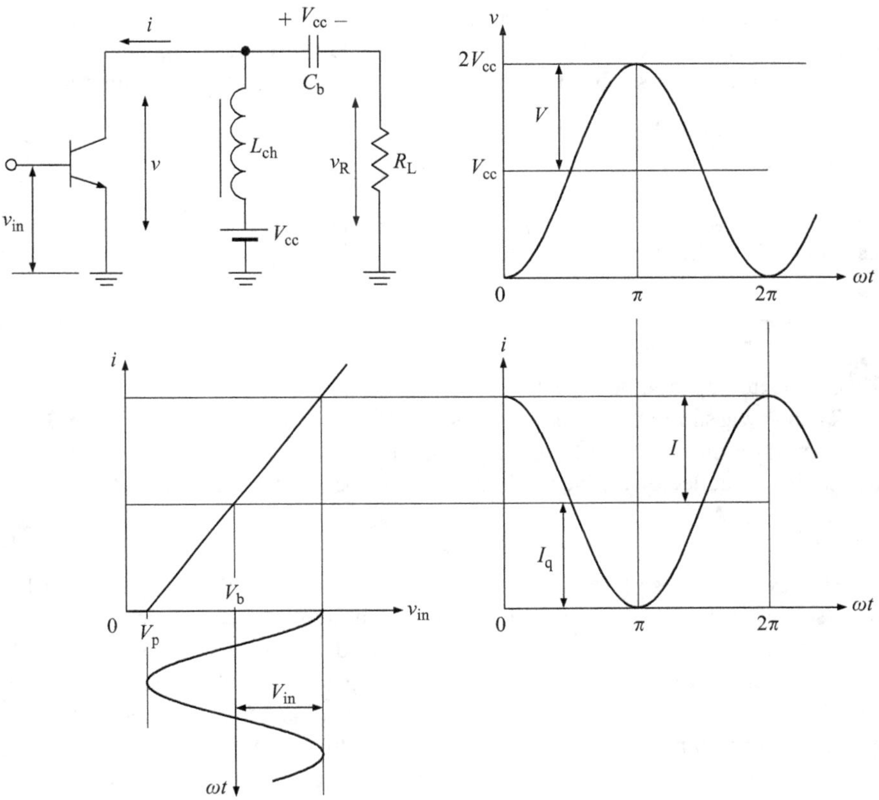

Figure 1.3 Voltage and current waveforms in Class-A operation

smaller output voltage v. Thus, for a purely real load impedance when $Z_L = R_L$, the collector voltage v is shifted by 180° relatively to the input voltage v_{in} and can be written as

$$v = V_{cc} + V\cos(\omega t + 180^\circ) = V_{cc} - V\cos\omega t \tag{1.19}$$

where V is the output voltage amplitude.

Substituting (1.18) into (1.19) yields:

$$v = V_{cc} - \left(i - I_q\right)R_L \tag{1.20}$$

where $R_L = V/I$, and (1.20) can be rewritten as

$$i = \left(I_q + \frac{V_{cc}}{R_L}\right) - \frac{v}{R_L} \tag{1.21}$$

which determines a linear dependence of the collector current versus collector voltage. Such a combination of the cosine collector voltage and current waveforms is known as a Class-A operation mode. In practice, because of the device nonlinearities, it is necessary to connect a parallel *LC* circuit with resonant frequency equal to the operating frequency to significantly suppress any possible harmonic components.

Circuit theory prescribes that the collector efficiency η can be written as

$$\eta = \frac{P}{P_0} = \frac{1}{2}\frac{I}{I_q}\frac{V}{V_{cc}} = \frac{1}{2}\frac{I}{I_q}\xi \tag{1.22}$$

where $P_0 = I_q V_{cc}$ is the dc output power, $P = IV/2$ is the power delivered to the load resistance R_L at the fundamental frequency f_0, and

$$\xi = \frac{V}{V_{cc}} \tag{1.23}$$

is the collector voltage peak factor.

Then, by assuming the ideal conditions of zero saturation voltage when $\xi = 1$ and maximum output current amplitude when $I/I_q = 1$, from (1.22) it follows that the maximum collector efficiency in a Class-A operation mode is equal to

$$\eta = 50\% \tag{1.24}$$

However, as it also follows from (1.22), increasing the value of I/I_q can further increase the collector efficiency. This leads to a step-by-step nonlinear transformation of the current cosine waveform to its pulsed waveform when the amplitude of the collector current exceeds zero value during only a part of the entire signal period. In this case, an active device is operated in the active region followed by the operation in the pinch-off region when the collector current is zero, as shown in Figure 1.4. As a result, the frequency spectrum at the device output will generally contain the second-, third-, and higher-order harmonics of the fundamental frequency. However, owing to high quality factor of the parallel resonant *LC* circuit, only the fundamental-frequency signal flows into the load, while the short-circuit conditions are fulfilled for higher-order harmonic components. Therefore, ideally the collector voltage represents a purely sinusoidal waveform with the voltage amplitude $V \leq V_{cc}$.

Equation (1.8) for the output current can be rewritten through the ratio between a quiescent current I_q and a current amplitude I as

$$\cos\theta = -\frac{I_q}{I} \tag{1.25}$$

As a result, the basic definitions for nonlinear operation modes of a power amplifier through half the conduction angle θ can be introduced as follows:

- when $\theta > 90°$, then $\cos\theta < 0$ and $I_q > 0$, corresponding to Class-AB operation;
- when $\theta = 90°$, then $\cos\theta = 0$ and $I_q = 0$, corresponding to Class-B operation; and
- when $\theta < 90°$, then $\cos\theta > 0$ and $I_q < 0$, corresponding to Class-C operation.

The periodic pulsed output current $i(\omega t)$ is represented as a Fourier-series expansion by (1.11), where the dc current component is a function of θ in the

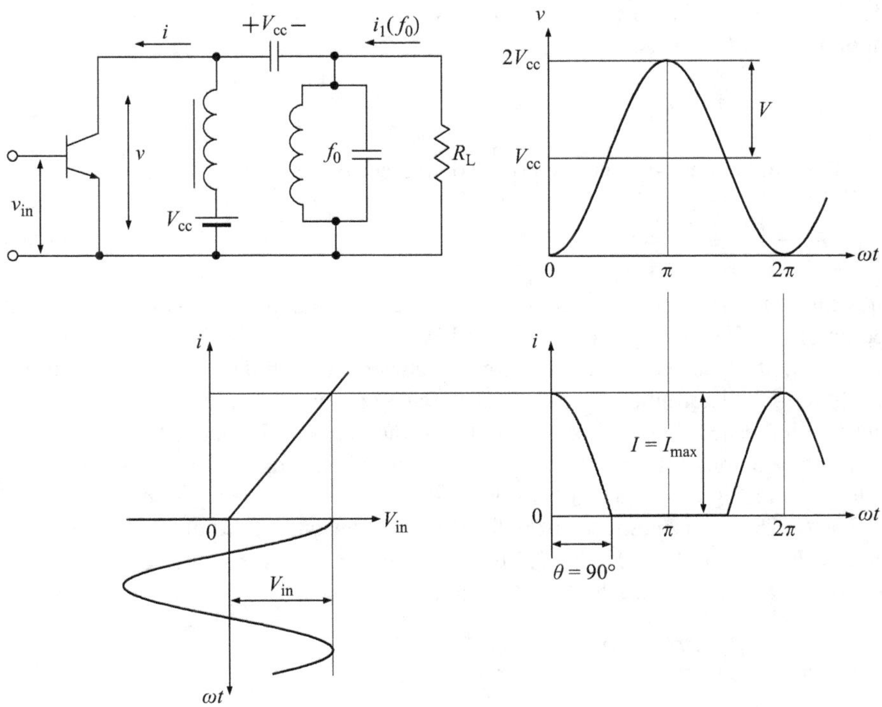

Figure 1.4 Voltage and current waveforms in Class-B operation

operation modes with $\theta < 180^\circ$, in contrast to a Class-A operation mode where $\theta = 180^\circ$ and the dc current is equal to the quiescent current during the entire period.

The collector efficiency of a power amplifier with parallel resonant circuit, biased to operate in a nonlinear mode with certain conduction angle, can be obtained by

$$\eta = \frac{P_1}{P_0} = \frac{1}{2}\frac{I_1}{I_0}\xi = \frac{1}{2}\frac{\gamma_1}{\gamma_0}\xi \tag{1.26}$$

which is a function of θ only, where P_1 is the output power at fundamental frequency and

$$\frac{\gamma_1}{\gamma_0} = \frac{\theta - \sin\theta\cos\theta}{\sin\theta - \theta\cos\theta} \tag{1.27}$$

The vacuum-tube Class-B power amplifiers were defined as those which operate with a negative grid bias such that the anode current is practically zero with no excitation grid voltage, and in which the output power is proportional to the square of the excitation voltage [5]. If $\xi = 1$ and $\theta = 90^\circ$, then from (1.26) and

(1.27) it follows that the maximum collector efficiency in a Class-B operation mode is equal to

$$\eta = \frac{\pi}{4} \cong 78.5\% \tag{1.28}$$

The fundamental-frequency power delivered to the load $P_L = P_1$ is defined as

$$P_1 = \frac{VI_1}{2} = \frac{VI\gamma_1(\theta)}{2} \tag{1.29}$$

showing its direct dependence on the conduction angle 2θ. This means that reduction in θ results in lower γ_1, and, to increase the fundamental-frequency power P_1, it is necessary to increase the current amplitude I. Since the current amplitude I is determined by the input voltage amplitude V_{in}, the input power P_{in} must be increased. The collector efficiency increases with reduced value of θ as well and becomes maximum when $\theta = 0°$, where the ratio γ_1/γ_0 is maximal, as follows from Figure 1.3(a). For instance, the collector efficiency η increases from 78.5% to 92% when θ reduces from 90° to 60°. However, it requires increasing the input voltage amplitude V_{in} by 2.5 times, resulting in lower values of the power-added efficiency (*PAE*), which is defined as

$$PAE = \frac{P_1 - P_{in}}{P_0} = \frac{P_1}{P_0}\left(1 - \frac{1}{G_p}\right) \tag{1.30}$$

where

$$G_p = \frac{P_1}{P_{in}} \tag{1.31}$$

is the operating power gain.

The vacuum-tube Class-C power amplifiers were defined as those that operate with a negative grid bias more than sufficient to reduce the anode current to zero with no excitation grid voltage, and in which the output power varies as the square of the anode voltage between limits [5]. The main distinction between Class B and Class C is in the duration of the output current pulses, which are shorter for Class C when the active device is biased beyond the cutoff point. It should be noted that, for the device transfer characteristic ideally represented by a square-law approximation, the odd-harmonic current coefficients $\gamma_n(\theta)$ are not equal to zero in this case, although there is no significant difference between the square-law and linear cases [6]. To achieve the maximum anode (collector) efficiency in Class C, the active device should be biased (negative) considerably past the cutoff (pinch-off) point to provide the sufficiently low conduction angles [7].

In order to obtain an acceptable trade-off between a high-power gain and a high power-added efficiency in different situations, the conduction angle should be chosen within the range of $120° \leq 2\theta \leq 190°$. If it is necessary to provide high collector efficiency of the active device having a high-gain capability, it is necessary to choose a Class-C operation mode with θ close to 60°. However, when the input power is limited and power gain is not sufficient, a Class-AB operation mode is recommended with small quiescent current when θ is slightly greater than 90°. In the latter case, the linearity of the power amplifier can be significantly improved.

1.2 Load line and output impedance

The graphical method of laying down a load line on the family of the static curves representing anode current against anode voltage for various grid potentials was already well known in the 1920s [8]. If an active device is connected in a circuit in which the anode load is a pure resistance, the performance may be analyzed by drawing the load line where the lower end of the line represents the anode supply voltage and the slope of the line is established by the load resistance, that is, the load resistance is equal to the value of the intercept on the voltage axis divided by the value of the intercept on the current axis.

In a Class-A operation mode, the output voltage v across the device anode (collector or drain) represents a sum of the dc supply voltage V_{cc} and cosine voltage across the load resistance R_L, and can be defined by (1.19). In this case, the power dissipated in the load and the power dissipated in the device is equal when $V_{cc} = V$, and the load resistance $R_L = V/I$ is equal to the device output resistance R_{out} [7]. In a pulsed operation mode (Class AB, B, or C) when the parallel LC circuit is tuned to the fundamental frequency, ideally the voltage across the load resistor R_L represents a cosine waveform. By using (1.7), (1.13), and (1.19), the relationship between the collector current i and the collector voltage v during a time period of $-\theta \le \omega t < \theta$ can be expressed by

$$i = \left(I_q + \frac{V_{cc}}{\gamma_1 R_L}\right) - \frac{v}{\gamma_1 R_L} \tag{1.32}$$

where the fundamental current coefficient γ_1 as a function of θ is determined by (1.16), and the load resistance is defined by $R_L = V/I_1$, where I_1 is the fundamental current amplitude. Equation (1.32) determining the dependence of the collector current on the collector voltage for any values of conduction angle in the form of a straight-line function is called the *load line* of the active device. For a Class-A operation mode with $\theta = 180°$ when $\gamma_1 = 1$, the load line defined by (1.32) is identical to the load line defined by (1.21).

Figure 1.5 shows the idealized active device output *I–V* curves and load lines for different conduction angles according to (1.32) with the corresponding collector and current waveforms. From Figure 1.5, it follows that the maximum collector current amplitude I_{max} corresponds to the minimum collector voltage V_{sat} when $\omega t = 0$, and is the same for any conduction angle. The slope of the load line defined by its slope angle β is different for different conduction angles and values of the load resistance, and can be obtained by

$$\tan\beta = \frac{I_{max}}{V(1 - \cos\theta)} = \frac{1}{\gamma_1 R_L} \tag{1.33}$$

from which it follows that greater slope angle β of the load line results in smaller value of the load resistance R_L for the same θ.

The load resistance R_L for the active device as a function of θ, which is required to terminate the device output to deliver the maximum output power to the load, can be written in a general form as

$$R_L(\theta) = \frac{V}{\gamma_1(\theta)I} \tag{1.34}$$

which is equal to the device equivalent output resistance R_{out} at the fundamental frequency [5]. The term "equivalent" means that this is not a real physical device resistance as in a Class-A mode, but its equivalent output resistance, the value of which determines the optimum load, which should terminate the device output to deliver maximum fundamental-frequency output power. The equivalent output resistance is calculated as a ratio between the amplitudes of the collector cosine voltage and fundamental-frequency collector current component, which depends on the angle θ.

In a Class-B mode when $\theta = 90°$ and $\gamma_1 = 0.5$, the load resistance R_L^B is defined as $R_L^B = 2V/I_{max}$. Alternatively, taking into account that $V_{cc} = V$ and $P_{out} = I_1 V/2$ for the fundamental-frequency output power, the load resistance $R_L^B = V/I_1$ can be written in a simple idealized analytical form with zero saturation voltage V_{sat} as

$$R_L^B = \frac{V_{cc}^2}{2P_{out}} \tag{1.35}$$

In general, the entire load line represents a broken line *PK* including a horizontal part, as shown in Figure 1.5. Figure 1.5(a) represents a load line *PNK* corresponding to a Class-AB mode with $\theta > 90°$, $I_q > 0$, and $I < I_{max}$. Such a load line moves from point *K* corresponding to the maximum output current amplitude I_{max} at $\omega t = 0$ and determining the device saturation voltage V_{sat} through the point *N* located at the horizontal axis v where $i = 0$ and $\omega t = \theta$. For a Class-AB operation, the conduction angle for the output current pulse between points N' and N'' is greater than 180°. Figure 1.5(b) represents a load line *PMK* corresponding to a Class-C mode with $\theta < 90°$, $I_q < 0$, and $I > I_{max}$. For a Class-C operation, the load line intersects a horizontal axis v in a point *M*, and the conduction angle for the output current pulse between points M' and M'' is smaller than 180°. Hence, generally the load line represents a broken line with the first section having a slope angle β and the other horizontal section with zero current i. In a Class-B mode, the collector current represents half-cosine pulses with the conduction angle of $2\theta = 180°$ and $I_q = 0$.

Now let us consider a Class-B operation with increased amplitude of the cosine collector voltage. In this case, as shown in Figure 1.6, an active device is operated in the saturation, active, and pinch-off regions, and the load line represents a broken line *LKMP* with three linear sections (*LK*, *KM*, and *MP*). The new section *KL* corresponds to the saturation region, resulting in a half-cosine output current

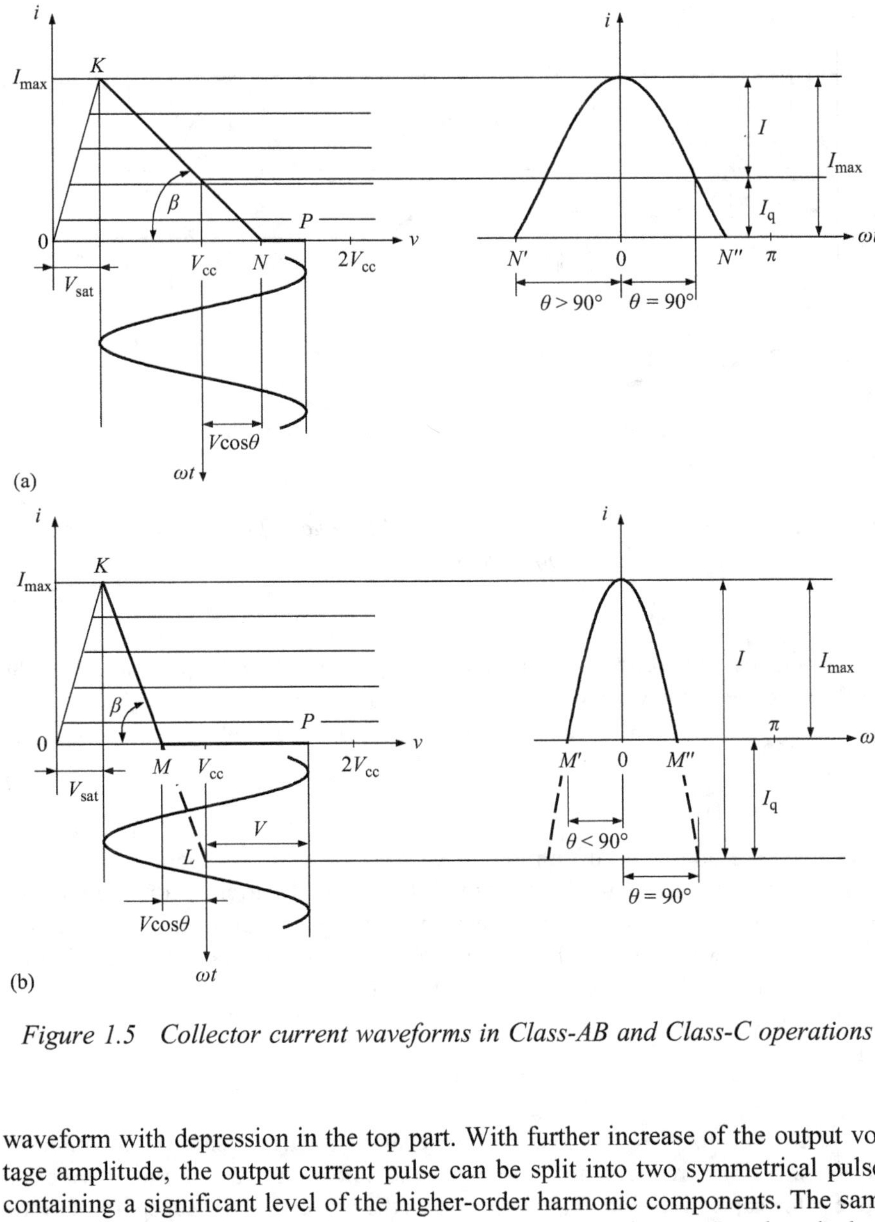

Figure 1.5 Collector current waveforms in Class-AB and Class-C operations

waveform with depression in the top part. With further increase of the output voltage amplitude, the output current pulse can be split into two symmetrical pulses containing a significant level of the higher-order harmonic components. The same result can be achieved by increasing a value of the load resistance R_L when the load line is characterized by smaller slope angle β.

The collector current waveform becomes asymmetrical for the complex load, the impedance of which represents the load resistance and capacitive or inductive reactance. In this case, the Fourier-series expansion of the output current given by

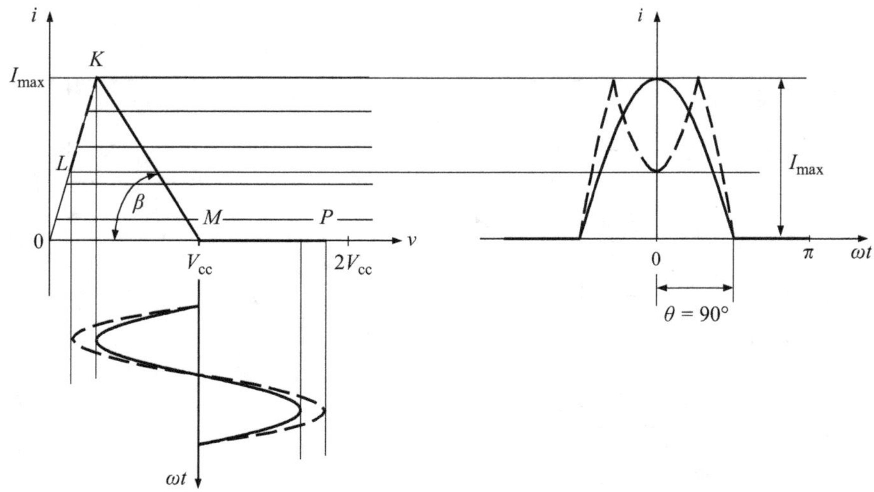

Figure 1.6 Collector current waveforms for the device operating in saturation, active, and pinch-off regions

(1.11) includes a phase for each harmonic component. Then, the output voltage at the device collector is written as

$$v = V_{cc} - \sum_{n=1}^{\infty} I_n |Z_n| \cos(n\omega t + \phi_n) \tag{1.36}$$

where I_n is the amplitude of nth output current harmonic component, $|Z_n|$ is the magnitude of the load-network impedance at nth output current harmonic component, and ϕ_n is the phase of nth output current harmonic component. If Z_n is zero for $n = 2, 3, \ldots$, which is possible for a resonant load network having negligible impedance at any harmonic component except the fundamental, (1.36) can be rewritten as

$$v = V_{cc} - I_1 |Z_1| \cos(\omega t + \phi_1) \tag{1.37}$$

As a result, for the inductive load impedance, the depression in the collector current waveform reduces and moves to the left-hand side of the waveform, whereas the capacitive load impedance causes the depression to deepen and shift to the right-hand side of the collector current waveform [9]. This effect can simply be explained by the different sign for the phase angle ϕ_1 in (1.37), as well as generally by the different phase conditions for fundamental and higher-order harmonic components composing the collector current waveform, and is illustrated by the different load lines for (a) inductive and (b) capacitive load impedances shown in Figure 1.7. Note that now the load line represents a two-dimensional curve with complicated behavior.

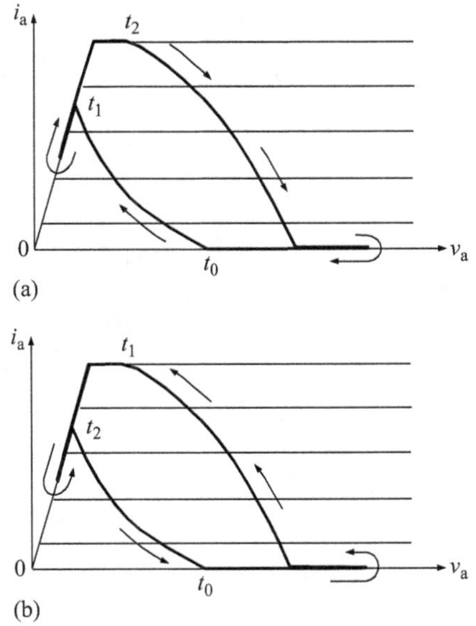

Figure 1.7 Load lines for (a) inductive and (b) capacitive load impedances

1.3 Classes of operation based upon finite number of harmonics

Figure 1.8(a) shows the block diagram of a generic power amplifier, where the active device (which is shown as a MOSFET device but can be a bipolar transistor or any other suitable device) is controlled by its drive and bias to operate as a multiharmonic current source or switch, V_{dd} is the supply voltage, and I_0 is the dc current flowing through the RF choke [10]. The load-network bandpass filter is assumed linear and lossless and provides the drain load impedance $R_1 + jX_1$ at the fundamental frequency and pure reactances X_k at each *k*th-harmonic component. For analysis simplicity, the load-network filter can incorporate the reactances of the RF choke and device drain-source capacitance which is considered voltage independent. Since such a basic power amplifier is assumed to generate power at only the fundamental frequency, harmonic components can be present generally in the voltage and current waveforms depending on class of operation. In a Class-AB, -B, or -C operation, harmonics are present only in the drain current. However, in a Class-F mode, a given harmonic component is present in either drain voltage or drain current, but not both, and all or most harmonics are present in both the drain voltage and current waveforms in a Class-E mode. The required harmonics with optimum or near-optimum amplitudes can be produced by driving the power amplifier to saturation. The analysis based on a Fourier-series expansion of the drain voltage and current waveforms shows that maximum achievable efficiency depends not upon the class of operation, but upon the number of harmonics

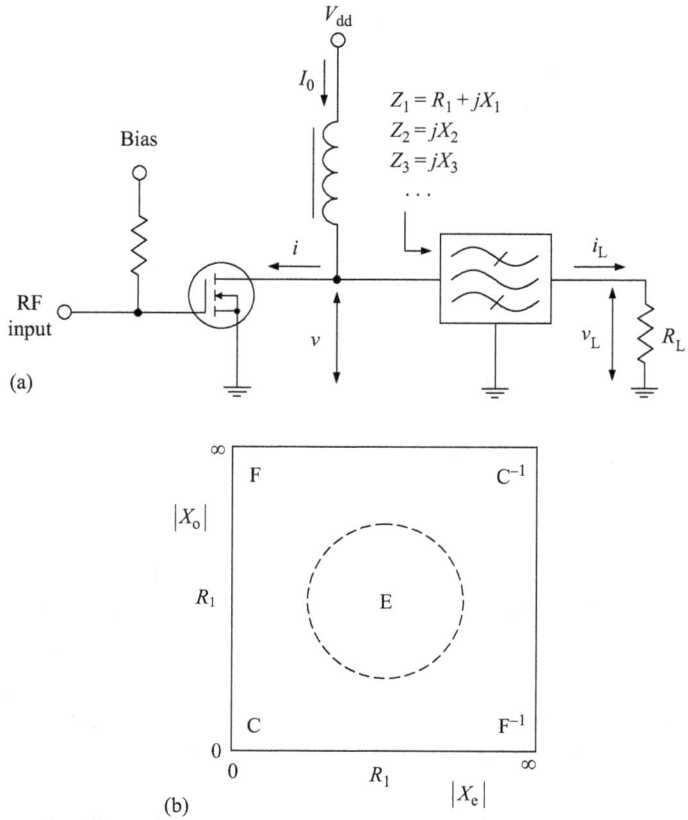

Figure 1.8 Basic power-amplifier structure and classes of amplification

employed [10,11]. For any set of harmonic reactances, the same maximum efficiency can be achieved by proper adjustment of the waveforms and the fundamental-frequency load reactance.

A mechanism for differentiating the various classes of power amplifier operation implemented with small numbers of harmonic components is shown in Figure 1.8(b) [10]. It is based on the relative magnitudes of the even (X_e) and odd (X_o) harmonic impedances relative to the fundamental-frequency load resistance R_1. In this case, the classes of operation can be characterized in terms of a small number of harmonics as follows:

- Class F: even-harmonic reactances are low and odd-harmonic reactances are high so that the drain voltage is shaped toward a square wave and drain current is shaped toward a half-sine wave;
- Inverse Class F (Class F^{-1}): even-harmonic reactances are high and odd-harmonic reactances are low so that the drain voltage is shaped toward a half-sine wave and drain current is shaped toward a square wave;

- Class C: all harmonic reactances are low so that the drain current is shaped toward a narrow pulse;
- Inverse Class C (Class C^{-1}): all harmonic reactances are high so that the drain voltage is shaped toward a narrow pulse and
- Class E: all harmonic reactances are negative and comparable in magnitude to the fundamental-frequency load resistance.

The transition from "low" to "comparable" occurs in the range from $R_1/3$ to $R_1/2$, whereas the transition from "comparable" to "high" similarly occurs in the range from $2R_1$ to $3R_1$. In this case, the circular boundary is for illustration only, and the point at which an amplifier transitions from one class to another is somewhat judgmental and arbitrary, as there is not an abrupt change in the mode of operation. All power amplifier degenerate to a Class-A operation when there is only a single (fundamental) frequency component. Class B is the special case of a pulsed operation with a conduction angle of 180°, which is represented by a half-sine current waveform based upon even harmonics. Class D can be considered as a push–pull Class-F power amplifier, in which the two active devices provide each other with paths for the even harmonics.

The transition from Class F to Class E and then to Class F^{-1} moves diagonally in Figure 1.8(b) by progressively increasing X_2 from zero to ∞ while decreasing X_3 from ∞ to zero so that $X_3 = 1/X_2$. In a Class F with $X_2 = 0$ and $X_3 = \infty$, the voltage is a third-harmonic maximum-power waveform, while the current is a second-harmonic maximum-power waveform. For $X_2 = X_3 = -1$, the voltage waveform leans leftward and the current waveform leans rightward, thus approximating the all-harmonic Class-E waveforms. Finally, when $X_2 = \infty$ and $X_3 = 0$, the power amplifier operates in an inverse Class F (Class F^{-1}). The transition from Class F to Class C moves down to the left-hand side of Figure 1.8(b) by setting X_2 at zero and progressively decreasing X_3 from ∞ to zero, and the waveforms remain almost unchanged for $X_3 \leq -3$. The explicit analytical expression for maximum achievable efficiency of finite-harmonic Class C with conduction angle $2\theta \rightarrow 0$ can be written as

$$\eta = \cos\left(\frac{\pi}{n+2}\right) \tag{1.38}$$

where n is a number of harmonics [12].

1.4 Mixed-mode Class C and nonlinear effect of collector capacitance

In contrast to the conventional Class-C power amplifiers with a parallel resonant circuit resulting in a sinusoidal collector voltage waveform, the so-called mixed-mode Class-C configuration with a series resonant circuit was widely although somewhat accidentally adopted for most VHF and UHF transistor power amplifiers, which could provide better efficiency performance and where it is easier to provide the drive and bias [13,14]. For low saturation resistance and significant nonlinear collector capacitance, it is difficult to maintain a sinusoidal collector voltage waveform. Instead, a nonlinear collector capacitance produces a voltage waveform containing harmonics in response to a sinusoidal current. As a result, the saturated bipolar transistor usually dominates the parallel-tuned circuit, flattening the collector voltage waveform [15]. Besides, it is also enough difficult in practice to implement the parallel-resonant circuit required for true Class-C operation in power amplifiers using either FET or bipolar devices, especially with a high-quality factor. There are several additional difficulties in implementing true Class-C operation in solid-state power amplifiers, especially at VHF and UHF in view of the device lead lengths and stray reactances, causing a significant effect at these frequencies.

Figure 1.9 shows the simplified schematic of a mixed-mode Class-C power amplifier with a series resonant circuit in a load network, which provides the near-sinusoidal collector current and pulsed collector voltage with pulse duration less than one-half the period, depending on the value of the collector capacitance. The level of a Class-C operation with corresponding conduction angle is defined by the value of the resistor in a base bias circuit, where the inductor value is chosen to maximize the operating power gain.

As an example, by using a 28-V MRF373A LDMOSFET device in a series-tuned Class-C power amplifier, whose simplified circuit schematic is shown in Figure 1.10(a), an output power of 50 W and a drain efficiency of 58% (by 10% lower than obtained with the idealized simulations) were achieved at 435 MHz [16]. Here, for the theoretical analysis, it was assumed that the transistor is driven so hard that its operation can be described by a switch, and the switch is turned on (closed) when the gate-source voltage is above the threshold voltage and turned off (open) when it is below. When the transistor is turned off, the total drain current i_d, which is a sum of the sinusoidal load current and dc supply current, charges or discharges the transistor drain-source capacitance C_{ds}. In this case, there are two basic power loss mechanisms: the transistor series loss, due to finite value of its

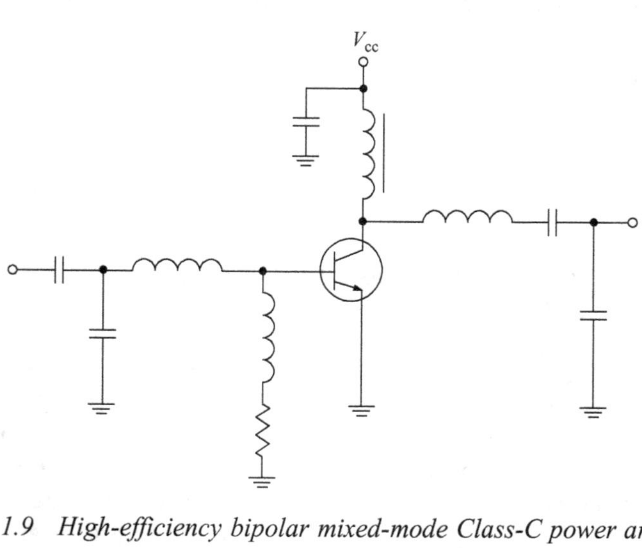

Figure 1.9 High-efficiency bipolar mixed-mode Class-C power amplifier

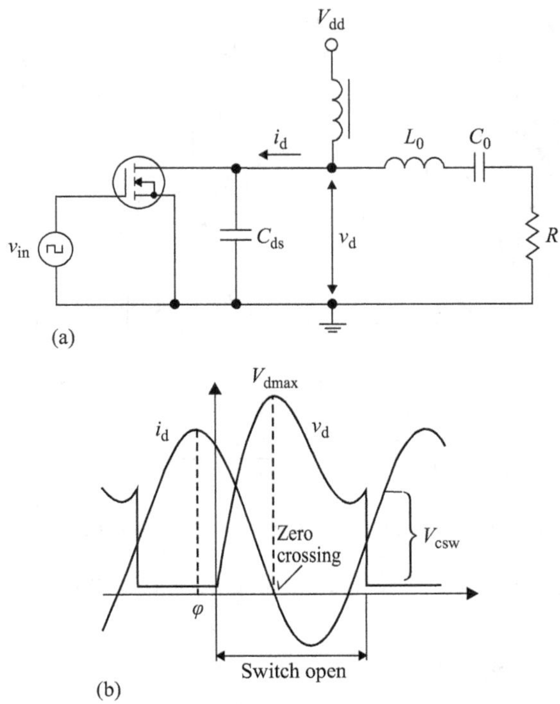

Figure 1.10 High-efficiency bipolar mixed-mode Class-C power amplifier

saturation voltage, and the switching loss that accompanies the switch turning on. The switching loss is determined by the value of the capacitor voltage V_{csw} obtained prior to the start of switching on. Figure 1.10(b) shows the drain voltage and current waveforms, where the switch starts turning off at zero time instant and the drain voltage $v_d(\omega t)$ achieves maximum value V_{dmax} when the drain current $i_d(\omega t)$ turns negative through zero-crossing point. Significantly higher efficiency was achieved using the Cree CGH40120F transistor when the output power of 115 W was achieved under pulsed conditions with the drain efficiency between 78.4% and 82.7% at the same operating frequency for the duty ratio varying from 36% to 25% [17].

Generally, the dependence of the collector capacitance on the output voltage represents a nonlinear function. To evaluate the influence of the nonlinear collector capacitance on electrical behavior of the power amplifier, consider the load network including a series resonant L_0C_0 circuit tuned to the fundamental frequency that provides open-circuit conditions for the second- and higher-order harmonic components of the output current and a low-pass L-type matching circuit with the series inductor L and shunt capacitor C, as shown in Figure 1.11(a). The matching circuit is needed to match the equivalent output resistance R, corresponding to the required output power at the fundamental frequency, with the standard load resistance R_L. Figure 1.11(b) shows the simplified output equivalent circuit of the bipolar power amplifier.

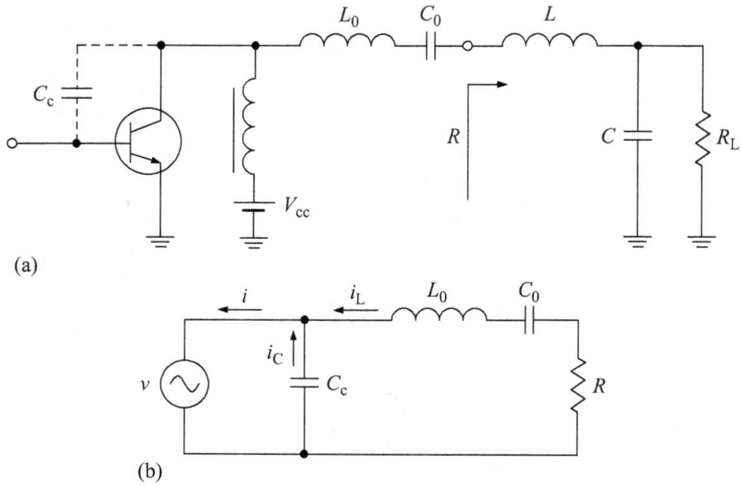

Figure 1.11 Circuit schematics of bipolar tuned power amplifier

The total output current flowing through the device collector can be written as

$$i = I_0 + \sum_{n=1}^{\infty} I_n \cos(n\omega t + \phi_n) \tag{1.39}$$

where I_n and ϕ_n are the amplitude and phase of the nth-harmonic component, respectively.

An assumption of a high-quality factor of the series resonant circuit allows the only fundamental-frequency current component to flow into the load. The current flowing through the nonlinear collector capacitance consists of the fundamental-frequency and higher-order harmonic components, which is written as

$$i_C = i_C \cos(\omega t + \phi_1) + \sum_{n=2}^{\infty} I_n \cos(n\omega t + \phi_n) \tag{1.40}$$

where I_C is the fundamental-frequency capacitor current amplitude.

The nonlinear behavior of the collector junction capacitance is described by

$$C_c = C_0 \left(\frac{\varphi + V_{cc}}{\varphi + v} \right)^{\gamma} \tag{1.41}$$

where C_0 is the collector capacitance at $v = V_{cc}$, V_{cc} is the supply voltage, φ is the contact potential, and γ is the junction sensitivity equal to 0.5 for abrupt junction.

As a result, the expression for charge flowing through collector capacitance can be obtained by

$$q = \int_0^v C(v)dv = \int_0^v \frac{C_0(\varphi + V_{cc})^{\gamma}}{(\varphi + v)^{\gamma}} dv \tag{1.42}$$

When $v = V_{cc}$, then

$$q_0 = \frac{C_0(\varphi + V_{cc})}{1-\gamma}\left[1 - \left(\frac{\varphi}{\varphi + V_{cc}}\right)^{1-\gamma}\right] \tag{1.43}$$

Although the dc charge component q_0 is a function of the voltage amplitude, its variations at maximum voltage amplitude normally do not exceed 20% for $\gamma = 0.5$ [18]. Then, assuming q_0 is determined by (1.43) as a constant component, the total charge q of the nonlinear capacitance can be represented by the dc component q_0 and ac component Δq as

$$\begin{aligned} q &= q_0 + \Delta q = q_0\left(1 + \frac{\Delta q}{q_0}\right) \\ &= q_0\frac{(\varphi + v)^{1-\gamma} - \varphi^{1-\gamma}}{(\varphi + V_{cc})^{1-\gamma} - \varphi^{1-\gamma}} \end{aligned} \tag{1.44}$$

Because $V_{cc} \gg \varphi$ in the normal case, from (1.44) it follows that

$$\frac{v}{V_{cc}} = \left(1 + \frac{\Delta q}{q_0}\right)^{\frac{1}{1-\gamma}} \tag{1.45}$$

where $q_0 \cong C_0V_{cc}/(1-\gamma)$.

On the other hand, the charge component Δq can be written using (1.39) as

$$\Delta q = \int i_C(t)dt = \frac{I_C}{\omega}\sin(\omega t + \phi_1) + \sum_{n=2}^{\infty}\frac{I_n}{n\omega}\sin(n\omega t + \phi_n) \tag{1.46}$$

As a result, substituting (1.46) into (1.45) yields:

$$\frac{v}{V_{cc}} = \left[1 + \frac{I_C(1-\gamma)}{\omega C_0 V_{cc}}\sin(\omega t + \phi_1) + \sum_{n=2}^{\infty}\frac{I_n(1-\gamma)}{n\omega C_0 V_{cc}}\sin(n\omega t + \phi_n)\right]^{\frac{1}{1-\gamma}} \tag{1.47}$$

After applying a Taylor-series expansion to (1.47), it is sufficient to be limited to its first three terms to reveal the parametric effect. Then, equating the fundamental-frequency collector voltage components results in

$$\begin{aligned} \frac{v_1}{V_{cc}} &= \frac{I_C}{\omega C_0 V_{cc}}\sin(\omega t + \phi_1) \\ &\quad + \frac{I_C I_2 \gamma}{(2\omega C_0 V_{cc})^2}\cos(\omega t + \phi_2 - \phi_1) \\ &\quad + \frac{I_2 I_3 \gamma}{12(\omega C_0 V_{cc})^2}\cos(\omega t + \phi_3 - \phi_2) \end{aligned} \tag{1.48}$$

Consequently, by taking into account that $v_1 = V_1\sin(\omega t + \phi_1)$, the fundamental voltage amplitude V_1 can be obtained from (1.48) as

$$\frac{V_1}{V_{cc}}=\frac{I_C}{\omega C_0 V_{cc}}\left[1+\frac{I_2\gamma}{4\omega C_0 V_{cc}}\cos(90^\circ+\phi_2-2\phi_1)+\frac{I_2 I_3\gamma}{12\omega C_0 V_{cc} I_C}\cos(90^\circ+\phi_3-\phi_2-\phi_1)\right] \tag{1.49}$$

Because a large-signal value of the abrupt-junction collector capacitance usually does not exceed 20%, the fundamental-frequency capacitor current amplitude I_C can be written in a first-order approximation as

$$I_C \cong \omega C_0 V_1 \tag{1.50}$$

As a result, from (1.49) it follows that, because of the parametric transformation due to the collector capacitance nonlinearity, the fundamental-frequency collector voltage amplitude increases by σ_p times according to

$$\begin{aligned}\sigma_p = 1 &+ \frac{I_2\gamma}{4\omega C_{0V_{cc}}}\cos(90^\circ + \phi_2 - 2\phi_1)\\ &+ \frac{I_2 I_3\gamma}{12(\omega C_0)^2 V_1 V_{cc}}\cos(90^\circ + \phi_3 - \phi_2 - \phi_1)\end{aligned} \tag{1.51}$$

where $\sigma_p = \xi_p/\xi$ and ξ_p is the collector voltage peak factor with parametric effect [9].

From (1.51) it follows that to maximize the collector voltage peak factor and consequently the collector efficiency for a given value of the supply voltage V_{cc}, it is necessary to provide the following phase conditions:

$$\phi_2 = 2\phi_1 - 90^\circ \tag{1.52}$$

$$\phi_3 = 3\phi_1 - 180^\circ \tag{1.53}$$

Then, for $\gamma = 0.5$,

$$\sigma_p = 1 + \frac{I_2}{8\omega C_0 V_{cc}} + \frac{I_2 I_3}{24(\omega C_0)^2 V_1 V_{cc}} \tag{1.54}$$

Equation (1.54) shows the theoretical possibility to increase the collector voltage peak factor by 1.1 to 1.2 times, thus achieving collector efficiency of 85% to 90%. Physically, the improved efficiency can be explained by the transformation of powers corresponding to the second- and higher-order harmonic components into the fundamental-frequency output power due to the collector capacitance nonlinearity. However, this becomes effective only in the case of the load network with a series resonant circuit (mixed-mode Class C), because it ideally provides infinite impedance at the second- and higher-order harmonics, unlike the load network with a parallel resonant circuit (true Class C) having ideally zero impedance at these harmonics.

1.5 Power gain and stability

Power amplifier design aims for maximum power gain and efficiency for a given value of output power with a predictable degree of stability. In order to extract the maximum power from a generator, it is a well-known fact that the external load should have a vector value which is conjugate of the internal impedance of the source [19]. The power delivered from a generator to a load, when matched on this basis, will be called the available power of the generator [20]. In this case, the power gain of the four-terminal network is defined as the ratio of power delivered to the load impedance connected to the output terminals to power available from the generator connected to the input terminals, usually measured in decibels, and this ratio is called the *power gain* irrespective of whether it is greater or less than one [21,22].

Figure 1.12 shows the basic block schematic of a single-stage power amplifier circuit, which includes an active device, an input matching circuit to match with the source impedance, and an output matching circuit to match with the load impedance. Generally, the two-port active device is characterized by a system of the immittance W-parameters, that is, any system of impedance Z-parameters, hybrid H-parameters, or admittance Y-parameters [23,24]. The input and output matching circuits transform the source and load immittances W_S and W_L into specified values between points 1–2 and 3–4, respectively, by means of which the optimal design operation mode of the power amplifier is realized.

The operating power gain G_P, which represents the ratio of power dissipated in the active load $\mathrm{Re}W_L$ to the power delivered to the input port of the active device, can be expressed in terms of the immittance W-parameters as

$$G_P = \frac{|W_{21}|^2 \mathrm{Re} W_L}{|W_{22} + W_L|^2 \mathrm{Re} W_{in}} \tag{1.55}$$

where

$$W_{in} = W_{11} - \frac{W_{12} W_{21}}{W_{22} + W_L} \tag{1.56}$$

is the input immittance and W_{ij} $(i, j = 1, 2)$ are the immittance two-port parameters of the active device equivalent circuit.

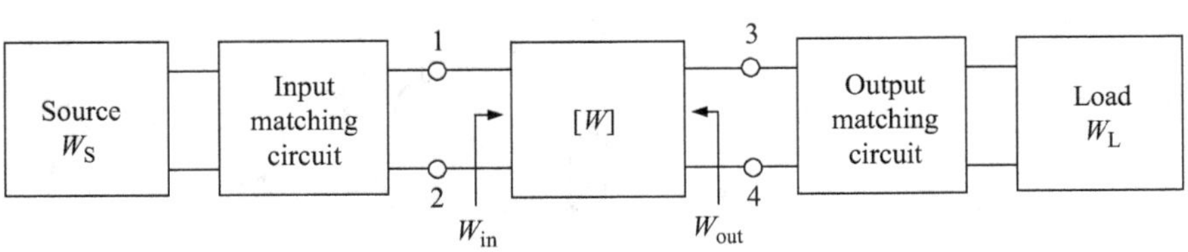

Figure 1.12 Block schematic of single-stage power amplifier

The transducer power gain G_T, which represents the ratio of power dissipated in the active load $\mathrm{Re}W_L$ to the power available from the source, can be expressed in terms of the immittance W-parameters as

$$G_T = \frac{4|W_{21}|^2 \mathrm{Re}W_S \ \mathrm{Re}W_L}{|(W_{11}+W_S)(W_{22}+W_L) - W_{12}W_{21}|^2} \tag{1.57}$$

The operating power gain G_P does not depend on the source parameters and characterizes only the effectiveness of the power delivery from the input port of the active device to the load. This power gain helps to evaluate the gain property of a multistage amplifier when the overall operating power gain $G_{P(\mathrm{total})}$ is equal to the product of each stage G_P. The transducer power gain G_T includes an assumption of conjugate matching of both the load and the source.

The simplified small-signal π-hybrid equivalent circuit of the bipolar transistor shown in Figure 1.13 provides an example for a conjugate-matched bipolar power amplifier. The impedance Z-parameters of the equivalent circuit of the bipolar transistor in a common-emitter configuration can be written as

$$\begin{aligned} Z_{11} &= r_b + \frac{1}{g_m + j\omega C_\pi} \\ Z_{12} &= \frac{1}{g_m + j\omega C_\pi} \\ Z_{21} &= -\frac{1}{j\omega C_c}\frac{g_m - j\omega C_c}{g_m + j\omega C_\pi} \\ Z_{22} &= \left(1 + \frac{C_\pi}{C_c}\right)\frac{1}{g_m + j\omega C_\pi} \end{aligned} \tag{1.58}$$

where g_m is the small-signal transconductance, r_b is the series base resistance, C_π is the base-emitter capacitance including both diffusion and junction components, and C_c is the feedback collector capacitance.

By setting the device feedback impedance Z_{12} to zero and complex conjugate-matching conditions at the input as $R_S = \mathrm{Re}Z_{in}$ and $L_{in} = -\mathrm{Im}Z_{in}/\omega$ and at the

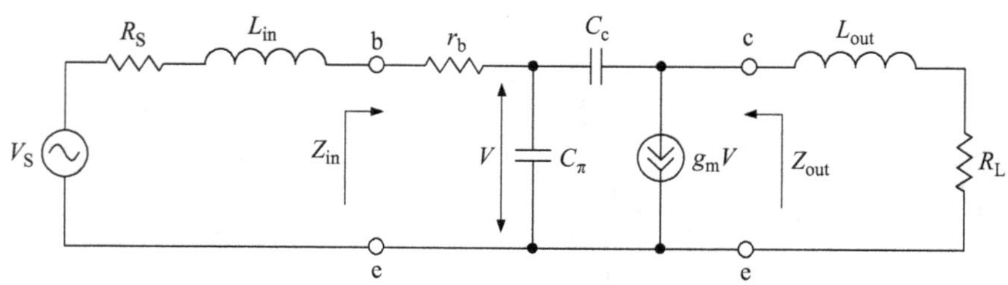

Figure 1.13 Simplified equivalent circuit of matched bipolar power amplifier

output as $R_L = \text{Re}Z_{\text{out}}$ and $L_{\text{out}} = -\text{Im}Z_{\text{out}}/\omega$, the small-signal transducer power gain G_T can be obtained by

$$G_T = \left(\frac{f_T}{f}\right)^2 \frac{1}{8\pi f_T r_b C_c} \tag{1.59}$$

where $f_T = g_m/2\pi C_\pi$ is the device transition frequency.

Figure 1.14 shows the simplified circuit schematic for a conjugate-matched FET (field-effect transistor) power amplifier. The admittance Y-parameters of the small-signal equivalent circuit of the FET device in a common-source configuration can be written as

$$\begin{aligned} Y_{11} &= \frac{j\omega C_{gs}}{1 + j\omega C_{gs} R_{gs}} + j\omega C_{gd} \\ Y_{12} &= -j\omega C_{gd} \\ Y_{21} &= \frac{g_m}{1 + j\omega C_{gs} R_{gs}} - j\omega C_{gd} \\ Y_{22} &= \frac{1}{R_{ds}} + j\omega\left(C_{ds} + C_{gd}\right) \end{aligned} \tag{1.60}$$

where g_m is the small-signal transconductance, R_{gs} is the gate-source resistance, C_{gs} is the gate-source capacitance, C_{gd} is the feedback gate-drain capacitance, C_{ds} is the drain-source capacitance, and R_{ds} is the differential drain-source resistance.

Since the value of the gate-drain capacitance C_{gd} is usually relatively small, the effect of the feedback admittance Y_{12} can be neglected in a simplified case. Then, it is necessary to set $R_S = R_{gs}$ and $L_{in} = 1/\omega^2 C_{gs}$ for input matching, whereas $R_L = R_{ds}$ and $L_{out} = 1/\omega^2 C_{ds}$ for output matching. Hence, the small-signal transducer power gain G_T can simply be obtained by

$$G_T\left(C_{gd} = 0\right) = MAG = \left(\frac{f_T}{f}\right)^2 \frac{R_{ds}}{4R_{gs}} \tag{1.61}$$

where $f_T = g_m/2\pi C_{gs}$ is the device transition frequency and MAG is the maximum available gain representing a theoretical limit on the power gain that can be achieved under complex conjugate-matching conditions.

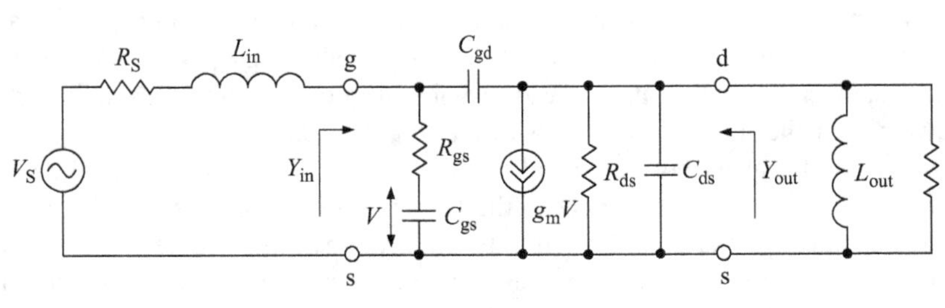

Figure 1.14 Simplified equivalent circuit of matched FET power amplifier

From (1.59) and (1.61), it follows that the small-signal power gain of a conjugately matched power amplifier for any type of the active device drops off as $1/f^2$ or 6 dB per octave. Therefore, $G_T(f)$ can readily be predicted at a certain frequency f, if a power gain is known at the transition frequency f_T, by

$$G_T(f) = G_T(f_T)\left(\frac{f_T}{f}\right)^2 \tag{1.62}$$

It should be noted that previous analysis is based upon the linear small-signal consideration when generally nonlinear device current source as a function of both input and output voltages can be characterized by the linear transconductance g_m as a function of the input voltage and the output differential resistance R_{ds} as a function of the output voltage. This is a result of a Taylor-series expansion of the output current as a function of the input and output voltages with maintaining only the dc and linear components. Such an approach helps to understand and derive the maximum achievable power amplifier parameters in a linear approximation. In this case, an active device is operated in a Class-A mode when one-half of the dc power is dissipated in the device, whereas the other half is transformed to the fundamental-frequency output power flowing into the load, resulting in a maximum ideal collector efficiency of 50%. The device output resistance R_{out} remains constant and can be calculated as a ratio of the dc supply voltage to the dc current flowing through the active device. In a common case for a complex conjugate-matching procedure, the device output immittance under large-signal consideration should be calculated using a Fourier-series analysis of the output current and voltage fundamental components. This means that, unlike a linear Class-A mode, an active device is operated in a linear region only part of the entire period, and its output resistance is defined as a ratio of the fundamental-frequency output voltage to the fundamental-frequency output current. This is not a physical resistance resulting in a power loss inside the device, but an equivalent resistance required to use for a conjugate matching procedure. In this case, the complex conjugate matching concept is valid when it is necessary first to compensate for the reactive part of the device output impedance and second to provide a proper load resistance resulting in a maximum power gain for a given supply voltage and required output power delivered to the load. Note that this is not a maximum available small-signal power gain which can be achieved in a linear operation mode, but a maximum achievable large-signal power gain that can be achieved for operation mode with a certain conduction angle. Of course, the maximum large-signal power gain is smaller than the small-signal power gain for the same input power, since the output power in a nonlinear operation mode also includes the powers at the harmonic components of the fundamental frequency.

Therefore, it makes more practical sense not to introduce separately the concepts of the gain match with respect to the linear power amplifiers and the power match in nonlinear power amplifier circuits since the maximum large-signal power gain, being a function of the conduction angle, corresponds to the maximum fundamental-frequency output power delivered to the load due to large-signal conjugate output matching. It is very important to provide a conjugate matching at both input and output device ports to achieve maximum power gain in a large-signal mode. In a Class-A mode, the maximum small-signal power gain ideally remains constant regardless of the output power level.

The transistor characterization in a large-signal mode can be done based on equivalent quasi-harmonic nonlinear approximation under the condition of sinusoidal port voltages [25]. In this case, the large-signal impedances are generally determined in the following manner. The designer tunes the load network (often by trial and errors) to maximize the output power to the required level using a particular transistor at a specified frequency and supply voltage. Then, the transistor is removed from the circuit and the impedance seen by the collector is measured at the carrier frequency. The complex-conjugate of the measured impedance then represents the equivalent large-signal output impedance of the transistor at that frequency, supply voltage, and output power. Similar design process is used to measure the input impedance of the transistor in order to maximize power-added efficiency of the power amplifier.

In early radio-frequency vacuum-tube transmitters, it was observed that the tubes and associated circuits may have damped or undamped oscillations depending upon the circuit losses, the feedback coupling, the grid and anode potentials, and the reactance or tuning of the parasitic circuits [26,27]. Various parasitic oscillator circuits such as the tuned-grid–tuned-anode circuit with capacitive feedback, Hartley, Colpitts, or Meissner oscillators can be realized at high frequencies, which potentially can be eliminated by adding a small resistor close to the grid or anode connections of the tubes for damping the circuits. Inductively coupled rather than capacitively coupled input and output circuits should be used wherever possible.

According to the immittance approach applied to the stability analysis of the active nonreciprocal two-port network, it is necessary and sufficient for its unconditional stability if the following system of equations can be satisfied for the given active device:

$$\mathrm{Re}[W_\mathrm{S}(\omega) + W_\mathrm{in}(\omega)] > 0 \tag{1.63}$$

$$\mathrm{Im}[W_\mathrm{S}(\omega) + W_\mathrm{in}(\omega)] = 0 \tag{1.64}$$

or

$$\mathrm{Re}[W_\mathrm{L}(\omega) + W_\mathrm{out}(\omega)] > 0 \tag{1.65}$$

$$\mathrm{Im}[W_\mathrm{L}(\omega) + W_\mathrm{out}(\omega)] = 0 \tag{1.66}$$

where $\mathrm{Re}W_\mathrm{S}$ and $\mathrm{Re}W_\mathrm{L}$ are considered to be greater than zero [28,29]. The active two-port network can be treated as unstable or potentially unstable in the case of the opposite signs in (1.63) and (1.65).

Analysis of (1.63) or (1.65) on extremum results in a special relationship between the device immittance parameters called the device stability factor:

$$K = \frac{2\mathrm{Re}W_{11}\,\mathrm{Re}W_{22} - \mathrm{Re}(W_{12}W_{21})}{|W_{12}W_{21}|} \tag{1.67}$$

which shows a stability margin indicating how far from zero value are the real parts in (1.63) and (1.65) if they are positive [29]. An active device is unconditionally stable if $K \geq 1$ and potentially unstable if $K < 1$.

When the active device is potentially unstable, an improvement of the power amplifier stability can be provided with the appropriate choice of the source and load immittances W_S and W_L. In this case, the circuit stability factor K_T is defined in the same way as the device stability factor K, taking into account of $\mathrm{Re}W_S$ and $\mathrm{Re}W_L$ along with the device W-parameters, and written as

$$K_T = \frac{2\mathrm{Re}(W_{11} + W_S)\mathrm{Re}(W_{22} + W_L) - \mathrm{Re}(W_{12}W_{21})}{|W_{12}W_{21}|} \tag{1.68}$$

If the circuit stability factor $K_T \geq 1$, the power amplifier is unconditionally stable. However, the power amplifier becomes potentially unstable if $K_T < 1$. The value of $K_T = 1$ corresponds to the border of the circuit unconditional stability. The values of the circuit stability factor K_T and device stability factor K become equal if $\mathrm{Re}W_S = \mathrm{Re}W_L = 0$.

For the active device stability factor $K > 1$, the operating power gain G_P has to be maximized. By analyzing (1.65) on extremum, it is possible to find optimum values $\mathrm{Re}W_L^o$ and $\mathrm{Im}W_L^o$ when the operating power gain G_P is maximal [30,31]. As a result:

$$G_{P\max} = \left|\frac{W_{21}}{W_{12}}\right| \Big/ \left(K + \sqrt{K^2 - 1}\right). \tag{1.69}$$

The power amplifier with an unconditionally stable active device provides a maximum power gain operation only if the input and output of the active device are conjugately matched with the source and load impedances, respectively. For the lossless input matching circuit when the power available at the source is equal to the power delivered to the input port of the active device, that is, $P_S = P_{in}$, the maximum operating power gain is equal to the maximum transducer power gain, that is, $G_{P\max} = G_{T\max}$.

Domains of the device potential instability include the operating frequency ranges where the active device stability factor is equal to $K < 1$. Within the bandwidth of such a frequency domain, parasitic oscillations can occur, defined by internal positive feedback and operating conditions of the active device. The instabilities may not be self-sustaining, induced by the RF drive power but remaining on its removal. One of the most serious cases of the power amplifier instability can occur when there is a variation of the load impedance. Under these conditions, the transistor may be destroyed almost instantaneously. However, even it is not destroyed, the instability can result in an increased level of the spurious emissions in the output spectrum of the power amplifier tremendously. Generally, the following classification for linear instabilities can be made [32]:

- Low-frequency oscillations produced by thermal feedback effects;
- Oscillations due to internal feedback;

- Negative resistance or conductance-induced instabilities due to transit-time effects, avalanche multiplication, etc.; and
- Oscillations due to external feedback as a result of insufficient decoupling of the dc supply, etc.

Therefore, it is very important to determine the effect of the device feedback parameters on the origin of the parasitic self-oscillations and to establish possible circuit configurations of the parasitic oscillators. Based on the simplified bipolar equivalent circuit shown in Figure 1.13, the device stability factor can be expressed through the parameters of the transistor equivalent circuit as

$$K = 2r_b g_m \frac{1 + \frac{g_m}{\omega_T C_c}}{\sqrt{1 + \left(\frac{g_m}{\omega C_c}\right)^2}} \tag{1.70}$$

where $\omega_T = 2\pi f_T$ [18,33].

At very low frequencies, the bipolar transistors become potentially stable and the fact that $K \to 0$ when $f \to 0$ in (1.70) can be explained by simplifying the bipolar equivalent circuit. In practice, at low frequencies, it is necessary to take into account the dynamic base-emitter resistance r_π and early collector-emitter resistance r_{ce}, the presence of which substantially increase the value of the device stability factor. This gives only one unstable frequency domain with $K < 1$ and low-boundary frequency f_{p1}. However, an additional region of possible low-frequency oscillations can occur due to thermal feedback where the collector junction temperature becomes frequently dependent, and the common-base configuration is especially affected by this [34].

The high-boundary frequency of a frequency domain of the bipolar transistor potential instability can be determined by equating the device stability factor K to unity as

$$f_{p2} = \frac{g_m}{2\pi C_c} \Bigg/ \sqrt{(2r_b g_m)^2 \left(1 + \frac{g_m}{\omega_T C_c}\right)^2 - 1}. \tag{1.71}$$

When $r_b g_m > 1$ and $g_m >> \omega_T C_c$, (1.71) is simplified to

$$f_{p2} \approx \frac{1}{4\pi r_b C_\pi}. \tag{1.72}$$

At higher frequencies, a presence of the parasitic reactive intrinsic transistor parameters and package parasitics can be of great importance in view of the power amplifier stability. The parasitic series emitter lead inductance L_e shown in Figure 1.15 has a major effect on the device stability factor. The presence of L_e leads to the appearance of the second frequency domain of potential instability at higher frequencies. The circuit analysis shows that the second frequency domain of potential instability can be realized only under certain ratios between the normalized parameters $\omega_T L_e/r_b$ and $\omega_T r_b C_c$ [18,33]. For example, the second domain does not occur for any values of L_e when $\omega_T r_b C_c \geq 0.25$.

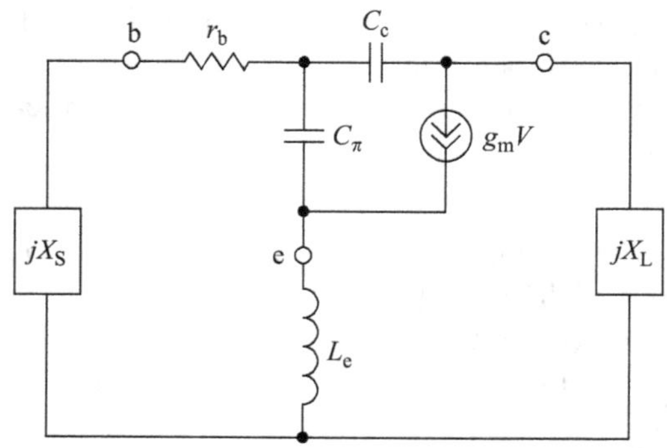

Figure 1.15 Simplified bipolar π-hybrid equivalent circuit with emitter lead inductance

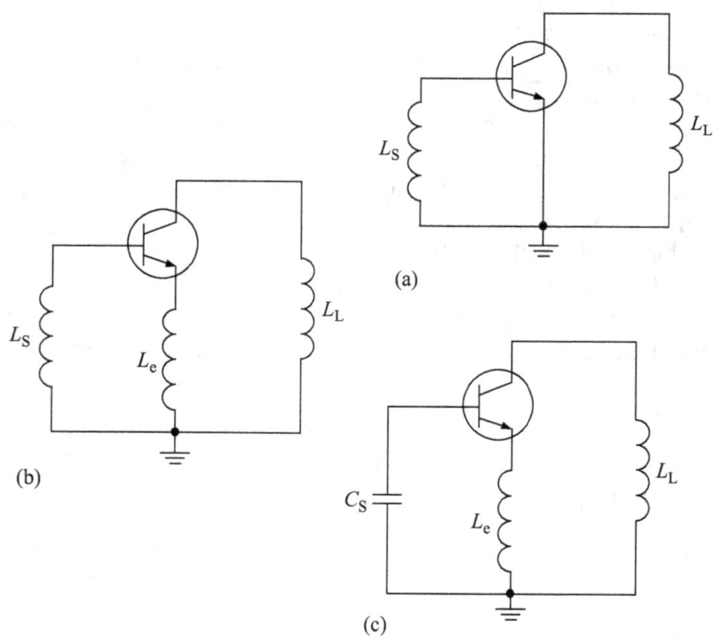

Figure 1.16 Equivalent circuits of parasitic bipolar oscillators

An appearance of the second frequency domain of the device potential instability is the result of the corresponding changes in the device feedback phase conditions and takes place only under a simultaneous effect of the collector capacitance C_c and emitter lead inductance L_e. If the effect of one of these factors is missing, the active device is characterized by only the first domain of its potential instability.

Figure 1.16 shows the potentially realizable equivalent circuits of the parasitic oscillators. If the value of a series-emitter inductance L_e is negligible, the parasitic oscillations can occur only when the values of the source and load reactances are positive, that is, $\text{Im}Z_S = X_S > 0$ and $\text{Im}Z_L = X_L > 0$. In this case, the parasitic oscillator shown in Figure 1.16(a) represents the inductive three-point circuit, where the inductive elements L_S and L_L in combination with the collector capacitance C_c form a Hartley oscillator. From a practical point of view, the more the value of the collector dc-feed inductance exceeds the value of the base-bias inductance, the more likely low-frequency parasitic oscillators can be created. It was observed that a very low inductance, even a short between the emitter and the base, can produce very strong and dangerous oscillations which may easily destroy a transistor [32]. Therefore, it is recommended to increase a value of the base choke inductance and to decrease a value of the collector choke inductance.

The presence of L_e leads to narrowing of the first frequency domain of the potential instability, which is limited to the high-boundary frequency f_{p2}, and can contribute to appearance of the second frequency domain of the potential instability at higher frequencies. The parasitic oscillator that corresponds to the first frequency domain of the device potential instability can be realized only if the source and load reactances are inductive, that is, $\text{Im}Z_S = X_S > 0$ and $\text{Im}Z_L = X_L > 0$, with the equivalent circuit of such a parasitic oscillator shown in Figure 1.16(b). The parasitic oscillator corresponding to the second frequency domain of the device potential instability can be realized only if the source reactance is capacitive and the load reactance is inductive, that is, $\text{Im}Z_S = -X_S < 0$ and $\text{Im}Z_L = X_L > 0$, with the equivalent circuit shown in Figure 1.16(c). The series emitter inductance L_e is an element of fundamental importance for the parasitic oscillator that corresponds to the second frequency domain of the device potential instability. It changes the circuit phase conditions so it becomes possible to establish the oscillation phase-balance condition at high frequencies. However, if it is possible to eliminate the parasitic oscillations at high frequencies by other means, increasing of L_e will result to narrowing of a low-frequency domain of potential instability, thus making the power amplifier potentially more stable, although at the expense of reduced power gain.

Similar analysis of the MOSFET power amplifier also shows two frequency domains of MOSFET potential instability due to the internal feedback gate-drain capacitance C_{gd} and series source inductance L_s [34]. Because of the very high gate-leakage resistance, the value of the low-boundary frequency f_{p1} is sufficiently small. For usually available conditions for power MOSFET devices when $g_m R_{ds} = 10$–30 and $C_{gd}/C_{gs} = 0.1$–0.2, the high boundary frequency f_{p2} can approximately be calculated from:

$$f_{p2} \approx \frac{1}{4\pi R_{gs} C_{gs}}. \tag{1.73}$$

It should be noted that power MOSFET devices have a substantially higher value of $g_m R_{ds}$ at small values of the drain current than at its high values. Consequently, for small drain current, the MOSFET device is characterized by a wider domain of potential instability. This domain is significantly wider than the same first domain of the potential instability of the bipolar transistor. The series source inductance L_s contributes to the appearance of the second frequency domain of the device potential instability. The potentially realizable equivalent circuits of the MOSFET parasitic oscillators are the same as for the bipolar transistor, as shown in Figure 1.16 [33].

Thus, to prevent the parasitic oscillations and to provide a stable operation mode of any power amplifier, it is necessary to take into consideration the following common requirements:

- Use an active device with stability factor $K > 1$;
- If it is impossible to choose an active device with $K > 1$, it is necessary to provide the circuit stability factor $K_T > 1$ by the appropriate choice of the real parts of the source and load immittances;
- Disrupt the equivalent circuits of the possible parasitic oscillators and
- Choose proper reactive parameters of the matching circuit elements adjacent to the input and output ports of the active device, which are necessary to avoid the self-oscillation conditions.

Generally, the parasitic oscillations can arise at any frequency within the potential instability domains for certain values of the source and load immittances W_S and W_L. The frequency dependences of W_S and W_L are very complicated and very often cannot be predicted exactly, especially in multistage power amplifiers. Therefore, it is very difficult to propose a unified approach to provide a stable operation mode of the power amplifiers with different circuit configurations and operation frequencies. In practice, the parasitic oscillations can arise close to the operating frequencies due to the internal positive feedback inside the transistor and at the frequencies sufficiently far from the operating frequencies due to the external positive feedback created by the surface mounted elements. As a result, the stability analysis of the power amplifier must include the methods to prevent the parasitic oscillations in different frequency ranges.

It should be noted that expressions in (1.63)–(1.69) are given by using the device immittance parameters that allow the power gain and stability to be calculated using the impedance *Z*- or admittance *Y*-parameters of the device equivalent circuit and to physically understand the corresponding effect of each circuit parameter, but not through the scattering *S*-parameters which are very convenient during the measurement procedure required for device modeling. Moreover, by using modern simulation tools, there is no need to even draw stability circles on a Smith chart or analyze stability factor across the wide frequency range since *K*-factor is just a derivation from the basic stability conditions and usually is a function of linear parameters, which can only reveal linear instabilities. Besides, it is difficult to predict unconditional stability for a multistage power amplifier because parasitic oscillations can be caused by the interstage circuits.

In this case, the easiest and most effective way to provide stable operation of the multistage power amplifier (or single-stage power amplifier) is to simulate the real part of the device input impedance $Z_{in} = V_{in}/I_{in}$ at the input terminal of each transistor across the entire frequency range as a ratio between the input voltage and current by placing a voltage node and a current meter, as shown in Figure 1.17(a).

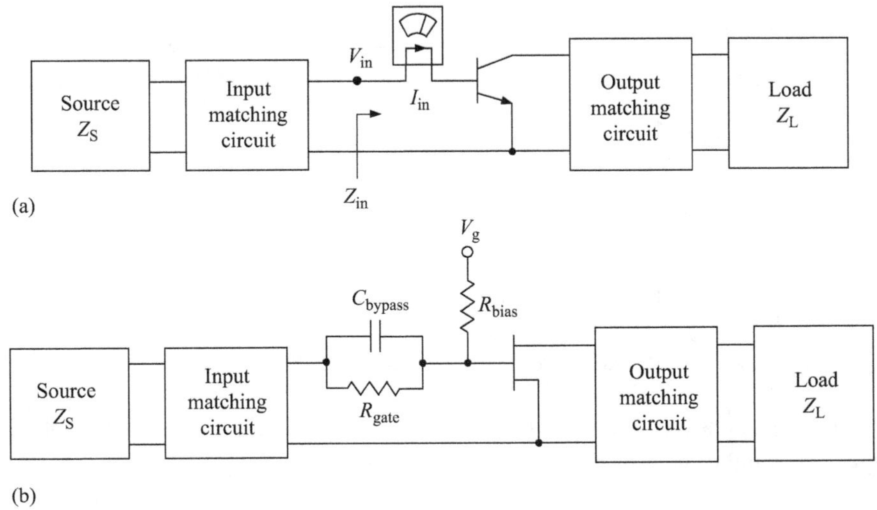

Figure 1.17 Single-stage power amplifiers with measured device input impedance

If ReZ_{in} < 0, then either a low-value series resistor must be added to the device base terminal as a part of the input matching circuit or a load-network configuration can be properly chosen to provide the resulting positive value of ReZ_{in}. In this case, not only linear instabilities with small-signal soft startup oscillation conditions but also nonlinear instabilities with large-signal hard startup oscillation conditions or parametric oscillations can be identified around operating region. Figure 1.17(b) shows the parallel RC stabilizing circuit with a bypass capacitor C_{bypass} connected in series to the input port of a GaN HEMT device [35]. In this case, using a stabilizing resistor R_{gate} and a low-value gate-bias resistor R_{bias} improves stability factor considerably at low frequencies without affecting the device performance at higher frequencies.

Figure 1.18 shows the example of a stabilized bipolar VHF power amplifier configured to operate in a zero-bias Class-C mode. Conductive input and output loading due to resistances R_1 and R_2 eliminate a low-frequency instability domain. The series inductors L_3 and L_4 contribute to higher power gain if the resistance values are too small, and can compensate for the capacitive input and output device impedances. To provide a negative-bias Class-C mode, the shunt inductor L_2 can be removed. The equivalent circuit of the potential parasitic oscillator at higher frequencies is realized by means of the parasitic reactive parameters of the transistor and external circuitry. The only possible equivalent circuit of such a parasitic oscillator at these frequencies is shown in Figure 1.16(c). It can only be realized if the series-emitter lead inductance is present. Consequently, the electrical length of the emitter lead should be reduced as much as possible, or, alternatively, the appropriate reactive immittances at the input and output transistor ports are provided. For example, it is possible to avoid the parasitic oscillations at these frequencies if the inductive immittance is provided at the input of the transistor and

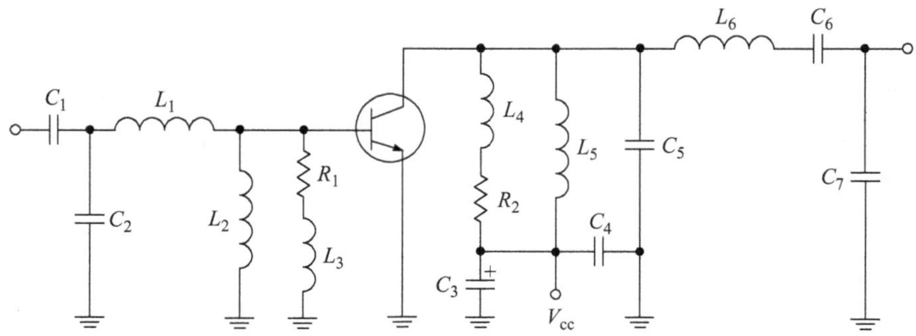

Figure 1.18 Stabilized bipolar Class C VHF power amplifier

capacitive reactance is provided at the output of the transistor. This is realized by an input series inductance L_1 and an output shunt capacitance C_5.

1.6 Impedance matching

In a common case, an optimum solution for impedance matching depends on the circuit requirements, such as the simplicity in practical realization, the frequency bandwidth and minimum power ripple, design implementation and adjustability, stable operation conditions, and sufficient harmonic suppression. As a result, many types of the matching networks are available which are based on the lumped elements and transmission lines. To simplify and visualize the matching design procedure, an analytical approach when all parameters of the matching circuits are calculated using simple analytical equations alongside with their Smith chart visualization can be used.

1.6.1 Basic principles

Impedance matching is necessary to provide maximum delivery to the load of the RF power available from the source by using some impedance matching network which can modify the load as viewed from the generator [36]. This means that generally, when the electrical signal propagates in the circuit, a portion of this signal might be reflected at the interface between the sections with different impedances. Therefore, it is necessary to establish the conditions that allow to fully transmitting the entire RF signal without any reflection. To determine an optimum value of the load impedance Z_L, at which the power delivered to the load is maximal, the equivalent circuit shown in Figure 1.19(a) can be considered.

In this case, the power delivered to the load can be defined as

$$P = \frac{1}{2} V_{in}^2 \mathrm{Re}\left(\frac{1}{Z_L}\right) = \frac{1}{2} V_S^2 \left|\frac{Z_L}{Z_S + Z_L}\right|^2 \mathrm{Re}\left(\frac{1}{Z_L}\right) \tag{1.74}$$

where $Z_S = R_S + jX_S$ is the source impedance, $Z_L = R_L + jX_L$ is the load impedance, V_S is the source voltage amplitude, and V_{in} is the load voltage amplitude.

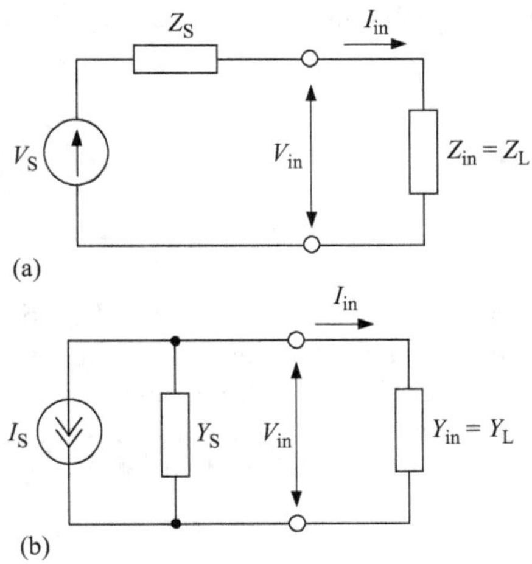

Figure 1.19 Equivalent circuits with (a) voltage and (b) current sources

Substituting the real and imaginary parts of the source and load impedances Z_S and Z_L into (1.74) yields:

$$P = \frac{1}{2} V_S^2 \frac{R_L}{(R_S + R_L)^2 + (X_S + X_L)^2} \tag{1.75}$$

If the source impedance Z_S is fixed, then it is necessary to vary the real and imaginary parts of the load impedance Z_L until maximum power is delivered to the load. To maximize the output power, the following analytical conditions in the form of derivatives with respect to the output power can be written:

$$\frac{\partial P}{\partial R_L} = 0 \qquad \frac{\partial P}{\partial X_L} = 0 \tag{1.76}$$

Applying these conditions and taking into consideration (1.75), the system of two equations can be obtained as

$$\frac{1}{(R_L + R_S) + (X_L + X_S)^2} - \frac{2R_L(R_L + R_S)}{\left[(R_L + R_S)^2 + (X_L + X_S)^2\right]^2} = 0 \tag{1.77}$$

$$\frac{2X_L(X_L + X_S)}{\left[(R_L + R_S)^2 + (X_L + X_S)^2\right]^2} = 0 \tag{1.78}$$

Simplifying (1.77) and (1.78) results in

$$R_S^2 - R_L^2 + (X_L + X_S)^2 = 0 \tag{1.79}$$

$$X_L(X_L + X_S) = 0 \tag{1.80}$$

By solving (1.79) and (1.80) simultaneously for R_S and X_S, one can obtain:

$$R_S = R_L \tag{1.81}$$

$$X_L = -X_S \tag{1.82}$$

or, in an impedance form:

$$Z_L = Z_S^* \tag{1.83}$$

where (*) denotes the complex-conjugate value [37]. Equation (1.83) is called an *impedance conjugate-matching condition*, and its fulfillment results in maximum power delivered to the load for fixed source impedance.

The *admittance conjugate-matching condition*, which are applied to the equivalent circuit shown in Figure 1.19(b) and given as

$$Y_L = Y_S^* \tag{1.84}$$

can be readily obtained in the same way.

Thus, the conjugate-matching conditions in a common case can be rewritten through the immittance parameters (representing any system of the impedance Z-parameters or admittance Y-parameters) as

$$W_L = W_S^* \tag{1.85}$$

The matching circuit is connected between the source and the input of an active device, as shown in Figure 1.20(a), and between the output of an active device and the load, as shown in Figure 1.20(b). For a multistage amplifier, the load represents an input circuit of the next stage. Therefore, the matching circuit, in this case, is connected between the output of the active device of the preceding amplifier stage and the input of the active device of the succeeding stage of the power amplifier, as shown in Figure 1.20(c). The main objective is to properly transform the load immittance W_L to the optimum device output immittance W_{out}, whose value is properly determined by the supply voltage, output power, device saturation voltage, and selected operation class to maximize the amplifier efficiency and output power.

It should be noted that (1.85) is given in a general immittance form without indication of whether it is used in a small-signal or large-signal application. In the latter case, this only means that the device immittance W-parameters are fundamentally averaged over large-signal swing across the device equivalent circuit parameters and that the conjugate-matching principle is valid in both the small-signal application and the large-signal application, where the optimum equivalent device output resistance (or conductance) at the fundamental frequency is matched to the load resistance (or conductance) and the effect of the device output reactive elements is compensated by the conjugate reactance of the load network. In addition, the matching circuits should be designed to provide the required voltage and current waveforms at the device output and the stability of operation conditions. The losses in the output matching circuits must be as small as possible to deliver the output power to the load with maximum efficiency. Finally, it is desirable that the matching circuit be easy to tune.

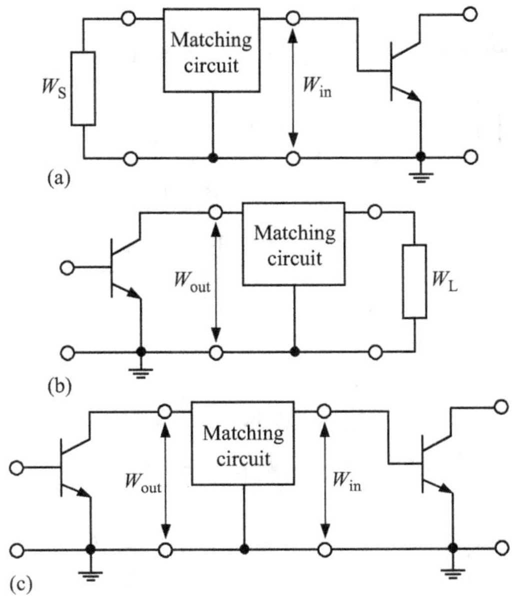

Figure 1.20 Matching circuit arrangements

1.6.2 Matching with lumped elements

Generally, there is a variety of configurations for matching networks to efficiently deliver signal from the source to the load, and application of any of these matching networks in the power amplifier depends on its class of operation, level of output power, operating frequency, frequency bandwidth, or level of harmonic suppression.

The lumped matching networks in the form of (a) *L*-transformer, (b) *Π*-transformer, or (c) *T*-transformer shown in Figure 1.21, respectively, have proved for a long time to be effective for power amplifier design [36]. The simplest and most popular matching network is the circuit in the form of the *L*-transformer. The transforming properties of this matching circuit can be analyzed by using the equivalent transformation of a parallel into a series representation of *RX* network.

Consider the parallel *RX* network shown in Figure 1.22(a), where R_1 is the real (resistive) part and X_1 is the imaginary (reactive) part of the network impedance $Z_1 = jX_1R_1/(R_1 + jX_1)$, and the series *RX* network shown in Figure 1.22(b), where R_2 is the resistive part and X_2 is the reactive part of the circuit impedance $Z_2 = R_2 + jX_2$. These two impedance networks (series and parallel) can be considered equivalent at some frequency if $Z_1 = Z_2$, resulting in

$$R_2 + jX_2 = \frac{R_1X_1^2}{R_1^2 + X_1^2} + j\frac{R_1^2X_1}{R_1^2 + X_1^2} \tag{1.86}$$

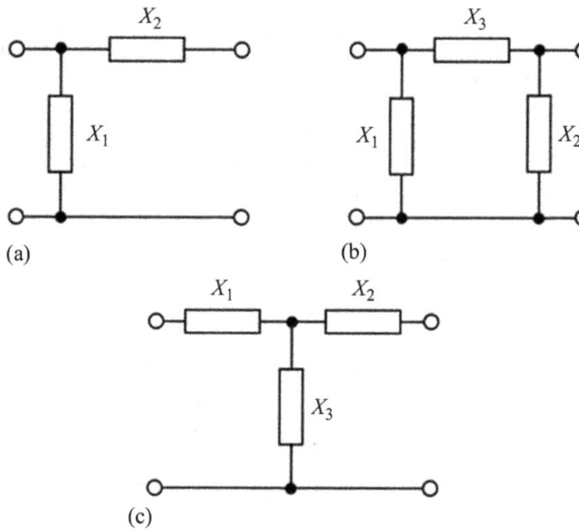

Figure 1.21 Matching networks in the form of (a) L-, (b) Π-, and (c) T-transformers

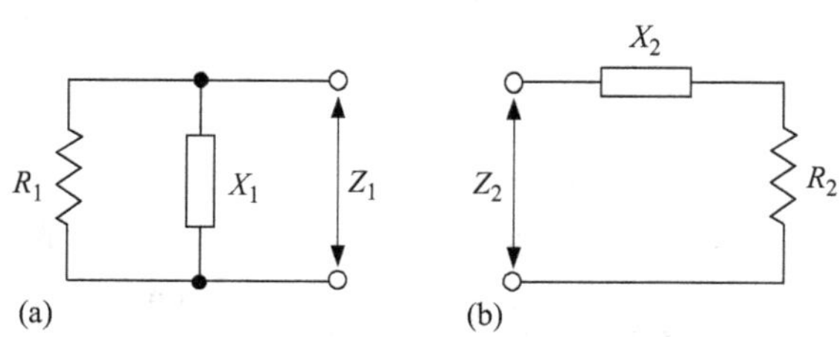

Figure 1.22 Impedance (a) parallel and (b) series equivalent networks

Equation (1.86) can be rearranged into two separate equations for real and imaginary parts as

$$R_1 = R_2(1 + Q^2) \tag{1.87}$$

$$X_1 = X_2(1 + Q^{-2}) \tag{1.88}$$

where $Q = R_1/X_1 = X_2/R_2$ is the network (or loaded) quality factor, which is equal for both series and parallel RX networks.

Consequently, if the reactive impedance or reactance $X_1 = -X_2(1 + Q^{-2})$ is connected in parallel to the series circuit composed of the resistance R_2 and reactance X_2, it allows the reactance of the series circuit to be compensated. In this case, the input impedance of such a two-port network, which is shown in Figure 1.23, will be resistive only and equal to R_1. Consequently, to transform the resistance R_1 into the other resistance R_2 at the certain frequency, it is sufficient to connect between them a two-port L-transformer with the opposite signs of the reactances X_1

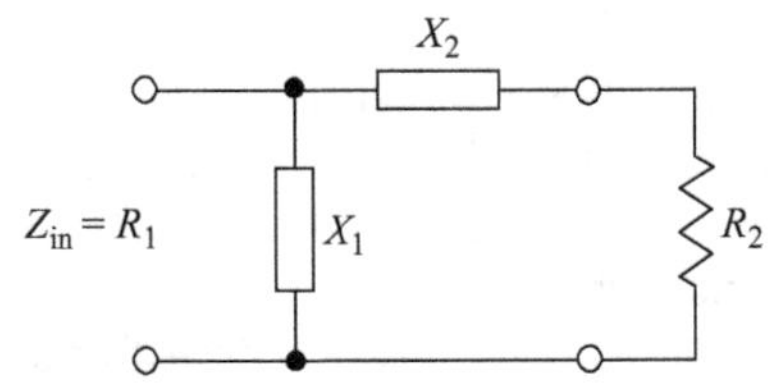

Figure 1.23 Input impedance of two-port network

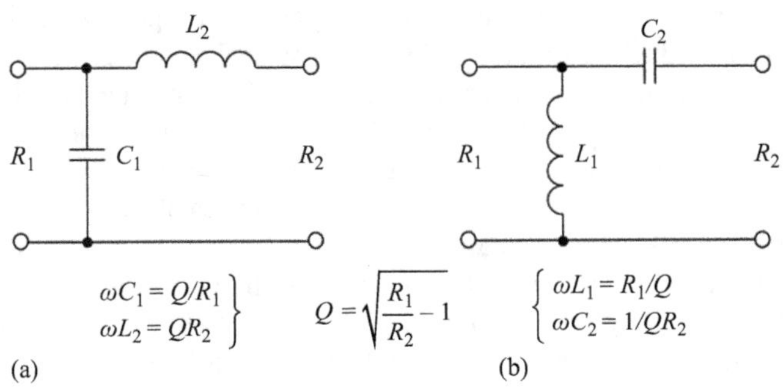

Figure 1.24 L-type matching circuits and relevant equations

and X_2, whose parameters can be easily obtained from the following simple design equations:

$$|X_1| = \frac{R_1}{Q} \tag{1.89}$$

$$|X_2| = R_2 Q \tag{1.90}$$

where

$$Q = \sqrt{\frac{R_1}{R_2} - 1} \tag{1.91}$$

is the loaded quality factor expressed through the resistances to be matched. Thus, to design a matching circuit with fixed resistances to be matched, first need to calculate the loaded quality factor Q according to (1.91) and then to define the reactive elements, according to (1.89) and (1.90).

Due to the opposite signs of the reactances X_1 and X_2, the two possible circuit configurations (one in the form of a low-pass filter section and another in the form of a high-pass filter section) with the same transforming properties can be realized, which are shown in Figure 1.24 together with the design equations. In practice, the single two-port L-transformers can be used as the input or interstage matching circuits in power amplifiers, where the requirements for the out-of-band suppression and harmonic control required for higher efficiency are not as high as for the output matching circuits. In this case, the main advantage of such an L-transformer is its simplicity when the only two reactive elements with fast tuning are needed. For larger values of $Q \geq 10$, it is possible to use a cascade connection of L-transformers, which allows wider frequency bandwidth and transformer efficiency to be realized.

The matching circuits in the form of (a) Π-transformer and (b) T-transformer can be realized by appropriate connection of two L-transformers, as shown in Figure 1.25. For each L-transformer, the resistance R_1 and the resistance R_2 are transformed to some intermediate resistance R_0 with the value of $R_0 < (R_1, R_2)$ for a Π-transformer and the value of $R_0 > (R_1, R_2)$ for a T-transformer. The value of R_0 is not fixed and can be chosen arbitrary depending on the frequency bandwidth. This means that, compared to the simple L-transformer with fixed parameters for the same ratio of R_2/R_1, the circuit parameters of the Π- or T-transformer can be different. However, they provide narrower frequency bandwidths due to higher quality factors because the intermediate resistance R_0 is either greater or smaller than each of the resistances R_1 and R_2. By taking into account the two possible circuit configurations of the L-transformer shown in Figure 1.24, it is possible to develop the different circuit configurations of the two-port transformers shown in Figure 1.25(a), where $X_3 = X_3' + X_3''$, and in Figure 1.25(b), where $X_3 = X_3' X''/(X_3' + X_3'')$. In this case, a Π-transformer with two shunt capacitors will represent a face-to-face connection of the two low-pass L-transformers without special requirement for the resistances R_1 and R_2, which means that the ratio R_1/R_2 can be greater or smaller than unity.

However, a Π-transformer with series capacitor representing a face-to-face connection of the high-pass and low-pass L-transformers, as shown in Figure 1.26(a), can only be used for impedance matching when $R_1/R_2 > 1$. The design equations for a high-pass section are written using (1.89)–(1.91) as

$$\omega L_1 = \frac{R_1}{Q_1} \tag{1.92}$$

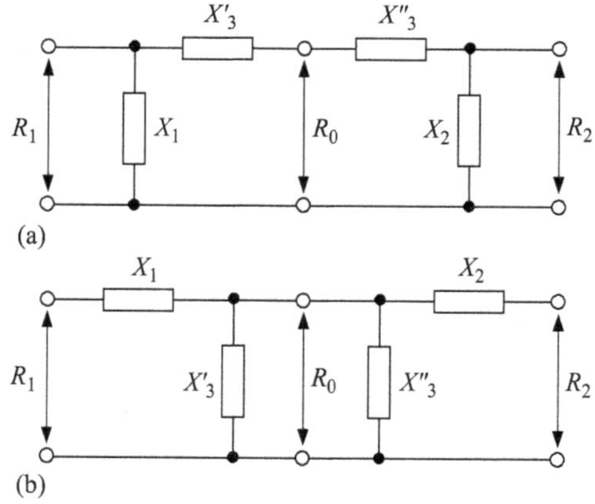

Figure 1.25 Matching circuits developed by connecting two L-transformers

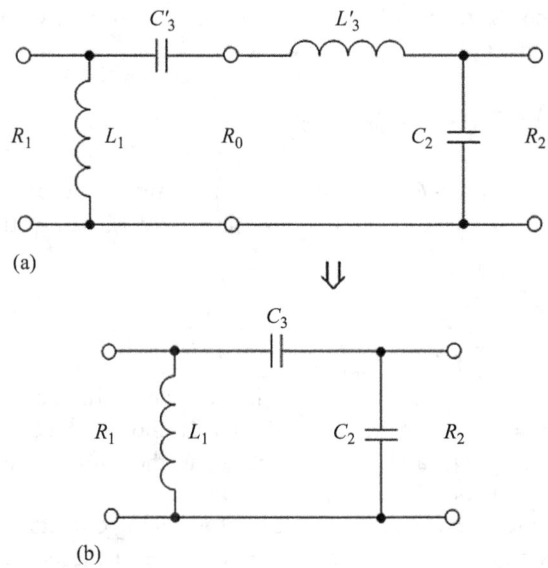

Figure 1.26 Circuit structures of Π-transformer with series capacitor

$$\omega C_3' = \frac{1}{Q_1 R_0} \tag{1.93}$$

$$Q_1^2 = \frac{R_1}{R_0} - 1 \tag{1.94}$$

where R_0 is the intermediate resistance.

Similarly, for a low-pass section:

$$\omega C_2 = \frac{Q_2}{R_2} \tag{1.95}$$

$$\omega L_3' = Q_2 R_0 \tag{1.96}$$

$$Q_2^2 = \frac{R_2}{R_0} - 1 \tag{1.97}$$

Since it is assumed that $R_1 > R_2 > R_0$, from (1.94) it follows that the loaded quality factor Q_1 of a high-pass L-transformer can be chosen from the condition:

$$Q_1^2 > \frac{R_1}{R_2} - 1 \tag{1.98}$$

Substituting (1.94) into (1.97) results in

$$Q_2 = \sqrt{\frac{R_2}{R_1}(1 + Q_1^2) - 1}. \tag{1.99}$$

Combining the reactances of two series elements (capacitor C_3' and inductor L_3') defined by (1.93) and (1.96), respectively, yields:

$$\omega L_3' - \frac{1}{\omega C_3'} = R_0(Q_2 - Q_1) = \frac{R_2(Q_2 - Q_1)}{(1 + Q_2^2)} \tag{1.100}$$

As a result, since $Q_1 > Q_2$, the total series reactance is negative, which can be provided by a series capacitance C_3, as shown in Figure 1.26(b), with a susceptance:

$$\omega C_3 = \frac{1 + Q_2^2}{R_2(Q_1 - Q_2)}. \tag{1.101}$$

On the other hand, if $Q_2 > Q_1$ when $R_2 > R_1 > R_0$, the total series reactance is positive, which can be provided by a series inductance L_3. In this case, it needs first to choose the value of Q_2 for fixed resistances R_1 and R_2 to be matched, then to determine the value of Q_1, and finally to calculate the values of the shunt inductance L_1, shunt capacitance C_2, and series inductance L_3.

At microwave frequencies, lumped-element matching circuits are very useful to provide a broadband operation and to reduce the size of the power amplifiers for miniaturization simultaneously. Figure 1.27 shows the circuit schematic of a broadband medium-power single-stage GaAs MESFET amplifier, where the matching circuits in the form of the Π-transformers with two inductors and one capacitor on each side of the transistor can cover the frequency bandwidth of 6 to 12 GHz [38]. Here, the shunt inductors L_2 and L_3 implemented with ribbons essentially affect the low-frequency gain response, whereas the series low-pass networks (L_1, C_1 and L_4, C_2) match the high end of the frequency bandwidth and selectively mismatch the device at the low end.

Now let us demonstrate a lumped matching circuit technique at very high frequencies to design a 150-W MOSFET power amplifier with a supply voltage of 50 V operating in a frequency bandwidth of 132 to 174 MHz ($\Delta f = 42$ MHz) and providing a power gain greater than 10 dB. In this case, the center bandwidth frequency is equal to $f_0 = \sqrt{132 \cdot 174} = 152\,\text{MHz}$. For example, from the datasheet on the MOSFET device it follows that the values of the input and output impedances at this frequency are $Z_{in} = (0.9 - j1.2)\ \Omega$ and $Z_{out} = (1.8 + j2.1)\ \Omega$, respectively, where the input reactance $\text{Im}Z_{in}$ is capacitive and the output reactance $\text{Im}Z_{out}$ is inductive due to inductive effect of both drain bondwire and output

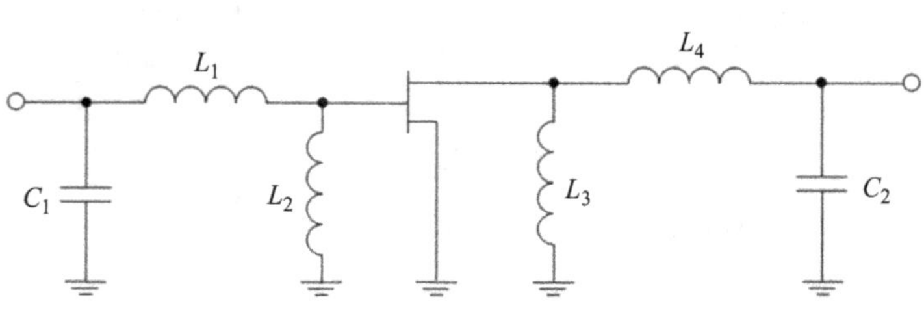

Figure 1.27 Schematic of broadband lumped-element FET amplifier

package lead. To cover the required frequency bandwidth, the low-Q matching circuits should be used that allows reduction of the in-band amplitude ripple and improvement of the input return loss. The value of a network quality factor for 3-dB bandwidth level must be less than $Q = f_0/\Delta f = 152$ MHz/42 MHz = 3.6. In this case, it is very convenient to design the input and output matching circuits using the simple L-transformers in the form of low-pass and high-pass filter sections connected in series [39].

Hence, to match the required low-value input series capacitive impedance to the standard 50-Ω source impedance within a frequency bandwidth of 27.6%, it needs to use at least three filter sections, as shown in Figure 1.28. From the negative imaginary part of the input impedance Z_{in} it follows that the input capacitance C_{in} at the operating frequency of 152 MHz is equal to approximately 873 pF. To compensate for this capacitive reactance at the center bandwidth frequency, it is sufficient to connect an inductance of 1.3 nH in series to the device input capacitance. When the device input capacitive reactance has been compensated, next step is to proceed with the design of the input matching circuit. To simplify the matching design procedure, it is best to cascade the L-transformers with equal value of Q [33]. Although equal-Q values are not absolutely necessary, this provides a convenient guide for both analytical calculation of the matching circuit parameters and the Smith chart graphical design.

In this case, the following ratio can be written for the input matching circuit:

$$\frac{R_1}{R_2} = \frac{R_2}{R_3} = \frac{R_3}{R_{in}} \tag{1.102}$$

resulting in $R_2 = 13$ Ω and $R_3 = 3.5$ Ω for $R_{source} = R_1 = 50$ Ω and $R_{in} = \text{Re}Z_{in} = 0.9$ Ω. Consequently, a loaded quality factor of each L-transformer according to (1.91) is equal to $Q = 1.7$. The elements of the input matching circuit using equations given in Figure 1.24 can be calculated as $L_1 = 31$ nH, $C_1 = 47$ pF, $L_2 = 6.2$ nH, $C_2 = 137$ pF, $L_3 = 1.6$ nH, and $C_3 = 509$ pF.

This equal-Q approach significantly simplifies the matching circuit design using the Smith chart. In this case, it is necessary first to plot two circles of equal-Q values on the Smith chart. The circle of equal Q is plotted taking into account that a ratio X/R or B/G must be the same for each point located at this circle. Then, each element of the input matching circuit can be readily determined, as shown in Figure 1.29. Each

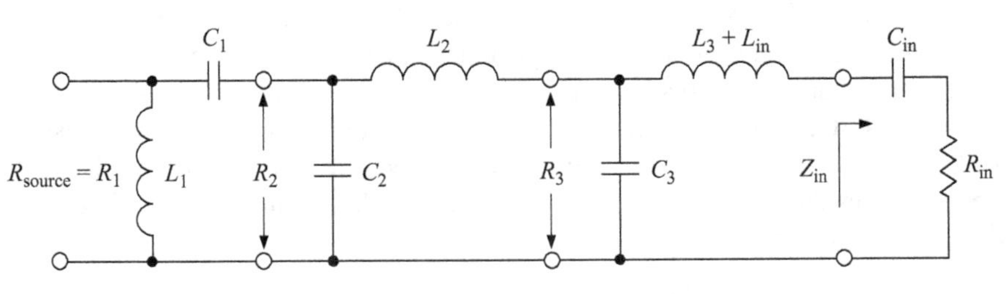

Figure 1.28 Complete broadband input matching circuit

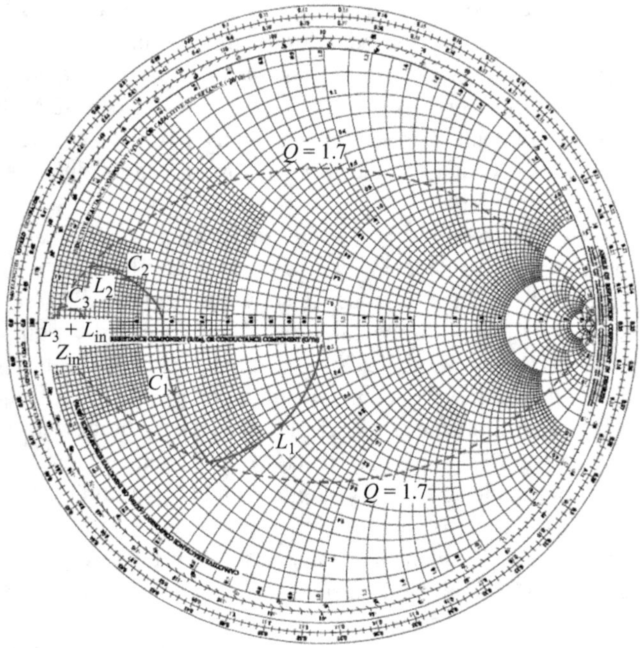

Figure 1.29 Smith chart with elements from Figure 1.28

trace for the series inductance and capacitance must be plotted as far as the intersection point with Q-circle, whereas each trace for the shunt capacitance and inductance should be plotted until the intersection with horizontal real axis.

1.6.3 Matching with transmission lines

At very high frequencies, it is very difficult to implement lumped elements with predefined accuracy in view of a significant effect of their parasitic parameters, for example, the parasitic interturn and direct-to-ground capacitances for lumped inductors and the stray inductance for lumped capacitors. However, these parasitic parameters can represent a part of a distributed LC structure such as a transmission line. In this case, for a microstrip line, the series inductance is associated with the flow of current in the conductor and the shunt capacitance is associated with the strip separated from the ground by the dielectric substrate. If the line is wide, the inductance is reduced but the capacitance is large. However, for a narrow line, the inductance is increased but the capacitance is small.

Figure 1.30 shows an impedance matching circuit in the form of a transmission-line transformer connected between the source impedance Z_S and load impedance Z_L. The input impedance as a function of the electrical length of the transmission line with arbitrary load impedance is defined as

$$Z_{in} = Z_0 \frac{Z_L + jZ_0 \tan\theta}{Z_0 + jZ_L \tan\theta} \tag{1.103}$$

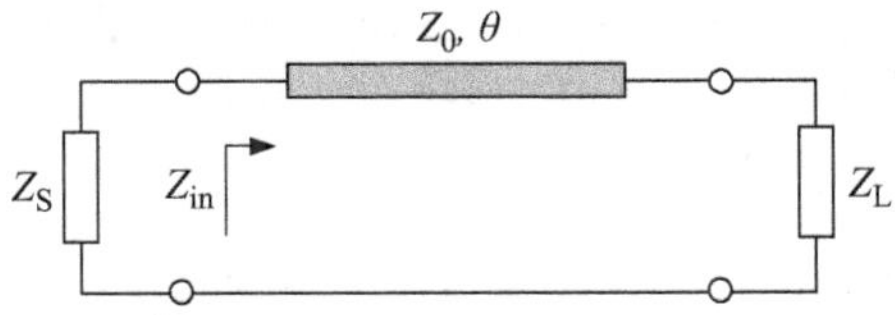

Figure 1.30 Transmission-line impedance transformer

where Z_0 is the characteristic impedance and θ is the electrical length of the transmission line [40,41].

For a quarter-wavelength transmission line when $\theta = \pi/2$ or 90°, the expression for Z_{in} in (1.103) reduces to

$$Z_{in} = Z_0^2/Z_L \tag{1.104}$$

from which it follows that, for example, a 50-Ω load can be matched to a 12.5-Ω source with the transmission-line characteristic impedance of 25 Ω.

Usually, such a quarter-wavelength impedance transformer is used for impedance matching in a narrow frequency bandwidth of 10% to 20%, and its length is chosen at the bandwidth center frequency. However, using a multisection quarter-wave transformer widens the bandwidth and expands the choice of the substrate to include materials with high dielectric permittivity, which reduces the transformer size. For example, by using a transformer composed of seven quarter-wavelength transmission lines of different characteristic impedances, whose lengths are selected at the highest bandwidth frequency, the power gain flatness of ±1 dB was achieved over a frequency range of 5 to 10 GHz for a 15-W GaAs MESFET power amplifier [42].

To provide a complex-conjugate matching of the input transmission line impedance Z_{in} with the source impedance $Z_S = R_S + jX_S$ when $R_S = \text{Re}Z_{in}$ and $X_S = -\text{Im}Z_{in}$, (1.103) can be rewritten as

$$R_S - jX_S = Z_0 \frac{R_L + j(X_L + Z_0 \tan\theta)}{Z_0 - X_L \tan\theta + jR_L \tan\theta} \tag{1.105}$$

Generally, (1.105) can be divided into two equations representing the real and imaginary parts as

$$R_S(Z_0 - X_L \tan\theta) - R_L(Z_0 - X_S \tan\theta) = 0 \tag{1.106}$$

$$X_S(X_L \tan\theta - Z_0) - Z_0(X_L + Z_0 \tan\theta) + R_S R_L \tan\theta = 0 \tag{1.107}$$

Solving (1.106) and (1.107) for the two independent variables Z_0 and θ yields:

$$Z_0 = \sqrt{\frac{R_S(R_L^2 + X_L^2) - R_L(R_S^2 + X_S^2)}{R_L - R_S}} \tag{1.108}$$

$$\theta = \tan^{-1}\left(Z_0 \frac{R_S - R_L}{R_S X_L - X_S R_L}\right) \tag{1.109}$$

As a result, the transmission line having the characteristic impedance Z_0 determined by (1.108) and the electrical length θ determined by (1.109) can match any source and load impedances when the impedance ratio gives a positive value inside the square root in (1.108).

In practice, to simplify the power amplifier designs at microwave frequencies, the simple matching circuits are very often used, which can include an L-transformer with a series transmission line as the basic matching section. It is convenient to analyze the transforming properties of this matching circuit by substituting the equivalent transformation of the parallel RX circuit to the series one. For example, R_1 is the resistance and $X_1 = -1/\omega C$ is the reactance of the impedance $Z_1 = jR_1X_1/(R_1 + jX_1)$ for the parallel RC circuit, and $R_{in} = \mathrm{Re}Z_{in}$ is the resistance and $X_{in} = \mathrm{Im}Z_{in}$ is the reactance of the input impedance $Z_{in} = R_{in} + jX_{in}$ for the loaded series transmission-line circuit shown in Figure 1.31. For a conjugate matching when $Z_1 = Z_{in}^*$, one can obtain:

$$\frac{R_1X_1^2}{R_1^2 + X_1^2} + j\frac{R_1^2X_1}{R_1^2 + X_1^2} = R_{in} - jX_{in}. \tag{1.110}$$

The solution of (1.110) can be rewritten in the form of two expressions for real and imaginary impedance parts by

$$R_1 = R_{in}(1 + Q^2) \tag{1.111}$$

$$X_1 = -X_{in}(1 + Q^{-2}) \tag{1.112}$$

where $Q = R_1/|X_1| = X_{in}/R_{in}$ is a quality factor equal for both parallel capacitive and series transmission-line circuits. From (1.103), the real and imaginary parts of the input impedance Z_{in} can be written as

$$R_{in} = Z_0^2R_2\frac{1 + \tan^2\theta}{Z_0^2 + (R_2 \tan\theta)^2} \tag{1.113}$$

$$X_{in} = Z_0 \tan\theta\frac{Z_0^2 - R_2^2}{Z_0^2 + (R_2 \tan\theta)^2} \tag{1.114}$$

From (1.114), it follows that an inductive input impedance (necessary to compensate for the capacitive parallel component) is provided when $Z_0 > R_2$ for $\theta < \pi/2$ and $Z_0 < R_2$ for $\pi/2 < \theta < \pi$. As a result, to transform the resistance R_1 into

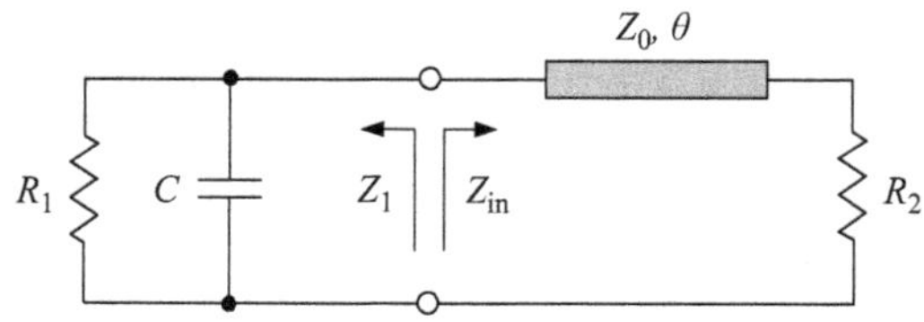

Figure 1.31 L-transformer with series transmission line

the other resistance R_2 at the given frequency, it is necessary to connect a two-port L-transformer with shunt capacitor and series transmission line between them. When one parameter (usually the characteristic impedance Z_0) is known in advance, the matching circuit parameters can be calculated from the following two equations:

$$C = \frac{Q}{\omega R_1} \tag{1.115}$$

$$\sin 2\theta = \frac{2Q}{\frac{Z_0}{R_2} - \frac{R_2}{Z_0}} \tag{1.116}$$

where

$$Q = \sqrt{\frac{R_1}{R_2}\left[\cos^2\theta + \left(\frac{R_2}{Z_0}\right)^2 \sin^2\theta\right] - 1} \tag{1.117}$$

is the loaded quality factor defined as a function of the resistances R_1 and R_2 and the parameters of the transmission line (characteristic impedance Z_0 and electrical length θ).

As it follows from (1.116) and (1.117), the electrical length θ can be calculated as a result of the numerical solution of a transcendental equation with one unknown parameter θ. In this case, it is more convenient to rewrite them in the implicit form:

$$\frac{R_1}{R_2} = \frac{1 + \left(\frac{Z_0}{R_2} - \frac{R_2}{Z_0}\right)^2 \sin^2\theta \cos^2\theta}{\cos^2\theta + \left(\frac{R_2}{Z_0}\right)^2 \sin^2\theta} \tag{1.118}$$

A Π-transformer can be realized by back-to-back connection of the two L-transformers, as shown in Figure 1.25(a), where the resistances R_1 and R_2 are transformed to some intermediate resistance R_0. In this case, to minimize the electrical length of the transmission line, the value of R_0 should be smaller than that of both R_1 and R_2, that is, $R_0 < (R_1, R_2)$. The same procedure for a T-transformer shown in Figure 1.26(b) gives the value of R_0 that is larger than that of both R_1 and R_2, that is, $R_0 > (R_1, R_2)$. Then, for a T-transformer, the two shunt adjacent capacitances are combined. For a Π-transformer, the two adjacent series transmission lines are combined into a single transmission line with a total electrical length.

Consider the design example of a broadband 150-W power amplifier for TV applications, which is required to operate over a frequency bandwidth of 470 to 860 MHz with a power gain of more than 10 dB at a supply voltage of 28 V [43]. In this case, it is convenient to use a high-power balanced LDMOS transistor as an active device, which is specially designed for UHF TV transmitters. Let us assume that the manufacturer states the input impedance for each transistor-balanced part as $Z_{in} = (1.7 + j1.3)\ \Omega$ at the center bandwidth frequency $f_0 = \sqrt{470 \times 860} = 635$ MHz. The input impedance Z_{in} represents a series combination of the input resistance and inductive reactance. To cover the required frequency bandwidth, the low-Q matching circuits should be used to reduce an in-band amplitude ripple and to improve an input return loss.

To achieve a 3-dB frequency bandwidth, the value of a network quality factor must be less than $Q = 635/(860 - 470) = 1.63$. Based on this value of Q, the next step is to define a number of matching sections. For example, for a single-stage input lumped matching circuit, the value of a quality factor Q is chosen as

$$Q > \sqrt{\frac{50}{1.7} - 1} = 5.33$$

which means that the entire frequency range can be appropriately cover using a multistage matching circuit only.

In this case, the device input quality factor is calculated as $Q_{in} = 1.3/1.7 = 0.76$, which is smaller than the required value of 1.63 to provide the broadband performance. It is very convenient to design the input matching circuit (as well as the output matching circuit) by using simple low-pass L-transformers composed of a series transmission line and a shunt capacitor each with a constant value of Q for both balanced parts of the active device. Then, these two input matching circuits are combined by inserting the shunt capacitors, whose values are reduced by two times, between the two series transmission lines.

To match the series input inductive impedance Z_{in} with the standard 50-Ω source, it is best to use three low-pass transmission-line L-transformers, as shown in Figure 1.32. In this case, the input resistance R_{in} can be assumed to be constant over the entire frequency range. At the center bandwidth frequency of 635 MHz, the input inductance is approximately equal to 0.3 nH. Taking this inductance into account, it is necessary to subtract the appropriate value of the electrical length θ_{in} from the total electrical length θ_3. Due to the short-length size of this transmission line when $\tan \theta_{in} \approx \theta_{in}$, a value of θ_{in} can be easily calculated as

$$\theta_{in} \cong \frac{X_{in}}{Z_0} = \frac{\omega L_{in}}{Z_0} \tag{1.119}$$

According to (1.117), there are two simple possibilities to provide an input matching using a technique with equal quality factors of the L-transformers. One option is to use the same values of the characteristic impedance for all transmission lines, and the other one is to use the same electrical lengths for all transmission lines.

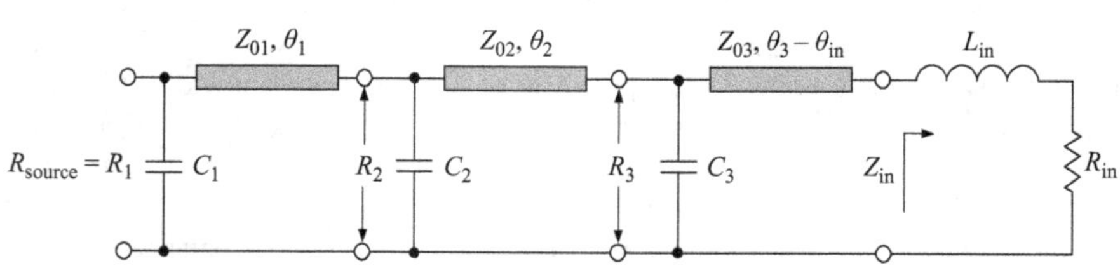

Figure 1.32 Complete broadband input matching circuit

When considering the first approach, which also allows direct use of the Smith chart, it is convenient to choose the value of the characteristic impedance as $Z_0 = Z_{01} = Z_{02} = Z_{03} = 50\ \Omega$. In this case, the ratio of the input and output resistances can be written as

$$\frac{R_1}{R_2} = \frac{R_2}{R_3} = \frac{R_3}{R_{in}} \tag{1.120}$$

which results in $R_2 = 16.2\ \Omega$ and $R_3 = 5.25\ \Omega$ for $R_{source} = R_1 = 50\ \Omega$ and $R_{in} = 1.7\ \Omega$. The values of the corresponding electrical lengths are determined from (1.118) as $\theta_1 = 30°$, $\theta_2 = 7.5°$, and $\theta_3 = 2.4°$.

To calculate the quality factor Q (equal for each L-transformer) from (1.117), it is enough to know the electrical length θ_1 of the first L-transformer. The remaining two electrical lengths θ_2 and θ_3 can be directly obtained from (1.116). As a result, the quality factor of each L-transformer is equal to a value of $Q = 1.2$. The values of the shunt capacitors using (1.115) are $C_1 = 6$ pF, $C_2 = 19$ pF, and $C_3 = 57$ pF.

For a constant Q, we can simplify significantly the design of the multisection matching circuit by using the Smith chart. After calculating the value of Q, it is necessary to plot a constant Q-circle on the Smith chart. Figure 1.33 shows the input matching circuit design using the Smith chart with a constant Q-circle, where the curves for the series transmission lines represent the arcs of the circles with

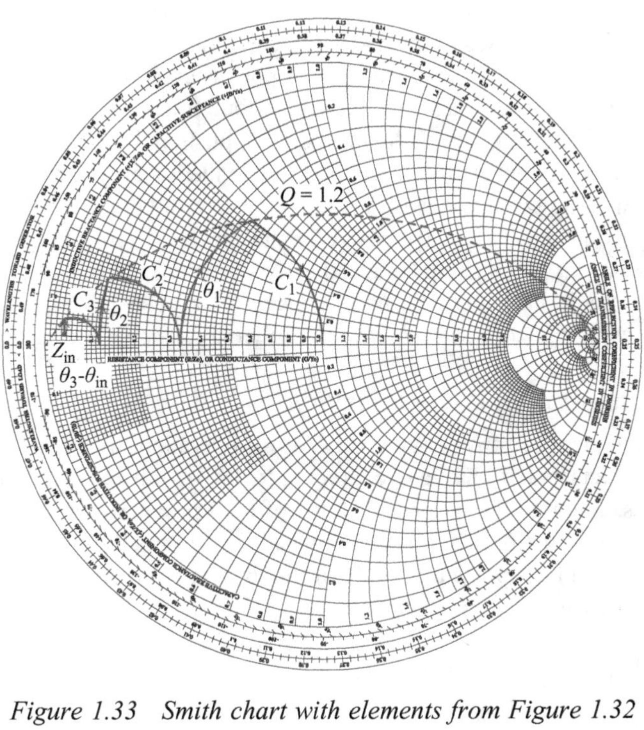

Figure 1.33 Smith chart with elements from Figure 1.32

center point at the center of the Smith chart. The capacitive traces are moved along the circles with the increasing susceptances and constant conductances.

Another approach assumes the same values of the electrical lengths $\theta = \theta_1 = \theta_2 = \theta_3$ and calculates the characteristic impedances of series transmission lines from (1.117) at equal ratios of the input and output resistances according to (1.120). Such an approach is more convenient in practical design, because, when using the transmission lines with standard characteristic impedance $Z_0 = 50\ \Omega$, the electrical length of the transmission line adjacent to the transistor input terminal is usually too short. In this case, it makes sense to set the characteristic impedance $Z_{01} = 50\ \Omega$ for the first transmission line only. Then, the value of $\theta = 30°$ is determined from (1.118). The subsequent calculation of Q from (1.117) for fixed θ and Z_0/R_2 yields $Q = 1.2$. The characteristic impedances of the remaining two transmission lines are then calculated from (1.116). Their values are $Z_{02} = 15.7\ \Omega$ and $Z_{03} = 5.1\ \Omega$ with the same values of the shunt capacitances.

1.7 Push–pull and balanced power amplifiers

Generally, if it is necessary to increase an overall output power of the power amplifier, several active devices can be used in parallel or push–pull configurations. In a parallel configuration, the active devices are not isolated from each other which requires a very good circuit symmetry, and output impedance becomes too small in the case of high output power. The latter drawback can be eliminated in a push–pull configuration, which provides increased values of the input and output impedances. In this case, for the same output power level, the input impedance Z_{in} and output impedance Z_{out} are approximately four times as high as that of in a parallel connection of the active devices since a push–pull arrangement is essentially a series connection. At the same time, the loaded quality factors of the input and output matching circuits remain unchanged because both the real and reactive parts of these impedances are increased by the factor of four. Very good circuit symmetry can be provided using balanced active devices with common emitters (or sources) in a single package. The basic concept of a push–pull operation can be analyzed by using the corresponding circuit schematic shown in Figure 1.34 [44].

1.7.1 Basic push–pull configuration

It is most convenient to consider an ideal Class-B operation, which means that each transistor conducts exactly half a cycle (equal to 180°) with zero quiescent current. Let us also assume that the number of turns of both primary and secondary windings of the output transformer T_2 is equal when $n_1 = n_2$ and the collector current of each transistor can be represented in the following half-sinusoidal form:
for the first transistor

$$i_{c1} = \begin{cases} +I_c \sin\omega t & 0 \le \omega t < \pi \\ 0 & \pi \le \omega t < 2\pi \end{cases} \tag{1.121}$$

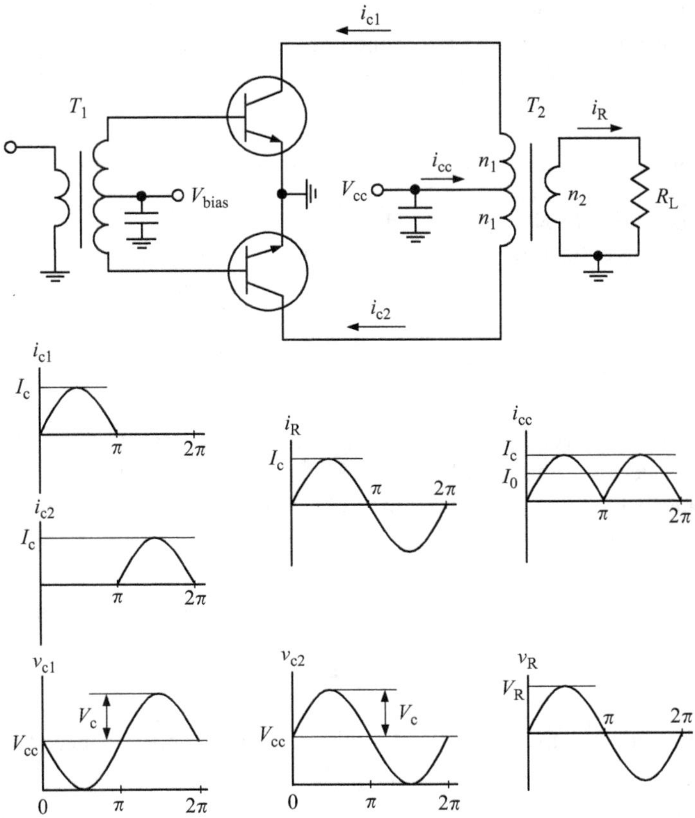

Figure 1.34 Basic concept of push–pull operation

for the second transistor

$$i_{c2} = \begin{cases} 0 & 0 \le \omega t < \pi \\ -I_c \sin \omega t & \pi \le \omega t < 2\pi \end{cases} \tag{1.122}$$

where I_c is the output current amplitude.

Being transformed through the output transformer T_2 with the appropriate phase conditions, the total current flowing through the load R_L is obtained as

$$i_R(\omega t) = i_{c1}(\omega t) - i_{c2}(\omega t) = I_c \sin \omega t \tag{1.123}$$

The current flowing into the center tap of the primary windings of the output transformer T_2 is the sum of the collector currents, resulting in

$$i_{cc}(\omega t) = i_{c1}(\omega t) + i_{c2}(\omega t) = I_c|\sin \omega t| \tag{1.124}$$

Ideally, the even-order harmonics being in phase are canceled out and should not appear at the load. In practice, a level of the second-harmonic component of

30–40 dB below the fundamental is allowable. However, it is necessary to connect a bypass capacitor to the center tap of the primary winding to exclude power losses due to even-order harmonics. The current $i_R(\omega t)$ produces the load voltage $v_R(\omega t)$ onto the load R_L as

$$v_R(\omega t) = I_c R_L \sin \omega t = V_R \sin \omega t \tag{1.125}$$

where V_R is the load voltage amplitude.

The total dc collector current is defined as the average value of $i_{cc}(\omega t)$, which yields

$$I_0 = \frac{1}{2\pi} \int_0^{2\pi} i_{cc}(\omega t) d\omega t = \frac{2}{\pi} I_c \tag{1.126}$$

The total dc power P_0 and fundamental-frequency output power P_{out} for the ideal case of zero saturation voltage of both transistors when $V_c = V_{cc}$ and taking into account that $V_R = V_c$ for equal turns of windings when $n_1 = n_2$ are calculated respectively from

$$P_0 = \frac{2}{\pi} I_c V_{cc} \tag{1.127}$$

$$P_{out} = \frac{I_c V_{cc}}{2} \tag{1.128}$$

Consequently, the maximum theoretical collector efficiency that can be achieved in a push–pull Class-B operation is equal to

$$\eta = \frac{P_{out}}{P_0} = \frac{\pi}{4} \cong 78.5\% \tag{1.129}$$

In a balanced circuit, identical sides carry 90° quadrature or 180° out-of-phase signals of equal amplitude. In the latter case, if perfect balance is maintained on both sides of the circuit, the difference between signal amplitudes becomes equal to zero in each midpoint of the circuit, as shown in Figure 1.35. This effect is called the *virtual grounding*, and this midpoint line is referred to as the *virtual ground*. The virtual ground, being actually inside the balanced transistor package having two identical transistor chips, reduces a common-mode inductance and results in better stability and usually higher power gain [45].

When using a balanced transistor, new possibilities for both internal and external impedance matching procedure emerge. For instance, for a push–pull operation mode of two single-ended transistors, it is necessary to provide reliable grounding for input and output matching circuits for each device, as shown in Figure 1.36(a). Using the balanced transistors simplifies significantly the matching circuit topologies, with the series inductors and shunt capacitors connected between amplifying paths, as shown in Figure 1.36(b), where the dc-blocking capacitors are not needed [46]. Such an approach can provide additional design flexibility when, for example, a two-stage monolithic push–pull X-band GaAs MESFET power

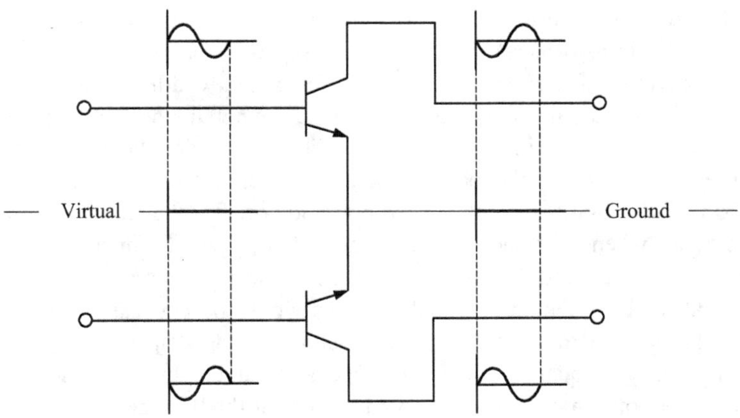

Figure 1.35 Basic concept of balanced transistor

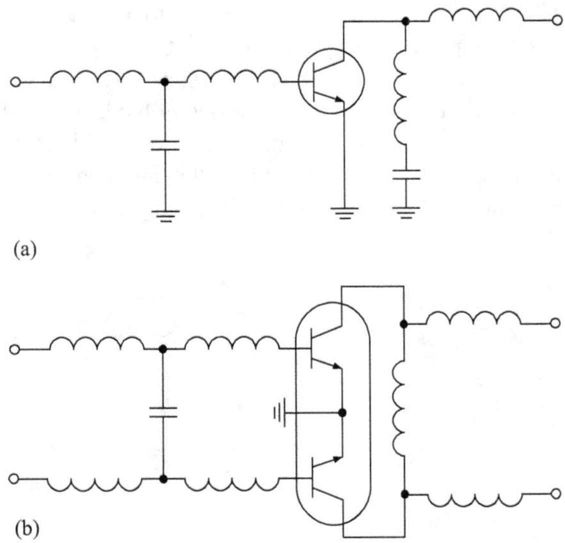

Figure 1.36 Matching technique for (a) single-ended and (b) balanced transistors

amplifier can be optimized for either small-signal, high-gain operation, or for large-signal power saturated operation by changing the lengths of the bondwires which form the shunt inductance at the drain circuits of each stage [47].

1.7.2 Baluns

For a push–pull operation of the power amplifier with a balanced transistor, it is necessary to provide the unbalanced-to-balanced transformation referenced to the ground both at the input and at the output of the power amplifier. The most suitable approach to solve this problem in the best possible manner at high frequencies and microwaves is to use the transmission-line baluns (balanced-to-unbalanced transmission-line transformers). The first transmission-line balun for coupling a single coaxial line having a quarter wavelength at the center bandwidth frequency to a push–pull coaxial line (or a pair of coaxial lines), which maintains perfect balance over a wide frequency range, was introduced and described by Lindenblad in 1939 [48,49].

Figure 1.37(a) shows the basic structure and equivalent circuit of a simple coaxial balun, where port *A* is unbalanced port and port *B* is the balanced port. To be a perfect balun when a balanced load is connected to port terminals *B*, the shield return current $I_2 - I_3$ would equal to I_1, which would ideally represent the delayed input current, and the output terminal voltages would be equal and opposite with respect to ground. In this case, if the characteristic impedance Z_0 of the coaxial transmission line is equal to the input impedance at the unbalanced end of the transformer, the total impedance from both outputs at the balanced end of the transformer will be equal to the input impedance. Hence, such a transmission-line transformer can be used as a 1:1 balun. The equivalent circuit for this coaxial balun demonstrates the basic drawback of this balun, when its inner conductor is shielded from ground having practically infinite impedance to ground, whereas the outer shield does have finite impedance to ground when a balun is placed above a printed circuit board. The presence of the lower ground plane creates a shunt short-circuit stub with the characteristic impedance Z_1 across one of the load and this converts the high-pass balun structure into a bandpass one. As a result, this stub has a dramatic effect on the balun performance, with the bandwidth being reduced to about an octave based on phase imbalance. One of the solutions is simply to raise the transmission

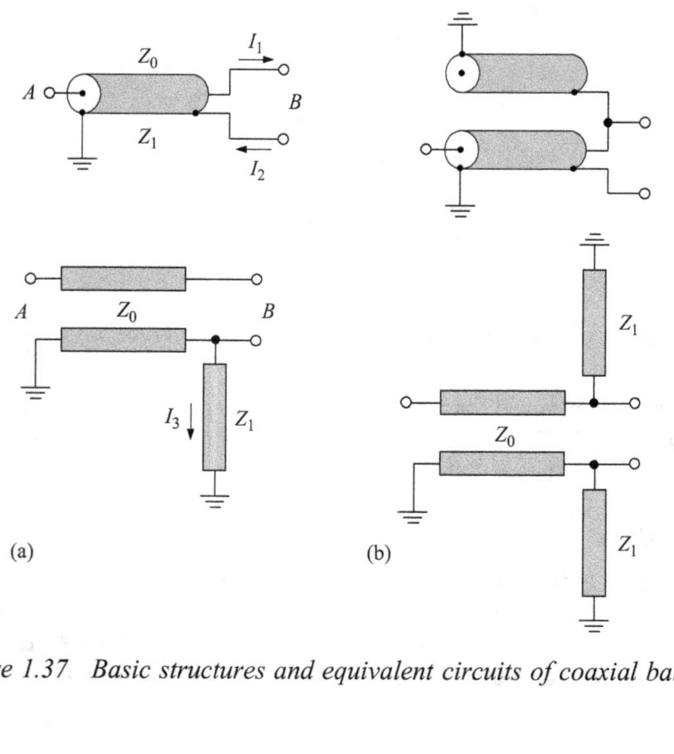

Figure 1.37 Basic structures and equivalent circuits of coaxial baluns

line above the printed circuit board as high as possible and make both conductors symmetrical with respect to lower ground plane. The other solution is to attach a compensating stub to the other load, as shown in Figure 1.37(b), which results in perfect amplitude and phase balance above the low-frequency cutoff region providing less than 1-dB insertion loss achieved from 5 MHz to 2.5 GHz [50].

Figure 1.38 shows the basic structure of a push–pull power amplifier with a balanced bipolar transistor including the input and output matching circuits. To extend operating frequency range to lower frequencies, the outside of the coaxial line of the balun can be loaded with low-loss ferrite core which acts as chokes to force equal and opposite currents in the inner and outer conductor and isolate the 180° output from the input ground terminal by creating a high and lossy impedance for Z_1. In this case, the measured *S*-parameters of the back-to-back connected baluns showed an insertion loss of about 0.5–0.6 dB and better than 20-dB return loss over 50–1000 MHz [51]. As an example, four broadband GaN HEMT power amplifier units were combined using such a low-loss coaxial balun that transforms an unbalanced 50-Ω load into two 25-Ω impedances that are 180° out of phase and each of the 25-Ω end is driven by a pair of the power amplifier units connected in parallel. A similar balun is used at the input to create the 180° out-of-phase input to the two pairs of the power amplifier units, resulting in over 100-W output power and higher than 60% drain efficiency across the frequency bandwidth of 100–1000 MHz. Lower cutoff bandwidth frequency can be provided to cover down to 10 MHz by adding lower frequency ferrites, but it may affect performance at high bandwidth frequencies. Generally, since ferrite has limited bandwidth, it is possible to use several ferrite cores to broader the frequency bandwidth. For example, by using a low-frequency ferrite core covering 1 MHz to 10 MHz, a medium-frequency ferrite core covering 10 MHz to 200 MHz, and a high-frequency ferrite core covering high frequencies above 200 MHz, the balun can cover 1 MHz to 2.5 GHz with the loss of 0.25 dB at low frequency and 1.3 dB at 2.5 GHz [52].

The miniaturized compact input unbalanced-to-balanced transformer shown in Figure 1.39 covers the frequency bandwidth up to an octave with well-defined rejection-mode impedances [53]. To avoid the parasitic capacitance between the outer conductor and the ground, the coaxial semirigid transformer T_1 is mounted atop microstrip shorted stub l_1 and soldered continuously along its length. The electrical length of this stub is usually chosen from the condition of $\theta \leq 90°$ on the

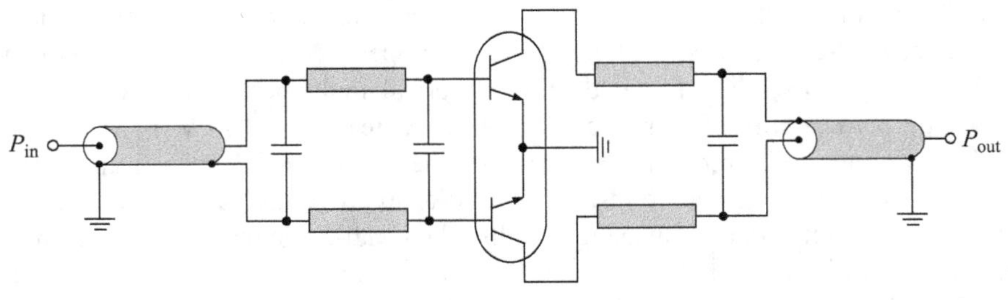

Figure 1.38 Push–pull bipolar power amplifier with input and output baluns

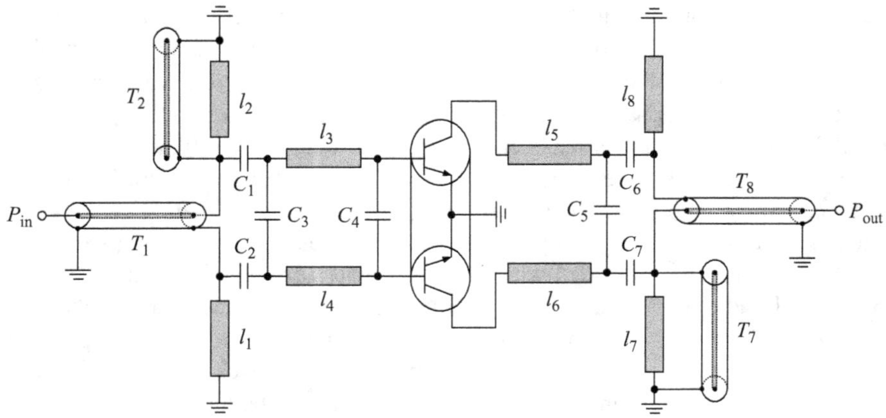

Figure 1.39 Push–pull power amplifier with compact balanced-to-unbalanced transformers

high bandwidth frequency depending on the matching requirements. To maintain circuit symmetry on the balanced side of the transformer network, another semi-rigid coaxial section T_2 with unconnected center conductor is soldered continuously along microstrip shorted stub l_2. The lengths of T_2 and l_2 are equal to the lengths of T_1 and l_1, respectively. Because the input short-circuited microstrip stubs provide inductive impedances, the two series capacitors C_1 and C_2 of the same value are used for matching purposes, thereby forming the first high-pass matching section and providing dc blocking at the same time. The practical circuit realization of the output matching circuit and balanced-to-unbalanced transformer can be the same as for the input matching circuit.

Figure 1.40 shows the circuit schematic of a broadband LDMOSFET power amplifier operating in a frequency range from 10 MHz to 512 MHz with a typical peak output power of 120 W, a power gain of more than 22.5 dB with gain flatness of 1.8 dB and a drain efficiency of more than 50% at a dc-supply voltage of 28 V with a quiescent current of 1 A [54]. Here, the input signal is divided into two 180° out-of-phase signals by a 50-Ω coaxial-cable 1:1 balun, whereas two 25-Ω coaxial-cable 4:1 transformers provide input impedance matching with a return loss better than 5 dB across the entire frequency range (15 dB below 250 MHz). Similarly, the output combining network consists of two 15-Ω coaxial-cable 4:1 transformers and a 50-Ω coaxial-cable 1:1 balun. Each balun and impedance transformer at the input and output circuits is incorporated in a cylindrical multiaperture low-loss ferrite core with permeability of more than 20 at frequencies below 100 MHz to provide high impedance to common-mode currents. The parallel *RC* feedback and input shunt 20-Ω resistor for each transistor is used to improve the gain flatness across the entire frequency range. By using broadband coaxial-cable input 4:1 transformer and input and output 1:1 baluns with ferrite cores, the push–pull GaN HEMT power amplifier based on two Cree CGH20025 devices demonstrated a *PAE* from 83% to 64% with a 30-W average output power and a power gain of 15±0.5 dB in a

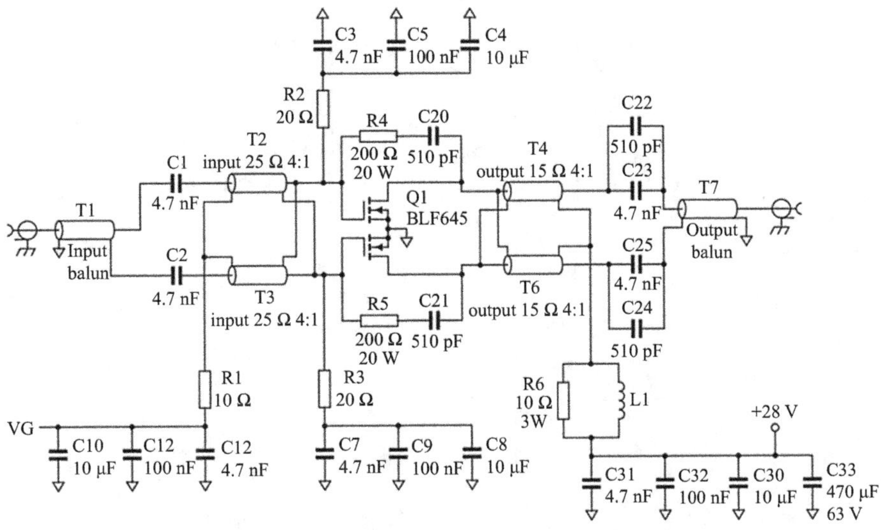

Figure 1.40 Schematic of 1-kW VHF-UHF push–pull LDMOSFET power amplifier

frequency bandwidth of 50–500 MHz at a dc-supply voltage of 28 V with a quiescent current of 500 mA [55]. The input matching circuit is configured so as to properly drive the two transistors in a push–pull switching mode, resulting in a *PAE* higher than 80% from 50 MHz to 300 MHz.

1.7.3 Balanced power amplifiers

Balanced amplifier technique using the quadrature 3-dB couplers for power dividing and combining represents an alternative approach to push–pull operation. Figure 1.41(a) shows the basic circuit schematic of a balanced amplifier where two power amplifier units of the same performance are arranged between the input splitter and output combiner, each having a 90° phase difference between coupled and through ports. The fourth port of each quadrature coupler must be terminated with a ballast resistor R_{bal}, which is equal to 50 Ω for a 50-Ω system impedance. The input signal is split into two equal-amplitude components by the first 90° hybrid coupler with 0° and 90° paths, then amplified, and finally recombined by the second 90° hybrid coupler. Owing to proper phase shifting, both signals in the load of the isolated port of the combiner are cancelled, and the load connected at the combiner output port sees the sum of these two signals.

The theory of balanced amplifiers was derived by Kurokawa with demonstration of the operating frequency bandwidth over 1.2 octaves using one-section distributed quarterwave 3-dB directional couplers [56]. For a wide frequency range, the main advantages of the balanced design are the improved input and output impedance matching, gain flatness, intermodulation distortion, and potential design simultaneously for minimum noise figure and good input match. As an example, a

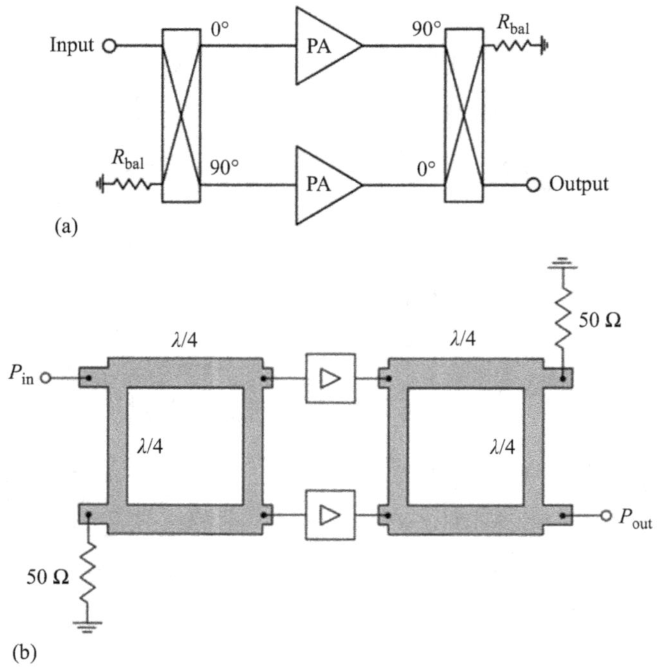

Figure 1.41 Schematics of balanced power amplifiers with quadrature hybrid couplers

four-stage balanced bipolar amplifier achieved a power gain of 20±0.5 dB and an input voltage standing wave ratio (*VSWR*) less than 1.2 across the octave frequency bandwidth from 0.8 to 1.6 GHz [57]. By extending the wide operating frequency range to higher frequencies, an output power of around 23 dBm with gain variations close to 1 dB over 4.5–6.5 GHz and 8–12 GHz was achieved for the balanced microstrip GaAs MESFET amplifiers [58,59].

To combine the output power from two or more transistors at microwaves, as shown in Figure 1.41(b) for two power amplifiers with standard 50-Ω input and output impedances, the branch-line 90° microstrip-line hybrid combiners are the most popular. In this case, the characteristic impedances of the transverse branches should be of 50 Ω, whereas the longitudinal branches must have the characteristic impedance of $50/\sqrt{2} = 35.4\,\Omega$. In practice, because the quarter-wavelength transmission line requirements, the 3-dB bandwidth of such a balanced amplifier based on two quadrature branch-line hybrids is limited to 10%–20%.

If the individual amplifiers with equal performance in the balanced pair are not perfectly matched at certain frequencies, then a signal in the 0° path of the coupler will be reflected from the corresponding amplifier and a signal in the 90° path of the coupler will be similarly reflected from the other amplifier. The reflected signals will again be phased with 90° and 0°, respectively, and the total reflected power as a sum of the in-phase reflected signals flows into the isolated port and dissipates on the ballast resistor R_{bal}. As a result, an input *VSWR* of the quadrature coupler does not depend on the equal load mismatch level. This gives a constant well-defined load to the driver stage, improving amplifier stability and driver power flatness across the operating frequency range. Generally, the stability factor of a balanced stage can be an order of magnitude higher than its single-ended equivalent, depending on the *VSWR* and isolation of the quadrature couplers. If one of the amplifiers fails or turned off, the balanced configuration provides a gain reduction of -6 dB only. Besides, the balanced structure provides ideally the cancellation in the load of the third-order products such as $2f_1 + f_2$, $2f_2 + f_1$, $3f_1$, $3f_2$, ..., and attenuation by 3 dB of the second-order products such as $f_1 \pm f_2$, $2f_1$, $2f_2$, ... In a microstrip implementation for octave-band power amplifiers, one of the most popular couplers for power dividing and combining is a 3-dB Lange hybrid coupler.

Figure 1.42(a) shows the microstrip single-section topology of a coupled-line directional coupler, which can be used for broadband power dividing and combining. Its electrical properties are described using a concept of two types of excitations for the coupled lines in TEM approximation. In this case, for the even mode, the currents flowing in the strip conductors are equal in amplitude and flow in the same direction. The electric field has even symmetry about the center line, and no current flows between two strip conductors. For the odd mode, the currents flowing in the strip conductors are equal in amplitude, but flow in opposite directions. The electric field lines have an odd symmetry about the center line, and a voltage null exists between these two strip conductors. An arbitrary excitation of the coupled lines can always be treated as a superposition of appropriate amplitudes of even and odd modes.

Therefore, the characteristic impedance for even excitation mode Z_{0e} and the characteristic impedance for the odd excitation mode Z_{0o} characterize the

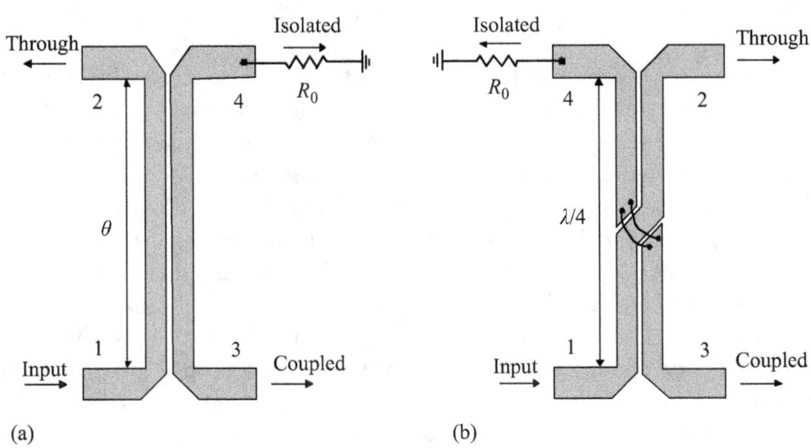

(a) (b)

Figure 1.42 Coupled-line directional couplers

coupled lines. For two coupled equal-strip lines used in a system with the characteristic impedance of Z_0, $Z_0^2 = Z_{0e}Z_{0o}$,

$$Z_{0e} = Z_0\sqrt{\frac{1+C}{1-C}} \tag{1.130}$$

$$Z_{0o} = Z_0\sqrt{\frac{1-C}{1+C}} \tag{1.131}$$

where the midband voltage coupling coefficient C of the directional coupler is defined as

$$C = \frac{Z_{0e} - Z_{0o}}{Z_{0e} + Z_{0o}} \tag{1.132}$$

where $C = 0$ for zero coupling and $C = 1$ for completely superposed transmission lines [60,61].

For a quarter-wavelength-long coupler, the voltage-split ratio K is defined as the ratio between voltages at ports 2 and 3 as

$$K = \left|\frac{S_{12}}{S_{13}}\right| = \frac{\sqrt{1-C^2}}{C} \tag{1.133}$$

from which it follows that equal voltage split between the output ports 2 and 3 can be provided with $C = 1/\sqrt{2}$ (or 3 dB).

If it is necessary to implement the output ports 2 and 3 at one side, it is possible to use a construction of a microstrip directional coupler with crossed bondwires, as shown in Figure 1.42(b). The strip crossover for a stripline directional coupler can be easily achieved with the three-layer sandwich. The microstrip 3-dB directional coupler fabricated on alumina substrate for idealized zero strip thickness should have the calculated strip spacing of less than 10 μm. Such a narrow value easily explains the great interest to the constructions of the directional couplers with larger spacing.

A popular way to increase the coupling between two edge-coupled microstrip lines is to use several parallel narrow microstrip lines interconnected with each other by the bondwires, as shown in Figure 1.43. For an interdigitated Lange coupler shown in Figure 1.43(a), four coupled microstrip lines are used, resulting in a 3-dB coupling over an octave or more bandwidth [62]. In this case, the signal flowing to the input port 1 is distributed between the output ports 2 and 3 with the phase difference of 90°. However, this structure is quite complicated for practical implementation when, for alumina substrate with $\varepsilon_r = 9.6$, the dimensions of a 3-dB interdigitated Lange coupler are $W/h = 0.107$ and $S/h = 0.071$, where W is the width of each strip and S is the spacing between adjacent strips.

Figure 1.43(b) shows the unfolded Lange coupler with four strips of equal length offering the same electrical performance but easier for circuit modeling [63]. The even-mode characteristic impedance Z_{e4} and odd-mode characteristic impedance Z_{o4}

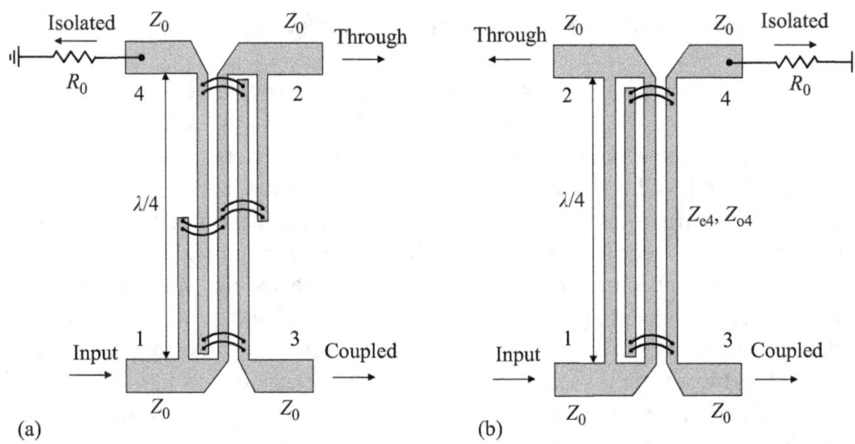

Figure 1.43 Lange directional couplers

of the unfolded Lange coupler with $Z_0^2 = Z_{e4}Z_{o4}$ in terms of the characteristic impedances of a two-conductor line (which is identical to any pair of adjacent lines in the coupler) can be obtained by

$$Z_{e4} = \frac{Z_{0o} + Z_{0e}}{3Z_{0o} + Z_{0e}} Z_{0e} \tag{1.134}$$

$$Z_{o4} = \frac{Z_{0e} + Z_{0o}}{3Z_{0e} + Z_{0o}} Z_{0o} \tag{1.135}$$

where Z_{0e} and Z_{0o} are the even- and odd-mode characteristic impedances of the two-conductor pair [64].

The midband voltage coupling coefficient C is given by

$$C = \frac{Z_{e4} - Z_{o4}}{Z_{e4} + Z_{o4}} = \frac{3\left(Z_{0e}^2 - Z_{0o}^2\right)}{3\left(Z_{0e}^2 + Z_{0o}^2\right) + 2Z_{0e}Z_{0o}} \tag{1.136}$$

The even- and odd-mode characteristic impedances Z_{0e} and Z_{0o} as functions of the characteristic impedance Z_0 and coupling coefficient C are determined by

$$Z_{0e} = Z_0\sqrt{\frac{1+C}{1-C}}\frac{4C - 3 + \sqrt{9-8C^2}}{2C} \tag{1.137}$$

$$Z_{0o} = Z_0\sqrt{\frac{1-C}{1+C}}\frac{4C + 3 - \sqrt{9-8C^2}}{2C} \tag{1.138}$$

For alumina substrate with $\varepsilon_r = 9.6$, the dimensions of such a 3-dB unfolded Lange coupler are $W/h = 0.112$ and $S/h = 0.08$, where W is the width of each strip and S is the spacing between the strips.

1.8 Transmission-line transformers and combiners

The transmission-line transformers and combiners can provide very wide operating bandwidths up to the frequencies of 3 GHz and higher [44,65]. They are widely used in matching networks for antennas and power amplifiers in the HF and VHF bands, and their low losses make them especially useful in high-power circuits [66,67]. Typical structures for transmission-line transformers and combiners consist of parallel wires, coaxial cables, or bifilar twisted wire pairs. In the latter case, the characteristic impedance can be easily determined by the wire diameter, the insulation thickness, and, to some extent, the twisting pitch [68,69]. For coaxial-cable transformers with correctly chosen characteristic impedance, the theoretical high-frequency bandwidth limit is reached when the cable length comes in order of a half wavelength, with the overall achievable bandwidth being about a decade. By introducing the low-loss high permeability ferrites alongside a good quality semirigid coaxial or symmetrical strip cable, the low-frequency limit can be significantly improved providing bandwidths of several or more decades.

The concept of a broadband impedance transformer consisted of a pair of interconnected transmission lines was first disclosed and described by Guanella [70,71]. Figure 1.44(a) shows a Guanella transformer system with transmission-line character achieved by an arrangement comprising one pair of cylindrical coils, which are wound in the same sense and are spaced a certain distance apart by an intervening dielectric. In this case, one cylindrical coil is located inside the insulating cylinder and the other coil is located on the outside of this cylinder. For the currents flowing through both windings in opposite directions, the corresponding flux in the coil axis is negligibly small. However, for the currents flowing in the same direction through both coils, the latter may be assumed to be connected in

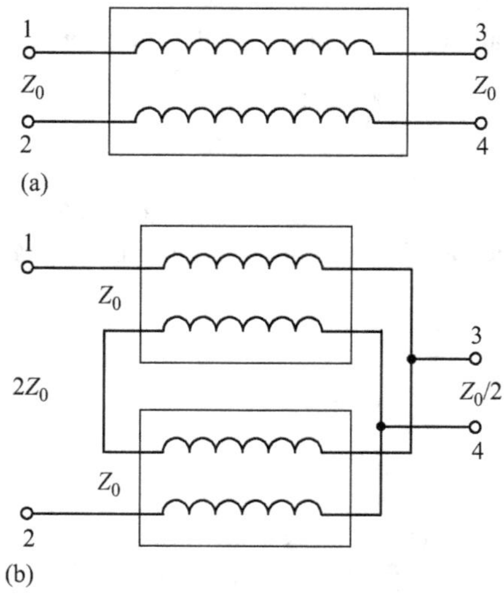

Figure 1.44 Schematic configurations of Guanella (a) 1:1 and (b) 4:1 transformers

parallel, and a coil pair represents a considerable inductance for such currents and acts like a choke coil. With terminal 4 being grounded, such a 1:1 transformer provides matching of the balanced source to unbalanced load. In this case, if terminal 2 is grounded, it represents simply a delay line. When terminals 2 and 3 are grounded, the transformer performs as a phase inverter.

A series-parallel connection of a plurality of coil pairs can produce a match between unequal source and load resistances. Figure 1.44(b) shows a 4:1 impedance (2:1 voltage) transmission-line transformer, where the two pairs of cylindrical transmission-line coils are connected in series at the input and in parallel at the output. For the characteristic impedance Z_0 of each transmission line, this results in two times higher impedance $2Z_0$ at the input and two times lower impedance $Z_0/2$ at the output. By grounding terminal 4, such a 4:1 impedance transformer provides impedance matching of the balanced source to the unbalanced load. In this case, when terminal 2 is grounded, it performs as a 4:1 *unun* (*un*balanced-to-*un*balanced transformer). With a series-parallel connection of n coil pairs with the characteristic impedance Z_0 each, the input impedance is equal to nZ_0 and the output impedance is equal to Z_0/n. Since voltages that have equal delays through the transmission lines are added, such a technique results in the so called *equal-delay* transmission-line transformers.

The simplest transmission-line transformer is a quarterwave transmission line, whose characteristic impedance is chosen to give the correct impedance transformation. However, this transformer provides a narrowband performance valid only around frequencies, for which the transmission line is odd multiples of a quarter wavelength. If a ferrite sleeve is added to the transmission line, common-mode currents flowing in both transmission-line inner and outer conductors in phase and in the same direction are suppressed and the load may be balanced and floating above ground [72,73]. If the characteristic impedance of the transmission line is equal to the terminating impedances, the transmission is inherently broadband. If not, there will be a dip in the response at the frequency, at which the transmission line is a quarter-wavelength long.

A coaxial cable transformer, whose physical configuration is shown in Figure 1.45(a) and equivalent circuit representation with polarity reversing is shown in Figure 1.45(b), consists of the coaxial line arranged inside the ferrite core, or wound around the ferrite core. Either end of the load resistor can be grounded, depending on the desired output polarity. The larger the core permeability, the fewer the turns required for a given low-frequency response and the larger the overall bandwidth. Due to its practical configuration, the coaxial cable transformer takes a position between the lumped and distributed systems. Therefore, at lower frequencies its equivalent circuit represents a conventional low-frequency transformer, as shown in Figure 1.45(c), whereas at higher frequency it is a transmission line with the characteristic impedance Z_0, as shown in Figure 1.45(d). The advantage of such a transformer is that the parasitic interturn capacitance determines its characteristic impedance, whereas this parasitic capacitance negatively contributes to the transformer frequency performance of the conventional wire-wound transformer with discrete windings by resonating with the leakage inductance that produces a loss peak.

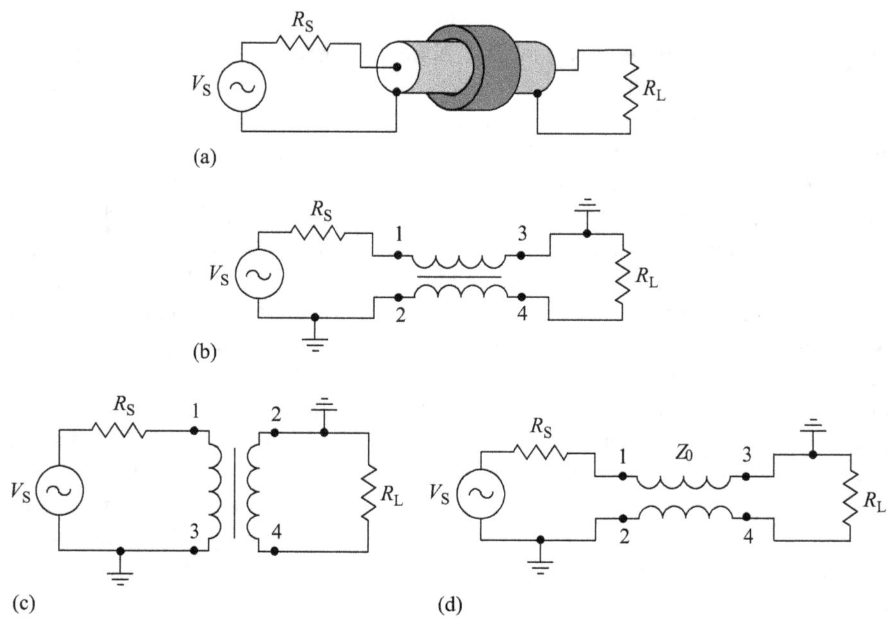

Figure 1.45 Schematic configurations of coaxial cable transformer

When $R_S = R_L = Z_0$, the transmission line can be considered a transformer with a 1:1 impedance transformation. To avoid any resonant phenomena, especially for complex loads, which can contribute to the significant output power variations, the length l of the transmission line, as a rule of thumb, is kept to no more than an eight of a wavelength λ_{min} at the highest operating frequency:

$$l \leq \frac{\lambda_{min}}{8} \tag{1.139}$$

where λ_{min} is the minimum wavelength in the transmission line corresponding to the high operating frequency f_{max}.

The low-frequency bandwidth limit of a coaxial cable transformer is determined by the effect of the magnetizing inductance L_m of the outer surface of the outer conductor. An approximation to the magnetizing inductance can be made by considering the outer surface of the coaxial cable to be the same as that of a straight wire (or linear conductor) which, at higher frequencies where the skin effect causes the current to be concentrated on the outer surface, would have the self-inductance of

$$L_m(\text{nH}) = 2l\left[\ln\left(\frac{2l}{r}\right) - 1\right] \tag{1.140}$$

where l is the length of the coaxial cable in cm and r is the radius of the outer surface of the outer conductor in cm [67]. If a toroid is used for the core, then high permeability of core materials results in shorter transmission lines.

An approach using the transmission line based on a single-bifilar-wound coil to realize a broadband 1:4 impedance transformation was introduced by Ruthroff [74,75]. In this case, by using a core material of a sufficiently high permeability, the number of turns can be significantly reduced. Figure 1.46(a) shows the circuit schematic of an unbalanced-to-unbalanced 1:4 transmission line transformer, where terminal 4 is connected to the input terminal 1. As a result, for $V = V_1 = V_2$, the output voltage is twice the input voltage, and the transformer has a 1:2 voltage step-up ratio. As the ratio of input voltage to input current is one-fourth the load voltage to load current, the transformer is fully matched for maximum power transfer when $R_L = 4R_S$, and the transmission-line characteristic impedance Z_0 is equal to the geometric mean of the source and load impedances:

$$Z_0 = \sqrt{R_S R_L} \tag{1.141}$$

where R_S is the source resistance and R_L is the load resistance. Figure 1.46(b) shows an impedance transformer acting as a phase inverter, where the load resistance is included between terminals 1 and 4 to become a 1:4 balun. This technique is called the *bootstrap effect*, which doesn't have the same high-frequency response as Guanella equal-delay approach because it adds a delayed voltage to a direct one [76]. The delay becomes excessive when the transmission line reaches a significant fraction of a wavelength.

Figure 1.47(a) shows the physical implementation of a Ruthroff 4:1 impedance transformer using a coaxial cable arranged inside the ferrite core. At lower frequencies, such a transformer can be considered an ordinary 2:1 voltage autotransformer.

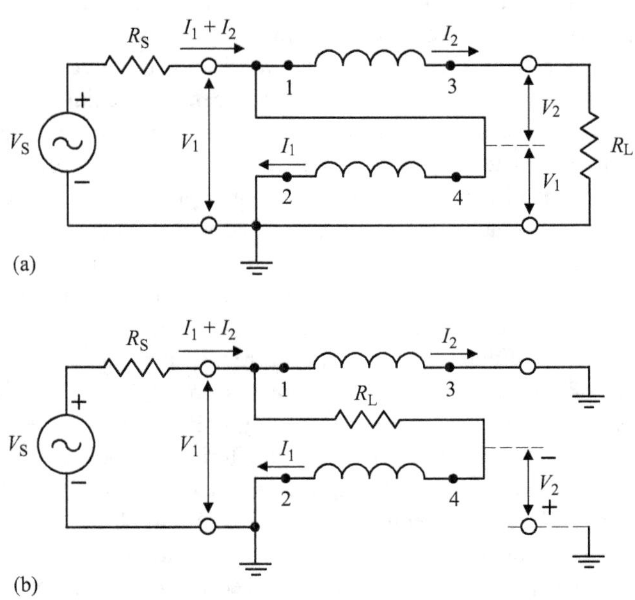

Figure 1.46 Schematic configurations of Ruthroff 1:4 impedance transformer

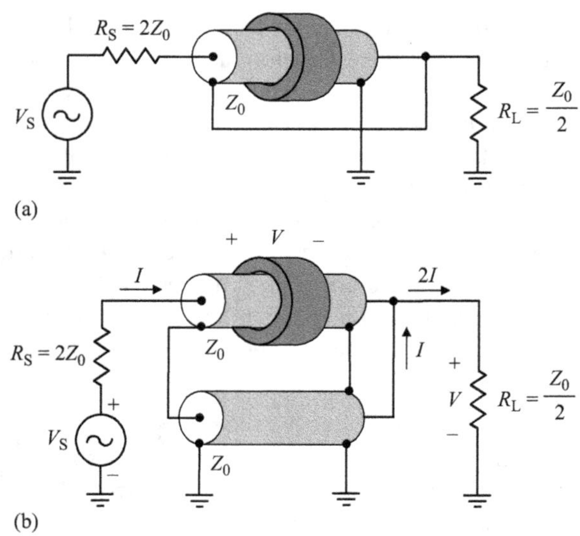

Figure 1.47 Schematic configurations of 4:1 coaxial cable transformer

To improve the performance at higher frequencies, it is necessary to add an additional phase-compensating line of the same length, as shown in Figure 1.47(b), resulting in a Guanella ferrite-based 4:1 impedance transformer. In this case, a ferrite core is necessary only for the upper line because the outer conductor of the lower line is grounded at both ends, and no current flows through it. A current I driven into the inner conductor of the upper line produces a current I that flows in the outer conductor of the upper line, resulting in a current $2I$ flowing into the load R_L. Because the voltage $2V$ from the transformer input is divided in two equal parts between the coaxial line and the load, such a transformer provides impedance transformation from $R_S = 2Z_0$ into $R_L = Z_0/2$, where Z_0 is the characteristic impedance of each coaxial line. To adopt this transmission-line transformer for microwave planar applications, the coaxial line can be replaced by a pair of stacked strip conductors or coupled microstrip lines [77,78].

Figure 1.48 shows similar arrangements for the 3:1 voltage coaxial cable transformers, which produce 9:1 impedance transformation. A current I driven into the inner conductor of the upper line in Figure 1.48(a) will cause a current I to flow in the outer conductor of the upper line. This current then produces a current I in the outer conductor of the lower line, resulting in a current $3I$ flowing into the load R_L. The lowest coaxial line can be removed, resulting in a 9:1 impedance coaxial cable transformer shown in Figure 1.48(b). The characteristic impedance of each transmission line is specified by the voltage applied to the end of the line and the current flowing through the line and is equal to Z_0. In a 0.18-μm CMOS process with six metal layers, a 1:9 transmission-line transformer with broadside-coupled and multiple-metal stacked transmission lines achieved a broadband impedance transformation from 5.0±0.1 Ω optimal load impedance of the power cell to 50-Ω load with a bandwidth of 4.4 to 6.6 GHz and an insertion loss of about 1 dB [79].

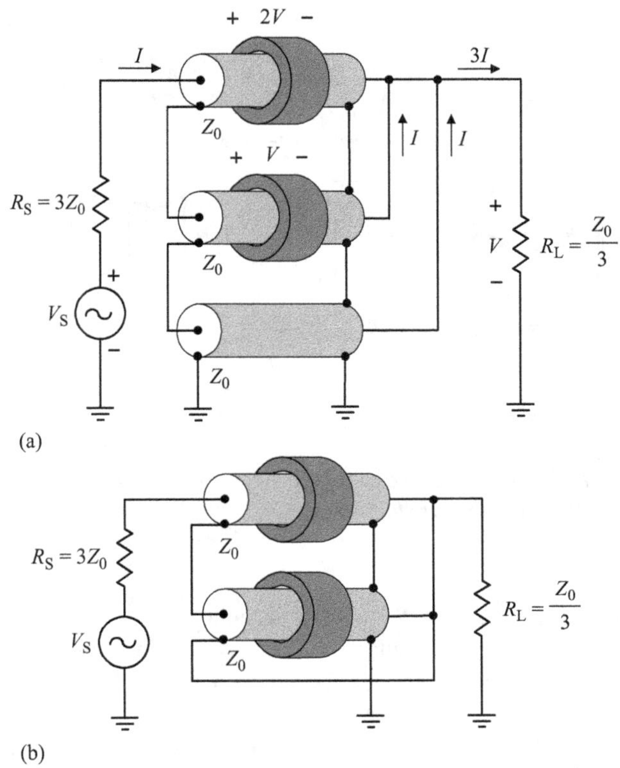

Figure 1.48 Schematic configurations of 9:1 coaxial cable transformer

In RF power amplifiers with very high output power, it is desirable to eliminate the need for output dc-blocking capacitors, because they must be able to handle large RF currents and should be of a multilayer chip type to minimize the series inductance and power loss. Figure 1.49 shows the simplified circuit schematic of a broadband high-power push–pull MOSFET amplifier, where the input and output impedance-transforming baluns T_1 and T_2 using low-loss ferrite core material provide high impedance ratios of 16:1 and 9:1, respectively, representing very practical ones for their simplicity and ease of implementation [80]. The impedance transformation is achieved by parallel connection of the outer conductors of the coaxial cables and series connection of their inner conductors. This design, in principle, resembles the multifilar-type transformer and must be considered a conventional (nontransmission line) transformer. In this case, the characteristic impedance of the coaxial transmission lines is not defined in the same manner as in transmission-line transformers, and therefore is not critical. For example, using a 50-Ω line in the output impedance transformer balun T_2 resulted in a 0.8-dB gain reduction across the band, plus an additional 0.3 dB at high-bandwidth frequencies compared to the unit made with 16-Ω cable. The input impedance-transforming balun T_1 is based on a 25-Ω coaxial cable, and the input high-impedance etched

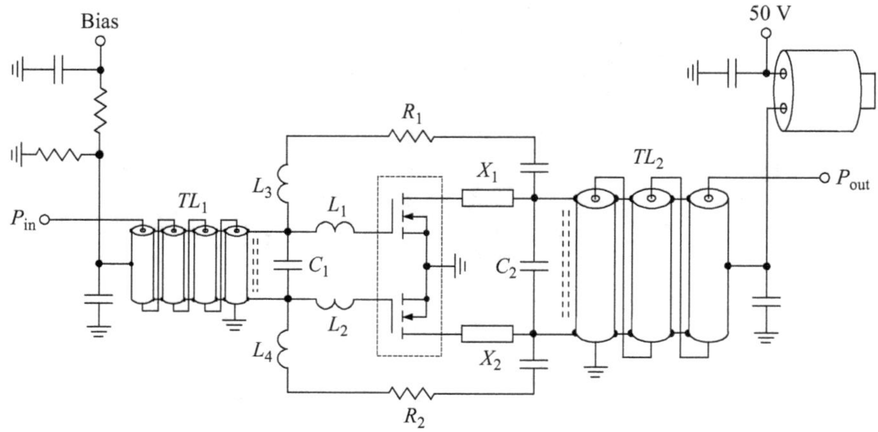

Figure 1.49 Schematic of high-power VHF MOSFET power amplifier with coaxial-cable transformers

lines, acting as inductors L_1 and L_2 with adjustable values, are necessary to compensate for the gate-source capacitances. The effect of the drain-source capacitances is partly compensated by the 12-Ω etched lines (X_1 and X_2) to increase efficiency by 6%–8% at 90 MHz, although at a cost of reduced power gain by approximately 0.5 dB. To improve the gain flatness across the entire bandwidth, the negative *RL* feedback for each transistor is introduced having reactance of about 8 Ω at the midband, where lower inductance values would result in increased feedback and a flatter gain response. Besides, the dummy resistor connected in parallel to the capacitor C_1 can be used to make the input impedance more resistive and to improve the input return loss. As a result, a 1-kW output power is achieved over the frequency bandwidth of 10–90 MHz with a power gain varying from minimum value of 11 dB at 90 MHz to almost 14 dB at midband with a drain efficiency varying from 60% at 10 MHz to 40% at 90 MHz at a supply voltage of 50 V with a quiescent current of 2 × 800 mA.

References

[1] A. A. Oswald, "Power Amplifiers in Trans-Atlantic Radio Telephony," *Proc. IRE*, vol. 13, pp. 313–324, 1925.

[2] L. E. Barton, "High Audio Power from Relatively Small Tubes," *Proc. IRE*, vol. 19, pp. 1131–1149, 1931.

[3] A. I. Berg, *Theory and Design of Vacuum-Tube Generators* (in Russian), Moskva: GEI, 1932.

[4] P. H. Osborn, "A Study of Class B and C Amplifier Tank Circuits," *Proc. IRE*, vol. 20, pp. 813–834, 1932.

[5] C. E. Fay, "The Operation of Vacuum Tubes as Class B and Class C Amplifiers," *Proc. IRE*, vol. 20, pp. 548–568, 1932.

[6] F. E. Terman and J. H. Ferns, "The Calculation of Class C Amplifier and Harmonic Generator Performance of Screen-Grid and Similar Tubes," *Proc. IRE*, vol. 22, pp. 359–373, 1934.

[7] L. B. Hallman, "A Fourier Analysis of Radio-Frequency Power Amplifier Wave Forms," *Proc. IRE*, vol. 20, pp. 1640–1659, 1932.

[8] C. E. Kilgour, "Graphical Analysis of Output Tube Performance," *Proc. IRE*, vol. 19, pp. 42–50, 1931.

[9] V. I. Kaganov, *Transistor Radio Transmitters* (in Russian), Moskva: Energiya, 1976.

[10] F. H. Raab, "Class-E, Class-C, and Class-F Power Amplifiers based upon a Finite Number of Harmonics," *IEEE Trans. Microwave Theory Tech.*, vol. MTT-49, pp. 1462–1468, 2001.

[11] F. H. Raab, "Maximum Efficiency and Output of Class-F Power Amplifiers," *IEEE Trans. Microwave Theory Tech.*, vol. MTT-49, pp. 1162–1166, 2001.

[12] A. Juhas and L. A. Novak, "Comments on Class-E, Class-C, and Class-F Power Amplifiers Based upon a Finite Number of Harmonics," *IEEE Trans. Microwave Theory Tech.*, vol. MTT-57, pp. 1623–1625, 2009.

[13] J. Vidkjaer, "A Computerized Study of the Class-*C*-Biased RF-Power Amplifier," *IEEE J. Solid-State Circuits*, vol. SC-13, pp. 247–258, 1978.

[14] H. L. Krauss, C. W. Bostian, and F. H. Raab, *Solid State Radio Engineering*, New York: John Wiley & Sons, 1980.

[15] B. E. Rose, "Notes on Class-D Transistor Amplifiers," *IEEE J. Solid-State Circuits*, vol. SC-4, pp. 178–179, 1969.

[16] J. Vidkjaer, "Series-Tuned High Efficiency RF-Power Amplifiers," *2008 IEEE MTT-S International Microwave Symposium Digest*, Atlanta, GA, 15–20 June, pp. 73–76.

[17] N. Le Gallou, J. Vidkjaer, and C. Poivey, "Suitability of GaN and LDMOS for 70%–82% Efficiency 120–200 W HPA Addressing Spaceborne P-band Radar Applications," *Proceedings of the 42nd European Microwave Conference*, Amsterdam, 29 Oct–1 Nov, pp. 691–694, 2012.

[18] V. M. Bogachev and V. V. Nikiforov, *Transistor Power Amplifiers* (in Russian), Moskva: Energiya, 1978.

[19] W. L. Everitt, "Output Networks for Radio-Frequency Power Amplifiers," *Proc. IRE*, vol. 19, pp. 725–737, 1931.

[20] H. T. Friis, "Noise Figure of Radio Receivers," *Proc. IRE*, vol. 32, pp. 419–422, 1944.

[21] S. Roberts, "Conjugate-Image Impedances," *Proc. IRE*, vol. 34, pp. 198–204, 1946.

[22] S. J. Haefner, "Amplifier-Gain Formulas and Measurements," *Proc. IRE*, vol. 34, pp. 500–505, 1946.

[23] R. L. Pritchard, "High-Frequency Power Gain of Junction Transistors," *Proc. IRE*, vol. 43, pp. 1075–1085, 1955.

[24] A. R. Stern, "Stability and Power Gain of Tuned Power Amplifiers," *Proc. IRE*, vol. 45, pp. 335–343, 1957.

[25] L. S. Houselander, H. Y. Chow, and R. Spense, "Transistor Characterization by Effective Large-Signal Two-Port Parameters," *IEEE J. Solid-State Circuits*, vol. SC-5, pp. 77–79, 1970.

[26] B. J. Thompson, "Oscillations in Tuned Radio-Frequency Amplifiers," *Proc. IRE*, vol. 19, pp. 421–437, 1931.
[27] G. W. Fyler, "Parasites and Instability in Radio Transmitters," *Proc. IRE*, vol. 23, pp. 985–1012, 1935.
[28] F. B. Llewellyn, "Some Fundamental Properties of Transmission Systems," *Proc. IRE*, vol. 40, pp. 271–283, 1952.
[29] D. F. Page and A. R. Boothroyd, "Instability in Two-Port Active Networks," *IRE Trans. Circuit Theory*, vol. CT-5, pp. 133–139, June 1958.
[30] J. M. Rollett, "Stability and Power Gain Invariants of Linear Two-Ports," *IRE Trans. Circuit Theory Appl.*, vol. CT-9, pp. 29–32, 1962.
[31] J. G. Linvill and L. G. Schimpf, "The Design of Tetrode Transistor Amplifiers," *Bell Syst. Tech. J.*, vol. 35, pp. 813–840, 1956.
[32] O. Muller and W. G. Figel, "Stability Problems in Transistor Power Amplifiers," *Proc. IEEE*, vol. 55, pp. 1458–1466, 1967.
[33] A. Grebennikov, *RF and Microwave Power Amplifier Design*, New York: McGraw-Hill, 2015.
[34] O. Muller, "Internal Thermal Feedback in Fourpoles, Especially in Transistors," *Proc. IEEE*, vol. 52, pp. 924–930, 1964.
[35] Application Note AN-010, *GaN for LDMOS Users*, Nitronex Corp., Durham, NC, USA, 2008.
[36] W. L. Everitt, "Output Networks for Radio-Frequency Power Amplifiers," *Proc. IRE*, vol. 19, pp. 725–737, 1931.
[37] S. Roberts, "Conjugate-Image Impedances," *Proc. IRE*, vol. 34, pp. 198–204, 1946.
[38] R. L. Camisa, J. B. Klatskin, and A. Mikelsons, "Broadband Lumped-Element GaAs FET Power Amplifier," *1981 IEEE MTT-S International Microwave Symposium Digest*, Los Angeles, CA, 15–19 June, pp. 126–128.
[39] A. Tam, "Network Building Blocks Balance Power Amp Parameters," *Microwaves & RF*, vol. 23, pp. 81–87, 1984.
[40] P. H. Smith, *Electronic Applications of the Smith Chart*, New York: Noble Publishing, 2000.
[41] D. M. Pozar, *Microwave Engineering*, New York: John Wiley & Sons, 2004.
[42] Y. Ito, M. Mochizuki, M. Kohno, H. Masuno, T. Takagi, and Y. Mitsui, "A 5–10 GHz 15-W GaAs MESFET Amplifier with Flat Gain and Power Responses," *IEEE Microwave and Guided Wave Lett.*, vol. 5, pp. 454–456, 1995.
[43] A. V. Grebennikov, "Create Transmission-Line Matching Circuits for Power Amplifiers," *Microwaves & RF*, vol. 39, pp. 113–122, 2000.
[44] H. L. Krauss, C. W. Bostian, and F. H. Raab, *Solid State Radio Engineering*, New York: John Wiley & Sons, 1980.
[45] L. B. Max, "Balanced Transistors: A New Option for RF Design," *Microwaves*, vol. 16, pp. 42–46, 1977.
[46] J. Johnson, "A Look Inside Those Integrated Two-Chip Amps," *Microwaves*, vol. 19, pp. 54–59, 1980.
[47] V. Sokolov and R. E. Williams, "Development of GaAs Monolithic Power Amplifiers in *X*-Band," *IEEE Trans. Electron Devices*, vol. ED-27, pp. 1164–1171, June 1980.

[48] N. E. Lindenblad, "Television Transmitting Antenna for Empire State Building," *RCA Rev.*, vol. 3, pp. 387–408, 1939.

[49] N. E. Lindenblad, "Junction between Single and Push–Pull Lines," U.S. Patent 2,231,839, February 1941 (filed May 1939).

[50] A. Riddle, "Ferrite and Wire Baluns with Under 1 dB Loss to 2.5 GHz," *1998 IEEE MTT-S International Microwave Symposium Digest*, vol. 2, Baltimore, MD, 7–12 June, pp. 617–620.

[51] K. Krishnamurthy, T. Driver, R. Vetury, and J. Martin, "100 W GaN HEMT Power Amplifier Module with >60% Efficiency over 100–1000 MHz Bandwidth," *2010 IEEE MTT-S International Microwave Symposium Digest*, Anaheim, CA, 23–28 May, pp. 940–943.

[52] A. K. Ezzedine and H. C. Huang, "10 W Ultra-Broadband Power Amplifier," *2008 IEEE MTT-S International Microwave Symposium Digest*, Atlanta, GA, 15–20 June, pp. 643–646.

[53] L. B. Max, "Apply Wideband Techniques to Balanced Amplifiers," *Microwaves*, vol. 19, pp. 83–88, 1980.

[54] Application Note 10953, "BLF645 10 MHz to 600 MHz 120 W Amplifier," Ampleon, 2015.

[55] D. Sardin and Z. Popovic, "Decade Bandwidth High-Efficiency GaN VHF/UHF Power Amplifier," *2013 IEEE MTT-S International Microwave Symposium Digest*, Seattle, WA, 2–7 June, pp. 1–3.

[56] K. Kurokawa, "Design Theory of Balanced Transistor Amplifiers," *Bell Syst. Tech. J.*, vol. 44, pp. 1675–1798, 1965.

[57] R. S. Engelbrecht and K. Kurokawa, "A Wideband Low Noise L-band Balanced Transistor Amplifier," *Proc. IEEE*, vol. 53, pp. 237–247, 1965.

[58] R. E. Neidert and H. A. Willing, "Wide-Band Gallium Arsenide Power MESFET Amplifiers," *IEEE Trans. Microwave Theory Tech.*, vol. MTT-24, pp. 342–350, 1976.

[59] K. B. Niklas, R. B. Gold, W. T. Wilser, and W. R. Hitchens, "A 12–18 GHz Medium-Power GaAs MESFET Amplifier," *IEEE J. Solid-State Circuits*, vol. SC-13, pp. 520–527, 1978.

[60] B. M. Oliver, "Directional Electromagnetic Couplers," *Proc. IRE*, vol. 42, pp. 1686–1692, 1954.

[61] E. M. T. Jones and J. T. Bolljahn, "Coupled-Strip-Transmission-Line Filters and Directional Couplers," *IRE Trans. Microwave Theory Tech.*, vol. MTT-4, pp. 75–81, 1956.

[62] J. Lange, "Interdigitated Stripline Quadrature Hybrid," *IEEE Trans. Microwave Theory Tech.*, vol. MTT-17, pp. 1150–1151, 1969.

[63] R. Waugh and D. LaCombe, "Unfolding the Lange Coupler," *IEEE Trans. Microwave Theory Tech.*, vol. MTT-20, pp. 777–779, 1972.

[64] W. P. Ou, "Design Equations for an Interdigitated Directional Coupler," *IEEE Trans. Microwave Theory Tech.*, vol. MTT-23, pp. 253–255, 1975.

[65] Z. I. Model, *Networks for Combining and Distribution of High Frequency Power Sources* (in Russian), Moskva: Sov. Radio, 1980.

[66] J. Sevick, *Transmission Line Transformers*, Norcross: Noble Publishing, 2001.

[67] C. Trask, "Transmission Line Transformers: Theory, Design and Applications," *High Frequency Electronics*, vol. 4, pp. 46–53, 2005 and vol. 5, pp. 26–33, 2006.

[68] E. Rotholz, "Transmission-Line Transformers," *IEEE Trans. Microwave Theory Tech.*, vol. MTT-29, pp. 148–154, 1981.

[69] J. Horn and G. Boeck, "Design and Modeling of Transmission Line Transformers," *Proc. 2003 IEEE SBMO/MTT-S Int. Microwave and Optoelectronics Conf.*, vol. 1, pp. 421–424.

[70] G. Guanella, "New Method of Impedance Matching in Radio-Frequency Circuits," *The Brown Boveri Rev.*, vol. 31, pp. 327–329, 1944.

[71] G. Guanella, "High-Frequency Matching Transformer," U.S. Patent 2,470,307, May 1949 (filed April 1945).

[72] R. K. Blocksome, "Practical Wideband RF Power Transformers, Combiners, and Splitters," *Proc. RF Technology Expo 86*, pp. 207–227, 1986.

[73] J. L. B. Walker, D. P. Myer, F. H. Raab, and C. Trask, *Classic Works in RF Engineering: Combiners, Couplers, Transformers, and Magnetic Materials*, Norwood: Artech House, 2005.

[74] C. L. Ruthroff, "Some Broad-Band Transformers," *Proc. IRE*, vol. 47, pp. 1337–1342, 1959.

[75] C. L. Ruthroff, "Broadband Transformers," U.S. Patent 3,037,175, May 1962 (filed May, 1958).

[76] J. Sevick, "A Simplified Analysis of the Broadband Transmission Line Transformer," *High Frequency Electronics*, vol. 3, pp. 48–53, 2004.

[77] M. Engels, M. R Jansen, W. Daumann, R. M. Bertenburg, and F. J. Tegude, "Design Methodology, Measurement and Application of MMIC Transmission Line Transformers," *1995 IEEE MTT-S International Microwave Symposium Digest*, vol. 3, Orlando, FL, 16–20 May, pp. 1635–1638.

[78] S. P. Liu, "Planar Transmission Line Transformer Using Coupled Microstrip Lines," *1998 IEEE MTT-S International Microwave Symposium Digest*, vol. 2, Baltimore, MD, 7–12 June, pp. 789–792.

[79] H. K. Chiou and H. Y. Liao, "Broadband and Low-Loss 1:9 Transmission-Line Transformer in 0.18-μm CMOS Process," *IEEE Electron Device Lett.*, vol. 31, pp. 921–923, 2010.

[80] H. O. Granberg, "New MOSFETs Simplify High Power RF Amplifier Design," *RF Design*, pp. 43–52, October 1986.

Chapter 2

Nonlinear active device modeling

Iltcho Angelov[1] *and Mattias Thorsell*[1]

2.1 Introduction: active devices

The chapter covers main characteristics of the devices used in power amplifiers (PAs) and provides basic knowledge about the specifications of these devices and the transistor models that are used in PA designs.

The Preface covers the basic characteristics of main semiconductor device types used in PA designs. These devices have been in use for many decades, the technology of which is well established and its production is repeatable with high yield. Now, these devices show really impressive results. They work up to the terahertz region and can deliver kilowatts of power at low radio frequencies (RF). Probably one of the most promising devices now is the GaN high-electron-mobility transistor (HEMT), and they yield high-power and high-frequency results in a number of applications. There are a plenty number of literature on these devices regarding how to design PA, and using these transistors has put a tremendous success in its field. A very small part of these references is listed here [1–87]. The models come in small signal (SS), large signal (LS) categories, and for all these devices, extraction procedures, software is available [11–60].

The main part of the device-modeling chapter is devoted to GaN [1–10], because now the GaN HEMT is the most promising device for PA designs. The GaN HEMT technology is improved and it is being used in multiple applications for generating power up to the millimeter waves. Most of the procedures for model extraction, which are described in this chapter for GaN HEMT, are directly applicable for other types of semiconductor devices as well. The procedures are common for all devices that should be measured and modeled in a similar way.

2.1.1 Semiconductor devices for PAs

The following are the main types of semiconductor devices used in today's PA:

1. GaAs, INP HEMT [11–60]
2. GaN HEMT [1–9,78–83]
3. CMOS FET [61–68]
4. HBT, BJT: SiGe, GaAs, InP, Si [69–72]

[1]Microwave Electronics Lab (MEL), Chalmers University of Technology, Gothenburg, Sweden

These devices have many common features, so all devices should be measured in a similar way, after procuring enough data, but keeping the device safe:

1. *IV* with minimum 10×10 matrix of control voltages.
2. Capacitance measurement, extraction and modeling with the same matrix of control voltages. The measurement can be performed with capacitance, impedance meter or using *S*-parameter measurements.
3. High-frequency *S*-parameter measurements for parasitic extraction, fit.
4. Power spectrum measurements. It is the most efficient, simple way to get a good quality, precise description of harmonics generated in the device at RF. Two sets of RF are used usually:
   (i) Low RF to precisely evaluate the nonlinearity of the current part.
   (ii) High-frequency RF to evaluate the nonlinearity of capacitance part.
5. LS evaluation. This can be load pool, power sweeps, and the best is to be combined with large signal vector network analyzer (LSVNA) measurements.

All these devices show some similar parameters:

1. The max current device can deliver in the range of I_{max} = 0.7–2 A/mm. The GaN devices can be designed to have very high I_{max}/mm in the range of 2–3 A, Figure 2.1.
2. The total device capacitance is in the range C_g = 0.7–1 pF/mm. The capacitance depends on the gate length (L_g)—device with smaller L_g will have less capacitance.
3. Very important parameter is the residual resistance of the device R_{on}, when the device delivers the maximum current.

For good GaAs, residual resistance R_{on} (resistance in the linear region, below the knee) is in the range of R_{on} = 0.2–0.5/mm. GaN devices now have higher R_{on}, but devices are improving.

4. The main parameters, I_{max}, I_g, C_g for field-effect transistor (FET) devices, scale well, linearly with the device size. The HBT, bipolar junction transistor (BJT) scale with the surface of the device.

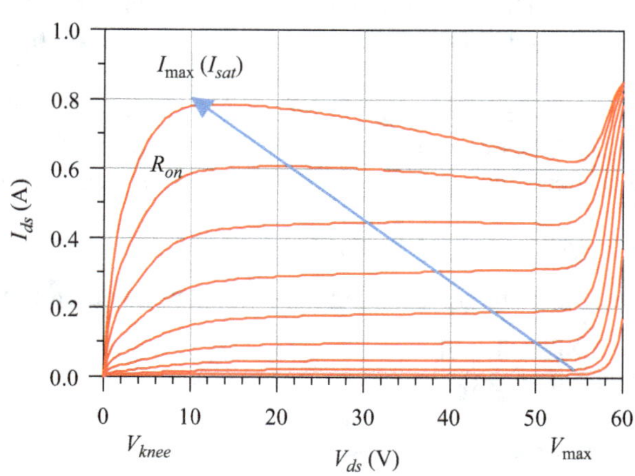

Figure 2.1 Typical I_{ds} vs. V_{ds}, stepping V_{gs}. Breakdown starts 55 V

These transistor parameters will determine to a very high degree what we can get from these devices in PA:

1. I_{max}, V_{knee} will determine output power.
2. R_{on}→output power, PAE.
3. V_{max} (breakdown at small I_{ds} (5%I_{max})). When the device is pumped, the drain voltage during waveform swing can reach many times the operating voltage. This will be discussed in the other chapters, but for simplicity, it can be considered that in Class F, operating swing can be 3.5 times higher than the operating voltage.
4. Drain voltage swing, which can be considered $V_{max}–V_{knee}$, will determine the power and load impedance.
   From linearity and matching point, it is always better to have large drain voltage swing and low I_{ds} current swing!
5. Gate voltage swing, and respective drain current, Figure 2.2 will be determined from the required input power pump. This is very important for high-frequency amplifiers, when the transistor gain is low. In this case, much more input power is needed to get the required output power P_{out} from the device. In addition, due to low gain, low PAE of the device will dissipate a lot of DC power and will be overheated; thus, these things should be very carefully considered.

2.1.2 GaAs FET and InP HEMT devices

Very extensive research was made on these devices and multiple papers and books are published. References [11,12,16,71,72] describe a little of it. The GaAs, InP devices show very high I_{sat}/mm, high F_t, F_{max}, not very high breakdown and so on.

The reliability is high, established technology, high yield in production, high F_t, F_{max} in the THz region.

They are very good for medium volume RF and mm wave MMIC: medium power PA, low-noise amplifier (AMP), mixers, multipliers.

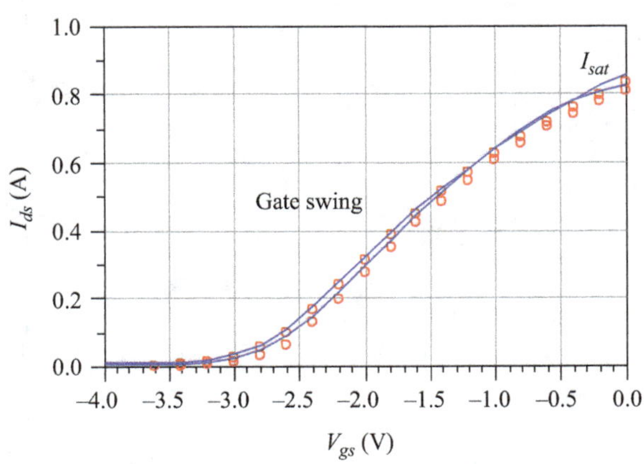

Figure 2.2 I_{ds} vs. V_{gs}

Figure 2.3 shows typical *IV* for a good quality $W_{tot} = 0.1$ mm, 2×50 μm devices, $L_g = 0.1$ μm process. The $I_{\max}$ is in the range of 0.8 A. The transconductance is bell shaped, and it is saturated at high V_{ds}, due to self-heating and mobility reduction.

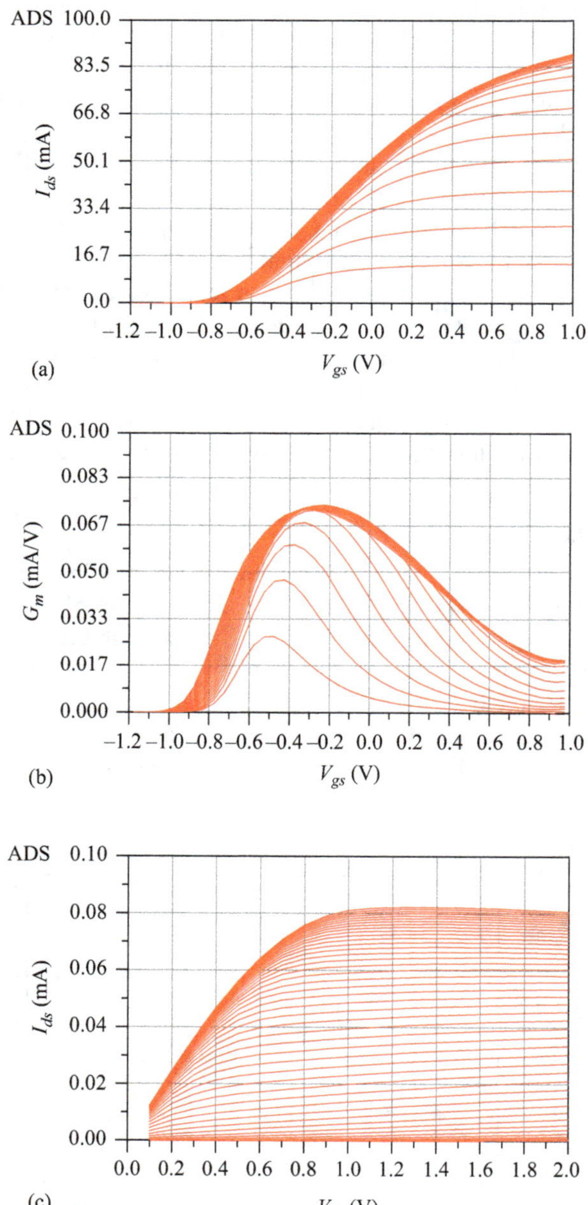

Figure 2.3 IV GaAs FET: (a) I_{ds} vs. V_{gs}; (b) G_m vs. V_{gs}; (c) I_{ds} vs. V_{ds}

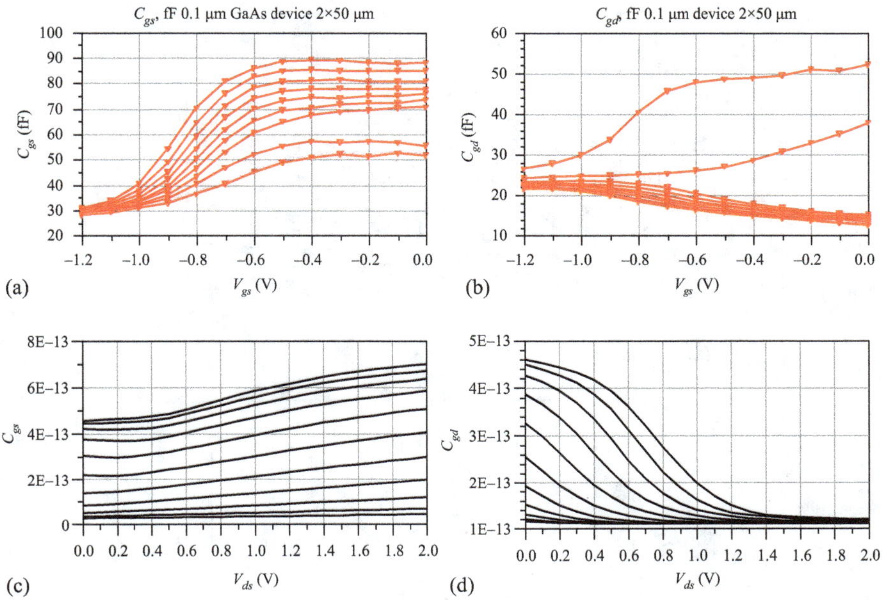

Figure 2.4 Capacitances GaAs FET: (a) C_{gs} vs. V_{gs}, V_{ds} parameter; (b) C_{gd} vs. V_{gs}, V_{ds} parameter; (c) C_{gs} vs. V_{ds}, V_{gs} parameter; (d) C_{gd} vs. V_{ds}, V_{gs} parameter

Figure 2.4 shows a typical C_{gs}–C_{gd} dependence on the voltages. It should be remembered that the primary controlled voltage for C_{gs} is V_{gs} and V_{gd} for C_{gd}. V_{ds} is the remote controlled voltage. The capacitances C_{gs}, C_{gd} are the minimum at the pinch off and gradually increase to the maximum. C_{gs} increases with the increase in V_{ds} and C_{gd} decreases at higher V_{ds}.

2.1.3 GaN HEMT devices

GaN devices are now becoming a main stream as PA design choice [78–83].

They exhibit high I_{sat}/mm, very high breakdown (kV), quite high F_t, $F_{\max}$. Resent results obtained, using these devices in PA, are extremely good. Recently, GaN devices became space qualified, and there are examples of high-efficiency communication amplifiers flying in space. There are multiple examples of these devices embedded in PA for radar and high power, high-reliability industrial applications. For example, a >10 kW DC-to-DC convertor is working in a Shinkansen bullet train (from 2017).

Still the technology is not so well established compared to that for CMOS, GaAs HEMT; the production yield is still improving.

For MMIC with a large number of transistors, the amplifier topology should be considered very carefully and the number of transistors should be minimized.

Figure 2.5 shows *IV* for a typical, good quality, general purpose 1 mm GaN HEMT. As with all FET, the G_m is peaking at some V_{gs}. For $V_{gs} = 0$, $V_{ds} > 0 > I_{ds} \geq 0$; HEMT is *normally* ON, i.e., this is a depletion mode device.

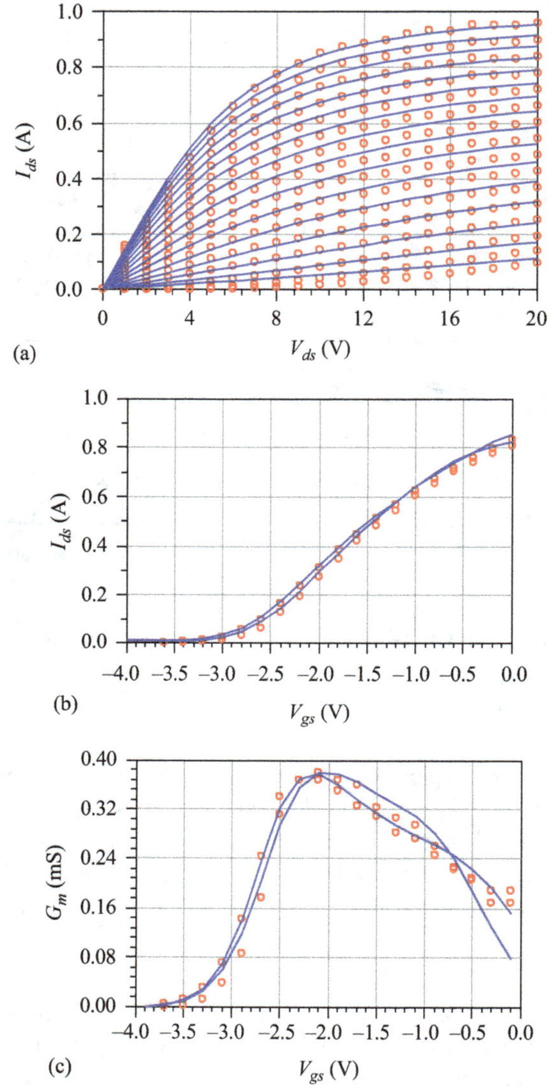

Figure 2.5 GaN IV measured points, model line: (a) I_{ds} vs. V_{ds} GaN 1 mm device, (b) I_{ds} vs. V_{gs}, (c) G_m vs. V_{gs}

The GaN capacitance behavior is similar to GaAs HEMT, Figure 2.6. The minimum is at the pinch off, and capacitance gradually increases with the increase in the control voltage. The inflection point of the voltage dependence is temperature dependent: this should be reflected in the capacitance models and considered in designs of PA or switching circuit. Some GaN HEMTs show a peak in C_{gs} and this

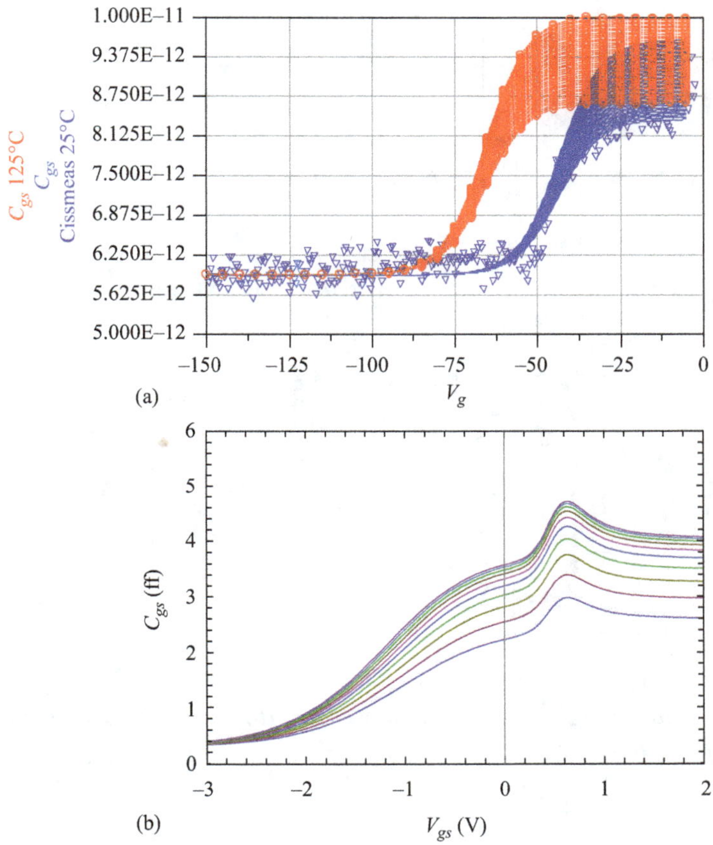

Figure 2.6 High voltage GaN HEMT: (a) C_{gs} high voltage GaN HEMT at 25°C and 125°C and (b) peaking C_{gs} GaN HEMT

should be modeled. The devices can easily handle hundreds of volts, and there are devices working well in the kV, kW range like the devices shown in Figure 2.7(a).

Now researchers are working on much higher power devices.

Figure 2.8 shows the high-frequency equivalent circuit of a typical GaAs, InP, GaN HEMT. It can be considered that we can split the equivalent circuit representation into two parts: linear and nonlinear.

Parasitic elements: R_g, R_{gd}, R_d, R_s, R_i, C_{ds}, L_g, L_d, L_s, layout elements, etc.

R_{del}, C_{del} shunting the gate control node V_{gsc} at high frequency:

1. Frequency-dependent gate control and delay.
2. Frequency-dependent R_s (SiC) nonlinear: I_{ds}, I_{gs}, I_{gd}, C_{gs}, C_{gd}→we need models.

Nonlinear parts are controlled by intrinsic voltages.

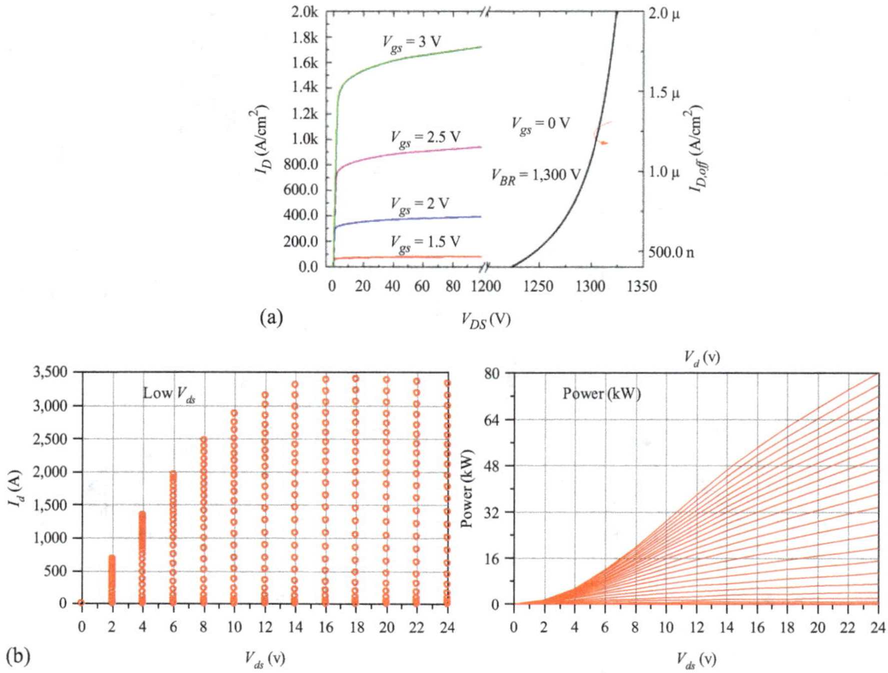

Figure 2.7 (a) IV high power 1.2 kV device and (b) very high power GaN FET IV and power

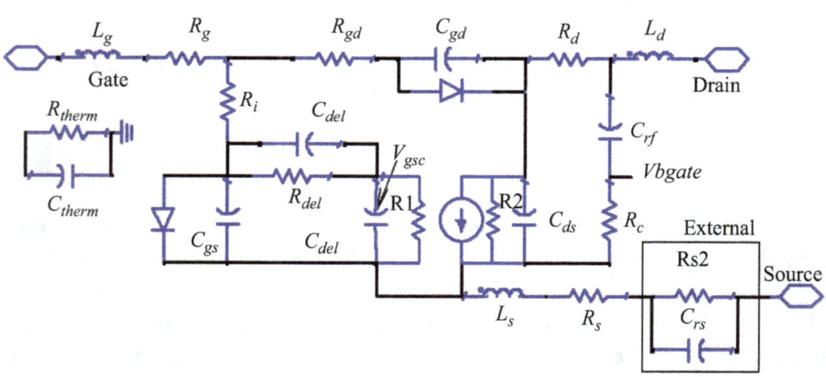

Figure 2.8 Equivalent circuit GaAs GaN HEMT

2.1.4 CMOS devices

Figure 2.9 shows *IV* for a typical 100 nm CMOS device. The *IV* is symmetrical, so should be the model.

CMOS devices for PA are enhancement (E)-type device, i.e., for $V_{gs} = 0$, $V_{ds}>0$, $I_{ds} = 0$, the device is normally OFF, Figure 2.10.

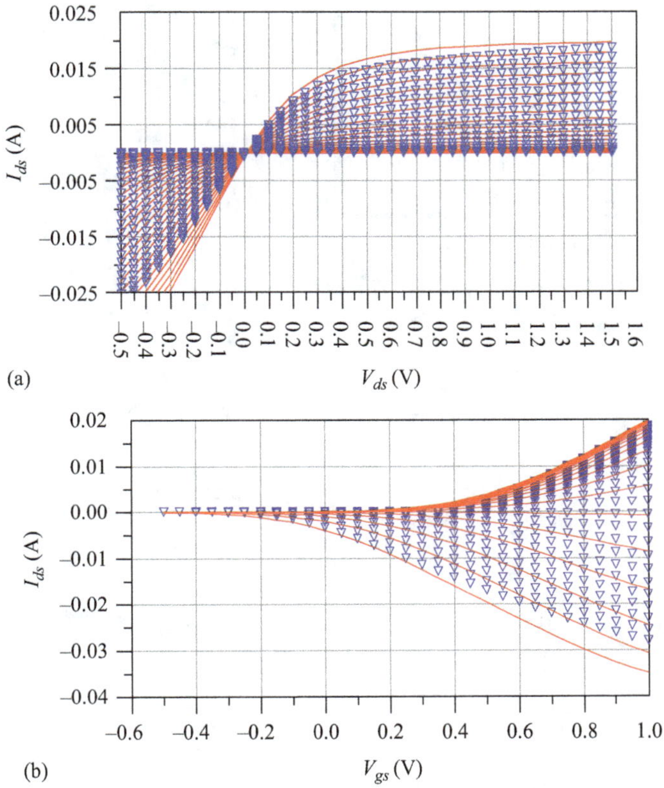

Figure 2.9 (a) CMOS I_{ds} vs. V_{ds} and (b) CMOS I_{ds} vs. V_{gs}

CMOS process is often combined with bipolar: BICMOS—these devices show very high reliability, technology is established, and the yield is very high.

They are very good for low-power mm MMIC, high-volume production [61–67], but the mask is expensive.

Power, PAE, noise is not very impressive, due to high-extrinsic R_s, R_d>R_{on}/mm.

Due to technology development, the device size is very small, L_g is in the range 7–100 nm. The CMOS devices show high F_t, $F_{\max}$ (intrinsic) in the range of THz. However, due to fundamental properties of the Si, lower mobility, etc., the extrinsic elements R_s, R_d, R_i are not that good as we can get from GaAs, InP HEMT. In addition, as we use the complete device (not the intrinsic), we will always have larger contribution in losses from R_s, R_d, R_i. An additional problem is the low breakdown of small gate length devices in Si CMOS. This is in a way fundamental—we shrink the device size and the breakdown will be worse, but when the material breakdown voltage is higher, we will get higher breakdown from the same L_g. We can use a parallel combination; combine many FETs to get more power from the low-voltage CMOS device. However, this has limits—we will need to transform very low impedance to 50 Ω, and the losses of the combing network will increase significantly.

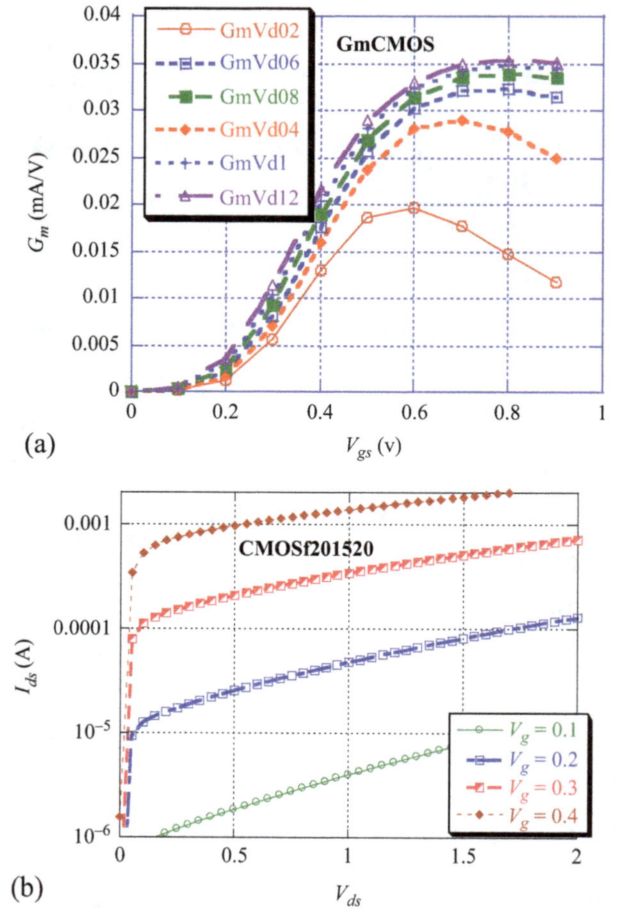

Figure 2.10 (a) CMOS G_m vs. V_{gs} and (b) CMOS I_{ds} vs. V_{ds}

The result is that the CMOS devices in reality will show lower power, PAE, and higher noise compared with GaAs, InP, GaN HEMT.

But for many applications we do not need extremely good circuit performance, but very high volume, low price in big volume.

This is a very good place for CMOS and in such high volume cases people will use CMOS. For example, 5G applications, circuits.

CMOS capacitances (Figure 2.11) are similar to ordinary FET capacitances C_{gs} and C_{gd}. The G_m is bell shaped (Figure 2.10(a)) as it is for all FET. It is important to notice that at low V_{ds}, low currents, the I_{ds} is exponentially dependent on V_{ds} (Figure 2.10(b)). As shown in Figure 2.10(b), the characteristic in logarithmic scale is linear. This can be important for some applications. The equivalent circuit of CMOS device, (Figure 2.12) is similar to GaAs FET, with multiple, additional coupling with the bulk: resistances, capacitances, diodes: R_{sb}, R_{db}, R_{gb}, C_{sb}, C_{db}, C_{gb}.

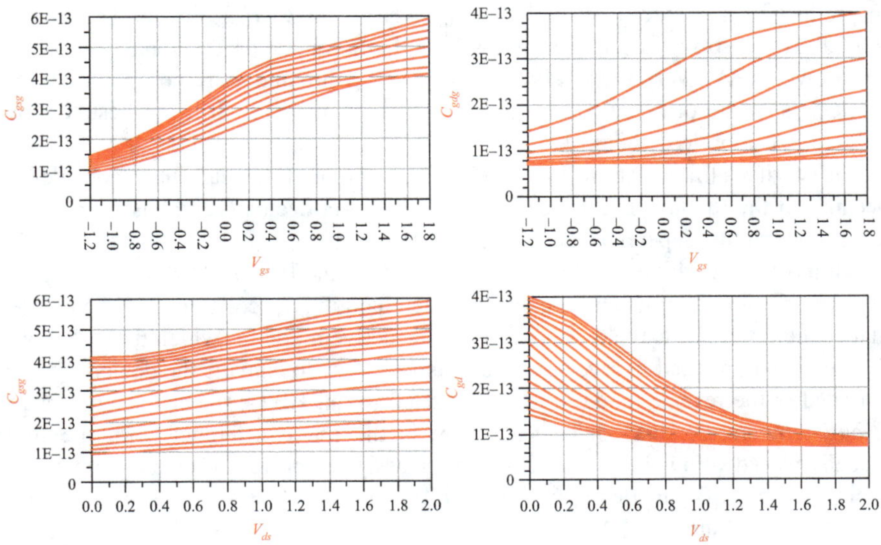

Figure 2.11 CMOS Cap

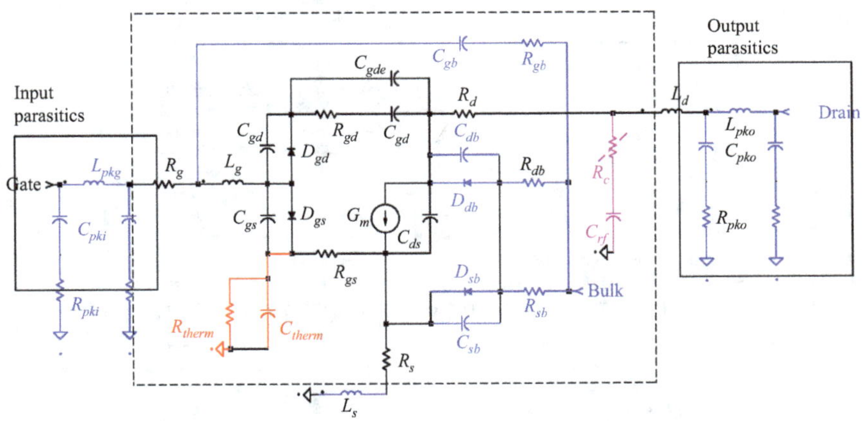

Figure 2.12 CMOS equivalent circuit

2.1.5 HBT devices

The bipolar point-contact transistor was invented in December 1947 at the Bell Telephone Lab. These were the first devices that were good enough and found application in many electronic products. The junction version known as the BJT, invented by Shockley in 1948, was the device of choice for three decades in the design of discrete and integrated circuits. Nowadays, the use of the BJT has declined in the favor of CMOS technology in the design of digital integrated circuits, but then, the HBT was invented and HBT devices are one of the best semiconductor devices available now. The technology is extremely well established—they have high yield in production and extremely high F_t, $F_{\max}$. HBT shows $F_{\max}$ in the THz region [69,70]. It is also very useful for mm MMIC, high-volume production.

Power—PAE very impressive due to low R_{on}/mm and extremely high gain.

G_m is in the order of Siemens even for small devices. Still the best for low voltage PA, mobile devices.

Linearity—linearization for the HBT PA should be carefully considered, because of the fundamental property of the HBT—collector I_c and base I_b currents are exponentially dependent on the control voltages.

Importantly, which should not be forgotten for the BJT, HBT devices, they are asymmetrical by design. This means that interchanging the collector and the emitter makes the transistor leave the forward normal active mode for which transistor was designed and operated in reverse mode. It is very easy to damage the device when is operated in the reverse mode. It is fundamental for all HBTs: the characteristics are *strong* temperature, bias dependent. Here is the main difference with the FET (usually biased with voltage sources): at high dissipated power, when the temperature is higher, the I_{ds}, G_m are decreasing. The FET device is getting some self-stabilization against temperature.

The HBT, when biased with a voltage source at the base, like in Figure 2.13(b) will start thermal runaway. The reason is fundamental as can be seen in Figure 2.14 from forward Gummel (FG) measurement when $V_c = V_b$. In the FG, we do not

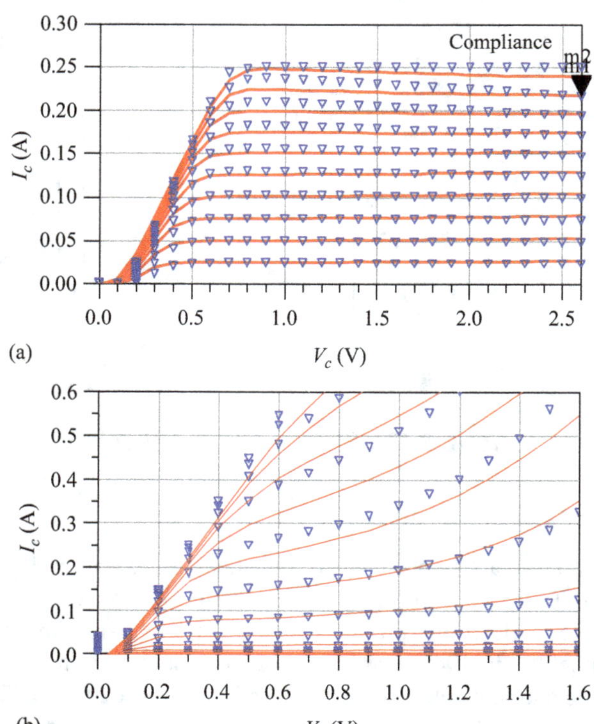

Figure 2.13 HBT IV: (a) I_{ce} vs. V_c, I_b parameter and (b) I_c vs. V_c, V_b parameter

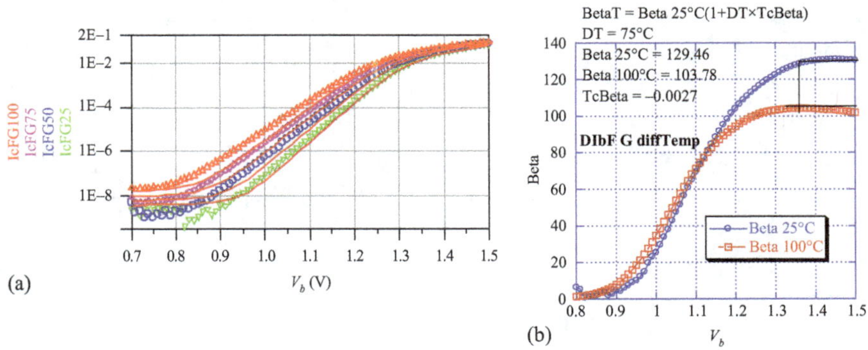

Figure 2.14 (a) HBT forward Gummel plot ($V_c = V_b$) and (b) HBT beta at different temperatures

dissipate power on the BC junction. At higher temperature FG, I_c and I_b characteristics move to the left when the temperature increases, i.e., when we bias with a voltage source (1.3 V, for example), the device will heat up and drive more current and I_c will exponentially increase. If the voltage-sweeping range is not set properly, the device will burn. The current-gain beta = I_c/I_b (Figure 2.14(b)) is bell shaped and also strongly temperature dependent. For this reason, the biasing range of voltages, compliance should be selected carefully. The biasing network should be designed to provide proper bias, correct the temperature dependence and insure device stability (Figure 2.14).

EC *HBT* is similar to FET EC [84]. It consists of linear and nonlinear parts. Linear parts are considered inductances, most of the resistances. Some of the resistances like R_{bin} are nonlinear. Nonlinear, bias and temperature dependent are the current part, capacitances C_{be}, C_{bc}, and the delay, respectively (Figure 2.15).

The BJT, HBT capacitances, C_{be}, C_{bc} can be considered consisting of two parts: depletion and diffusion. The capacitances C_{be} and C_{bc} show rapid increase and then decrease at high-forward bias around V_{bi} = 1.46 V, Figure 2.16 for GaAs: the peak is

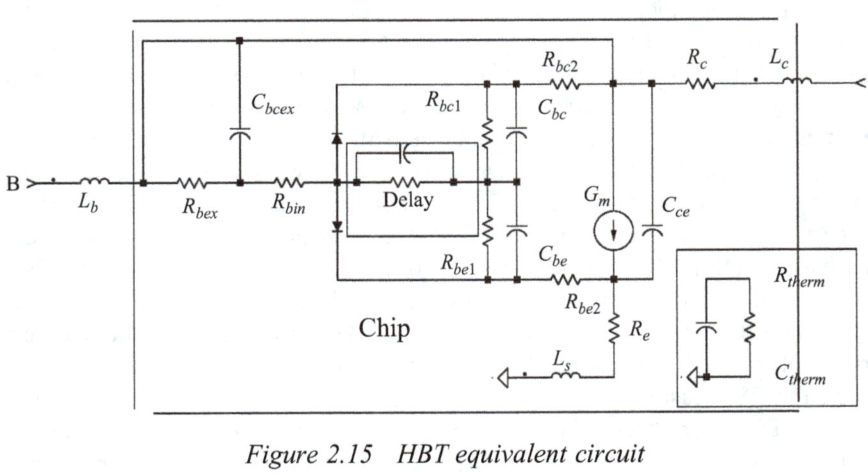

Figure 2.15 HBT equivalent circuit

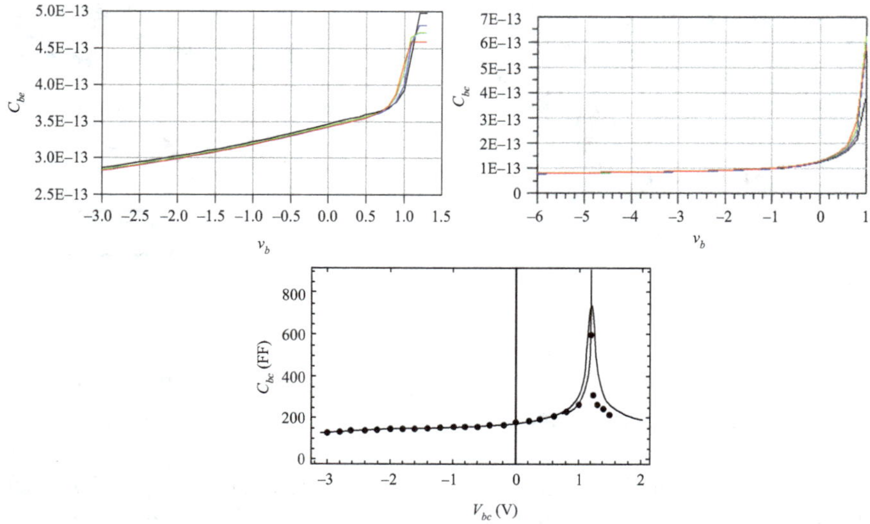

Figure 2.16 HBT C_{bc}: (a) C_{be} vs. V_b and (b) C_{bc} vs. V_{sbc}

at the built in voltage V_{bi} and then decline. When the currents start to flow via the junctions, mainly above V_{bi}, the diffusion part will start to work. The standard implementation of the junction capacitance $C_{dep} = C_{dep0}/(1 - V_{be}/V_{bi})^n$ will create convergence problems when the voltage swings around V_{bi} and above V_{bi}.

The accepted approach is to define the capacitance with two definitions—below and above V_{bi}, but this cannot solve the problem completely. So a proper, converging capacitance, charge model was developed and implemented. The converging, compact HBT model [84] is available in Verilog A (VA) for Advanced Design Station (ADS) and Spectre, CADENCE [53].

2.1.5.1 Importance of correct derivatives

Figure 2.17 shows measured and modeled I_{ds} vs. V_{gs}, V_{ds} above the knee, with respective measured and modeled harmonics. The device is pumped with a power of 0 dBm. Usually, two sets of frequencies are used; low RF (in the range of 0.1 GHz) to get the nonlinearities of the current source and high RF frequency, in the range of 5–10 GHz to get the capacitance part. The difference measured model current is very small, less than percent. However, there is quite a large difference in measured-modeled harmonics at RF. It is not certain how the DC fit influences the harmonics The extractions from the DC measurement I_{ds} harmonics are usually very noisy, and it is not possible to use them to get a better fit for harmonics products. In addition, there are some effects—gate current influence, dispersion, self-heating, onset of the breakdown. The influence of all these effects is difficult to see directly at the *IV*. For this reason, to get a good quality device model, the PS measurements must be considered. It is a very simple measurement: the DUT is actively biased, pumped with some power, enough to see the nonlinearities, the DC input control voltage is swept to evaluate the dependence

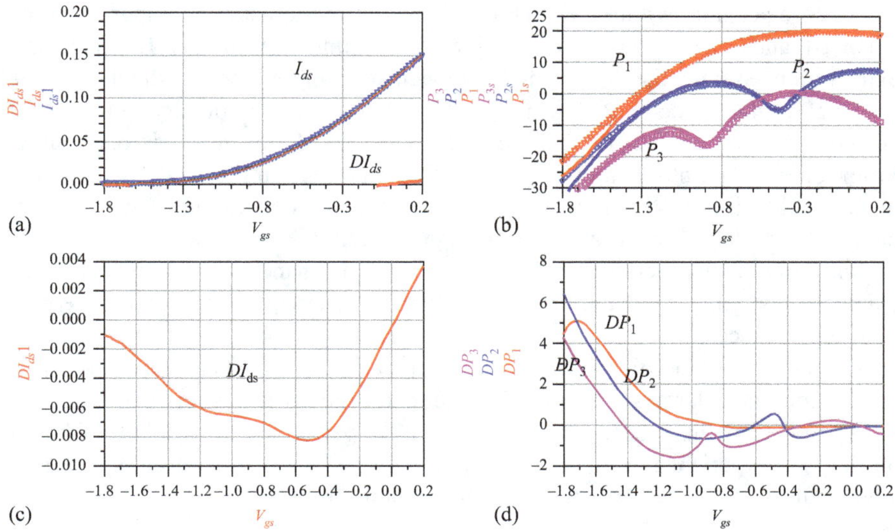

Figure 2.17 *(a) I_{ds} vs. V_{gs} GaN HEMT, (b) PS measured modeled, (c) difference measured-modeled, (d) difference measured-modeled PS*

of the harmonics on the input control voltage. The procedure can be applied for FET, CMOS HBT [47–49].

2.1.5.2 Model types

All these devices need models to be used practically from designers. Basically, there are 3 types of models:

1. Physical models (PM) used to design the device.
2. Measurements (Table)-based models (TBMs)-based on measured data from the device.
3. Empirical Equivalent Circuit Models (EECM). These are equivalent circuit models extracted from measured or simulated device data.

All model types have their place, use—we should use the right type for the specific application. We can mix and integrate model types to get the best from different types of models. In more details this is discussed in Section 2.5.2.

Important point is the accuracy of models. It is critical to get circuits working directly, with high yield, quickly, after they are designed. There are two main sources for discrepancy between designed and measured performance of the circuit:

1. Model accuracy.
2. Device, circuit manufacturing accuracy.

Model accuracy depends on the type of the model and is a result from device characteristics we want to describe and model complexity. The effects modeled, will effect number of parameters etc. Simulation speed, convergence are also important.

Device processing tolerance in a good foundry process is in the range of 5–10% device difference between the wafers. Within the wafer, (good foundry) devices are accurate within 1–5%. From this point of view is very important to be balanced and used a model which provide a similar accuracy. Every model extraction need time, the more complex models need more time to extract. There is no point trying to do model which is 1% accurate when foundry delivers the devices with 5% spread. In such situation 1–5% model accuracy is reasonable. From tolerance point of view, it is important to arrange in the model with statistical., yield facilities. When process is stable and device data is collected, statistical part of the model can be arranged, so designers should be able make yield analysis of their circuit. But even evaluating the circuit performance with slightly detuned bias voltages will give a hint on the stability of the circuit design on device processing tolerances.

High-power amplifiers (HPAs) that include GaN HEMTs have potential applications in radar systems, satellite communication systems, cellular base stations, etc. Efforts have been made to improve the performance of GaN FET based on the vast experience accumulated over the years [1–53]. Therefore, concurrent development of device and circuit-design methods are desired.

2.2 Sources of nonlinearity (I_{ds}, various G_m, R_d, R_{therm}, capacitances, breakdown)

Figures 2.18–2.23 show typical drain current I_{ds} dependences vs. gate–source (GS) voltage V_{gs} and drain-source voltage V_{ds} and first and second derivatives for GaN HEMT. Characteristics of the GaN HEMT are very similar to the characteristics of other FETs [1–11]. Device typically will pinch at GS voltages, V_{gs}, of −3 to −4 V, Figure 2.18. The knee is in the range of 4–6 V, V_{ds}, and depends on the quality of the Ohmic contacts, Figure 2.19. At some gate voltage, the derivative I_{ds} vs. V_{gs}, the transconductance G_m = diff(I_{ds} vs. V_{gs}) will show peak, Figure 2.20. Devices with good quality Ohmic contacts will have lower knee voltage for the same device size and give more power, better PAE. The device dissipates significant power, so to keep the device safe, measurement range should be split into two regions, Figure 2.19: high currents, low V_{ds} and high voltage, low current. The derivative I_{ds} vs. V_{ds}, or G_{ds}

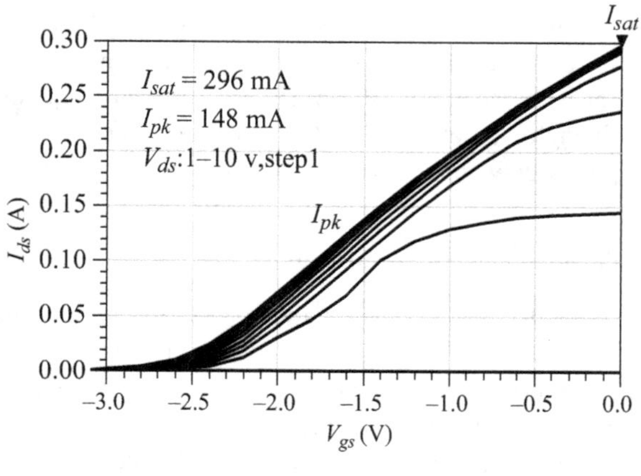

Figure 2.18 I_{ds} vs. V_{gs}, V_{ds} parameter GaN D-HEMT

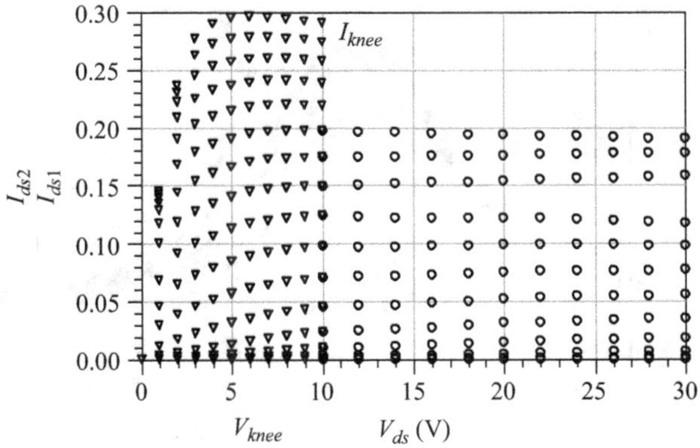

Figure 2.19 I_{ds} *vs.* V_{ds}, V_{gs} *parameter GaN HEMT*

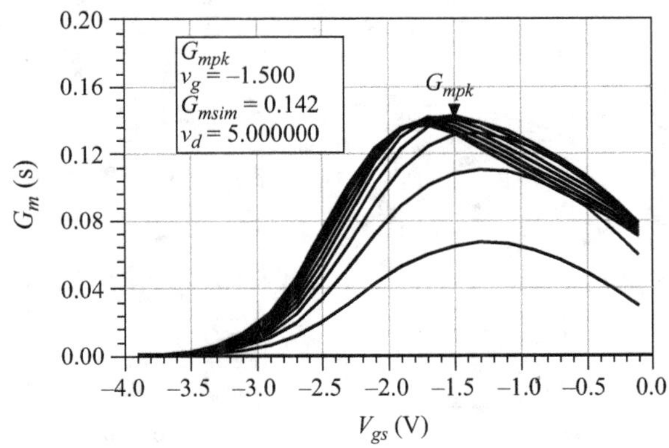

Figure 2.20 G_m *vs.* V_{gs} *GaN D-HEMT*

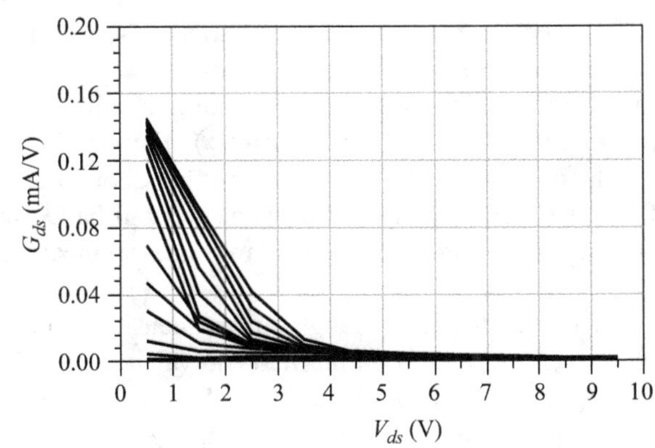

Figure 2.21 G_{ds} *vs.* V_{gs} *GaN HEMT*

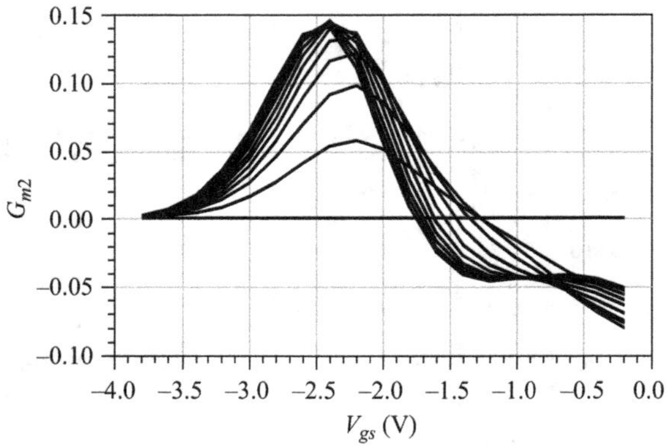

Figure 2.22 Second derivative of I_{ds}, GaN D-HEMT

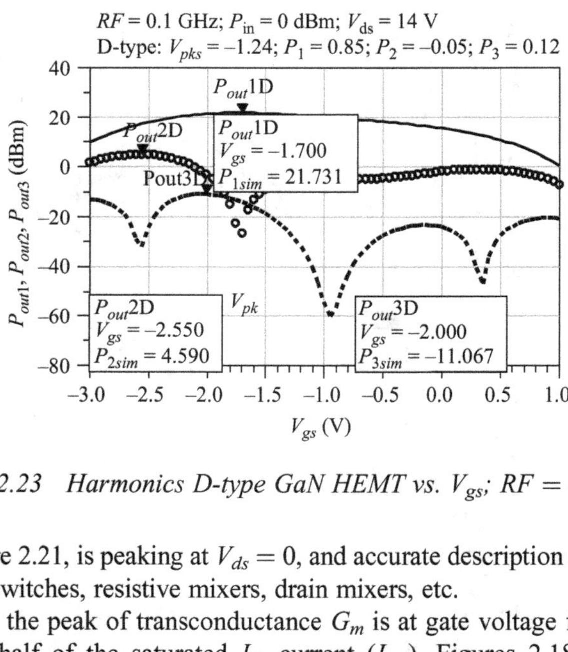

Figure 2.23 Harmonics D-type GaN HEMT vs. V_{gs}; RF = 0.1 GHz

shown in Figure 2.21, is peaking at $V_{ds} = 0$, and accurate description of this region is important for switches, resistive mixers, drain mixers, etc.

Normally, the peak of transconductance G_m is at gate voltage for which drain current I_{ds} is half of the saturated I_{ds} current (I_{sat}), Figures 2.18 and 2.20. As expected, at V_{gs} voltage for which G_m is maximum, the second derivative of drain current $G_{m2} = 0$, Figure 2.22. At this gate voltage, we will have minimum of the generated second harmonic, Figure 2.23, and in some application, this can be important.

Often GaN device operates at very high current density, gets hot and Ohmic contacts, and, respectively, R_d and R_s (simplified equivalent circuit 2.27) cannot be considered linear. Figure 2.24 shows experimental drain-resistance R_d and dissipated power P_{DC} vs. gate voltage V_{gs} (and respective I_{ds} current). Both drain R_d and source-resistance R_s change substantially when I_{ds}, P_{DC} (controlled by V_{gs}) increases. The effect, due to bias dependence of R_s and R_d, is accelerating, when the

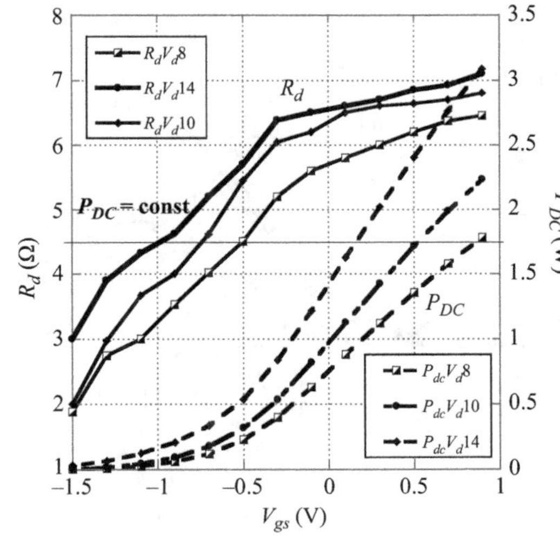

Figure 2.24 Nonlinear R_d GaN HEM

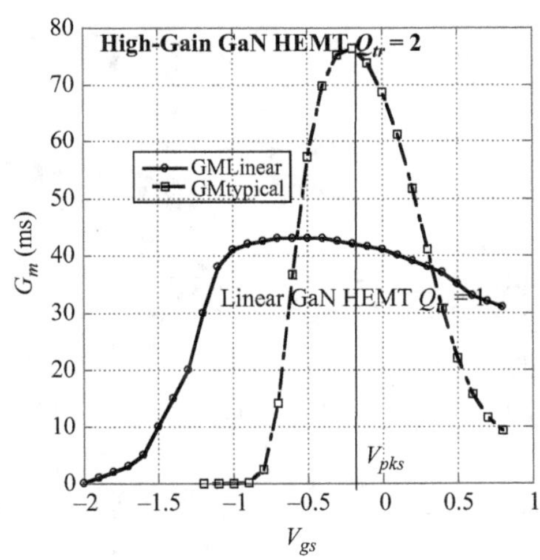

Figure 2.25 Low power GaN HEMT; typical G_m shapes

device is heated up at high current, high power. These two effects will substantially change the value of resistances R_s and R_d [73–77].

Very important parameter for the transistor operation is the ratio G_m/I_{pk}, the "quality" factor of transistor $P_1 = Q_{tr}$.

From gain point of view, device with high $P_1>2$, Figure 2.25, operating at the same I_{ds} will have higher G_m, more gain.

From linearity point of view, if P_1 is small, <0.5, device will be linear but will have low G_m, respectively, lower gain for the same device size, current.

Depending on the device design, HEMT can be depletion (D), Figures 2.20–2.23, or E-type, Figures 2.26 and 2.27.

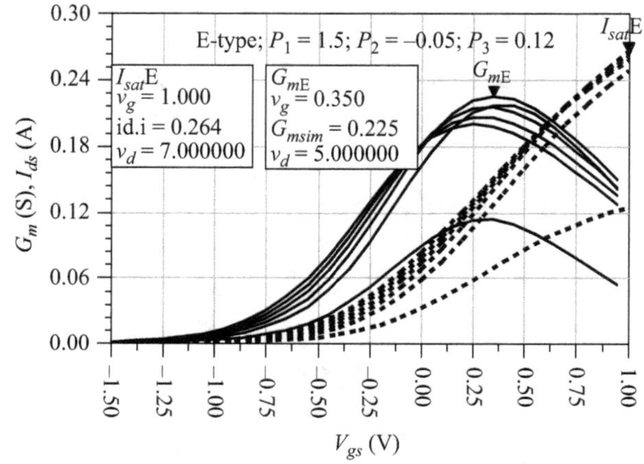

Figure 2.26 Enhancement GaN HEMT IV

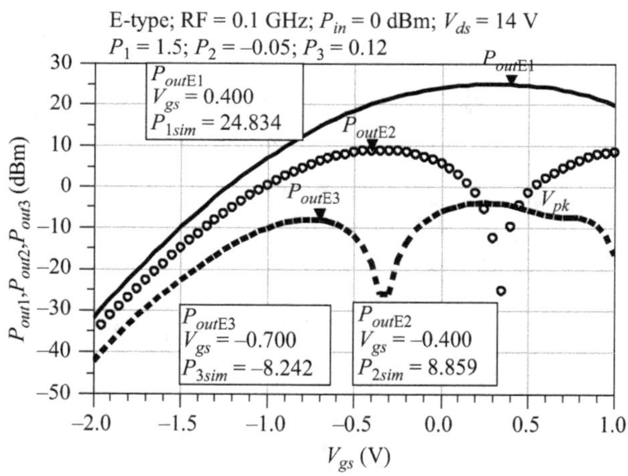

Figure 2.27 Harmonics E-type GaN HEMT vs. V_{gs}; RF = 0.1 GHz; P_{in} = 0 dBm

Depletion FET usually show peak of the transconductance in the negative range of V_{gs} (-3 to -1) and will pinch off at V_{gs}:-4 to -6 V. D-type FET are very good for high reliability, space and linear application, because, when input of the device is pumped hard, the rectified gate current is minimized.

The E-type of devices are usually pinched off at $V_{gs} = 0$ or close, transconductance shows peak at positive V_{gs}. To open the channel and reach saturated I_{sat}, V_{gs} should be increased to $V_{gs} = +1$ (+2 V for some). At high-positive gate voltage, usually, there is gate current I_{gs}. The enhancement HEMT usually provides higher P_1, more gain, but will generate more harmonics and, when pumped with large input power, can exhibit rectified gate current. For some applications like high-order multipliers, generating high-power harmonics, these features of E-mode FET are useful. In a similar way, in a low-noise stage, we need more gain at small current, and devices with high P_1 will perform very well. But if E-device is pumped hard, gate part will start to rectify, so this should be accounted for, minimized. Otherwise, the excess gate current can sacrifice the reliability. It is usually considered that rectified gate current I_{gs} should be less than 0.1 mA/mm for reliable operation.

For example, if we look at the harmonics at their maxima, keeping the input power constant $P = 0$ dBm, we will find the following: depletion FET generate: $P_{out\text{D}1} = 21.7$ dBm, $P_{out\text{D}2} = 4.6$ dBm, $P_{out\text{D}3} = -11$ dBm.

Enhancement FET generate: $P_{out\text{E}1} = 24.8$ dBm, $P_{out\text{E}2} = 8.86$ dBm, $P_{out\text{E}3} = -8.2$ dBm.

D-type: $V_{pks} = -1.24, P_1 = 0.85, P_2 = -0.05, P_3 = 0.12$; *E-type*: $V_{pks} = 0.4$, $P_1 = 1.5, P_2 = -0.05, P_3 = 0.12$

In this case, E-type HEMT will have more gain, more harmonics, but if saturated current, resistances are the same; they will produce about the same saturated power.

Important sources of the nonlinearity are FET capacitances such as GS C_{gs} and gate–drain (GD) C_{gd}. Typical bias dependences of C_{gs}, C_{gd} are shown in Figures 2.28–2.31.

When FET gate is in the middle between the gate and drain distance, C_{gs}, C_{gd} dependences vs. V_{gs} are identical. When V_{gs} is increased, the C_{gs}, C_{gd} increase, saturate and reach the maximum. When drain voltage increases, C_{gs} increases, and for V_{ds} voltages above the knee, C_{gs} is rather constant, Figure 2.29.

When the device is biased close to the pinch off $V_{gs} = -4.5$ V, Figure 2.28, the capacitance C_{gs} is the minimum. Dependence of C_{gd} vs. V_{gs} is similar to C_{gs} dependence, Figure 2.29, C_{gd} is the minimum at the pinch off. Normally, the minimum of the capacitance is in the order of $C_{gs\max}$, $C_{gd\max}/10$, i.e., capacitances are quite nonlinear. In some devices, the C_{gs}, C_{gd} change vs. gate voltages V_{gs} can be very sharp, so it is certain that working at high frequency, when gate voltage swings, harmonics will be generated from the capacitance part. GD capacitance (C_{gd}) plays a major role on the device performance. It makes parasitic feedback,

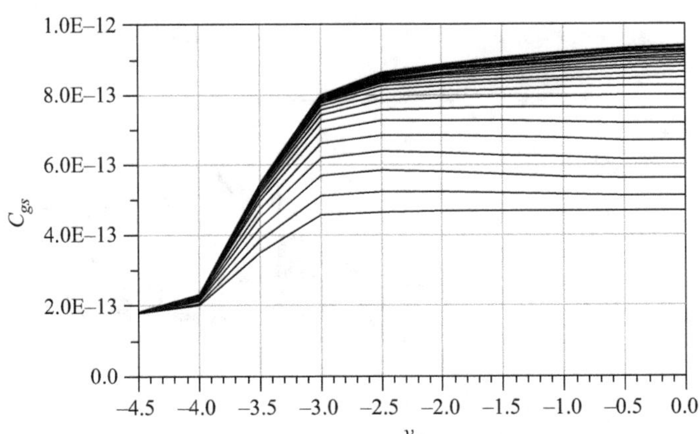

Figure 2.28 C_{gs} vs. V_{gs} GaN HEMT

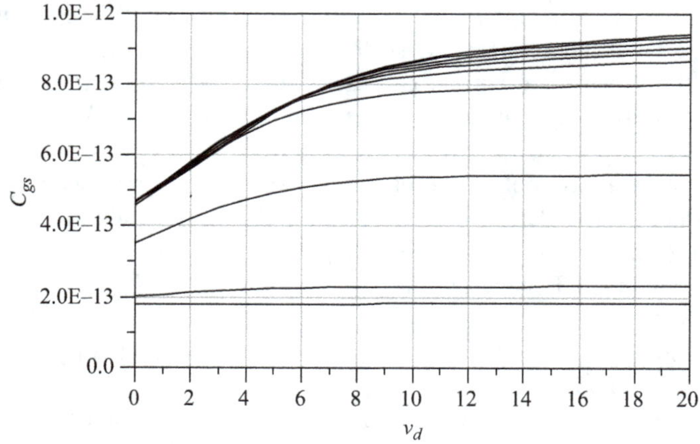

Figure 2.29 C_{gs} *vs.* V_{ds} *GaN HEMT*

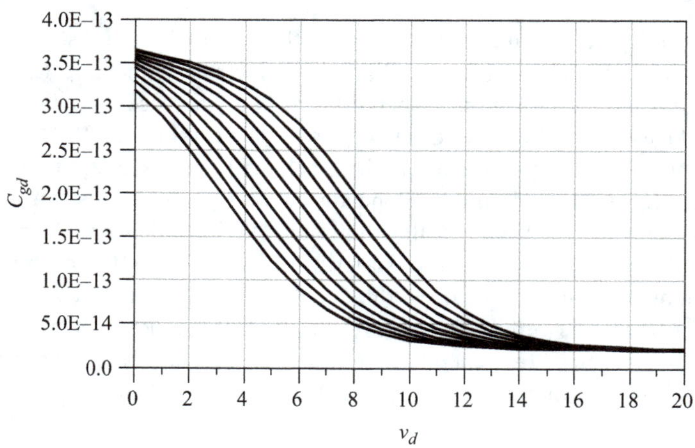

Figure 2.30 C_{gd} *vs.* V_{ds} *GaN HEMT*

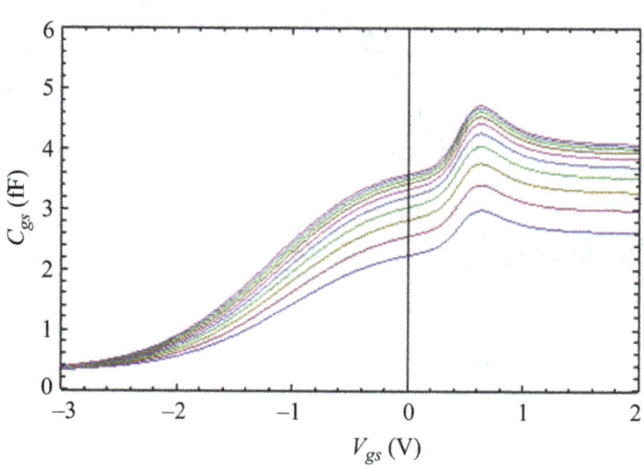

Figure 2.31 *Peaking* C_{gs}*, some GaN HEMTs*

decreases the gain, output power and generates harmonics when V_{gd} swing is larger. The C_{gd} is the maximum at low V_{ds}, Figure 2.30. When drain voltage increases, C_{gd} decreases. At the knee, many devices have very characteristic "crossing" C_{gd} vs. V_{ds}; for positive V_{gs}, the C_{gd} capacitance can be smaller.

The C_{gs}, C_{gd} bias dependences become much more complicated when GaN HEMT is "equipped" with field plates [1–10]. Field plates are arranged to distribute the electric field in GD, GS regions to improve the breakdown. Field plates also improve the reliability of the device, the dispersive characteristics, knee walkout, etc. Depending on the field-plate configuration, we will get different dependencies of C_{gs}, C_{gd} vs. voltages, different harmonics and intermodulation products will be generated. In this case, and also when the gate is not in the middle, it is certain that C_{gs}, C_{gd} will not be identical at $V_{ds} = 0$. Field plates give certain benefits, but as usual, we pay with something—in this case, we increase the capacitances, we pay with the bandwidth we can get. So compromise should be made. Modeling is becoming more complicated.

Quite often, to improve the gate leakage, increase the breakdown, a thin oxide layer is deposited under the gate (MIS FET). This improves the GS breakdown, but often the device becomes more dispersive. Devices with oxide under the gate often show peak in C_{gs}, which can be very sharp, Figure 2.31. A sharp change of capacitance can make unwanted increase of generated harmonics, so this should be modeled, respectively.

The GaN HEMTs can deliver very high power, power density per mm device size, so devices are pushed to the operational limits to get the maximum. When voltages, V_{ds}, V_{dg}, are increased above some certain threshold voltage, V_{tr}, there is an exponential increase of the drain current, as shown in Figure 2.32. Devices working at low RF application, 1–3 GHz, can have breakdown in the range of 30–100 V, millimeter wave GaN devices show breakdown in the range of 15–20 V, because mm wave devices are smaller, thinner channel, etc. The GaN devices for very high-power switching application can be designed to handle >5 kV and current in kA. The GS junction is usually weaker and can start to show I_{gs} breakdown already at V_{gs} −3 to −4 V, Figure 2.33. If the I_{ds} breakdown phenomenon is not reflected in the transistor models, designers can easily get world records in their amplifier simulation, to be disappointed later in the experimental evaluation.

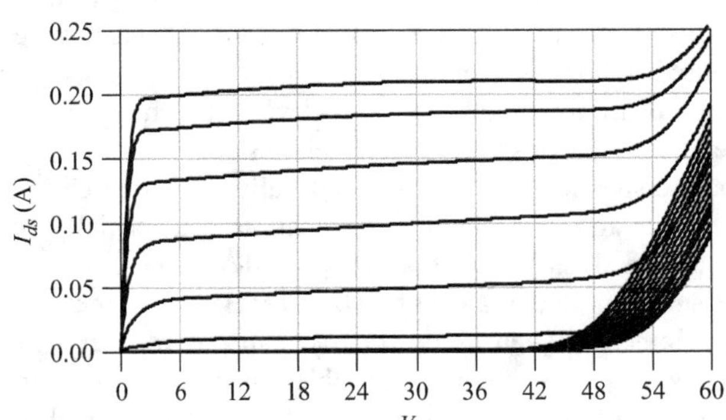

Figure 2.32 I_{ds} *vs.* V_{ds} *GaN HEMT in breakdown*

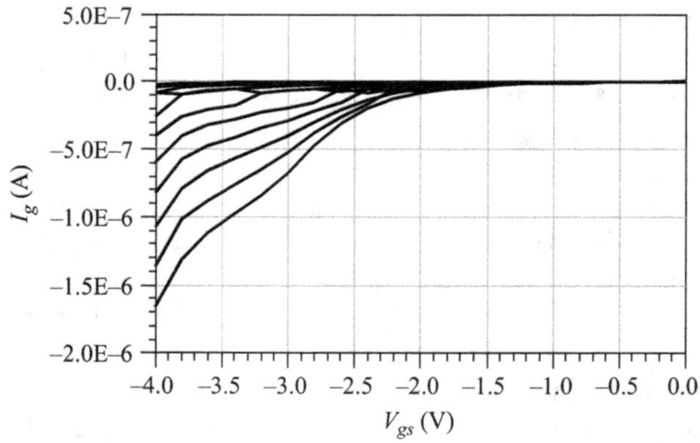

Figure 2.33 Experimental I_{gs} vs. V_{gs} in the breakdown

In addition, the sharp change of both I_{ds}, I_{gs} can create unwanted harmonics. So, for high power, realistic design, breakdown should be evaluated and modeled, respectively.

2.3 Memory effects

Every new technology, starting with GaAs, GaN devices now, show substantial dispersion problems, memory effects, hysteresis, etc. [37–44]. For example, when measuring GaN device stepping V_{ds} up or down, I_{ds} vs. V_{ds} (V_{gs} parameter), we can get different values for the current like in Figure 2.34(a). Similar is the effect when measuring I_{ds} vs. V_{gs}, V_{ds} parameter. The difference can be larger than 10% for one and the same set of V_{gs}, V_{ds}. This is one of the reasons to use in the measurements a separate sweep vs. V_{gs} stepping V_{ds} and vs. V_{ds}, stepping V_{gs}.

Part of trapping memory effects can be found this way. Much better way to evaluate trapping is to use pulse *IV* measurements like in Figure 2.34(b), low-frequency VNA or LSVNA measurements [37,46,92–100]. As can be seen in Figure 2.34(b), GaN devices show very slow recovery after the switching, comparing with a similar size LDMOS device. After some time, when technology matures, parasitic, unwanted dispersive effects are usually reduced. This is expected for GaN, in order to be used in communication system with advanced modulation schemes. The additional problem now is that the GaN device can change more than 10% in time, even after a simple load-pull measurement test.

The trapping dispersion phenomenon is usually concentrated in the frequency region between 1 Hz and 10 MHz; so to evaluate this, a specially designed setup, with properly designed bias tees, bias supply, should be used. Low-frequency impedance measurement in the range of 1 Hz–10 MHz, *S*-parameter measurements, pulsed *IV* measurements, when used properly, can provide very important information for the device quality. Additional problem is that the dispersion phenomena manifest itself in the same frequency region where is the thermal constant of the chip or packaged device TauTherm = Rtherm*Cthermal. To make things even

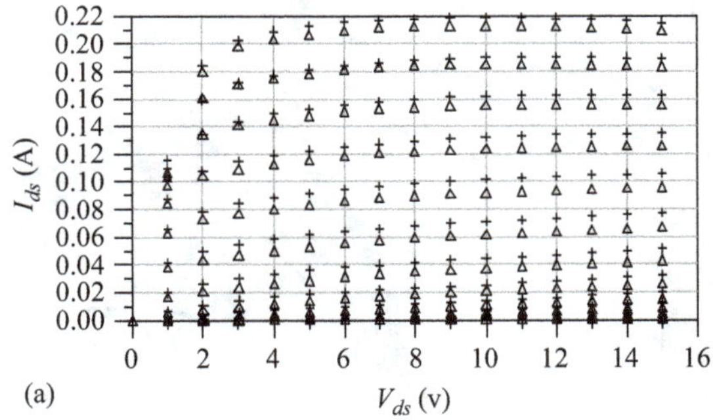

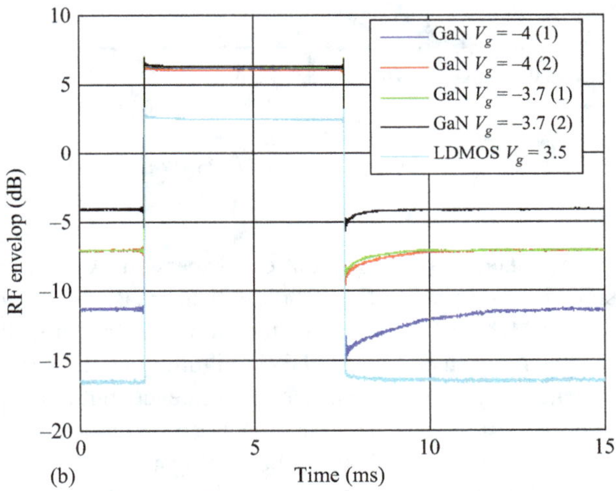

Figure 2.34 *(a) I_{ds} vs. V_{ds}, V_{gs} parameter stepping up with V_{ds} triangles, stepping V_{ds} down-cross. (b) Pulse switching characteristics GaN of different vendors and LDMOS*

more complicated, dispersion is strongly temperature dependent and so is thermal resistance. So things get mixed up and it is difficult to separate effects, difficult to model. To sort these effects, several measurements, such as CW *IV*, pulsed *IV*, *S*-parameter, pulsed *S*-parameter, LSVNA, shown in Figures 2.38–2.41, should be done at different temperatures. Exact description in this region is not always needed; it is difficult to measure and model precisely. For this reason, two border cases are usually considered: low frequency (DC) and high-frequency RF. This simplified solution was possible to tolerate: device is rarely operated directly in this frequency region, 1 Hz to 10 MHz. But now, when modern communication systems use very wide modulation bandwidth, this transient region becomes more important. The low-frequency region is also important for correct analysis of oscillators, because traps contribute directly to the phase noise of oscillators. Above the frequency at which dispersive effects are to some degree settled, device parameters stabilize to their high-frequency state (usually above 5–20 MHz). The change from low-to-high frequency state of parameters can be quite substantial. For example, Figure 2.35 shows the measured (points) and modeled (lines) DC transconductance

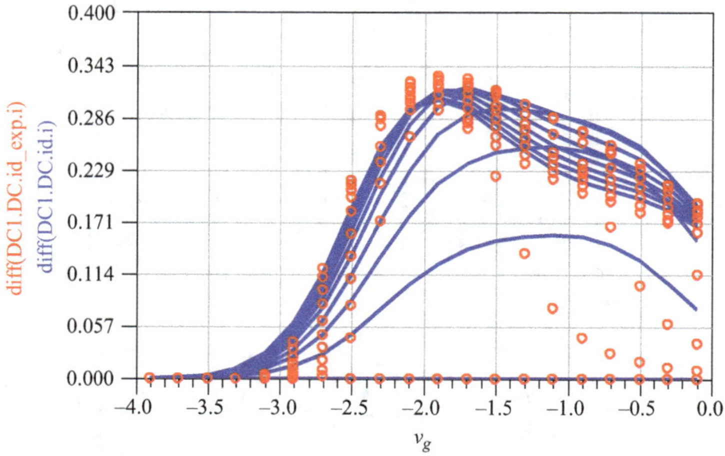

Figure 2.35 DC G_m; GaAs model

using GaAs device model, which is without G_m dispersion. At DC, low-frequency G_m fit is "pessimistic" and predict G_m 10% less than measured. Unfortunately, at high frequency, 200 MHz, *S*21, gain predicted from the "pessimistic" dispersion free model, is 5 dB higher than measured *S*21, Figure 2.36. This reduction of the gain at high frequency is due to traps, dispersion influence, not to device saturation from high input power. In the SS case, this effect can be modeled very accurately with the back-gating approach [37–44]. The back-gating is negative feedback, which returns part of the output RF signal in antiphase at the gate. Modeling the back-gating effect is physical; it was evaluated in multiple designs, starting from early GaAs transistors. The same back-gating approach is implemented now in CAD tools to model the SS dispersion of GaN HEMT [53].

In addition to the SS manifestation of the traps, some devices can exhibit an effect named "knee walkout" (Figure 2.37). This effect can be evaluated precisely using active load sweep (Figures 2.37–2.43), Thorsell from Chalmers [94,96] and Tasker [91,92] from Cardiff University. In this method, the load is purely active and is swept between 50 and 300 Ω (Figure 2.37). Applying correct input power at the gate, we get enough voltage–current swing at the output of the device, and we can precisely describe RF dynamic swinging of I_{ds} in the knee region. The solid line, Figure 2.37 shows the *IV* characteristics I_{dscw} vs. V_{ds} stepping V_{gs}, measured at DC. The DC knee is in the range of $V_{kneeDC} = 7$ V for high currents at $V_{gs} = +1$ V. The points on Figure 2.37 are recovered from RF measurements I_{dsRF} at 3 GHz. The RF knee at high currents, for the same $V_{gs} = +1$ V is $V_{kneeRF} = 12$ V. We see that, at RF, the I_{ds} current cannot reach the DC values at the knee. This knee walkout will decrease the output power, reduce PAE and should be accounted for. In this case, the device is biased at 20 V, but during the RF swing, the RF V_{ds} voltage can reach 32 V, which is enough to activate some traps. Usually the trapping, knee walkout, starts above certain drain voltage: the traps need energy (voltage) to get excited. Sometimes, at different drain voltages, different traps with

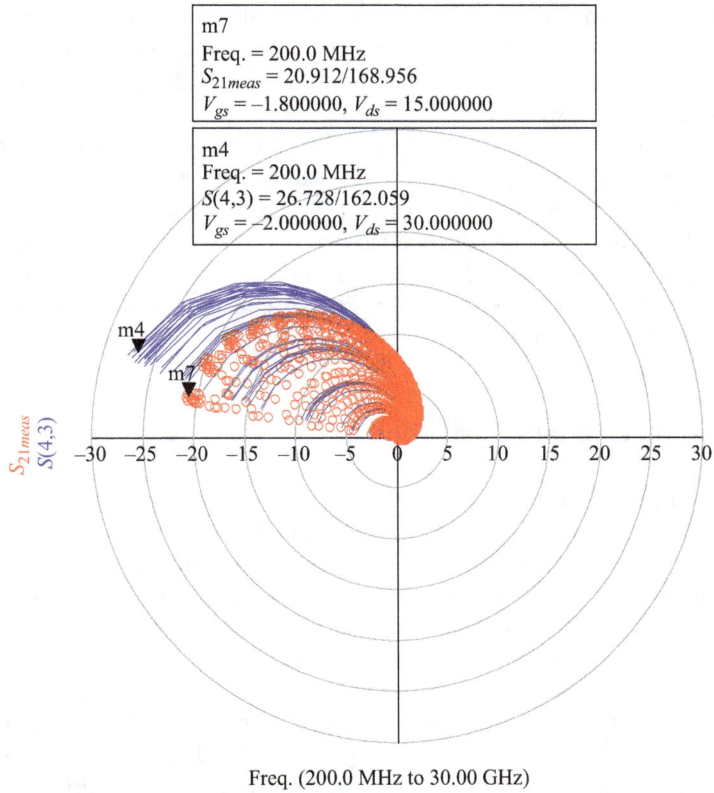

Figure 2.36 Simulated Mag S21: 3 dB higher with GaAs model

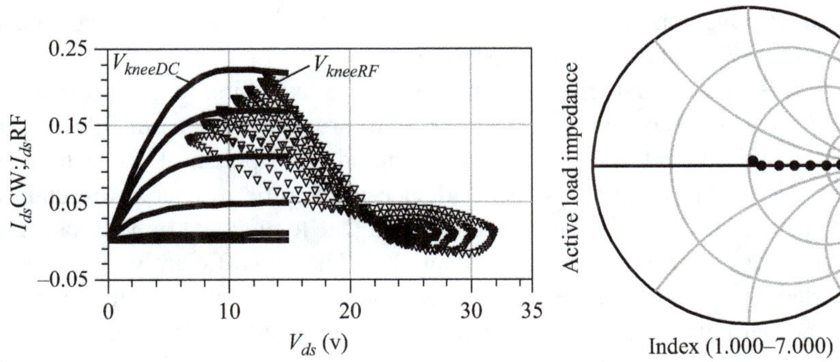

Figure 2.37 I_{ds} vs. V_{ds} GaN HEMT. Points RF swing, line-CW I_{ds} at different active load impedances

different time constants are excited. We are in a situation: we try to model problem instead of helping to solve the problem. Even if we have perfect models, they will not get the situation better from user's point of view: dispersion influence can be critical, many practical parameters, like output power (P_{out}), power added efficiency (PAE), low frequency noise (LFN), IMD products in the modulation bandwidth are worse for highly dispersive device. Some application still should be considered to use the old, well-working device-LDMOS until the GaN devices are better suited for these specific applications.

The good news is that GaN devices are getting better. We have seen this tendency with the GaAs, initially GaAs were also very dispersive. In a similar way, GaN devices are getting better. Now, for good-quality GaN devices, the knee walkout is less than 10% when operated in the specified drain voltage range. Seems logical: we better spent more time to evaluate and understand the problem in order to help device designers to fix the device problems, instead of investing most of the time to make ECM for the problem.

2.4 Nonlinear characterization

The standard procedure used for many years when characterizing transistors is to perform DC measurements to get the *IV* characteristics and use a VNA to obtain the *S*-parameters at each operating bias point. These measurements are either performed under CW condition, or under pulsed condition to reduce the thermal effects. These characterization methods work very well as long as the transistors behave linearly. For a linear operating device, the *S*-parameters will correctly predict the response to any applied stimuli (if stimuli are small with respect to the operating point values). However, the response due to an LS excitation is not characterized. Hence, effects such as harmonic generation, gain compression and intermodulation distortion are not included. The nonlinearities in the device are classified as odd and even, where the odd-order nonlinearities cause in-band distortion, and the even-order nonlinearities generate a low-frequency response around DC. The response at DC, due to an LS excitation, could give information about memory effects such as self-heating and trapping. These effects can significantly affect device performance and are not visible in *IV* and *S*-parameter measurements.

It is necessary to characterize the DUT under LS excitation to get information about the nonlinear distortion. The VNA is therefore no longer a suitable instrument, and a replacement is needed. The magnitude and phase of the generated harmonics due to the nonlinearity of the device need to be measured to obtain a complete characterization of the device operation. Due to practical limitations, only a limited number of harmonics can be measured, and hence sufficiently many needs to be included. Such a measurement system, capable of measuring the magnitude and phase of the harmonics is presented in Lott [86]. The system is limited to CW measurements and utilizes a mixer-based approach to measure the RF spectrum. A sampling-based measurement system is introduced in den Broeck and Verspecht [87], enabling the simultaneous measurement of multiple harmonics. The sampler based instrument is extended in as van den Broeck [88] to enable modulated and pulsed measurements by measuring a limited bandwidth around each harmonic. The mixer-based approach is further developed in Blockley *et al.* [89] to allow for modulated measurements. There are advantages and disadvantages with the sampler- and mixer-based measurement systems, and the two systems are compared in

Van Moer and Gomme [90]. The conclusion is that none of the two methods can be said to be superior to the other. All nonlinear measurements in this thesis are carried out by using a sampler-based instrument, the LSNA (MT4463, Maury Corp./NMDG NV). This allows for modulated measurements with up to 20 MHz modulation bandwidth.

The nonlinear measurement systems allow for new characterization methods. The voltage and current waveforms of a transistor can be measured under load-pull conditions to realize different amplifier classes of operation by waveform engineering [91]. Furthermore, the waveforms allow the study of dispersion effects such as knee walkout in GaN-based HEMTs [93]. These characterizations take a significant amount of time since several different load impedance combinations need to be measured to obtain a complete characterization. A new active load-pull setup is presented in Thorsell and Andersson [94], allowing for very fast load-pull measurements with waveform acquisition. The possibility of controlling the load impedance in a fast, or a dynamic way, also enables new characterization methods. Amplifier topologies such as dynamic load modulation [95] can be directly synthesized at the transistor terminals, as shown in Thorsell, Andersson, and Fager [96].

2.4.1 Active load-pull

The commonly used *S*-parameters are not valid when driving the DUT with LSs, such as in power amplifiers, due to the nonlinear distortion. The output power and efficiency of a transistor depends upon input power, bias, and source and load impedances at all harmonics containing power due to the excitation signal. It is therefore necessary to characterize the performance of the transistor vs. all these parameters to find the optimum combination for the desired application. The input power and bias are easily controlled, but the load and source impedances need a system capable of changing the reflection coefficient at the input and output of the DUT.

The classical approach to control the load impedance is to use mechanical impedance tuners [97]. The mechanical tuners are high-precision mechanical instruments consisting of a slabline and at least one metal probe. The tuner is either controlled manually or automatically. Automatic tuners use stepper motors to control the distance between the slabline and the probe, and also the position of the probe along the slabline. The magnitude of the reflection coefficient presented to the device is controlled by changing the height of the probe, and the angle of reflection coefficient is set by the position of the probe along the slabline. The mechanical tuner is therefore very sensitive to mechanical disturbance, and very high repeatability in the stepper motors are needed for accurate measurements. Furthermore, the speed of the mechanical tuner is limited by the speed of the stepper motors.

The load–reflection coefficient (Γ_L) is defined as the ratio between the outgoing voltage wave (b_2) and the reflected (due to the load) voltage wave (a_2), according to

$$\Gamma_L = \frac{a_2}{b_2}$$

Hence any load–reflection coefficient may be synthesized by controlling the amplitude and phase of a_2. This can be achieved by injecting a_2 rather than reflecting as done with the mechanical tuner [98]. There are two commonly used methods to control the injected signal, either the open loop approach [98], or the closed loop approach [99], illustrated in Figure 2.38.

The closed-loop active load-pull setup reuses the outgoing b_2 wave; hence, it achieves perfect phase coherence between the active injection and the drive signal on the input of the DUT. The load impedance is set by controlling the amplitude and phase of the decoupled signal before it is injected back toward the DUT. There is no need to measure the injected wave to determine the load impedance if the loop gain is known by a pre-calibration. The downside is that the closed loop setup suffers from significant instability problems due to potential oscillations within the closed loop.

The open-loop active load-pull setup injects an additional amplitude and phase-controlled signal rather than reusing the generated signal from the DUT. This removes the problems with instability related to the closed loop. The injected signal is either generated by an auxiliary signal source or by splitting input signal of the DUT. The phase coherence between the drive signal and the injected signal could therefore be a problem depending on the used configuration. The control of the load impedance is more difficult compared to the closed-loop setup. The outgoing b_2 wave needs to be measured to know how to set the amplitude and phase of the injected a_2. This requires additional measurement capabilities and increases both the complexity of the setup and the measurement time.

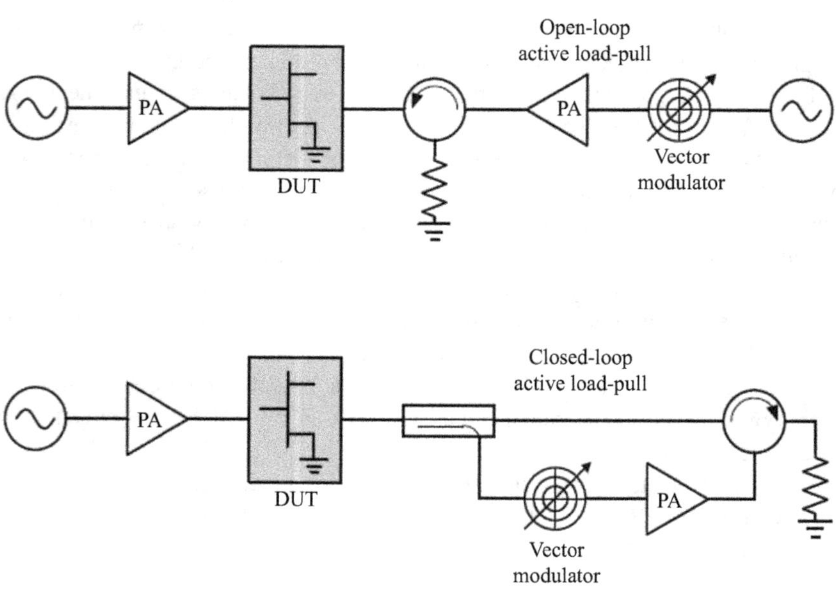

Figure 2.38 Comparison between open- and close-loop active load-pull setups

The amplitude and phase of a signal can be controlled in different ways by using, e.g., attenuators, phase shifters, variable gain amplifiers, *I*/*Q* mixers, vector modulators or vector signal generators. Depending on the setup, one or more of these are suitable. The vector modulator is used in the setup in this chapter, and a benefit compared to an *I*/*Q* mixer is the absence of local oscillator (LO) leakage.

2.4.2 Fast active load-pull

There is an increased interest in using active injection to control the load impedances rather than using mechanical tuners. This has led to several commercial suppliers of active load-pull systems. However, none of the commercially available systems offers both fast measurements and waveform acquisition capabilities. Both these requirements can be solved by using an LSNA. The LSNA is capable of doing multi-harmonic modulated measurements; hence, it fulfills the basic requirements of a fast multi-harmonic active load-pull system. The control signals to the vector modulators need to be synchronized with the LSNA samplers to avoid spectral leakage. The procedure for enabling these synchronized measurements is presented in Andersson *et al.* [100]. The complete multi-harmonic active load-pull setup capable of doing waveform acquisition is shown in Figure 2.39. The illustrated setup is configured for load-pull on two harmonics but can easily be extended. Only one signal source is used, and the signal is divided to generate both the drive signal and the injected signals on the load side. The higher order harmonics are generated using frequency doublers and triplers. The samplers in the LSNA and the trigger are controlled using a digital pattern generator (DPG). The DPG also generates the control signals to the vector modulators, enabling fast control of the load impedance, and synchronized measurements.

The fast-varying load impedance increases the measurement speed but adds an additional problem. The drain current depends on the load impedance; hence, the current varies fast and the reading obtained from the bias supply is a time average of the drain current. An oscilloscope is therefore included in the setup, with a

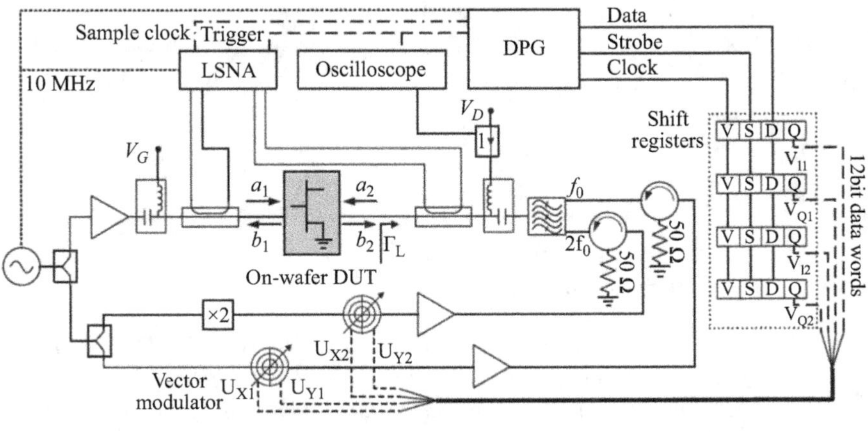

Figure 2.39 The multi-harmonic active load-pull setup capable of doing waveform acquisition

current probe connected to the drain bias line to measure the dynamic drain current. The oscilloscope is triggered by the DPG, and the measurement software combines the data from both the oscilloscope and the LSNA to reconstruct the voltage and current waveforms.

2.4.3 Nonlinear characterization using active load-pull

The combination of nonlinear measurements and multi-harmonic load-pull gives, for example, detailed information about the device operation in different amplifier topologies. The extrinsic load lines are measured, and the intrinsic waveforms can be determined by de-embedding the passive components enclosing the nonlinear core of the DUT [101].

The multi-harmonic active load-pull setup is used to measure GaN-based HEMTs to illustrate the possibilities that these load-dependent nonlinear characterizations provide. First, a standard fundamental load-pull measurement is carried out, shown in Figure 2.38. The measurement takes 21 ms, and 39 different impedance states are measured, covering almost the entire Smith chart. The drain current is measured using the current probe and oscilloscope, and its load dependence is clearly shown. This is to date the fastest setup for controlled load-pull measurements with complete waveform acquisition.

The importance of accurate control of the second harmonic is illustrated in Figure 2.41. The fundamental load impedance is kept constant at the impedance corresponding to the maximum output power. The phase of the load–reflection coefficient at the second harmonic is swept around the Smith chart with a fixed

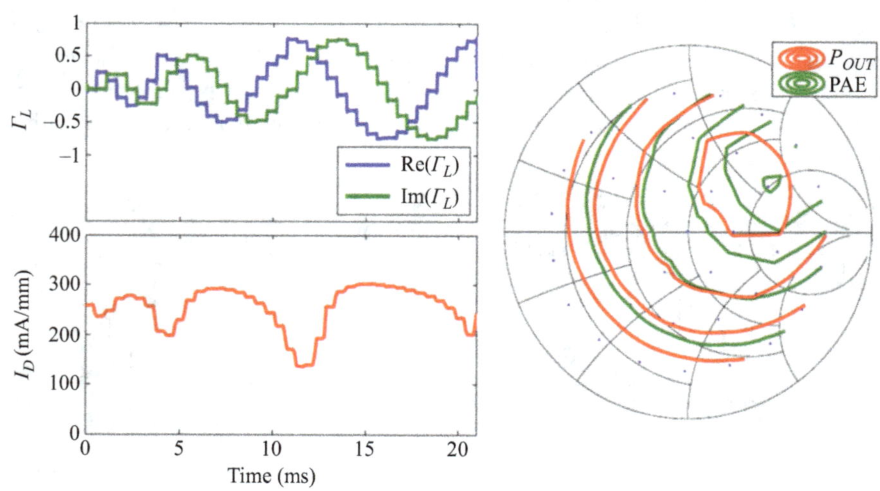

Figure 2.40 Active load-pull measurements on a GaN-based HEMT. Top: the measured modulated load reflection with 39 different states within 21 ms. Bottom: the measured drain current during the load-pull characterization. Right: output power and PAE contours

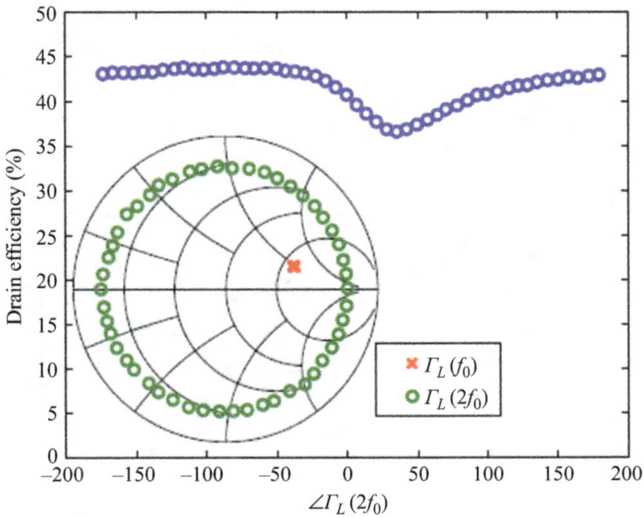

Figure 2.41 Power-added efficiency vs. phase of the second harmonic load impedance. Insert shows the measured load impedances at the fundamental and second harmonic

magnitude of 0.8. The efficiency is very sensitive to the angle of Γ_{2f_0}, and a dip occurs at around 30°. This clearly illustrates the importance of full control over the impedances at the higher order harmonics when performing load-pull measurements.

The waveform measurement capabilities are also very useful for device comparison studies. GaN-based HEMTs suffer from low-frequency dispersion due to trapping. The trapping is due to surface states, and a high-quality surface passivation is needed to reduce the influence of these charges. It is therefore necessary to evaluate the influence of different passivations on the RF performance. Low-quality surface passivations cause knee-walkout effects, increasing the on-resistance at RF, and hence reducing the output power. The active load-pull setup is very useful for knee-walkout analysis since multiple impedances are measured simultaneously and the entire knee region of the transistor is instantly characterized. A load-pull along the real axis in the Smith chart is carried out, and the output current and voltage waveforms are measured. The result is illustrated in Figure 2.42 for a 4 × 200 μm GaN HEMT.

The real axis load-pull is performed at different drain voltages, and the output waveforms are then compared to get a measure of the knee-walkout effect at RF. The knee-walkout effects on four different GaN HEMTs (three transistors from commercial GaN foundries and one transistor from Chalmers) are shown in Figure 2.43. It can be seen that the knee walkout is still a problem for some manufacturers, limiting their device performance.

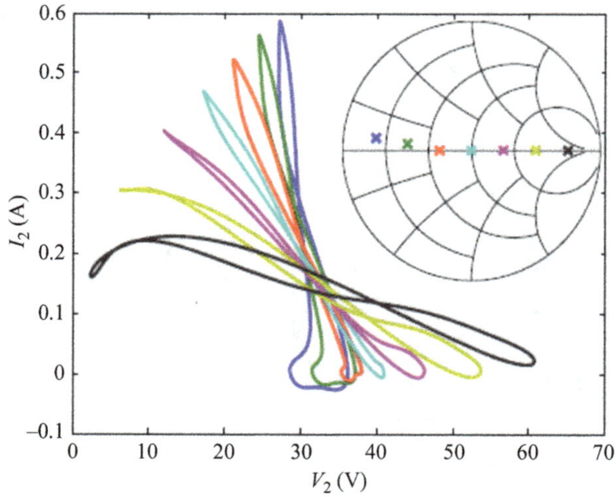

Figure 2.42 The output current vs. voltage waveforms for load impedances marked in the Smith chart for a 4 × 200 μm GaN HEMT

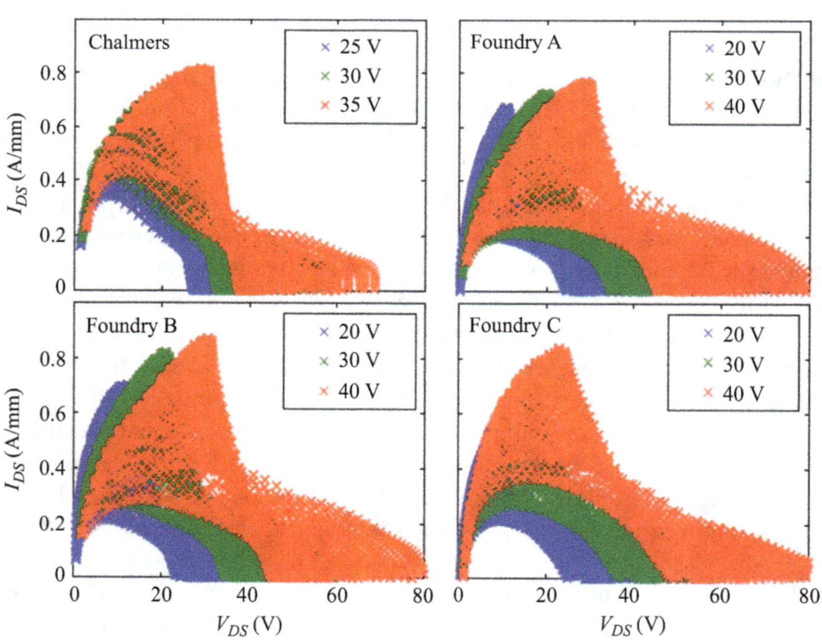

Figure 2.43 Study of the knee walkout on four different manufacturers

2.5 Small/Large signal compact models

2.5.1 Small-signal equivalent circuit models

Approximately at the same time, several very good works on the SS FET models and extraction appeared and this extraction procedure is in wide use today [21–27]. Their EC approach is based on the device physics; it is simple, easy to understand and very accurate. For good quality FET, the SS model extracted this way is accurate within 2%–10% with the measurements. Some additional optimization will make SS model accurate with 1%.

This is the first step when doing the LS model (LSM) to extract directly the SS model. The extraction is rather simple, and when data are organized in a proper way, the extraction can be done automatically, using directly the software tool:

$$[Y^i] = \begin{bmatrix} Y_{11}^i & Y_{12}^i \\ Y_{21}^i & Y_{22}^i \end{bmatrix} = \begin{bmatrix} \dfrac{jC_{gs}\omega}{1+jR_iC_{gs}\omega} + \dfrac{jC_{gd}\omega}{1+jR_{gd}C_{gd}\omega} & -\dfrac{jC_{gd}\omega}{1+jR_{gd}C_{gd}\omega} \\ \dfrac{G_m e^{-j\omega\tau}}{1+jR_iC_{gs}\omega} - \dfrac{jC_{gd}\omega}{1+jR_{gd}C_{gd}\omega} & gd + jC_{ds}\omega + \dfrac{jC_{gd}\omega}{1+jR_{gd}C_{gd}\omega} \end{bmatrix} \tag{2.1}$$

$$C_{gd} = -\frac{\Im(Y_{12}^i)}{\omega} \times \left[1 + \left(\frac{\Re(Y_{12}^i)}{\Im(Y_{12}^i)}\right)^2\right] \tag{2.2}$$

$$C_{ds} = \frac{1}{\omega} \times \Im(Y_{22}^i + Y_{12}^i) \tag{2.3}$$

$$C_{gs} = \frac{1}{\omega} \times \frac{1}{\Im(1/(Y_{11}^i + Y_{12}^i))} \tag{2.4}$$

$$R_i = \Re\left(\frac{1}{Y_{11}^i + Y_{12}^i}\right) \tag{2.5}$$

$$R_{gd} = -\frac{\Re(Y_{12}^i)}{\Im(Y_{12}^i)} \times \frac{1}{\left[1 + (\Re(Y_{12}^i)/\Im(Y_{12}^i))^2\right]} \tag{2.6}$$

$$g_d = \Re(Y_{22}^i) + \Re(Y_{12}^i) \tag{2.7}$$

$$G_m = \left|\left(\frac{Y_{21}^i - Y_{12}^i}{Y_{11}^i + Y_{12}^i}\right)\right| \times \frac{1}{\Im(1/(Y_{11}^i + Y_{12}^i))} \tag{2.8}$$

$$\tau = -\frac{1}{\omega} \times \left[\arg\left(\frac{Y_{21}^i - Y_{12}^i}{Y_{22}^i + Y_{12}^i}\right) + \frac{\pi}{2}\right] \tag{2.9}$$

The simplified SS equivalent circuit including extrinsic elements is shown in Figure 2.44. The equivalent circuit consists of the parasitic resistances, inductances, G_m, capacitances C_{gs}, C_{gd}, C_{ds}, etc. When extrinsic elements are de-embedded, the equivalent circuit further simplifies, Figure 2.45. Extrinsic inductances LG, LD, LS and resistances RG, RD, RS might require some tuning after the extraction, to make intrinsic parameters as frequency independent as possible, Figure 2.46.

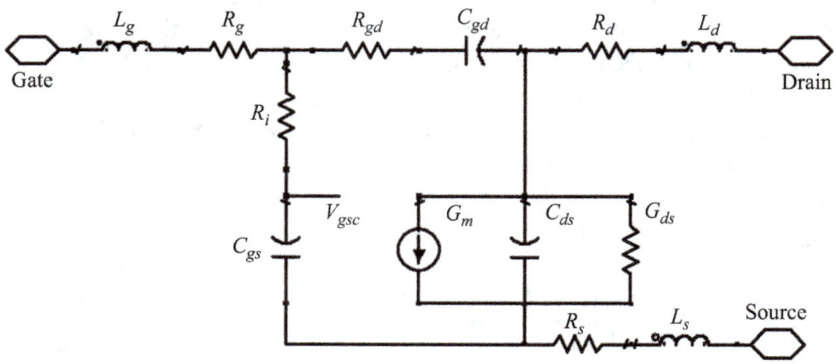

Figure 2.44 Small-signal equivalent circuit

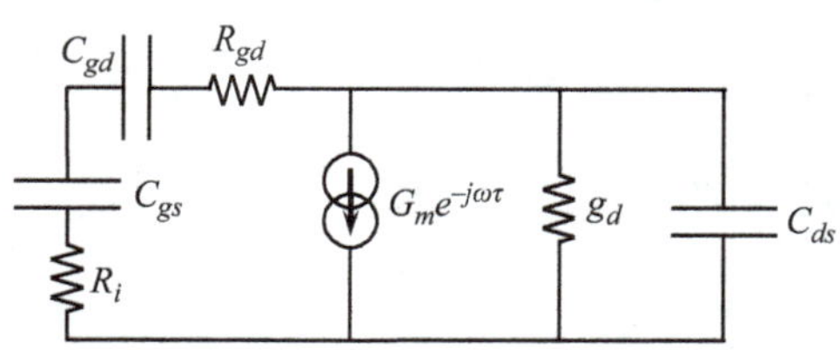

Figure 2.45 Intrinsic small signal equivalent circuit

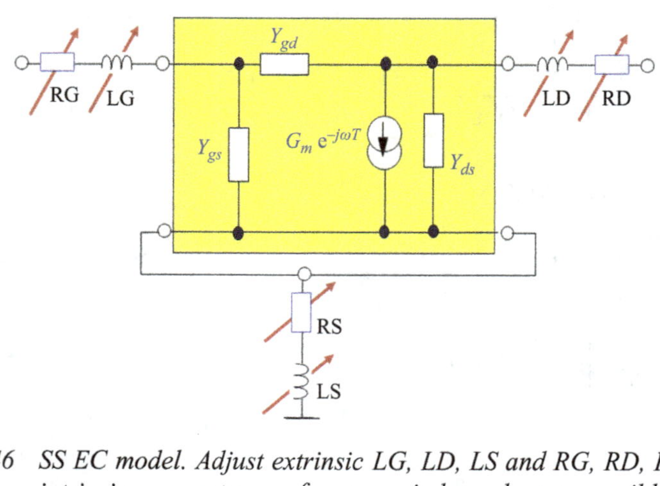

Figure 2.46 SS EC model. Adjust extrinsic LG, LD, LS and RG, RD, RS to make intrinsic parameters as frequency independent as possible

The first step in making SS model is to determine the parasitic inductances, resistances. At high frequency, the dominant player is the inductances, and they can be precisely adjusted at highest measured frequency. Inductances L_s, L_g, L_d, R_i and R_g are frequency independent; they should be tuned and fixed, monitoring that intrinsic parameters are as frequency independent as possible. The first step is to extract device parameters for $V_{ds} = 0$, so named cold FET measurement. For the GaAs devices extraction was simpler , very rarely we push the device to extremes, resistances are considered bias independent, so it was possible to extract R_s; R_d from cold FET measurement, fixed them.

But for GaN this is not the case—for GaN HEMT, R_s and R_d are bias, temperature dependent. For this reason, extraction from measured data for cold FET $V_{ds} = 0$ is just the starting value. The correct R_s–R_d bias dependence should be extracted at different temperatures, bias, etc. Several very good works are published on the topic of how to account the bias dependence of resistive parameters of the FET [25–29]. Usually, direct extraction works well for most parameters except R_i and τ. For these parameters, using *only* *S*-parameter information is easy to get significant error, up to 200%.

When parasitics are known, capacitance shape can be extracted using the approach published in [21–29] from multi-bias *S*-parameters. The capacitance extraction is usually done at low frequency, where the impedance of the capacitances is dominant and error from the L_s, L_g and L_d is reduced. The optimum frequency for capacitance extraction is in the range 3–10 GHz, depending on the device size. The impedance of the capacitor at the extraction frequency should be comparable with the impedance of the line. Figures 2.47 and 2.48 show extracted bias dependence C_{gs}, C_{gd} for a typical GaN device.

2.5.2 Large-signal compact models

When the SS model and extraction were established and implemented in the CAD tools, the next step was to increase the complexity, integrate the SS ECM into bias-dependent LSM. Some elements of the equivalent circuit are bias dependent and

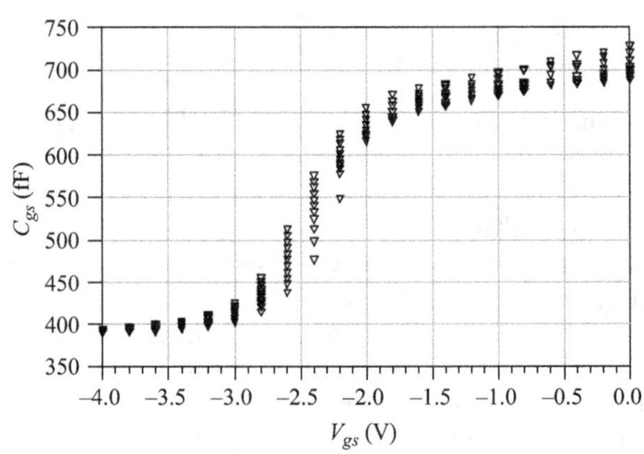

Figure 2.47 Measured C_{gs} vs. V_{gs}, V_{ds} step1v

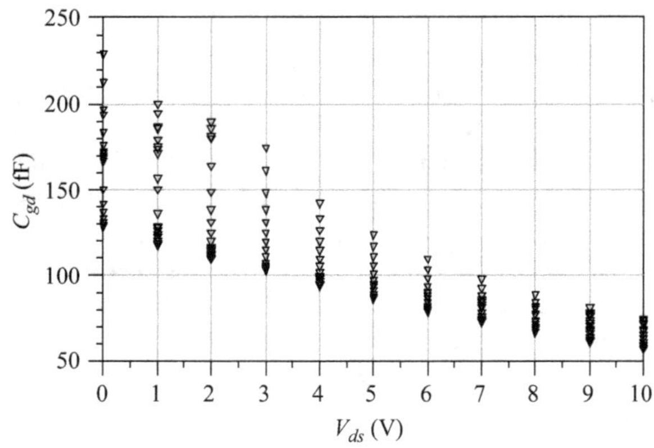

Figure 2.48 Measured C_{gd} vs. V_{ds}, V_{gs} step0.2v

extended, LS equivalent circuit approach was the simplest way to increase the performance of the CAD tools, complexity of the device models, i.e., including bias dependence of these elements. For example, capacitances, resulting from multi-bias *S*-parameter measurements like Figures 2.47 and 2.48, can be used. The LSM provides a possibility of doing accurately more complicated tasks to design nonlinear circuits like power amplifiers, mixers, oscillator multipliers, etc. At first, *IV*, capacitance bias-dependent characteristics were added to simulated model parameters. The SS *S*-parameters model parameters are generated directly from the LS ECM. It is natural to expect that *S*-parameters generated from the LSM, with small input power, should be equal to the *S*-parameters generated from the SS equivalent circuit. Gradually, LSM complexity increased, due to need to perform various simulations, like small signal *S*-parameter simulations, harmonic balance, transient, thermal analysis and so on. The number of model parameters is increasing respectively.

Today, in use are three main sets of LSMs:

1. Technology CAD, computer-aided design for semiconductor manufacturing technology (TCAD), i.e., physical LSMs (PLSMs)
2. Measurement-based LSMs (MBLSMs)
3. Equivalent circuit empirical LSMs (ECLSMs)

The first type of models, TCAD, PLSM [2–15,55,59] are very important in the device design stage. Now, TCAD tools have become very accurate, the devices are designed properly, and usually work quite close to simulations after manufacturing. Today, TCAD design is a key to creating good-quality, reliable devices.

In PMs, TCAD involving device/physical parameter calculations are too slow to be used directly in circuit simulations. Further, these models do not converge well during RF simulation, because of the complicated and exhaustive calculation for some boundary regions [11,14–16]. Therefore, simple EECM constructed using device/physical parameters can be very useful. Such mixed models may help improve the device performance to optimize for specific applications.

The measurement-based MBLSM [30–33] are very accurate in the measurement range (use around 1,000 bias points)—they are exact at the measured points. When used properly, without changing measurement condition (e.g., thermal resistance, thermal capacitance), they produce very accurate results and should be used when possible. Bias interpolation is possible, but they can be used *only in the measured frequency range*. Extending simulation frequency out of the measurement frequency usually is not working properly. With this type of models, it is difficult to handle process tolerances, scaling the device size. The measured data sets are very large, even simple *S*-parameter simulation can be slow; this model type will have difficulties to handle complicated dispersive devices. This is because dispersive effects such as temperature, process dependent, dispersion are not scaling well. When something is changed—for example, thermal resistance, the device will behave in a different way.

The ECLSM [17–20,34–37,48–54] use bias-dependent equivalent circuit; their accuracy is in the range of 1%–10%, can be easily extended out of the measurement bias and frequency range. Good ECLSM is normally extracted from 100 to 200 bias points. It easy to scale up ten times device size with reasonable accuracy, to change thermal environment, process tolerance implementation is straight forward. It was experimentally verified that ECLSM can be extracted at low frequency (up to 65 GHz) and used successfully to design circuits up to 220 GHz [56–58]. Additional benefit of using ECLSM is that they provide feedback for the process quality and help one to adjust the device manufacturing process. This is one of the reasons to make them popular.

ECLSM are widely used in circuit design now. They are fast, converge well and give an accurate fit for DC and RF characteristics [48–54]. However, such circuit models are not suitable for use in TCAD device design directly, because there is no explicit relationship between the model parameters and the device/ physical parameters.

Obviously depending on the application, respective type of models should be used, so all model types should be considered to make the picture complete. The best is to mix different types of models to combine their good features. For example, it is possible to mix PM with EC model. Some work is already done and more will follow [54,55,59]. The combined model is fast, can be used to design circuits to evaluate which is the best device structure for specific application. In this case, device physical parameters are used in the ECM equations for the current, capacitance, etc., and very encouraging results are already obtained, and we will see more results coming.

We can mix measurement based and ECLSM and combine their best features—accuracy, speed, possibility of extending out of the measurement frequency range, implement tolerances, etc. The mixed model can be constructed in such a way that part of the model is measurement based (difficult to model part) and the other, conventional part, is equation based.

Some work is already done and more will follow [54,83].

We will focus more on the ECLSM as they are used more frequently and the knowledge, approach for this model type is very similar for the MBLSM. For a good quality ECLSM, it is a key moment to use equations, an approach that follows from the device physics.

The LS HEMT models [11,13–15,35,36,45,48–53] are highly accurate models whose convergence is sufficiently high to be used in circuit-design simulation. However, as in the case of every empirical device model, models are used for reproducing the measured (or TCAD simulated) shape of the *I–V* curves, G_m, capacitance characteristics, etc., rather than for device design.

It is possible to modify the EECM to obtain a direct connection with the device/physical parameters. The model parameters are similar to the parameters used in device design. Furthermore, the number of model parameters for the nonlinear components (I_{ds}, C_{gs} and C_{gd}) can be reduced, because the same physical parameters are used to describe the *I–V* and capacitance characteristics. In the recent time, this approach of mixing physical–empirical models is becoming popular; it was shown that good agreement between physical parameters and measured *IV*, Cap, *S*-parameters, LS performance is obtained.

2.5.3 FET ECLSM model

Every LSM FET consists of the following parts:

1. Equivalent circuit of the device consisting of linear and nonlinear parts. Linear are usually considered parasitic resistances (except GaN), inductances, parasitic capacitances. Nonlinear are considered GS, GD diode, I_{ds} current source, capacitances C_{gs}, C_{gd}. A very good start for the LSM is to start with the SS equivalent circuit, like Figures 2.44–2.46. To treat multiple effects, normally more complicated, complete equivalent circuit is used in simulators, like Figures 2.8 and 2.65.
2. Equations for the nonlinear parts: I_{ds}, I_{gs} currents, capacitances.
3. Thermal, self-heating description.
4. Noise part: RF and low frequency noise part.
5. Breakdown description which can be part of the I_{ds}, I_{gs} equations.
6. Device scaling.

This LSM structure is common for all ECLSM.

In following parts, examples, illustrating how to construct and extract the LSM, will be given using the GaN FET model available in the CAD tools [48–54]. Many of the procedures described, including elements, equations, parameters, are general for all LSMs.

2.5.3.1 I_{ds} equations

The bases for Si, GaAs, GaN FET operation are two dependencies—the carrier velocity and carrier concentration, Figures 2.49 and 2.50. Their bias and temperature dependencies will be the main factors which will determine the transistor behavior. The I_{ds} vs. V_{gs} dependence is similar to the carrier concentration dependence vs. gate voltage, Figure 2.18, and corresponding modeling function should be selected. Generally, the solution of the Schrödinger and Poisson equation is ***erf*** type of functions, Eq4a, but ***erf*** function is usually not available in circuit simulators. Direct extraction of model using error function is not possible. That is why it can be replaced with other suitable modeling function, like tanh, I_{ds2} which is accurate enough to model FET. A pocket calculator can be used to extract model parameters from measured data I. Replacement is quite accurate, as can be seen from Eq4b, comparing the Tailor series expansion for the three cases:

1. I_{ds1}: erf(x);
2. I_{ds2}: tanh(x);
3. I_{ds3}: tanh(sinh(x)).

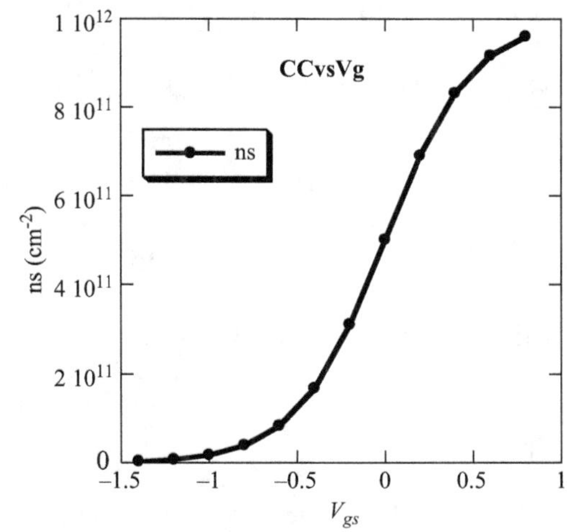

Figure 2.49 Variation of the 2DEG (ns) sheet densities

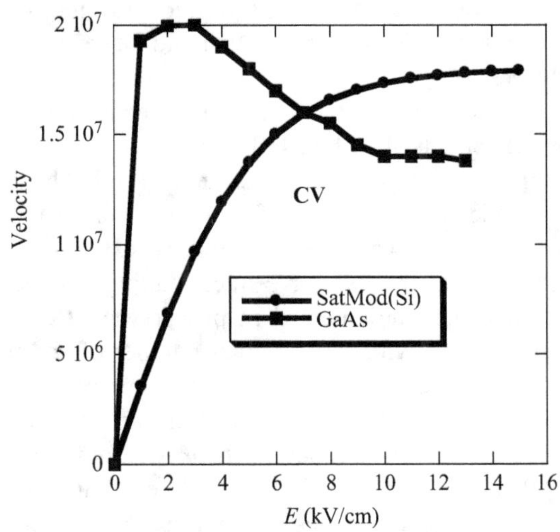

Figure 2.50 Velocity vs. electric field for GaAs and Si

The difference in the spectral content is very small:

1. The constant DC part and first harmonics are equal.
2. The difference in the third terms in the version I_{ds3} (with dual hyperbolic function) is specially arranged to provide faster pinching and Gm shape closer to rectangular for devices for which P_1 is small.

The parameter $P_1 < 1$ in many GaN, SiC devices. There is small difference is in the fifth terms for I_{ds1} and I_{ds2} (the standard Ids model.):

$$\operatorname{erf}(x) = \frac{2}{\sqrt{\pi}} \sum_{n=0}^{\infty} \frac{(-1)^n \cdot x^{2n+1}}{n!(2n+1)} = \frac{2}{\sqrt{\pi}} \left(x - \frac{x^3}{3} + \frac{x^5}{10} - \frac{x^7}{42} + \frac{x^9}{216} - \cdots \right) \tag{2.10}$$

1. $$I_{ds1}ef = 1 + \mathrm{erf}(\Psi_1)/(2/\mathrm{Sqrt}[\pi]); \Psi_1 = P_1(V_{gs} - V_{pks})$$
$$I_{ds1}ef = I_{pk} + I_{pk} \times P_1 \times V_{gs} - \frac{1}{3}\left(I_{pk} \times P_1^3\right)V_{gs}^3 + \frac{1}{10}\left(I_{pk} \times P_1^5\right)V_{gs}^5 \quad (2.11)$$

2. $$I_{ds2} = I_{pk}(1 + \tanh(\Psi_2)); \Psi_2 = P_1(V_{gs} - V_{pks}); \mathrm{GaAs}$$
$$I_{ds2} = I_{pk} + I_{pk} \times p_1 \times V_{gs} - \frac{1}{3}\left(I_{pk} \times P_1^3\right)V_{gs}^3 + \frac{2}{15}\left(I_{pk} \times p_1^5\right)V_{gs}^5 \quad (2.12)$$

3. $$I_{ds3} = I_{pk}(1 + \tanh(\Psi_3)); \Psi_3 = P_1(\sinh(V_{gs} - V_{pks})); \mathrm{GaN, SiC}$$
$$I_{ds3} = I_{pk} + I_{pk} \times P_1 \times V_{gs} - \frac{1}{6}\left(I_{pk} \times P_1^3\right)V_{gs}^3 + \frac{1}{40}\left(I_{pk} \times P_1^5\right)V_{gs}^5 \quad (2.13)$$

In GaAs, InP, GaN FET and other devices, at some electric field (V_{ds}, V_{gs}), we can observe maximum of the carrier velocity and respective transconductance, Figure 2.50. Depending on the EPI structure, we can have a variety of cases with quite different shapes of I_{ds}, G_m, G_{ds} vs. V_{gs}, V_{ds} comparing with Si, GaAs. The different shapes of G_m for GaN HEMT will produce different harmonic contents. This means that in GaN FET models, we should have respective parameters describing these differences.

There are some general requirements for the selection of the modeling functions in the empirical models. In FET, HBT, device characteristics can be considered depending on two voltages: $I = f_1(V_{gs})$ and $f_2(V_{ds})$. The best solution from extraction and user understanding point of view is to make both parts f_1 and f_2 completely independent—this will greatly simplify extraction. However, when follows from device physics that we have intercoupling between the $f_1(V_{gs})$ and $f_2(V_{ds})$ parts, this should be implemented in proper way. Because this coupling follows from device physics, with a very small number of additional parameters, the model will accurately describe the behavior. When proposed modeling function is correct, and the device is ideal, we should obtain linear function for the reverse extracted argument Ψ of respective functions f_1 *or* f_2 from the measured data. The derivative in Ψ will be equal to the first derivative, as in Figure 2.51. If from the reverse extraction we get two values of the argument, as this is shown for the example function P_{si2} (i.e., we have a $\partial\psi^2/\partial Vgs^2 = 0$), this is an indication that our choice for modeling function is not very good. The selected function P_{si2} will work in the simulations but will create problems in the extraction. For example, if the function we guess is $\Psi = Ax^2$, this will work well in the simulation. But obviously, there is a problem in the reverse extraction, because the same value of y can be produced by two values of the argument $\psi = \pm\sqrt{y/A}$. This is valid also for the sub-functions, responsible for the intercoupling between f_1 and f_2 (Figure 2.52).

Usually, the doping profile in the EPI layer is quite complicated. This will make the carrier concentration dependence on the gate voltage (and respective G_m vs. V_{gs}) quite complicated. For this reason, the I_{ds} model structure should be able to reflect a variety of cases. During the years, many different I_{ds} models were created and some found their place in CAD tools [7–14,54–60]. First, pioneering LSMs

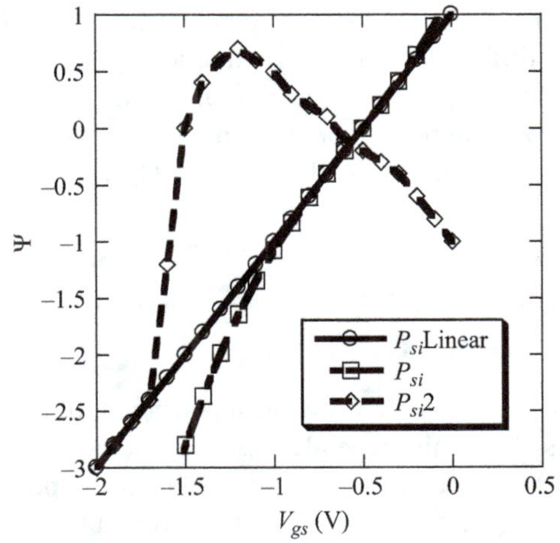

Figure 2.51 Extracted Ψ function for AlGaAs. GaAs MODFET vs. V_{gs}

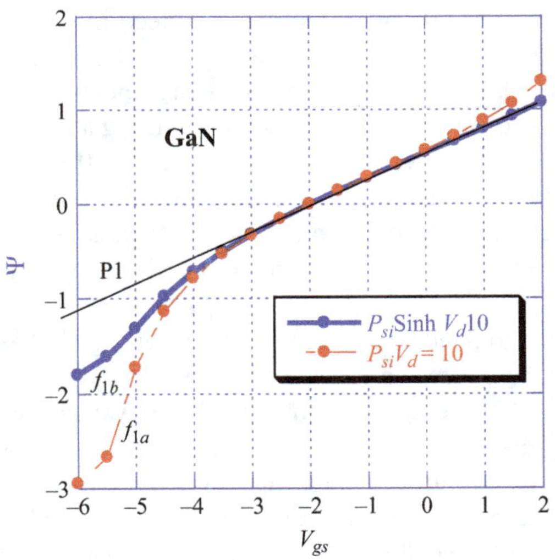

Figure 2.52 Extracted argument function GaN

were created when computers' simulators were very slow, and they needed to sacrifice the accuracy to get something calculated with reasonable speed. For this reason, many of the first models used polynomials to fit certain regions of the characteristics. To make a model more global, the characteristics were often divided into regions. For each region, the most efficient polynomial modeling approach was used. Later, when simulators became faster, more complicated functions, equations started to be used by using modeling functions that follow from the device physics. The simulations became computer resource intensive, but models became more global, more accurate from a physical point of view. A model that is in general use, implemented in all CAD tools is the model [48–53]. This model will be used to illustrate different modeling, extraction problems, which need to be solved. Many of the model parameters commonly exist in all compact models, so the shown extraction procedures can be used, modified, extended for other models, and similar results can be obtained with those models as well.

$$P_1 > 1; f_{1a}(V_{gs}) = 1 + \tanh[P_1 \cdot V_{gs}]; \Psi_{1a}(V_{gs}) = \operatorname{arctanh}[(I_{ds}/I_{pk0}) - 1]; \text{GaAs,GaN}$$

$$P_1 < 1; f_{1b}(V_{gs}) = 1 + \tanh[\sinh(P_1 V_{gs})]; \Psi_{1b}(V_{gs}) = \operatorname{arcsinh}[\operatorname{arctanh}[(I_{ds}/I_{pk0}) - 1]]; \text{GaN,SiC}$$

The tanh is used now in many FET LSMs as replacement of ***erf***. It is used to model the I_{ds} vs. V_{gs}, I_{ds} vs. V_{ds} drain dependences and the capacitances C_{gs}, C_{gd}. This approach follows from device physics and describes the dependence of the carrier concentration vs. V_{gs} correctly [14,48,59]. The current and capacitances should have the same modeling function, because the capacitances and I_{ds} current origin from the same gate charge.

Quite often the devices are not ideal; we need some flexibility to tune the model. It seems logical that a complex model is more likely to be accurate. This is correct, within limits, because we should always keep in mind that there are processing tolerances and there is no sense making model 1% accurate when process tolerances are 10%. The representation of the Ψ argument as a power series (APS) will give a possibility of fitting variety of devices. Fitting a polynomial function is a rather simple task, but even in this case, parameters of the Ψ should be selected properly. For example, when we have a negative second term in Ψ, we should always add positive third term and so on. This will exclude the possibility of a local maximum and dual argument reading at some set of voltages and provide required trimming.

It was found that to handle variety of cases, I_{ds} vs. V_{gs} and G_m vs. V_{gs} [54–60], V_{ds} above the knee three terms in the power series: P_1, P_2 and P_3 are enough to get a typical accuracy of 5% measurement model. The power series of the argument Ψ is centered at V_{pk}—the gate voltage for maximum transconductance. The arrangement with polynomial argument helps one to track variety of I_{ds} vs. V_{gs} gate dependences. One of the main model parameters $P_1 = G_m/I_{pk}$ is like the quality factor of the device. It connects (define) the maximum G_m at peak, directly define bell shape of the transconductance. Many FET devices show bell type of G_m. The higher order parameters like P_2 will introduce asymmetry of the I_{ds} vs. V_{gs}, Figure 2.53, influence the second harmonic. The parameter P_3 will trim drain current at gate voltages close to pinch off and influence the third harmonic. As it follows from experimental data, some of parameters like V_{pk}, P_1 are, in addition, bias, temperature dependent, and in order to have global model, model complexity needs to be increased [54–61].

The model parameters α_r and α_s change the I_{ds} slope at small drain voltages, Figure 2.54, parameter DV_{pk} describes the change of V_{pk} due to the drain-voltage influence. A good fit in the area of small or negative drain voltages can be important

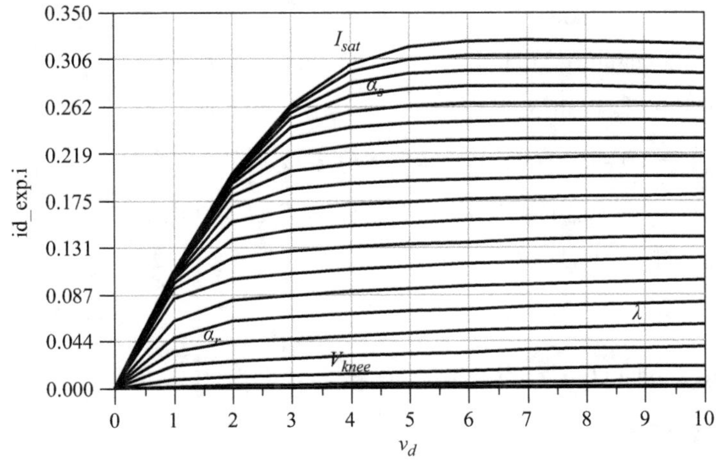

Figure 2.53 I_{ds}, G_m vs. V_{gs}

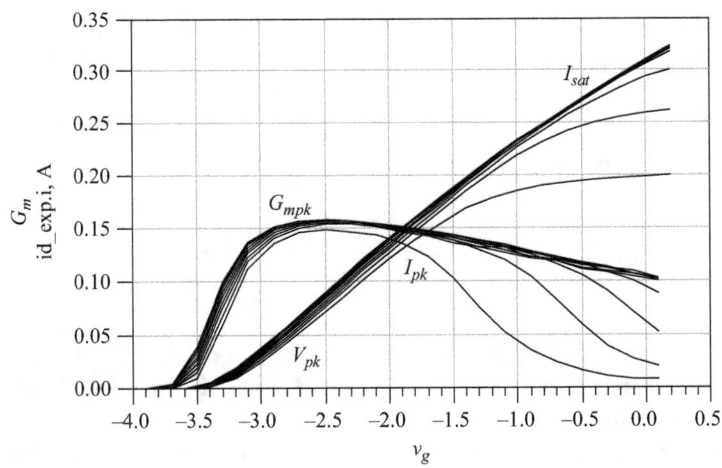

Figure 2.54 I_{ds} vs. V_{ds}

for circuits working at low V_{ds} like resistive mixers, switches, etc. The model parameters are rather independent in adjusting I_{ds}. For example, α_r will influence the drain current at small V_{ds}, small currents, parameter α_s will influence the drain current at small V_{ds}, and high currents, close to the knee, Figure 2.53. Above knee, the slope of I_{ds} vs. V_{ds} is adjusted with parameter λ. This parameter should be extracted above the knee in the linear part, at small currents, to avoid the influence from the self-heating. Breakdown modeling, Figure 2.32, if required, can be treated with parameters V_{tr}, L_{sb0} and V_{sb2} [48,53]:

I_{ds} drain current equations [54–60]

Original model I_{ds} model $(I_{ds\ mod} = 0)$:

$$\begin{aligned}
I_{ds} &= I_{pk0}\left(1 + \tanh\left(x_{1p}\right)\right)\tanh\left(\alpha_p \times V_{ds}\right)\left(1 + \lambda\ V_{ds} + L_{sb0}\exp\left(V_{dg} - V_{tr}\right)\right)\\
P_{1m} &= P_1\left(1 + B_1/\cosh(B_2\ V_{ds})\right)\\
V_{pkm} &= V_{pks} - DV_{pks} + DV_{pks}\tanh(\alpha \times V_{ds}) - V_{sb2}\left(V_{dg} - V_{tr}\right)^2\\
x_{1p} &= P_{1m}\left(\left(V_{gs} - V_{pkm}\right) + P_2\left(V_{gs} - V_{pkm}\right)\right)^2 + P_3\left(V_{gs} - V_{pkm}\right)^3\\
\alpha_p &= \alpha_r + \alpha_s\left(1 + \tanh\left(x_{1p}\right)\right)
\end{aligned} \tag{2.14}$$

Symmetric model $(I_{dsmod} = 1)$:

$$\begin{aligned}
I_{ds} &= 0.5\left(I_{dsp} - I_{dsn}\right)\\
I_{dsp} &= I_{pk0}\left(1 + \tanh\left(x_{1p}\right)\right)\left(1 + \tanh\left(\alpha_p\ V_{ds}\right)\right)\left(1 + \lambda_P\ V_{ds} + \lambda_{P_1}\exp\left(V_{ds}/V_{kn} - 1\right)\right)\\
I_{dsn} &= I_{pk0}\left(1 + \tanh(x_{1n})\right)\left(1 - \tanh(\alpha_n\ V_{ds})\right)\left(1 - \lambda_n\ V_{ds} - \lambda_{n_1}\exp\left(V_{ds}/V_{kn} - 1\right)\right)\\
x_{1n} &= P_{1m}\left(\left(V_{gd} - V_{pkm}\right) + P_2\left(V_{gd} - V_{pkm}\right)\right)^2 + P_3\left(V_{gd} - V_{pkm}\right)^3\\
\alpha_n &= \alpha_r + \alpha_s(1 + \tanh(x_{1n}))\\
\lambda_n &= \lambda + L_{vg}(1 + \tanh(x_{1n}))\\
\lambda_P &= \lambda + L_{vg}\left(1 + \tanh\left(x_{1p}\right)\right)\\
\lambda_{n_1} &= \lambda_1 + L_{vg}(1 + \tanh(x_{1n}))\\
\lambda_{P_1} &= \lambda_1 + L_{vg}\left(1 + \tanh\left(x_{1p}\right)\right)
\end{aligned} \tag{2.15}$$

The equations are shown as they appear in the CAD tools.

Important parameters like λ, $I_{pk0} = I_{sat}/2$, V_{pks}, P_1, α can be extracted directly from the measurement and later refined.

Many of these parameters are typical for all FET types, Table 2.1. For example, transconductance parameter P_1 for MESFETs is typically P_1 = 1.2–1.5,

Table 2.1 Typical model parameters for various device types

Parameter	**MESFET**	**HEMT**	**High-gain HEMT**	**Linear HEMT**	**SiC**	**GaN**	**LDMOS**
I_{chan} (A/mm)	0.3–0.6	0.3–0.6	0.17–0.25	0.3–0.6	0.35	1.5	0.75
P_1	1.1–1.5	2–3	4.5–5.5	1.5	0.1	0.3	2
V_{pk}	−0.5	−0.2	+0.05	−1.4	−9	−3	3.5
V_{knee}	0.75	0.75	0.75	0.75	9	4	3
α_s	1.3–1.5	2–2.5	3.7	1.5	0.14	0.4	1.5
Cap (pF/mm)	1	1	1.0	1.3	0.8	0.7	0.6

$P_1 = 2$ to >4 for the HEMT, $P_1 = 0.3$–1.5 for GaN, $P_1 =$ for 2 for LDMOS, etc. High value of P_1 will produce higher gain for the same current, which is good for low-noise and high-gain applications. But if P_1 is very large, the gate voltage swing (input power) can be limited and this will influence the linearity and intermodulation characteristics. Transistors with low P_1 like MESFET, GaAs HEMT specially designed for linear applications, SiC and GaN FET, will have better intermodulation properties, but lower gain. This means that some compromise should be made if we want to have high efficiency, high power and linear amplifier. Depending on the application, we can select the best P_1 for our application. Nowadays, the physical simulators are accurate, fast enough and can help one to optimize the device structure for specific application.

Some basic data are given in Table 2.1 for different FET devices. Capacitance (Cap/mm) depends on the gate length; devices with smaller gate length will have smaller Cap/mm.

Normally, we operate the devices at positive drain voltages, and it seems obvious that there is no need to look at negative V_{ds}. When drive level is small, this is correct, but when the device is used as power amplifier, switch or mixer, the instantaneous drain voltage is swinging into the negative V_{ds} region, i.e., the drain current model should describe properly I_{ds} at negative V_{ds} even if the device is biased with positive V_{ds}. Often, in the circuit simulators, the model switching at negative V_{ds} is arranged in a simple way. When the drain voltage V_{ds} is positive, the gate voltage V_{gs} controls the drain current. When V_{ds} is negative, the control voltage is switched to V_{gd}, and I_{ds} current is calculated from the same equation with the opposite sign (I_{ds} is negative). If the device is symmetrical, this is correct. But at the switching point $V_{ds} = 0$ will be a singularity and the derivative of I_{ds} is not defined. As a consequence, it will be more difficult for the HB to converge and the results of the simulations can be wrong in the vicinity of $V_{ds} = 0$. A solution to this is a continuous, single model equation for I_{ds} valid for all control voltages from $-\infty$ to $+\infty$.

For cases like switches and resistive mixers applications, operating at low and negative V_{ds} (as in Figure 2.62), the drain current equation (2.15) is symmetrical and is composed of two sources, I_{dsp} and I_{ds} and, which are controlled, respectively, by V_{gs} and V_{gd} [52,53].

There are cases with which the device has very complicated I_{ds} vs. V_{gs}, V_{ds} dependences, and it is very difficult to obtain a good correspondence between the model and measurements. In this case, the power series can be replaced with a data set calculated from measured data [54], i.e., combining both the empirical ECMs with TBMs or using the measurement-based models. Mixed empirical-table approach allows one to combine and use the best from both types of models. The empirical model is serving as an envelope for the TBM and the problem with spline function selection, out of the measurement region extension, and convergence are solved. This is because, correct spline functions, i.e., FET model equations are used as a spline. The derivatives are continuous and correct, and the model will converge well. The linear extrapolation out of the measured data range will be adequate, because the empirical model will limit the solution. The model will be limited and valid out of the measured range, because the data set is naturally limited by using the measured data for the extraction.

Often there is a spread of parameters and it is important to give the users some flexibility to tune basic model parameters in the empirical or mixed empirical TBM. For example, there are always some tolerances in G_m, pinch-off voltage, thermal resistance, etc., and the model can be arranged in such way that the users, without making complete measurement and extraction, can change the required parameter. This can be done with a proper arrangement of the mixed empirical TBM. The mixed empirical TBM can be arranged to access the basic parameters I_{pk}, V_{pk}, P_1, capacitances, etc., combining benefits of the empirical and the TBMs. The LSM is extracted for a typical device, but later it would be possible to modify environment parameters like R_{therm}, temperature, trace the process tolerances, etc.

2.5.3.2 Gate current

Quite often, we forget that FET devices have gates and ignore that FET can exhibit significant gate current when biased at positive V_{gs} gate voltages, Figure 2.55, or when pumped with RF signal. When the transistor is used as a low noise or an SS amplifier, the bias point is usually selected close to V_{gs} voltage for maximum transconductance. In this case, the gate current for normal device is rather small (well below 1 μA) and can be ignored. If device is showing gate current larger than 1 μA, this might degrade the noise characteristics. The problem will arise when the device is used as a power amplifier, mixer, oscillator, or switch. Then, at the device input is applied significant input power, the dynamic gate voltage can reach high-positive values and drive large gate current I_{gs}. The rectified gate current can easily reach 1 mA/mm device size for high input power. If the transistor model does not take into account the I_{gs} part and the designed amplifier is in practical use, the users can be surprised by the low reliability of the amplifier. This is because, large rectified I_{gs} current will accelerate migration processes in the gate metal and reduce the reliability. That is why, it is important to extract and use the gate current model of the FET in order to verify in advance that rectified gate current is within the manufacturer specifications. Another reason why users do not like gate models in CAD tools: the gate current I_{gs} dependence vs. V_{gs} is exponential and this can create problems in HB convergence, especially when a large number of harmonics are considered. We will always have this problem with the gate current when transistor is driven with high-input power (P_{in}>10 dBm).

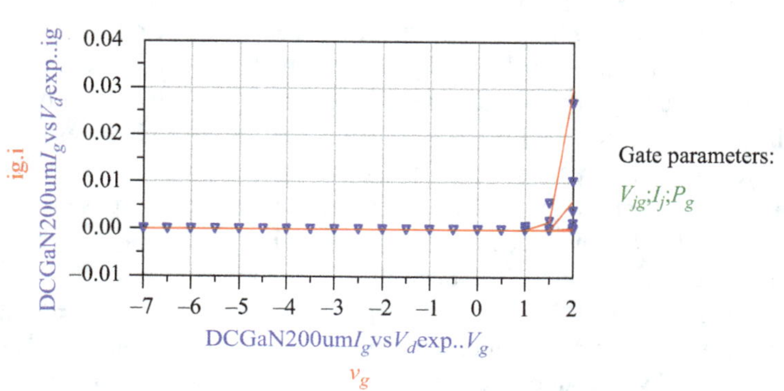

Figure 2.55 I_{gs} *vs.* V_{gs} V_{ds} *as a parameter*

Practically, the high-input drive will produce rectified gate current which can sacrifice the reliability when device is used. This can turn to severe problem in radar application, where leaking RF pulse power can be substantial and use of additional resistance, limiting the gate current, is nearly mandatory:

In the standard diode equation,

$$I_{gs} = I_s\left(\exp\left(\frac{V_{gs}}{V_{tr}\,N_e}\right) - 1\right) \tag{2.16}$$

I_s extracted when the diode is biased at $V_{gs} = -\infty$, i.e., at very small current and high negative $V_{gs} = -\infty$, for which we do not operate the device.

We can change the reference (extracting) point, rearranging the standard Schottky diode equation. In the new coordinate system, parameters are taken at the typical operating point. This can be the inflection point of the dependence I_{gs} vs. V_{gs} characteristics. For GaAs, this built in voltage is typically $V_{jg} = 0.8$ V, for GaN device: $V_{jg} = 0.9$ V. The exponent can be limited with a limited function to improve the convergence like in (2.18):

$$\begin{aligned} I_{gs} &= I_j(\exp(P_{be}) - \exp(P_{be0})) \\ P_{be} &= P_{be1}\left(V_{gs} - V_j\right), P_{be0} = -P_{be1}\left(V_j\right) \end{aligned} \tag{2.17}$$

$$\begin{aligned} P_{be1} &= q_e/K_b\,T_{ambK}\,N_{e1} = 1/V_{tr}\,N_{e1} \cong 38.695/N_{e1}, \\ I_{gs} &= I_j\left(\exp\left(P_{be1}\tanh\left(V_{gs} - V_j\right)\right) - \exp(P_{be1}\tanh\left(P_{be0}\right))\right) \end{aligned} \tag{2.18}$$

where q_e is the electron charge, K_b is the Boltzmann constant, N_{e1} is the ideality factor, I_j is measured I_{gs} at V_{jg} [52].

2.5.3.3 Charge equations [54–60]

Capmod = *0:* linear type implementation $C_{gs} = C_{gsp}$; $C_{gd} = C_{gdp}$; $C_{gd} = C_{gdpe}$; parameters work

Capmod = *1:* bias-dependent capacitance type implementation

Phi1 $= P_{10}+P_{11}\times V_{gsc}+P_{111}\times V_{ds}$; Phi2 $= P_{20}+P_{21}\times V_{ds}$

Phi3 $= P_{30}-P_{31}\times V_{ds}$; Phi4 $= P_{40}+P_{41}\times V_{gDC}-P_{111}\times V_{ds}$

$C_{gs} = C_{gsp}+C_{gs0}(1+\tanh(\text{Phi1}))\,(1+\tanh(\text{Phi2}))$

$C_{gd} = C_{gdp}+C_{gd0}((1-P_{111}+\tanh(\text{Phi3}))\,(1+\tanh(\text{Phi4}))+2\times P_{111})$

Capmod = *2:* bias-dependent charge type implementation

$L_{c1} = \ln(\cosh(\text{Phi1}))$; $L_{c10} = \ln(\cosh(P_{10}+P_{111}\times V_{ds}))$

$Q_{gs} = C_{gsp}\times V_{gsc}+C_{gs0}(\text{Phi1}+L_{c1}-Q_{gs0})\,(1+\tanh(\text{Phi2}))/P_{11}$

$Q_{gs0} = P_{10}+P_{111}\times V_{ds}+L_{c10}$

$L_{c4} = \ln(\cosh(\text{Phi4}))$

$L_{c40} = \ln(\cosh(P_{40}-P_{111}\times V_{ds}))$

$Q_{gd} = C_{gdp}\times V_{gDC}+C_{gd0}(\text{Phi4}+L_{c4}-Q_{gd0})\,(1-P_{111}+\tanh(\text{Phi3}))/P_{41}$

$Q_{gd0} = P_{40}-P_{111}\times V_{ds}+L_{c40}$

With $P_{111} = 0$ (high voltage C_{gs} vs. V_{ds} linear increase and C_{gd} crossing at the knee switched off the Phi1, Phi4 can be rewritten)

Phi1 $= P_{11}(V_{gsc}+P_{10}/P_{11})$; Phi4 $= P_{41}(V_{gDC}+P_{40}/P_{41})$

Capacitance parameters for symmetrical device without field plates and for charge conservation should be

$$P_{11} = P_{41}, P_{21} = P_{31}; P_{10} = P_{40}; P_{20} = P_{30}$$

That is, we can reduce the number of parameters in half. Even if model by structure is charge conservative, using wrong coefficients will make the model non-charge conservative.

Often, the inflection point V_{pks} for the current (max transconductance) coincide *with inflection point voltage for the* capacitance C_{gs}, C_{gd}, i.e., $P_{10} = -V_{pks}/P_{11}$, the derivative parameter for capacitances P_{11} is usually similar to derivative parameter P_1 or within 10%. This can give very good starting values for parameters describing the gate dependence of capacitances: using current parameters P_1, V_{pks},

$$P_{11} = P_1; P_{10} = -V_{pks}/P_1;$$

The drain voltage sensitivity for the capacitance is similar to drain-voltage sensitivity for the current, i.e., parameter a_s.

As result, we end up with the following starting capacitance parameters:

$$P_{11} = P_{41} = P_1; P_{10} = P_{40} = -V_{pks}/P_1; P_{20} = P_{30}; P_{21} = P_{31} = \alpha_s$$

The equations for parameters are entered directly on the circuit page:

$$P_{10} = -V_{pks}/P_1; P_{11} = P_1; P_{21} = \alpha_s$$
$$P_{20} = -0.0974072 \text{ opt } \{-0.1 \text{ to } 0.2\}$$

Temperature equations:

$$I_{pk0} = I_{pk0}(1+TcI_{pk0}(\text{Temp}-T_{nom}))$$
$$P_1 = P_1(1+TcP_1(\text{Temp}-T_{nom}))$$
$$L_{sb0} = L_{sb0}(1+TcL_{sb0}(\text{Temp}-T_{nom}))$$
$$C_{gs0} = C_{gs0}(1+TcC_{gs0}(\text{Temp}-T_{nom}))$$
$$C_{gd0} = C_{gd0}(1+TcC_{gd0}(\text{Temp}-T_{nom}))$$

Critically, most important parameters of the LSM, connect, use directly measured parameters or TCAD results from physical simulations like resistances, *IV* and capacitances bias dependences. This simplifies modeling and extraction. The model equations are with continuous derivatives, without poles from – to + infinity, without switching or conditioning. The original model $I_{dsmod} = 0$ was optimized to work in the saturation region for $V_{ds} > V_{knee}$ and V_{gs} for the peak of the transconductance. For saturated V_{ds} and $V_{gs} = V_{pk0}$ the function, and the drain current is $I_{ds} = I_{pk}$ by definition. The parameter $P_1 = G_m/I_{pk}$ will automatically define exact FET transconductance G_m at this point.

Parameters V_{pk0}, I_{pk}, $P_1 = G_m/I_{pk}$ are taken directly from the measurements, and as result, the extraction is very simple, i.e., five parameters $>I_{pk}$, V_{pk0}, P_1 at saturated V_{ds}, λ and α. The model and derivatives are strictly defined at V_{pk0} and in the vicinity of V_{pk0} where the maximum of the transconductance occurs. The transconductance is automatically generated with a bell shape; definition provides a typical global accuracy better than 10% with five parameters. Later the model was extended to a model of more effects, and the number of parameters was increased.

The parameter α_s, α_r together with R_d (and all DC line resistances in the measurement setup) will define the slope of I_{ds} vs. V_{ds} at small drain voltages $V_{ds}<V_{knee}$. The parameter λ will define the slope of I_{ds} vs. V_{ds} at high $V_{ds}>V_{knee}$ and is extracted at small currents to avoid the influence of the self-heating. These two parameters are common for many FET models.

2.6 The large-signal model extraction

Even if the model is probably one of the most compact models, the number of parameters is significant: for the GaAs model, 90 for the GaN FET Model: effects that need to be handled are huge. For this reason, it is good to follow certain order in the measurements and extraction. In the extraction procedure, the order shown is general and can be used for every FET model.

The work is more efficient if the following sequence is followed:

1. Extraction of ***On*** resistance.
2. I_{gs} parameter extraction and fit.
3. The I_{ds} model extraction and fit.
4. Thermal resistance R_{therm} fit.
5. *S*-parameter, capacitance extraction and fit.
6. *S*-parameter fit including dispersion part.
7. Nonlinear RF fit: power spectrum; waveform fit.
8. Model evaluation: load-pull, power sweeps.

*2.6.1 Extraction of on-resistance (*R$_{\text{on}}$*)*

Nonlinear models for currents and capacitances are controlled by intrinsic voltages. We need R_s, R_d to account for the voltage drop. R_s and R_d can be extracted from *S*-parameter measurements, but DC extraction provides a very good starting point, and limit the range of parameters for optimizations for R_s and R_d. It should be considered that extraction from *S*-parameters cold measurements, i.e., $V_{ds}=0$, is not good for GaN.

As shown Figure 2.24, R_s and R_d are bias, temperature dependent.

When device is biased in the linear region, low V_{ds}, Figures 2.56 and 2.57, the equivalent circuit is quite simple, Figure 2.58.

The channel-resistance R_{ch} is controlled by the gate voltage:

Resistance R_{off} is high—I_{ds} is minimum, device is pinched.

The channel resistance R_{ch} minimum, I_{ds} is maximum at positive V_{gs}, Figures 2.57 and 2.59.

The gate in the middle between the GD distance can be considered $R_s=R_d=R_{ch}$ when the channel is open. The extraction of R_{on} is done using I_{ds} vs. V_{gs} data for low V_{ds} in the linear part of the *IV*, below the knee.

V_{gs} should be large enough to open fully the channel (for GaN typically $V_{gs}=+1$ V). Very common mistake: the resistance in the bias lines is not measured. The resistance in the bias lines is in the order 0.8–3 Ω, when device is bias

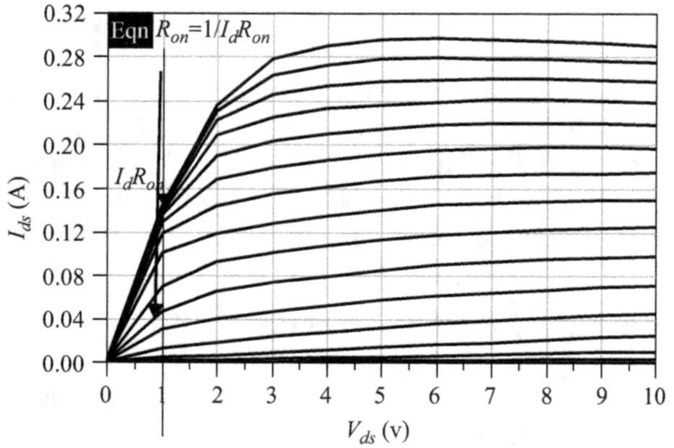

Figure 2.56 I_{ds} *vs.* V_{ds}, V_{gs} *parameter;* $I_{ds}\,R_{on} = 0.144$, $R_{on} = 1/I_{ds}\,R_{on} = 6.923\ \Omega$

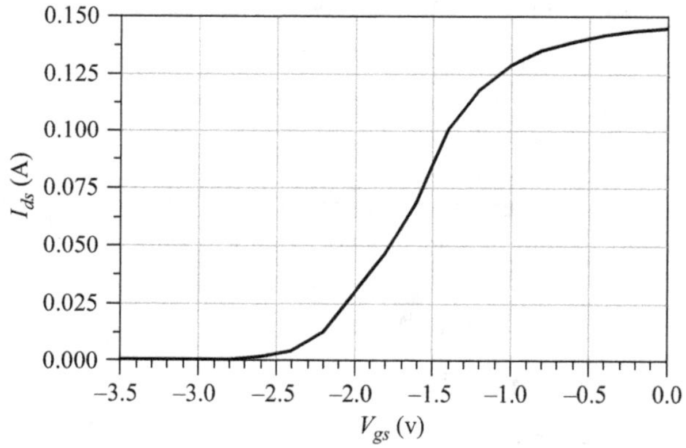

Figure 2.57 I_{ds} *vs.* V_{gs}, $V_{ds} = 1$, *GaN HEMT*

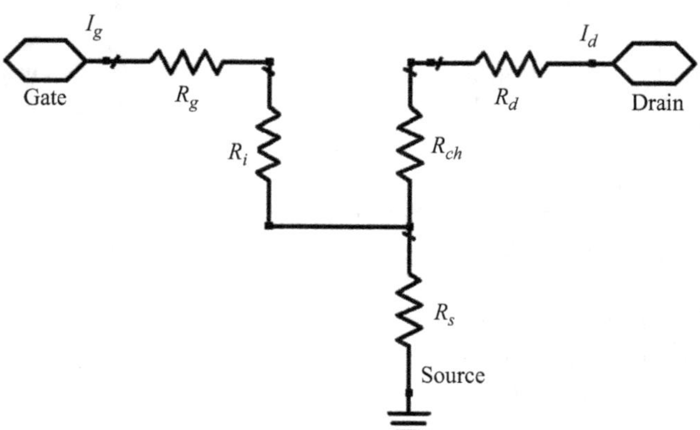

Figure 2.58 *DC measurements of* R_{ds}, R_s

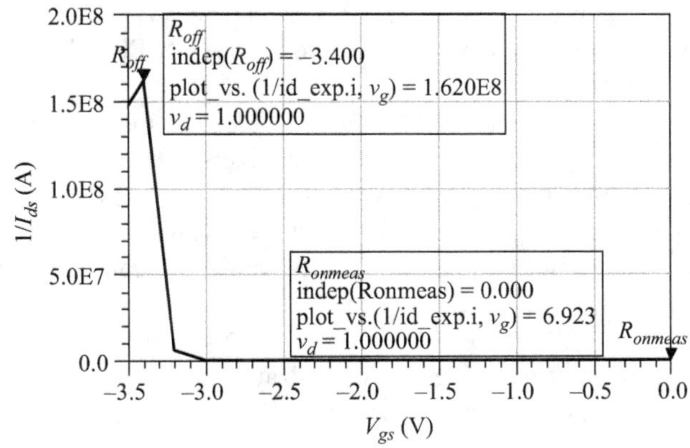

Figure 2.59 Channel resistance of GaN HEMT vs. V_{gs}

via VNA, so for high current devices, this will contribute significant errors in the voltages, S-parameters, extracted R_{on} and other parameters. There is no way to separate from IV measurements the influence of resistance of the bias line R_{biasL} and R_{on} resistance, if R_{biasL} is not measured or compensated. It is the simplest way to measure R_{biasL} between the bias supply and probes and use later in the simulations. A better way is to use bias sensing (probes with Kelvin connectors) and compensate for the voltage drop in the bias lines.

For GaN HEMT, drain voltage V_{ds} for R_{on} extraction should be in the range 1 to <2 V. For GaAs, V_{ds} can be selected $V_{ds} = 0.1$ to <0.4 V.

The total R_{ds} (R_{on}) $= V_{ds}/I_{ds}$ is calculated from measured I_{ds}.

R_{on} measurement is very safe; it is done at very low V_{ds} when the device does not dissipate power. This is the first measurement to be done on every FET device. The dependence of R_{on} vs. V_{gs}:

$$R_{dsOn} = R_s + R_d + R_{ch}.$$

gives important information for the quality of the device.

For $V_{gs} = +1$, the R_{dsOn}—resistance, is equal to

$$R_{dsOn} = R_s + R_d + R_{ch\min}.$$

When the device is pinched off, the $R_{dsOn} = R_{off}$, i.e., R_{ch} is the maximum.

The R_{off} gives preliminary information on how the device will pinch at high V_{ds}. Usually, a device with poor R_{off} will not behave properly at high V_{ds} showing substantial leakage but might be usable at low V_{ds}, low RF power. A good-quality device will show R_{off}>1 MΩ. The device that is not pinching properly will have lower breakdown and not very good characteristics as switch, power amplifier, resistive mixer, etc. The minimum R_{on} resistance gives information for the quality of the Ohmic contacts, recess etching, channel doping, the quality of the gate controlling I_{ds}, pinch-off characteristics, etc.

ON-Resistance is a very important parameter, which should be recorded and monitored on PCM samples on the wafer. Due to problems in the device manufacturing, R_{dsOn} resistance in some devices degrades (i.e., R_s, R_d increase) and the output power of the amplifier and PAE drop. For GaN FET the good number is R_{on}<0.4 Ω/mm devise size.

Normally, it is considered that the ratio R_{off}/R_{on} should be >100 (preferably >1,000), but for some applications, like resistive mixers, 10 can be sufficient. Though we should not expect to get very impressive, low-conversion loss from this resistive mixer or good performance from the switch, if device with low ration R_{off}/R_{on} is used.

For a Gate voltage in the middle of source–drain, we can consider

$$R_s = R_s = R_{ch} = R_{ds}/3.$$

These are very good starting values for optimization.

Roughly R_d, R_s can be estimated as

HEMT GaAs R_{on}/mm. Smaller device will have higher R_s, R_d.

Device R_s and R_d usually scale quite well with the respective distances GS, L_{sg}; GD, L_{sd}, with the total device size W_{tot}:

$$R_{sw} = 0.2;\ R_{dw} = 0.2$$

$$R_s = 0.02 + R_{sw}/W_{tot};\ R_d = 0.02 + R_{dw}/W_{tot}$$

$$W_{tot} = N_{fing} \times W_{fing}/1{,}000$$

HEMT GaN R_{on}/mm:

$$R_{dw} = 0.35;\ R_{sw} = 0.35$$

$$R_s = 0.01 + R_{sw}(0.2 + L_{sg}/3)/W_{tot};\ R_d = 0.05 + R_{dw}(0.2 + L_{gd}/3)/W_{tot}$$

R_g, R_i, as a start, can be considered equal to R_s, later refined from *S*-parameter measurements and fit.

But as R_{on}, R_s, R_d, R_g are process, foundry dependent, it is recommended to find correct scaling rules for the respective device measuring minimum three to four different device sizes.

A different method to extract R_s, R_d is also the fly-back method [2]. The method requires injecting high current in the gate to get reasonable signal quality. The method was invented for a bipolar transistor, where it is natural to inject current in the base, but injecting high current in the FET gate might not always be applicable. The method is based on injecting high current I_{gs} in the gate and measuring the voltage on the drain side. When drain current is 0, while measuring V_{dmeas}, we measure the voltage drop on R_s, Figure 2.58. This will give us the value of R_s:

$$R_{meas} = R_s + R_{ch};\ R_s = V_{dmeas}/I_{gs}$$

Flipping drain and source we can get $R_d = V_{dmeas}/I_{gs}$.

If the measurement is done at DC, considering $R_s = 1\ \Omega$, we need I_{gs} in the range of 5–10 mA to be in the range of sensitivity 5–10 mV. If the measurement is done at RF, we can get higher sensitivity, but nevertheless we need high gate current, which not all devices can tolerate.

The fly-back method should be used with extra care or should not be used for devices with oxides under the gate. Forcing large gate current in these devices might destroy the oxide insulation. In these cases, extraction of R_{on} resistance should be made at highest possible positive V_{gs}.

2.6.2 I_{gs} *parameter extraction and fit*

In all LSMs, when R_s, R_d are extracted, it is recommended to continue with I_{gs} extraction. This is because, for forward bias devices and E-type devices, I_{gs} will influence the intrinsic voltages. In addition, I_{gs} will influence $S11$-parameters. Therefore, it is important to extract I_{gs} part, have the gate current model working, before starting to do *IV*, *S*-parameters fit.

The gate current extraction is rather simple:

1. The gate current model parameter I_j is taken using directly measured current I_{gs} at V_j (forward gate biased device), Figure 2.55. Typically for GaN, $V_{jg} = 0.9$ V.
2. The slope I_{gs} vs. V_{gs} is adjusted using the parameter P_g (i.e., ideality factor). Typically, P_g is in the range of 12–15 (decades of current per volt). The P_g is connected with the Schottky diode ideality parameter NE. For GaN HEMT, a good start is $P_g = 15$. The best way to evaluate and adjust the slope P_g is to use ***abs***(I_{gs}) plotted in the log scale.

It is important to consider that at high I_{gs} currents, R_s, R_g, R_i will influence the slope and they will not influence the slope at low currents. So, they R_s, R_g, R_i should be extracted first. When I_{gs} current is small <0.1 mA, the influence from R_s, R_g is minimal.

2.6.3 *Drain* I_{ds} *current extraction and fit*

Extracting the I_{ds} current part of the model is a very important part of making good-quality FET LSM. Many of the steps are general for all I_{ds} FET current models. The drain current is to be measured in a wide range of biases sweeping both V_{gs} and V_{ds} as Figures 2.18 and 2.19. Typically, we will need grid of minimum 10 V_{gs} and 5–10 V_{ds}, depending on the voltage and power range of the transistor. The *IV* measurements are very fast, so this will not slow down measurement, extraction. When measurements and extraction are done properly, we can expect that at low frequency, where the contribution from reactive components (capacitance and inductances) is small, the model will be very good. In high-frequency applications, when the *IV* model is accurate in the knee region, we can expect that the output power P_{load} will be predicted with good accuracy.

The I_{ds} model extraction can be more reliable, faster if the following procedure is followed:

1. Extraction of λ slope I_{ds} vs. V_{ds} at high V_{ds}, above the knee, Figure 2.60.
2. Extraction and fit $I_{pk0}(I_{sat}/2)$, $P_1 = G_m$, V_{ds} is at the knee, Figure 2.61.
3. Extraction, Fit V_{pk}. Changing V_{pk} will shift I_{ds} vs. V_{gs} curve, Figure 2.61.
4. Fitting I_{ds}–V_{ds}, linear part below the knee: parameters α_s and α_r.
5. Thermal resistance R_{therm} fit.
6. Extraction, fit of P_2 (second derivative) and P_3 (third derivative) parameters fitting the G_m vs. V_{gs} curve.

R_d will change the slope of I_{ds} vs. V_{ds} at low V_{ds}, R_s will change the transconductance.

Repeat the procedure 1–5, because some parameters are interdependent.

At this point, we should get typically <10% fit global. It is very important to get a very good fit for I_{ds} at the knee, because this will determine the accuracy of the predicted max power. It is easy to get I_{ds} current at the knee I_{knee} within 1%, which will give a typical accuracy of the output power P_{load}, better than 1 dB. After the preliminary extraction and fit, it is recommended to continue with the following:

7. Global *IV* optimization: typically, the global fit is <2%–5%.
8. Breakdown parameter extraction, fit (optional).
9. Followed by global *IV* optimization, if needed.

2.6.3.1 Extraction fitting of λ

Parameter λ is common for many LSMs, for GaAs, GaN, InP, CMOS FET and reflects the influence of the drain voltage on the channel opening at high V_{ds}, Figures 2.53 and 2.60. When electrical field, due to V_{ds}, is becoming high, it will produce additional opening of the channel, i.e., we will observe an increase of the I_{ds} vs. V_{ds} at high V_{ds}. For GaAs, CMOS device λ is quite high: λ = 0.05–0.2, for

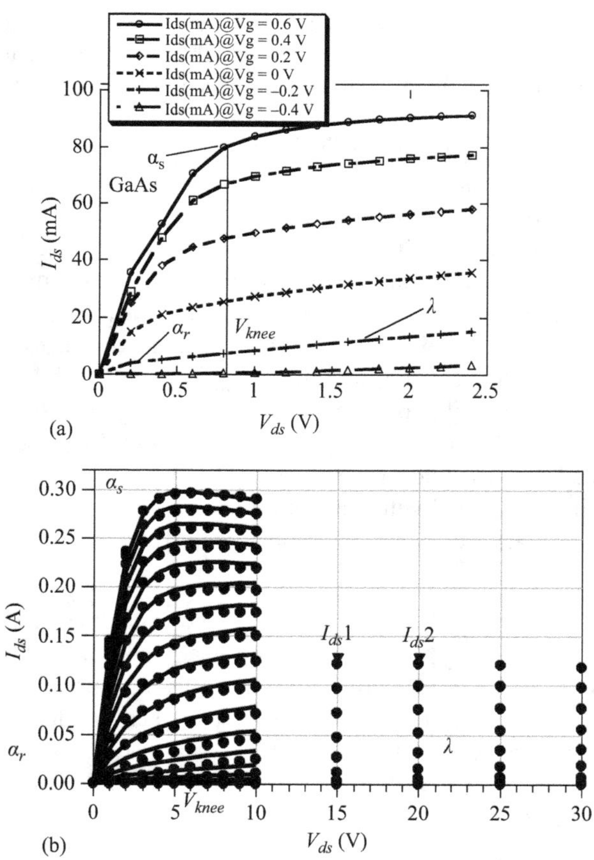

Figure 2.60 I_{ds} vs. V_{ds} fitting the I_{ds} at the knee region α_s, α_r, λ above the knee: (a) GaAs and (b) GaN

millimeter wave GaAs, InP HEMTs we can find even higher $\lambda = 0.4$–0.7. For GaN devices, this influence is not strong, especially for high voltage GaN devices. Normally, λ for GaN devices is in the order of $\lambda = 0.001$–0.04. Similar to GaAs HEMT, millimeter wave GaN HEMT can exhibit higher $\lambda = 0.05$–0.1.

This parameter should be extracted from the slope at small currents, high V_{ds}, above the knee, Figure 2.60.

For the extraction, two measurements I_{ds1}, I_{ds2} for the same V_{gs} are used:

$\lambda \sim (I_{ds2} - I_{ds1})/(I_{ds1} \times \delta\_V_{ds})$.
Assume $\delta\_V_{ds} = 1$; $\lambda = ((I_d\ V_{ds2}/I_d\ V_{ds1}) - 1)$.

For GaAs devices, extraction should be done in the range of 1–3 V and for GaN $V_{ds} = 10$–15 V, small currents. At high currents, high dissipated power self-heating will decrease the current and lower the slope. At very high power, the slope will become negative. Unfortunately, some researchers try to fit this negative slope, arranging new parts of I_{ds} equations, by manipulating the parameters of the nonlinear current source, which is not correct. The negative slope at higher dissipated power (above 0.15–0.2 W) should be handled by the self-heating part of the LSMs. To get correct information, avoiding the self-heating influence, especially for very large devices, we need good-quality pulse measurements. But good-quality pulsed *IV* measurement system is quite expensive. To get correct information, avoiding the self-heating influence, especially for very large devices, we need good-quality pulse *IV* measurements setup (PIVMS). But when evaluating the quality of a new GaN device, it is recommended to start with pulsed *IV* measurement. When the knee walkout is less than 10%, it can be considered that the device is good for some applications. For a very high-power large device, or a high current or voltage device, pulse measurements can be the only option. In such cases, it is recommended to start with a small device that can be measured, modeled in detail accurately and then continue with the large device making measurements which are possible, reasonable. The same *IV* and other information can be obtained from the waveform analysis using LSVNA.

For small power devices, it is the easiest way to define λ at V_{gs} voltages close to Gmpeak, V_{ds} above saturation.

In this case, the equation for the current simplifies to

$I_{ds2} = I_{ds1}(1+\lambda)$ and λ is found directly.

2.6.3.2 $I_{ds} = f(V_{gs})$ parameter extraction FIT

Extraction of I_{pk0}, V_{pks}, P_1, DV_{pks}

The first step in I_{ds} (V_{gs}) part of model extraction is to find the I_{pk} for which we have maximum transconductance and extract the argument function Ψ.

For GaN, usually, there are two cases: high gain $P_1>1$ and low gain devices $P_1<1$:

1. $P_1>1$: GaAs, GaN: I_{ds} vs. V_{gs} dependence can be considered $f_{1a}(V_{gs}) = 1+\tanh[P_1\ V_{gs}]$.
   Direct extraction is done $\Psi_{1a}(V_{gs}) = \text{arctanh}[(I_{ds}/I_{pk0})-1]$
2. $P_1<1$: GaN and SiC; I_{ds} vs. V_{gs} dependence is $f_{1b}(V_{gs}) = 1+\tanh[\sinh(P_1\ V_{gs})]$.
   $\Psi_{1b}(V_{gs}) = \text{arcsinh}[\text{arctanh}[(I_{ds}/I_{pk0})-1]]$

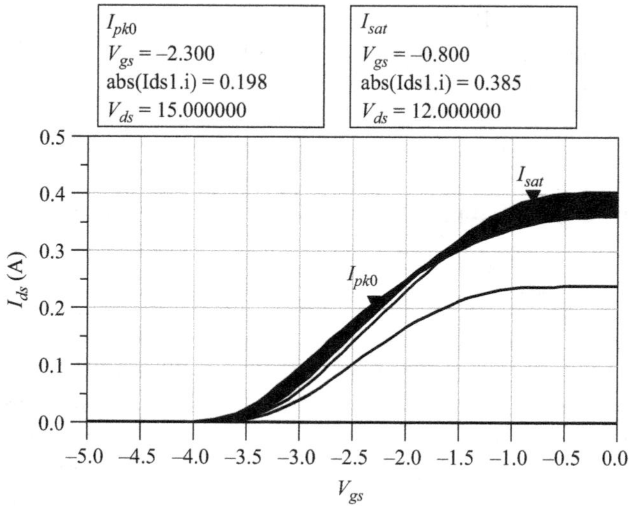

Figure 2.61 I_{ds} *vs.* V_{gs} *high power device:* $I_{pk} = I_{sat}/2 = 0.2$ *A,* $V_{pks} = -0.5$

When $P_1 = 1$, devices are quite linear and the current directly corresponds to G_m, gain. When linearity is important, designing, using such devices with $P_1 = 1$ can be considered.

In the first case, large $P_1 > 1$, by using tanh as main modeling function, the shape of the transconductance is closer to bell shape, similar to what we usually find in GaAs devices.

In the second case, small P_1, by using tanh[sinh], the shape is closer to rectangular, the device is pinching faster. Tanh Sinh helps one to make the pinching faster, a single term P_1 is enough to make good fit in a wide range of V_{gs} voltages. The fit for harmonics is also better, using tanh(sinh) for devices with low P_1.

I_{pk0} can be taken as $I_{sat}/2$—saturated drain current, Figures 2.18 and 2.54 for V_{ds} at the knee, channel fully opened, i.e., $V_{gs} = 0.9$ V. The explanation: usually the maximum of the transconductance occurs in the middle of I_{ds} vs. V_{gs} dependence, $I_{pk0} = I_{dsknee}/2$. In this example, $I_{sat} = 0.3$ A, so $I_{pk0} = 0.15$ is a very good starting point. The gate voltage for this current is $V_{gs} = -0.5$. At higher V_{ds}, the self-heating and breakdown will start to influence the *IV*. *Note:* A typical value for I_{sat} for GaAs device is I_{sat}_GaAs = 0.3–0.8 A/mm, for GaN can be higher: I_{sat}_GaN = 0.7–2 A/mm. V_{pks} is the V_{gs} gate voltage for I_{pk0}.

A common mistake is to forget that this refers to intrinsic voltage. Due to this, it is possible to select a wrong value for the I_{pk0} current for peak transconductance. In a similar way, if the transconductance is with a complex shape, it is difficult to decide where the peak of G_m is situated. In this case, the current should be set to $I_{sat}/2$ and fitted with the derivatives parameters, P_2, P_3. In addition, the complex G_m shape can be due to various reasons: temperature and bias dependence for R_s, R_d, early breakdown influence, gate leakage, etc. It is very important to find the reason for the shape of transconductance and modeled, respectively. *IV* measurements at different temperatures can help to determine the influence of the temperature.

Parameters DV_{pks}, B_1, B_2 will determine the change of P_1 when V_{ds} increases from low voltages to V_{ds}, well above the knee (P_1 usually decreases and saturates).

DV_{pks} refer to intrinsic change of V_{pk}, and it is smaller than extrinsic change DV_{pk} we observe, due to the voltage drop on R_s.

2.6.4 I_{ds} parameter extraction model fit low V_{ds}

Fitting I_{ds}–V_{ds}, linear part below the knee:

The slope I_{ds} vs. V_{ds}, for low V_{ds} is dependent on the gate voltage. The modeling is done with the alpha parameters α_r and α_s.

For device biased at high V_{gs} (all carriers are collected), the slope is max-α_s.

Parameter α_s is the slope I_{ds} vs. V_{ds} at high currents, low V_{ds}.

At high currents in the knee region, several parameters have combined influence on I_{ds}; R_d, I_{pk0} and to some degree R_s. When R_d, R_s are accurately extracted from R_{on} measurements, α_s can be fitted adjusting the slope in the area close to the knee, fitting carefully the saturated I_{sat} (current at the knee). It is very important to fit I_{sat} with high accuracy: 1%–2% can be easily achieved. To a large degree, this accuracy will determine the accuracy of the predicted output power. A good starting point for α_s is the value of P_1. The explanation: the sensitivity, derivatives of I_{ds} vs. controlling voltages V_{gs}, V_{ds} are similar below the knee.

For GaN devices, α_s is in the range of 0.2–1, for GaAs devices α_s is in the range of 1–3.

α_r is the slope at small currents and small V_{ds}. A typical value is in the range of 0.01–0.2. At small currents, small V_{ds} voltages derivative I_{ds} vs. V_{ds} is small.

For drain voltage above the knee voltage V_{knee}, Figure 2.59, and gate voltage $V_{gs} = +0.6$ V, the drain current of the FET will saturate and reach the maximum channel current. This maximum channel current depends on the material structure, doping profile, etc. For GaAs FET, the maximum channel current is 0.3–0.5 A/mm and for material structures like GaN, the maximum channel current can be as large as 1.6 A/mm.

There are cases in which the device is operated at low V_{ds} or swinging from negative to positive V_{ds}, for example, in switching applications. In this case, symmetrical I_{ds} model should be used [51–53,59,60]. Figure 2.62 shows the I_{ds} vs. V_{gs} dependence, stepping drain voltage V_{ds} from negative to positive. As can be seen, the device is not completely symmetrical, and this is in part due to the shift of

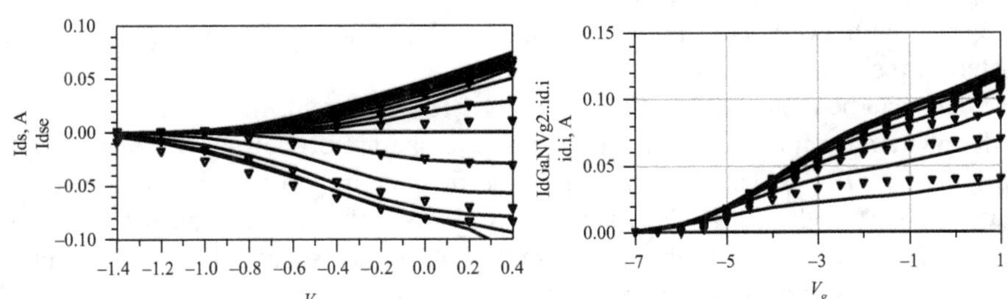

Figure 2.62 GaAs, GaN symmetrical model measured and modeled I_{ds} vs. V_{gs}

intrinsic V_{pk} when V_{ds} is negative. The G_m shape will change when the drain voltage is around $V_{ds} = 0$, the ration $P_1 = G_m/I_{pk}$ is higher at low V_{ds} for V_{ds} below the knee. The increase of P_1 at low V_{ds} is described with the parameter B_1, $P_{1m} = P_1(1+B_1/\cosh(B_2\times V_{ds}))$.

How fast P_1 decreases when V_{ds} is high is determined by parameter B_2 (typically in the range of 3–4).

If the drain voltage is further increased, the I_{ds} breakdown can occur.

If during the RF swing, V_{ds} reaches the breakdown, this will lead to reduced output power, PAE and low reliability. So, operating in the breakdown region should be avoided. The breakdown model should be set; otherwise, unrealistically high power can be simulated from the amplifier. Typically, high-power devices are biased for high-efficiency operation, i.e., at high voltages, low currents like Class AB or Class C. A properly constructed load line will keep the devices away from the breakdown area; the device will be switched from high voltage and low current to high currents and low voltages (close to the V_{knee}). If this is the case, there is no sense to spend much time making very detailed model for high voltage, high currents and accurate breakdown model. The breakdown model should be simple enough to mark that the device starts the breakdown. Only if the device will operate in the breakdown area, it is worth spending time to make very detailed and accurate breakdown model.

2.6.5 Self-heating modeling thermal resistance R_{therm} fit

When device is dissipating power, larger than 150–200 mW, we will start to see the influence of the self-heating.

This will be manifested with changes in the *IV* and capacitance part. The self-heating is modeled with thermal sub-circuit consisting the *RC* network, Figures 2.65 and 2.93, with thermal resistance, thermal capacitance, respective temperature coefficients for parameters influencing the *IV*, capacitance. Model without these key parameters is not good for modeling devices dissipating more than 0.2 W.

When junction temperature increases, the transconductance, I_{ds}, will drop, due to the reduction of the mobility and carrier concentration at high temperature. This can be clearly seen in the area where the device dissipates high power: high currents, high voltages. The slope I_{ds} vs. V_{ds} is becoming negative, Figure 2.63. If we make pulse *IV* measurements with a short pulse, we will not see this negative slope, as in Figure 2.64. For GaN FET and high-power GaAs FET resistances, R_s, R_d will also increase with the temperature. Due to the reduction of the mobility at high temperature, we will also see capacitances change. One of the first things we need to verify with every new process is to find the temperature coefficients of the material, resistances, capacitances, device, etc.

This can be done making *IV*, *S*-parameter (capacitance) measurements at three different temperatures and extract the main temperature coefficients TcP_1, TcI_{pk}, TcR_s, TcC_{gs}, TcC_{gd}. These temperature coefficients are very similar for different materials GaAs, GaN, InP. They are in the order of 0.001–0.003. This seems small, but as the junction is very tiny, it gets quite hot: we see substantial effect. A good starting point for temperature coefficients for GaN is

$TcP_1 = -0.002$; $TcI_{pk} = -0.002$; $TcC_{gs0} = +0.002$; $TcC_{gd0} = +0.002$; $TcR_s = 0.005$, i.e., according to Murphy's law, things get worse under pressure, when heat up.

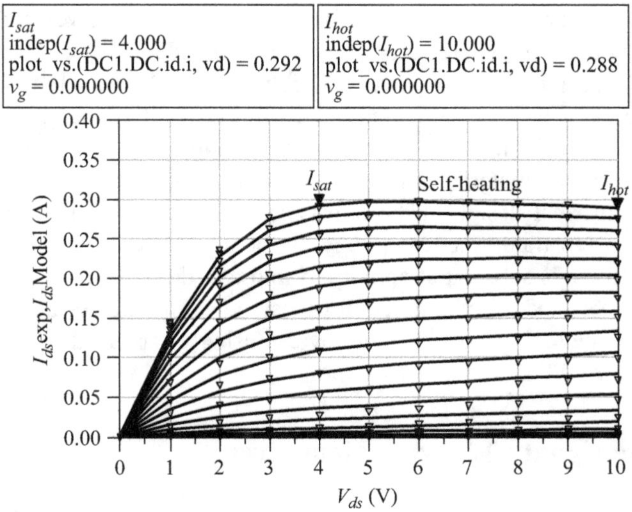

Figure 2.63 I_{ds} *GaN FET measured points model lines*

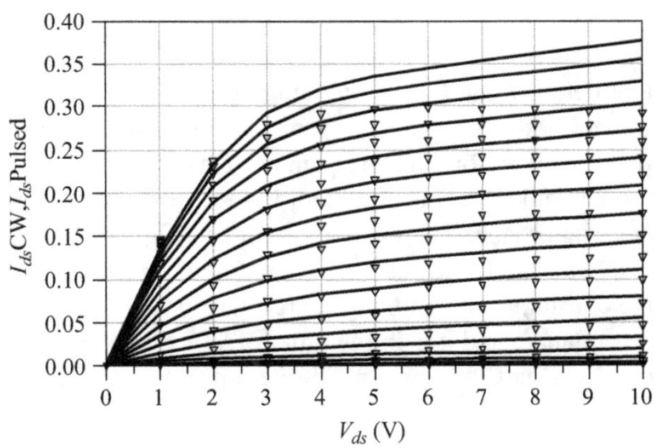

Figure 2.64 I_{ds} *GaN FET measured CW, pulsed IV lines*

The thermal resistance R_{therm} can be found directly making infrared (IR) measurements biasing the device for high power. From IR, measured temperature increase DT_j and known power P_{DC}, thermal resistance is calculated directly $R_{therm} = DT_j/P_{DC}$.

IR measurement system is not always available, so with a reasonable accuracy, R_{therm} can be found using the following procedure:

1. Extracting *IV* temperature coefficient from *IV* measurement at different temperatures (2 minimum).
2. Fitting the model accurately at I_{sat} at the knee.
3. Using R_{therm} to adjust the negative slope for high currents: I_{hot}.

Quite often due to large device size, high dissipated power and dispersive effects, the modeled *IV* characteristics are far from ideal, Figure 2.19. The self-heating will decrease the drain current at high dissipated power [64], Figure 2.63. The decrease of I_{ds} at high dissipated power will critically depend on the thermal resistance R_{therm} of the mounting package, so for high-power devices, it is important to select proper material with a high thermal conductivity for heat sink, to make a good thermal design of the transistor, i.e., using properly placed thermal shunts and thin substrate. The technology for the new GaN and SiC devices is very promising as there is substantial activity to improve these devices. We can expect that the *IV* curves and all parameters for these new, high-power devices will gradually become better, compared with devices using established technologies like CMOS and GaAs FET.

2.7 Large signal FET equivalent circuit

The LS-equivalent circuit with reactive components included is rather standard, Figure 2.65. In all LSMs, most of the elements are considered linear, and the nonlinear (bias dependent) are considered I_{gs}, I_{ds}, capacitances C_{gs}, C_{gd}, for GaN also R_s, R_d. The difference between the simple SS equivalent circuit Figures 2.44–2.46 and LS equivalent circuit Figure 2.65 are GS, GD diodes: drain current source I_{ds}, bias-dependent capacitances, thermal, dispersion part. They are described in more detail in the following paragraphs.

Large devices do not react so quickly, so to account for this, network R_{del}, C_{del} is implemented to account for high-frequency delay, high-frequency dispersion, high-frequency knee walkout. This is in addition to the dispersion modeling using back-gating ***Vbgate*** approach. The delay capacitance C_{del} shunts the input control node at high-frequency and reduces the output signal, respectively, at high frequency. The time constant is set with the charging resistor R_{del}. C_{del} is typically the gate footprint capacitance, and it is in the order of 1–2 fF and R_{del} is in the order of 10 Ω to 10 kΩ and will not influence much the input impedance.

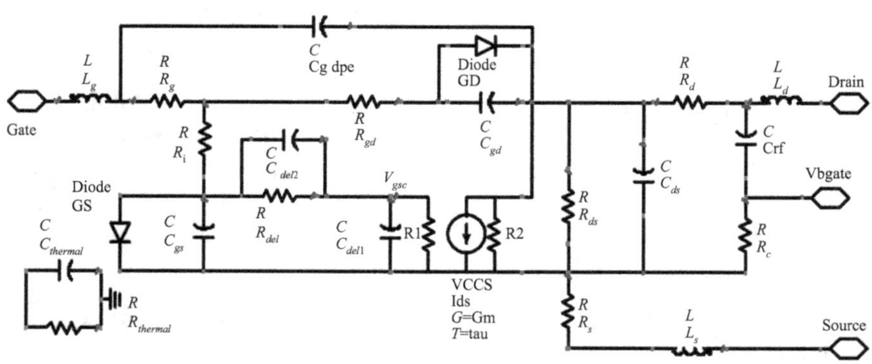

Figure 2.65 FETLS equivalent circuit

2.8 Capacitances and capacitance models' implementation in simulators

In multiple extraction and physical simulations on different FET structures, it was evaluated that the main device capacitances are bias dependent on both voltages $C_{gs} = f(V_{gs}, V_{ds})$ and $C_{gd} = f(V_{gd}, V_{ds})$, Figures 2.28–2.30. A consequence of the dependence of the capacitances on the remote voltage is that we need additional charge control element which is called transcapacitance [30–33].

This behavior is normal to expect, the problem is how to implement this in the circuit simulators. The charge implementation in CAD tools, charge conservation problem is very old; several good works are devoted on the topic and propose solutions [13,30–34].

Traditionally, total gate FET charge has been modeled by two nonlinear charges: GS Q_{gs} and GD through charge Q_{gd}, though for symmetrical cases, single charge definition Q_g should work. Possible implementation is to implement the charges using directly division by current proposed by Maas [13].

The main reason for implementation with two charges is that device is with three terminals, we can have different terminations at GS–GD junctions, device geometry GS–GD can be different, different resistive parts R_i, R_{gd}, different C_{gs}, C_{gd} dependences. This is even more important for GaN HEMTs where the distance GS–GD is normally different, with different configuration, doping, field plates, etc. This makes the dependence on the remote voltage even more complicated. There are several ways to implement the gate charges into two individual components: division by capacitances and division by charge [1].

As FET devices have both gate-to-source capacitance C_{gs} and gate-to-drain C_{gd}, it seems natural to use them directly. In this case,

$$I_g = \frac{\partial Q_g}{\partial t} = \frac{\partial Q_g}{\partial V_{gs}}\frac{dV_{gs}}{dt} + \frac{\partial Q_g}{\partial V_{gd}}\frac{dV_{gd}}{dt} \tag{2.19}$$

and

$$I_s = \frac{\partial Q_g}{\partial t} = \frac{\partial Q_g}{\partial V_{gs}}\frac{dV_{gs}}{dt} \tag{2.20}$$

$$I_d = \frac{\partial Q_g}{\partial V_{gd}}\frac{dV_{gd}}{dt} \tag{2.21}$$

where I_s and I_d are reactive parts of the source and drain current and

1. $C_{gs} = \dfrac{\partial Q_g}{\partial V_{gs}}; \quad C_{gd} = \dfrac{\partial Q_g}{\partial V_{gd}}$

2. $\dfrac{\partial V_{gs}}{\partial t}; \quad \dfrac{\partial V_{gd}}{\partial t}$

3. $I(\text{gsi}, \text{si}) = \dfrac{\partial V_{gs}}{\partial t} C_{gs}; \quad I(\text{gdi}, \text{di}) = \dfrac{\partial V_{gd}}{\partial t} C_{gd}$

That is, when using the capacitance implementation,

1. calculate capacitances,
2. calculate time derivative,
3. calculate currents for the intrinsic nodes.

When we use capacitances in the implementation, the currents I_s and I_d depend only on the time derivative of their own terminal voltage and not on the changes in any remote voltage. The resulting SS-equivalent circuit is completely consistent with the LS-equivalent circuit and requires no transcapacitances.

Second option is to divide the gate charge Q_g into two independent charges. Then,

$$Q_g = Q_{gs} + Q_{gd}$$

where both Q_{gs} and Q_{gd} are functions of V_{gs} and V_{gd}. Differentiating Q_g with respect to time gives

$$I_g = I_s + I_d$$

$$I_s = \frac{\partial Q_{gs}}{\partial t} = \frac{\partial Q_{gs}}{\partial V_{gs}}\frac{dV_{gs}}{dt} + \frac{\partial Q_{gs}}{\partial V_{gd}}\frac{dV_{gd}}{dt} \tag{2.22}$$

$$I_d = \frac{\partial Q_{gd}}{\partial t} = \frac{\partial Q_{gd}}{\partial V_{gs}}\frac{dV_{gs}}{dt} + \frac{\partial Q_{gd}}{\partial V_{gd}}\frac{dV_{gd}}{dt} \tag{2.23}$$

In this case, the reactive source and drain currents result from both capacitances and transcapacitances, and both definitions charge and capacitance are not equivalent.

A common approach to implement the charge part of every transistor model is to directly use the charge approach. In this case, the current of the capacitance is easy to calculate by taking the time derivative of the charge, i.e., multiplying by $j\omega$. This operation is very reliable, because making the derivative will always produce only one solution. This works very well with capacitance which depends only on their own terminal voltage. The problem with all FET transistors is that the gate capacitance depends on the two controlling voltages. When we multiply by $j\omega$, we are making in fact the full derivative of the charge and the result is not correct if the charge is obtained as integrating the capacitance equation by the terminal voltage. It is obvious that partial (considering the remote part constant) and full derivatives are different.

This can be shown with the case of the capacitance model given in Section 2.5.3.3 by integrating the C_{gs} capacitance by the terminal voltage V_{gs} and assuming that V_{ds} part is constant. If the ordinary charge approach is used, multiplying by $j\omega$ will obviously bring different results, i.e., if we need to compensate the difference due to the partial derivative, we need an extra term—the transcapacitance [13,30–34].

In some advance simulators, for the internal, compiled models, the partial derivatives of the charges are calculated analytically using the respective terminal voltage. This way the problem is solved in a better way—the CAD tool is making the derivative vs. respective terminal voltage, considering the remote voltage constant. This way, the capacitances calculated by both methods should be identical. This is the way charges, capacitance, SS analysis are implemented in some simulators.

In the capacitance implementation case, when the device is pumped with very small power, the SS capacitance calculated from SS analysis will coincide with the capacitance calculated in the HB. But when the device is pumped with power, the medium capacitance will change and will be different from what is calculated from the SS analysis. In the HB, a DC current will appear via the capacitance. The consequence is that the model is not charge conservative, the solution will become path dependent and the HB of the simulator will have difficulties to converge [13,30–34].

A problem that should be accounted in the capacitance, charge implementation is the boundary condition problem. The capacitance depends on two terminals, so we need to know what will be the capacitances (charge) when the capacitance terminal is shorted and there is voltage on the remote terminal.

For example, when the GS junction is shorted, $V_{gs} = 0$ and the charge $Q_{gs} = 0$. The charge Q_{gs} should be 0, independent of remote voltage V_{ds}. But the capacitance C_{gs} will continue to exist. Similar is the situation with the GD junction. This puts additional constraints, boundary conditions, for the charge definitions. For these reasons, most circuit simulators use separate Q_{gs}, Q_{gd}—we have different boundary conditions on the GS, GD part, different resistances R_i, R_{gd}. We take into account different boundary conditions using charges Q_{gs0} and Q_{gd0}. As it can be seen from Figures 2.70–2.73, when $V_{gs} = 0$, the charge $Q_{gs} = 0$, and when $V_{gd} = 0$, the charge $Q_{gd} = 0$ independent of the remote voltage V_{ds}.

Figures 2.66–2.69, show the typical shape of the simulated C_{gs} and C_{gd} capacitances. When the device is symmetrical, for $V_{ds} = 0$, capacitances C_{gs} and C_{gd} are equal. For gate voltages close to pinch off, capacitances C_{gs} and C_{gd} have their minimum values C_{gspi} and C_{gdpi}, and this should be used in the capacitance models to define the capacitance at the pinch-off. Increasing V_{gs} will increase C_{gs} and C_{gd}. Generally, when V_{ds} increases, C_{gs} will increase and saturate at voltages around

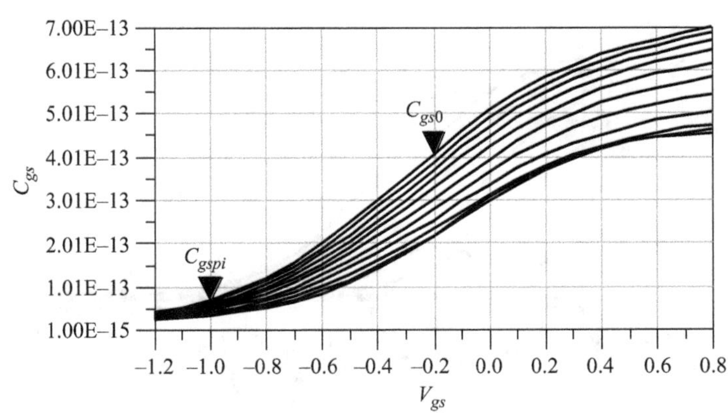

Figure 2.66 Charge Q_{gs} vs. V_{gs}, V_{ds} parameter

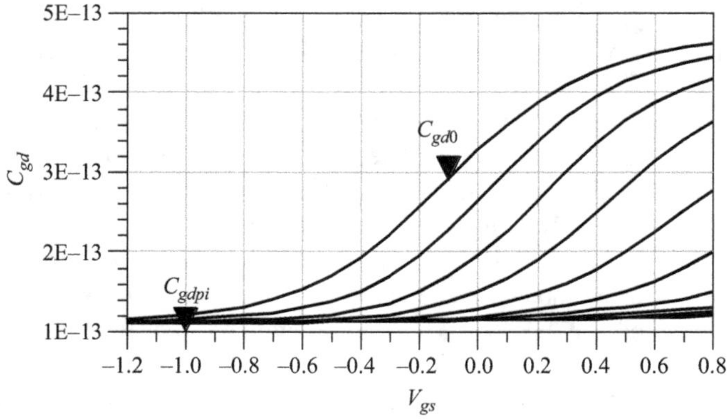

Figure 2.67 Charge Q_{gs} vs. V_{ds}, V_{gs} parameter

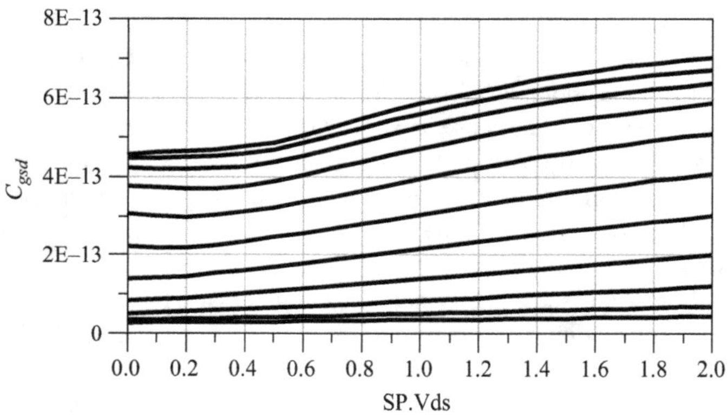

Figure 2.68 Charge Q_{gd} vs. V_{gd}, V_{ds} parameter

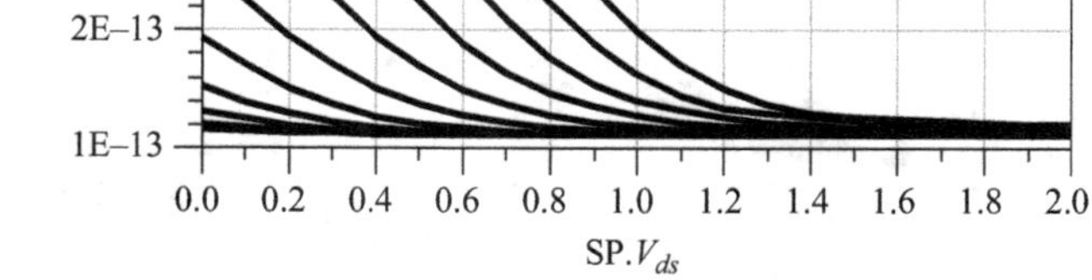

Figure 2.69 Charge Q_{gd} vs. V_{ds}, V_{gs} parameter

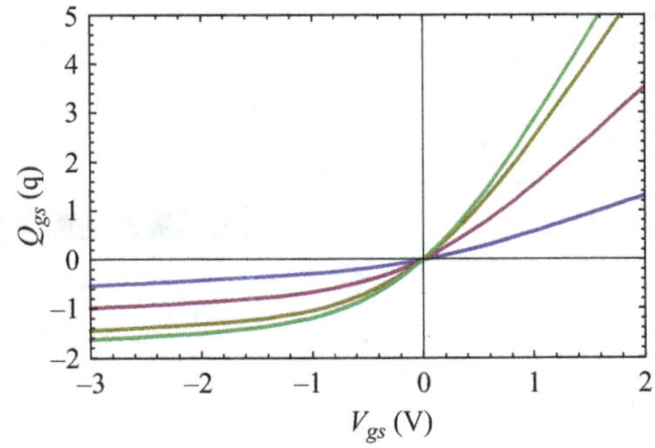

Figure 2.70 C_{gs} *vs.* V_{gs}, V_{ds} *parameter*

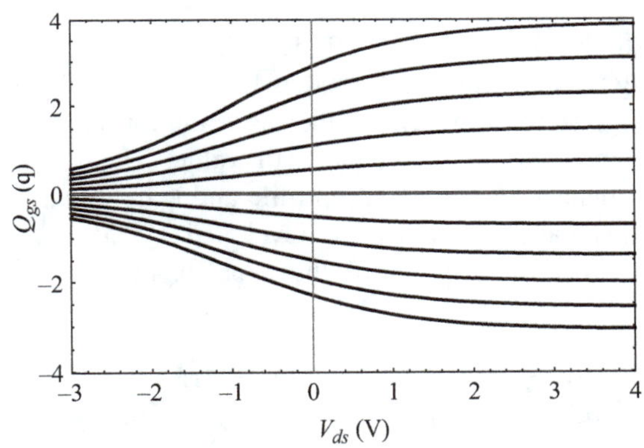

Figure 2.71 C_{gd} *vs.* V_{gs}, V_{ds} *parameter*

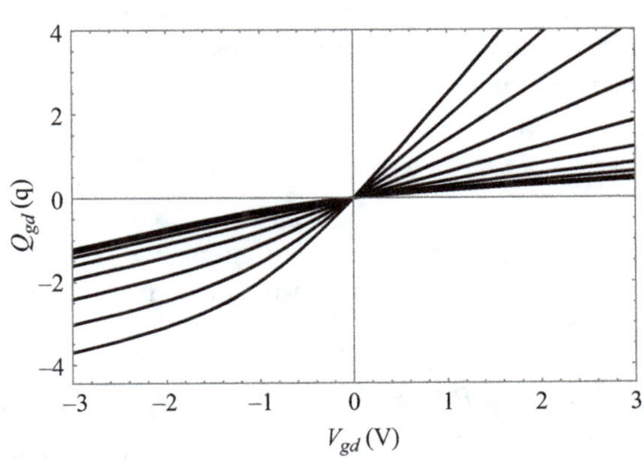

Figure 2.72 C_{gs} *(F) vs.* V_{ds}, V_{gs} *parameter*

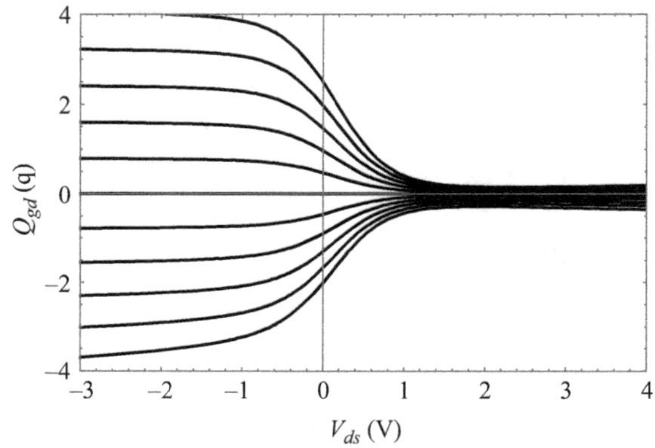

Figure 2.73 C_{gd} (F) vs. V_{ds}, V_{gs} parameter

$V_{ds} = 2$ V. The high voltage, large FET devices can show linear change of the C_{gs} vs. V_{ds} at high V_{ds}, Figure 2.68, and these problems should be addressed. The capacitance C_{gd} decreases when V_{ds} increases and for drain voltage around the V_{knee}, the capacitance C_{gd} settles to a small value. The shape of capacitance dependencies will depend on the doping profile and material and in some specific cases, a special capacitance model can be developed.

A reasonably good description of the capacitance shape for FET can be obtained by [48–53]

$$\psi_1 = P_{10} + P_{11}\, V_{gs} + P_{111}\, V_{ds}; \qquad \psi_2 = P_{20} + P_{21}\, V_{ds}$$
$$\psi_3 = P_{30} - P_{31}\, V_{ds}; \qquad \psi_4 = P_{40} + P_{41}\, V_{gd} - P_{111}\, V_{ds}$$
$$C_{gs} = C_{gsp} + C_{gs0}(1 + \tanh[\psi_1])\,(1 + \tanh[\psi_2])$$
$$C_{gd} = C_{gdp} + C_{gd0}(1 + \tanh[\psi_3])\,(1 - \tanh[\psi_4])$$
$$C_{gd} = C_{gdp} + C_{gd0}(1 - P_{111} + \tanh[\psi_3])\,(1 + \tanh[\psi_4] + 2\,P_{111})$$

Independent of the implementation (capacitance or charge) and the type of model, in order to have the capacitance model charge conservative, it is important to fulfill the following basic requirement:

$$\frac{\partial C_{gs}}{\partial V_{gd}} = \frac{\partial C_{gd}}{\partial V_{gs}}$$

In our case, using the charge, capacitance definitions, we get

$Q_{tot} = Q_{gs} + Q_{gd}$—Gate charge is conservative, path independent, because partial cross derivatives are equal

$$df_1 = D[C_{gs1}, V_{gDC}] = -C_{g0}\, P_{21}\, \mathrm{sech}[P_{21}(-V_{gDC} + V_{gsc})]^2 (1 + \tanh[P_{11}\, V_{gsc}])$$
$$df_2 = D[C_{gd1}, V_{gsc}] = -C_{g0}\, P_{21}\, \mathrm{sech}[P_{21}(-V_{gDC} + V_{gsc})]^2 (1 + \tanh[P_{11}\, V_{gDC}])$$

This means that the equations for the capacitances C_{gs} and C_{gd} should be symmetrical and model coefficients should be selected properly when $P_{11} = P_{41}$ and

$P_{22} = P_{33}$. The consequences of non-charge conservation can be non-convergence in the HB. A good test for the consistency of the capacitance models is to simulate the *S*-parameters in the SS case and *S*-parameters simulated in the LS case with HB, but with very small input power. If this difference is small, this means that the capacitance, charge model is correct and implemented properly.

In general case, the charges are

$$\begin{aligned} Q_{gs} &= \int C_{gs} \partial V_{gs} = C_{gsp}\, V_{gs} + C_{gs0}\left(\Psi_1 + L_{c1} - Q_{gs0}\right)(1 + \tanh[\Psi_2])/P_{11} \\ L_{c1} &= \mathrm{Log}[\cosh(\psi_1)] \quad L_{c10} = \mathrm{Log}[\cosh(P_{10} + P_{111}\, V_{ds})] \end{aligned} \tag{2.24}$$

$$Q_{gs0} = P_{10} + P_{111}\, V_{ds} + L_{c10} \tag{2.25}$$

$$\begin{aligned} Q_{gd} &= \int C_{gd} \partial V_{gd} \\ &= C_{gdp}\, V_{gd} + C_{gd0}\left(\Psi_4 + L_{c4} - Q_{gd0}\right)\,(1 - P_{111} + \tanh[\Psi_3]))/P_{41} \end{aligned} \tag{2.26}$$

$$\begin{aligned} Q_{gd0} &= P_{40} + P_{111}\, V_{ds} + L_{c40} \\ L_{c4} &= \mathrm{Log}[\cosh(\psi_4)] \qquad L_{c40} = \mathrm{Log}[\cosh(P_{40} - P_{111}\, V_{ds})] \end{aligned} \tag{2.27}$$

The functions for capacitances, charges and their derivatives are symmetrical and defined from $-\infty < V_{gs}, V_{gd}, V_{ds} < +\infty$.

Generally, the most circuit simulators use either charge approach or capacitance approach. Figures 2.74 and 2.75 show some results for C_{gs} capacitance and charge implementation. As can be seen from Figure 2.74, there is a very small difference mainly in the pinch-off area. The difference in C_{gd} is calculated both ways—charge and capacitance is smaller.

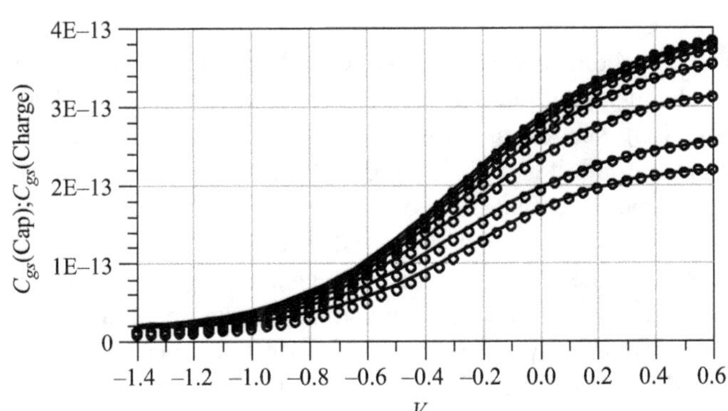

Figure 2.74 *C_{gs}, capacitance implementation-lines C_{gs} charge implementation points*

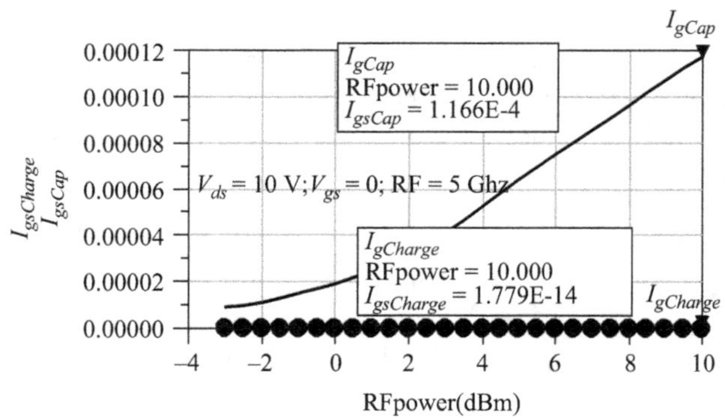

Figure 2.75 DC component of the C_{gs} capacitance current in charge $I_{gCharge} = 1e{-}14$; I_{gCap} implementation $= -1e{-}4$

The difference between both implementations is small in *S*-parameters, but the main problem is that in the capacitance implementation, device when pumped will show a DC current from the capacitance, Figure 2.75. This unphysical DC current via the capacitance is the same order as the Schottky gate diode current, so if this is critical, the capacitance implementation should be avoided. The DC component for charge implementation models will be determined from the accuracy of the HB, behave better in HB and converge much better.

It is important to know that *always there should be some small difference* in the calculated Cap, *S*(*Y*)-parameters depending on the implementation type—capacitance or charge, even if the same model parameters for the capacitances are used. The origin of this difference in the calculated *S*-parameters on the implementations is very well described by Maas [13]. As a consequence, it is important to keep the same tape of the model in extraction and later in the circuit simulations, because this small difference can be accounted when fitting the *S*-parameters with the selected capacitance model.

2.9 GaN implementation specifics

GaN FET show different characteristics depending on the channel thickness, Al content in the channel, field plates, gate structure, passivation, etc. This results in a variety of *IV* and G_m shapes, pinch-off voltages, SS and LS characteristics. When technology is not settled, it is probably more efficient and faster to work with table-based (typ. 1,000 measurement points) or empirical TBMs (typ. 100 measurement points). When technology is settled, empirical models (typ. 100 measurement points) can be very useful, as they allow easy implementation of process variations, thermal conditions and to extend the model far beyond the measurement range in voltages and frequency. The ETB model can be used in the transactional period or in the case when very complicated characteristics should be modeled accurately, keeping the model structure simple and using table for the complicated characteristics.

Generally, for GaN FET, the transconductance /mm is lower, channel current is higher and ratio G_m/I_{pk} is lower in comparison with ordinary GaAs and InP HEMT.

When GaN technology is in advance state, matured and device reliable, the CW and pulsed characteristics are very similar, i.e., dispersive effects are rather small. Comparison with ordinary FET shows that the basic GaN FET model is similar to the available GaAs FET model and respective Verilog implementation. For some GaN FET with small dispersion and small difference between pulsed and CW characteristics, high $P_1 = (G_m/I_{pk}) > 0.5$ ratio, the default model can be applied with a good accuracy. But as the GaN devices from different manufacturing processes differ, there are some specific issues for the GaN FET (device is pushed to the current and breakdown limits). Some of these issues are addressed in the Verilog A GaN FET model, equivalent circuit, as in Figure 2.65, implemented in the circuit simulators [53].

High power, large GaN devices can show additional problems:

1. The gate control is delayed and reduced at high frequency:
   Large, high-power devices do not respond immediately at RF, $\tau = C_{del} \times R_{del}$. C_{del} = 2–3 fF is the capacitance of the gate footprint, R_{del} = 2 kΩ (channel resistance).
2. Higher R_s and R_d/mm for GaN and SiC FET in comparison with GaAs FET.
3. R_d, R_s are bias and temperature dependent.
   Trew *et al.* [74,75], Inoue [76], Thorsell [73]
   When cold values for R_s and R_d are used, unrealistically high output power and PAE will be predicted.
4. Dynamic self-heating model for GaN FET model-must! R_{therm}, C_{therm}, thermal coefficients for current, capacitances should be set properly. LSM without these parameters is not accurate for $P > 200$ mW.
5. Breakdown important for high-power devices.
6. Often, GAN devices are leaky. The leakage is temperature dependent.
7. Current slump—in some cases at RF, we do not reach the CW DC knee I_{ds} values.
8. (Back-gate feedback) voltage will change effective V_{gs} at RF.
9. Keep device safe $< P_{\max}$ mounting quality is critical.

GaN model implementation summary:

The new, different things in the GaN FET model in comparison with the default GaAs models are as follows:

1. Modified psi function (tanh(sinh(Ψ))) I_{ds} model 3.
   This will produce the shape closer to rectangular using one term of the power series P_1. Very small addition of the P_3–R_d term can produce even sharper pinch-off and rectangular shape. In addition, by using double hyperbolical function tanh (sinh), the fit for the harmonics is better when ratio $P_1 = G_m/I_{pk}$ is low < 0.5, which is often the case for SiC FET and some high voltage GaN FET.
2. Capacitance model which can provide peaking of C_{gs}, if the device show this effect, Cap model 3, 4.
   The capacitances are implemented as charge-Cap model4 (to provide charge conservation) and capacitance Cap model 3 to keep the model compatible with simulators with capacitive.

In this implementation, it is considered that some of GaN devices show peak of C_{gs} at about -2 to 1 V. This is arranged with the peaking parameter part of P_{222}, C_{gs}, MJC is the standard depletion coefficient, m will define how large can be the peak capacitance. The default $P_{222} = 0$ will switch of the peaking part transferring to the standard capacitance models 1 and 2.

$$y = (((V_{gsc}+2.0)/P_{10})-1.0);\ MJC = 0.5$$
$$C_{gsdepl} = C_{gsdepl} = ((m+y^2)^{(-1-\mathrm{MJC})})\ (m+(1-2\times\mathrm{MJC})y^2)$$

The shape of the caps and position of the peak can be changed if required. If the device is very different, for example, the peak of the cap is at -10 V; this can be arranged modifying part $((V_{gsc}+2.0)/P_{10})$ to $((V_{gsc}+10.0)/P_{10})$.

3. Temperature and bias dependent R_d, R_s—this is very important when the device is pushed hard.
   Typical $TcR_s = 0.003$ which is used for both R_s and R_d. This will affect the knee region and change the junction temperature in time within the limits of the thermal constant.
   Bias dependence of R_d, R_s is assumed similar to the current dependence I_{ds}.
   $R_{d1} = R_d + R_{d2}(\text{tanh\_psi})$
   R_d is the bias independent part of the total R_{d1}; R_{d2} will define the bias dependent part of R_d, it is user-defined value, default 0.
4. Enhanced dispersion modeling:
   (i) Back gate approach V_{bg}.
   (ii) Delay circuit to model the knee walkout.

   Backgate V_{bg} influence:

$$V_{pkm} = V_{pks} - Dv_{pks} + Dv_{pks} \times \tanh(\alpha_s \times V_{ds}) - V_{bg} - V_{sb2}(V_{dg} - V_{tr})(V_{dg} - V_{tr}) \tag{2.28}$$

   Delay circuit, Figure 2.65, represents the slowed down reaction of I_{ds} on the gate control voltage at RF. It is arranged as $R_{del}\times C_{del}$. Typical values are $C_{del} = 1$–2 fF (the gate footprint of the gate capacitance) and $R_{del} = 1$–5 kΩ—the channel charging resistance. At high frequency C_{del} shunts directly the control V_{gs} input. This circuit can be used to model the high-frequency reduction of the output power and knee walkout [56–58].
5. The I_{ds} channel breakdown was available in the GaAs model with parameters—$L_{sb0}\_T \times \text{limexp}(V_{dg}-V_{tr})$. New for GaN HEMT: the breakdown exponent can be adjusted with parameter E_{bd}.

$$L_{sb0}{}^{-}T \times \text{limexp}(E_{bd}(V_{dg} - V_{tr}))$$

6. In the GaN model, devices are pushed to the limits, and the GS and GD Schottky junction breakdown is also included.
   This is arranged with breakdown voltages for the respective junctions V_{bdgs} (typ. 10), V_{bdgd} (typ. 100), the exponent slope P_{bdg} (typ. 0.5), and breakdown coefficient K_{bdgate}. Setting $K_{bdgate} = 0$ will switch off the Schottky junction breakdown (default = 0).

$$I_{gs1} = \exp(P_g \times \tanh(\_v3\text{-}V_{jg})) - \exp(-P_g \times \tanh(V_{jg}))$$
$$I_{gs} = I_j(I_{gs1} - K_{bdgate} \times 0.001 \times I_{gs2}) + 0.00000000000001$$

$$I_{gd1} = (\exp((P_g)\tanh(\_v2-V_{jg}))) - \exp(-P_g \times \tanh(V_{jg}))$$
$$I_{gd} = I_j(I_{gd1} - K_{bdgate} \times 0.001 \times I_{gd2}) + 0.000000000000001$$

7. Many functional changes made by Tiburon to improve the stability including small numbers e−13 added to improve numerical stability.

An interesting feature of Verilog A is that some result of simulations can be monitored, trace the CAD tool calculations. This can be arranged using the command Desc, for example, displaying calculated capacitance C_{gs}(×Desc = "Cgssim"×). It should be considered that, if too many parameters are displayed, this will slow the simulations.

2.10 Implementation of complex G_m shape

When a device has some specifics features, the model equations can be changed using the Verilog model as mainframe. For example, highly asymmetrical G_m shape or G_m asymmetrical with two peaks of the transconductance, Figure 2.76, can be easily implemented in Verilog A or symbolically defined device (SDD). But this will require adding two more parameters: V_{pks2} of the second peak and the splitting current parameter AA.

$$I_{ds} = I_{pk0}T(1 + \mathrm{AA} \times \mathrm{tanp} + (1 - \mathrm{AA})\mathrm{tanp}_{11})$$
$$\times \tanh(\alpha_p \times V_{ds})(1 + \lambda_p \times V_{ds} + L_{sb0} \times \exp(E_{bd}(V_{ds} - V_{tr})))$$
$$V_{pkm} = V_{pks} - DV_{pks} + DV_{pks} \times \tanh(\alpha_s \times V_{ds}) - V_{bg} - V_{sb2}(V_{dg} - V_{tr})^2$$
$$V_{pkm2} = V_{pks2} - DV_{pks} + DV_{pks} \times \tanh(\alpha_s \times V_{ds}) - V_{bg} - V_{sb2}(V_{dg} - V_{tr})^2$$
$$x_{11} = P_1T((V_{gs} - V_{pkm2}) + P_2(V_{gs} - V_{pkm2})^2 + P_3(V_{gs} - V_{pkm2})^3)$$
$$\mathrm{tanp}_{11} = \tanh(x_{11})$$

In a similar way, other changes can be arranged, but it should be considered: it might be more efficient to use measurement-based model, empirical TBM instead of increasing complexity, parameters to extract for the empirical model?

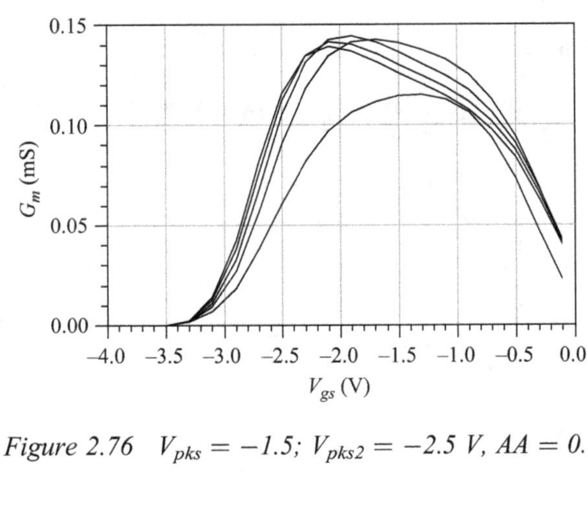

Figure 2.76 $V_{pks} = -1.5$; $V_{pks2} = -2.5$ V, AA = 0.1

2.11 Breakdown phenomena

For high-power devices, it is a key requirement to organize properly the measurements to keep the device safe. The solution is to use dual region measurements and simulations, Figure 2.77:

1. High I_{ds}, low V_{ds}
2. Low I_{ds}, high V_{ds}—the selection should be below $P_{\max}$. When measurement is properly designed, it should cover the load line!

At very high voltages, I_{gs} can start breakdown, like in Figures 2.32, 2.78 and 2.79.

Resistance $R_{meas} = 1\ \text{k}\Omega$ defines the current in the measurement path, $R_{con} = 1\ \text{M}\Omega$ defines the injected current in the connected path. Injected current should be <0.1 mA/mm for safety.

I_{ds} breakdown is the most critical for the prediction of the output power, Figure 2.32. The Schottky diodes breakdown, Figures 2.78 and 2.79 will influence reliability, noise, harmonics but will not contribute so much to the output power.

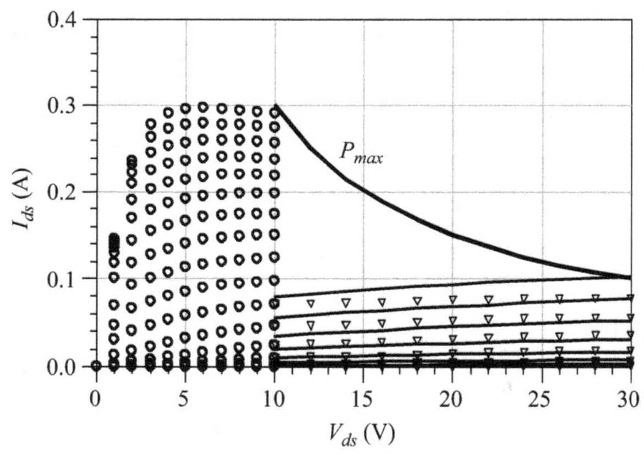

Figure 2.77 I_{ds} vs. V_{ds}

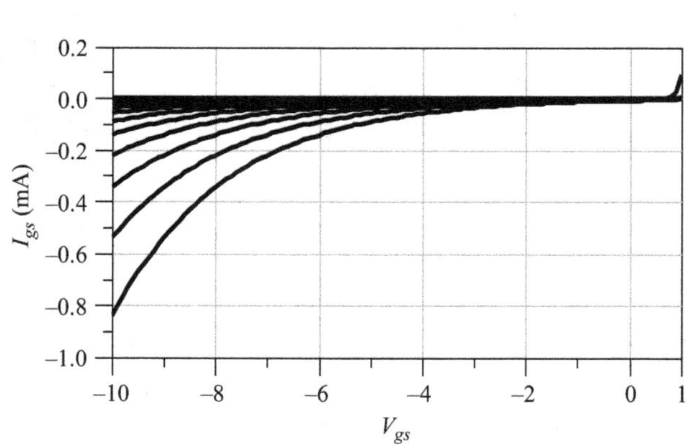

Figure 2.78 I_{gs} breakdown GaN HEMT

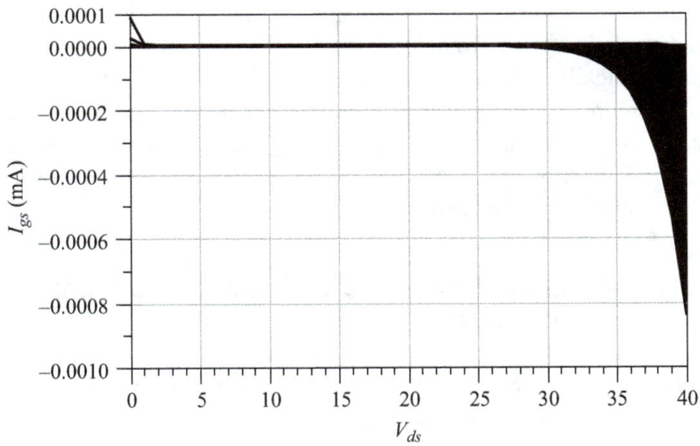

Figure 2.79 I_{gd} *breakdown, GaN HEMT*

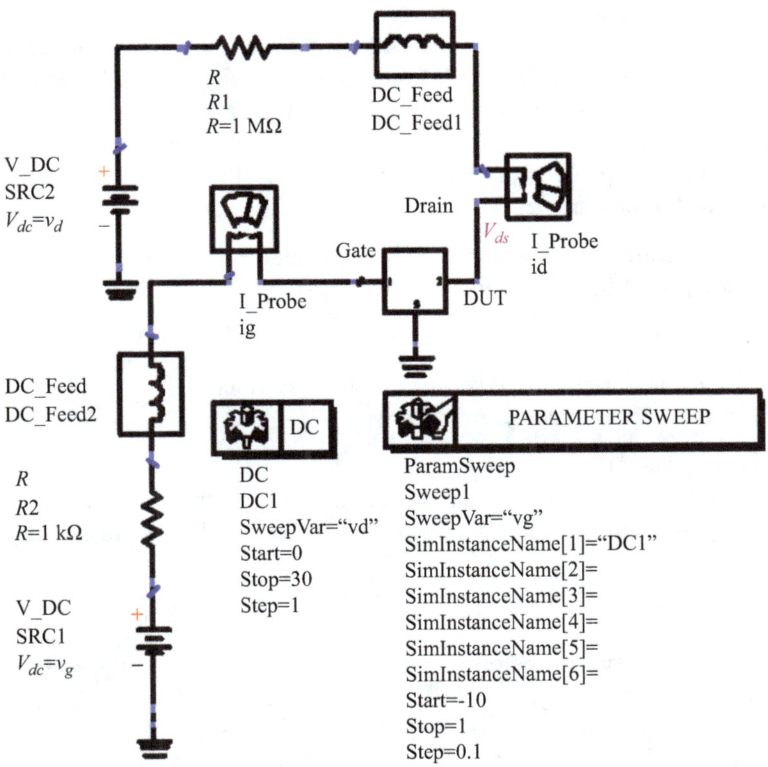

Figure 2.80 Gate–source breakdown measurement setup

To measure the breakdown is recommended to set carefully the compliance levels, to keep device, probes safe. Often compliance is not fast enough, so it will help if limiting resistors are used for breakdown measurements like in Figure 2.80. These resistances will set the max current and should be selected according to the device size. In the measurement branch, lower value resistance should be used: 50–1 kΩ. The free terminal should be grounded with a resistor to prevent current to flow in this direction, typically >1 MΩ. Resistance cabling should be with low inductance to avoid oscillations. The same resistances should be used later to simulate and model the breakdown. Using this approach, a unique value of the breakdown parameters is easily extracted. For very high-power devices, it should be considered that the breakdown is also temperature dependent.

2.12 Large-signal model evaluation: power-spectrum measurements and fit

For LS and harmonics modeling, we need correct derivatives, self-heating, dispersion, memory effects, complicate the picture. A problem is DC data for I_{ds} harmonics are often noisy. Most simple solution to evaluate harmonics is to use power spectrum measurement. More complicated, complete evaluation is done using load-pull, LSVNA waveform analyses, combined LSVNA-load-pull or LSVNA power spectrum.

The PS measurement is very universal, easy to set, fast, easy to repeat at different places—standard 50 Ω, simple equipment is required. The PS measurement can be done using set up similar to that shown at Figure 2.81.

In the PS measurements P_{in}, RFfreq, V_{ds} constant are constant.

To evaluate the modeling task more accurately and easily, the problem can be split into two PS measurements:

1. Low freq. RF PS> evaluation of I_{ds} current source.
2. High freq. RF PS> capacitances evaluation.

Equipment needed:

1. RF source, scalar NA or power meter+diplexer filter.
2. RF source, spectrum analyzer (SA).

Bias tees, attenuators.

PS using SAs, the solution is simpler and faster to arrange, there is no need for diplexer, not as accurate as using a Power meter, but usually accurate enough.

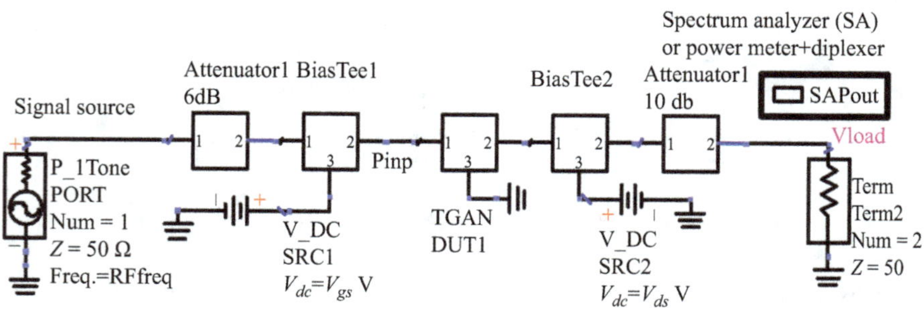

Figure 2.81 Power spectrum setup

The device is biased in active mode $V_{ds} = 10$–15 V for GaN, $V_{ds} = 2$–3 V (GaAs).

Low RF frequency is used to evaluate the nonlinearities of the I_{ds} and high RF to evaluate capacitance nonlinearities. Definition: low RF–high RF will depend on the device size. For devices size <0.5 mm, low RF can be considered low RF = 100 MHz. High RF = 3 GHz is good to evaluate the capacitances. Input power should be enough to drive the device in nonlinear mode but avoiding hard compression. For a medium-sized device (W_{tot} = 200 μm), P_{in} = 0 dBm is a good start.

Procedure for PS:

1. Set the frequency, calibrate respective input power (P_{in} = 10 dBm for SiC and GaN; P_{in} = 0 dBm for GaAs) at the device terminal—DUT for fundamental and harmonics (cable losses might be different).
2. Replace DUT with the thru, connect input output, calibrate losses of output cables, diplexer, calculate corrections of the SA, etc. Keep attenuators directly at the bias tees, close to DUT.
3. Measure 1, 2, 3 harmonics sweeping V_{gs} (10pts), V_{ds} parameter: V_{ds} = 1; 14 V, 24 V ... for GaN, SiC.

   V_{ds} = 0.1; 2 V, 3 V ... for GaAs.

When measurement is done, the first parameter which should be refined is the V_{pks}. This is the gate voltage V_{gs} for which we observe the minimum of the second harmonic. At this gate voltage, G_m is at the peak, second harmonic 0, i.e., we will observe minimum of the second harmonic. V_{pks} is usually the only parameter needed to tune. It should be noted that V_{pks} refers to the intrinsic parameters, so externally we will see voltage shift that will depend on the current. In the example: intrinsic $V_{pks} = -0.5$ V, but we observe minimum at $V_{gs} = -0.58$ V, Figure 2.82. Changing V_{pks} to $V_{pks} = -0.3$ V will shift the minimum, Figure 2.83.

Parameter P_1 will influence the gain and the output power, and like quality factor will change the bell characteristics of the first and higher harmonics. It should be considered that minimum of the second is at the maximum of the first, minimum of the thirds will be at the maximums (inflection points) for the second

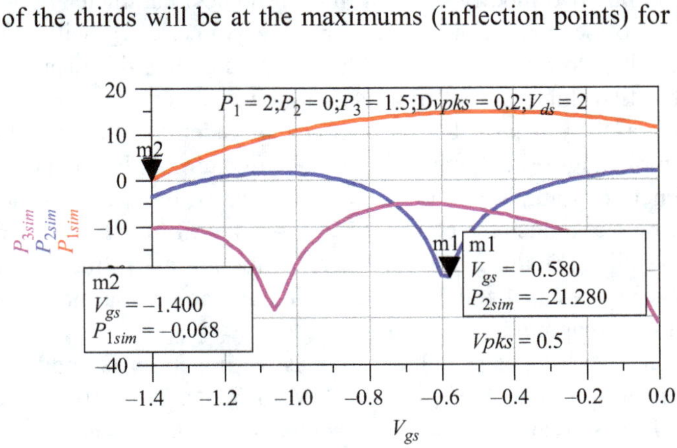

Figure 2.82 $V_{pks} = -0.5$ V; $P_1 = 2$; $V_{ds} = 2$ V; GaAs

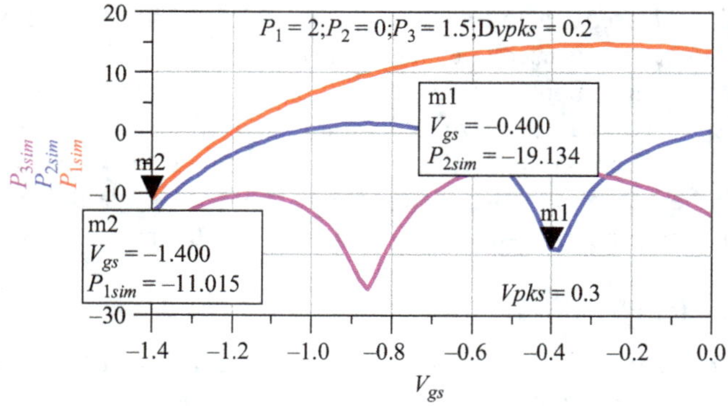

Figure 2.83 $V_{pks} = -0.3$ V; $P_1 = 2$; $V_{ds} = 3$ V; GaAs

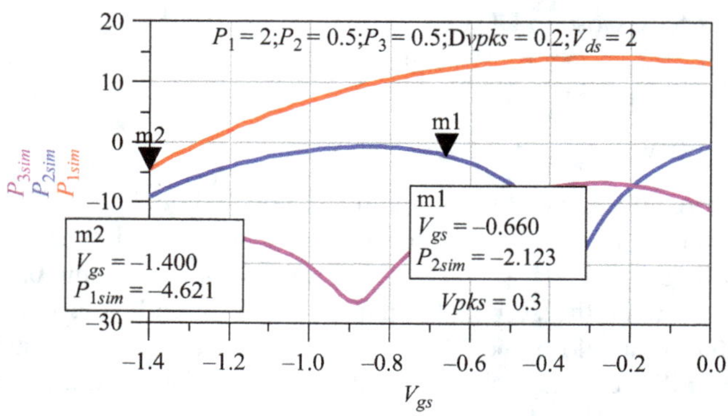

Figure 2.84 $P_1 = 2$; $P_3 = 0.5$

harmonic and so on, i.e., increasing P_1 will not only shrink the gate span for the first harmonic but also modify the span for harmonics. Parameter P_2 will change the magnitude of second harmonic and create asymmetry, P_3 will influence the harmonics for V_{gs} close to the pinch off and manipulate the third harmonic. When tuning P_3, the first thing we will find is as follows: increasing P_3 will reduce the first harmonic at the pinch off but will not change much the power at the peak.

For example, Figure 2.84, when we increase first term of harmonic series in the argument function from $P_1 = 2$ to $P_1 = 3$, we change the shape of the first harmonic at the output, reduce the power at $V_{gs} = -1.4$ V from $P_{1\text{sim}} = -4.68$ dBm to $P_{1sim} = 19.58$ dBm (Figure 2.85). The other harmonics are influenced too. When modifying P_3, Figures 2.84 and 2.86, we manipulate the first harmonic close to the pinch off and the magnitude of the third harmonic.

For devices working as switches and resistive mixers, it is good to have harmonic data measured at low V_{ds} ($V_{ds} = 1$ V for GaN, $V_{ds} = 0.1$ V for GaAs), Figure 2.87. This is because the harmonic content at low V_{ds} might be very different from the harmonic contents for $V_{ds} > V_{knee}$ in the saturated region. Quite often the ratio transconductance/current, the shape of the transconductance at low V_{ds} is

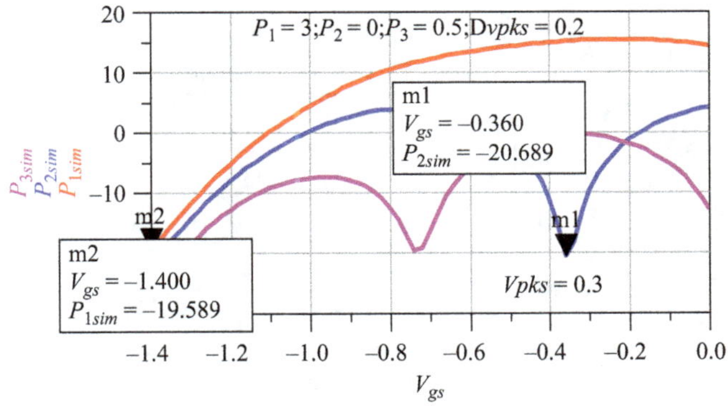

Figure 2.85 $P_1 = 3$; $P_3 = 0.5$

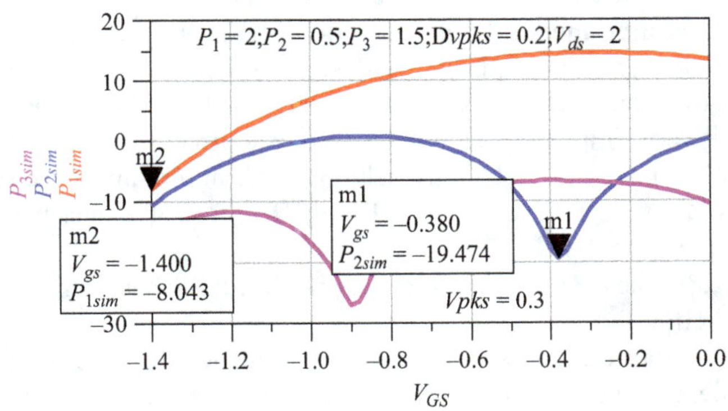

Figure 2.86 $P_1 = 2$; $P_2 = 0.5$; $P_3 = 1.5$; $V_{ds} = 2$ V

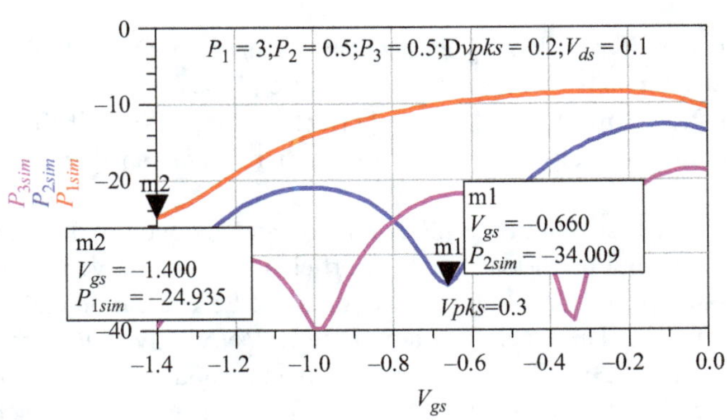

Figure 2.87 $V_{ds} = 0.1$ V; $DV_{pks} = 0.2$ V; $P_1 = 3$; $P_2 = 0.5$; $P_3 = 0.5$

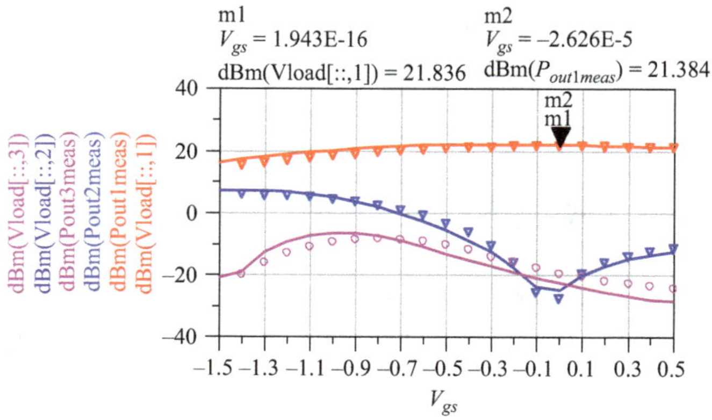

Figure 2.88 Power spectrum X-band measured and modeled GaN HEMT

quite different. When it is known in advance that device will be operated as switch or in resistive mode, the optimization, model refinement in this region will help one to get very good quality designs to determine exactly the shape, contribution of harmonics, improve the accuracy of IMD prediction, etc., keeping the model very simple, but accurate. The most correct refinement of DV_{pk} can be done first fitting the model V_{pks}, position of the minimum of the second harmonic in the active mode, and later refining D_{vpk} from PS measurements with low V_{ds}, as in Figure 2.87.

Figure 2.88 shows the measured and modeled PS spectrum for a typical 0.5 mm GaN HEMT

Usually initial tuning will give correspondence for the first harmonic 0.2–1 dB, depending on the output power level, good fit for I_{ds} DC part, harmonics—important for PAE, IMD products. If required, an optimization sequence can be arranged.

2.13 LSVNA measurement and evaluation

It is much better, if possible, to look not only for the magnitude of harmonics, scalar PS measurements, but also at the harmonic phases. From this point of view, the PS evaluation using LSVNA is the perfect solution. We will fit not only the correct magnitude but also the phases of the harmonics, so that we get correct waveforms.

The PS is the first step of the LS evaluation and is done in 50-Ω environment. But, normally the optimum load for high efficiency, HPAs is very different from 50 Ω, so it is highly desirable to further continue the LS evaluation making LSVNA load-pull evaluation. The block diagram of the LSVNA active LP system is shown in Figures 2.38 and 2.39. The amplifier at the output should provide enough power to compensate circulator and cable losses (2–3 dB). The circulator separates the injected and outgoing wave, terminating b_2 in a 50-Ω load. This gives full control of Γ_L seen by the DUT at f_0.

Combined LSNA and load-pull measurements are the best approach for LS evaluation of a high-power GaN device. The load impedance is swept and output power, waveforms, impedance data are recorded. The waveform fit will give very important information for the accuracy of the *IV*, capacitance models, breakdown, knee walkout; it is invaluable tool to evaluate the GaN devices. Figures 2.89–2.91 show some typical results for HP devices using active load-pull and LSVNA.

When the waveforms fit for the current, voltages are good, Figure 2.91; a very good fit for the load impedance can be obtained, Figure 2.90. Such a good fit is difficult to obtain using standard, load-pull setup. If in addition, some CAD optimization of the model is made using the LSVNA load-pull data, an accuracy for the LP in the range of 1° can be obtained.

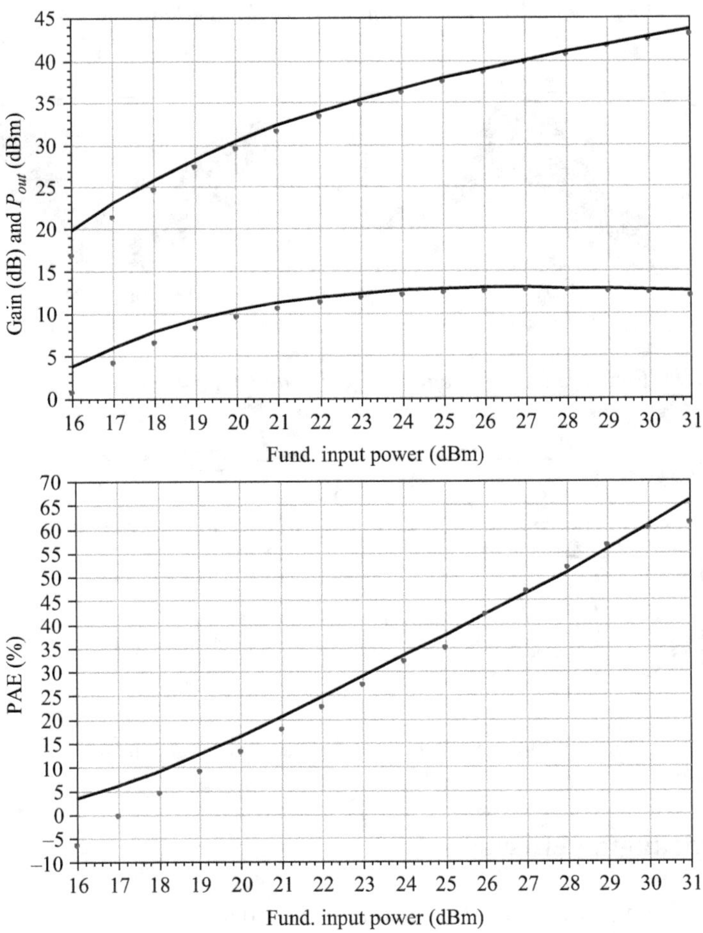

Figure 2.89 Measured, modeled RF and PAE

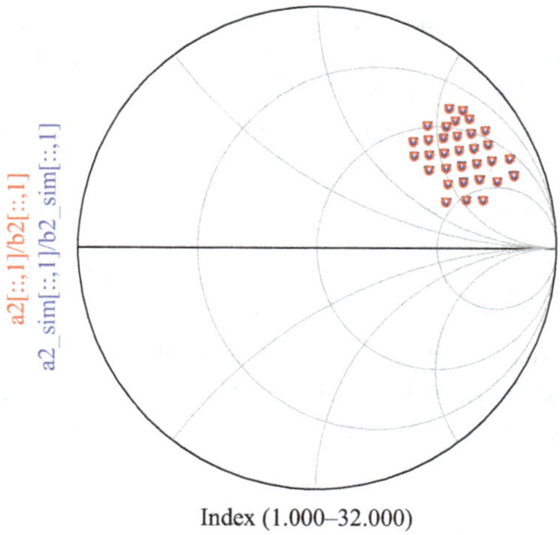

Figure 2.90 Measured, simulated load impedances C band

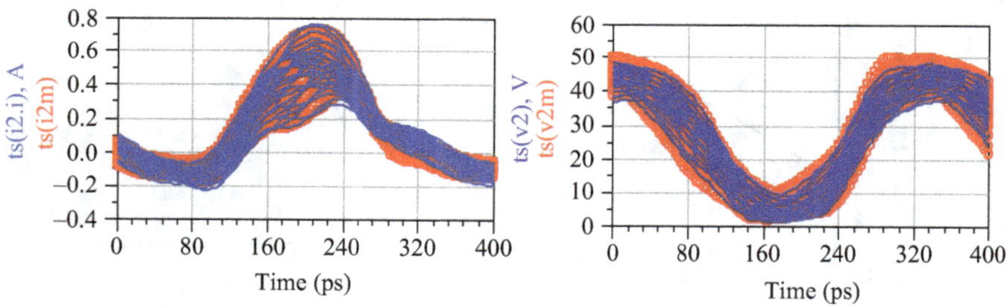

Figure 2.91 Waveforms I_{ds} (i2) and V_{ds} (v2) vs. time

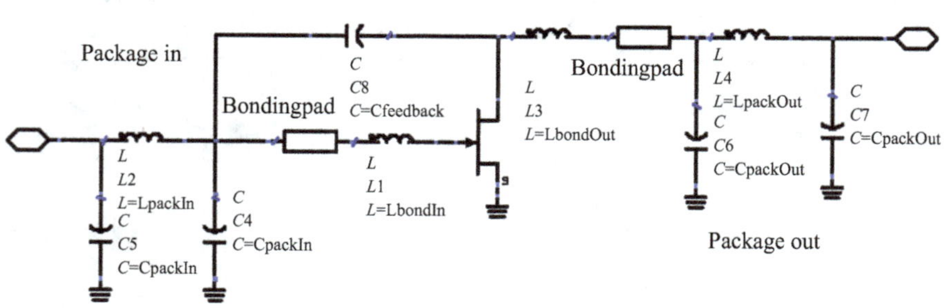

Figure 2.92 Simplified package equivalent circuit

2.14 Packaging effects

Very large, high-power devices operating at low frequency are often inserted in hermetical packages to improve handling, reliability.

Each of individual gate, source, and drain fingers are bonded separately or in groups to reduce the parasitic connecting inductance. To improve the matching, often additional matching capacitances, lines are used at the input–output, and they should be implemented in the model. Additional problem is that the input–output inductances are mutually coupled, and this should also be included in the passive part of the model. The model accuracy of the passive part of the package is *very critical* to obtain good LS simulation results from package device [102–108].

Seems natural to split the large-packaged transistor into individual cells, each bonded with its own bonding parameters, mutually coupled inductances. Modeling of the passive part can be done using an advanced electromagnetic (EM) simulator. Experimental evaluation of the passive, bonding part can be done replacing the device with thru, short. This evaluation step is mandatory to obtain good ECM for passive part of the packaged FET. Splitting the device into multiple parallel cells, making distributed model, will produce accurate *S*-parameters but most probably can create problems in the high-power HB evaluation. It was proven experimentally that the multiple cell packaged transistor can be modeled in a simplified way using single LSM for the whole packaged device or use minimum cells if device, power is very large [106,107]. This approach will converge in harmonic balance better, compared with the distributed case with multiple LSMs for fingers connected in parallel. It is shown, comparing simulation with *S*-parameter and load-pull measurement, that the concentrated model provides similar accuracy compared to the distributed model approach. The discrepancies are identified as being caused by losing the information for the phase-shift change due to lateral dimension of the package when deducing the concentrated model. This, however, is no significant drawback of the concentrated model. It can be overcome by slightly fine-tuning its model parameters inductances and package capacitances.

The simplified modeling approach, Figure 2.92, will work properly if the device is not very large. If the device is very large, we might see additional problem: the temperature distribution will be different between the fingers, parts of the transistor. In such case, it might need to use more complicated distributed ECM of the packaged transistor, using several parallel LSMs (3–5) with different thermal resistance, bonding inductances.

2.15 Self-heating modeling implementation GaN

The GaN FET model [53] is implemented in Verilog A with two self-heating arrangements:

1. the three-terminal version and
2. the four-terminal version.

Default is the four terminals, user-accessed thermal node, Figure 2.93. This version gives the user possibility of arranging complicated thermal circuits with several thermal constants. This can be useful for packaged devices where thermal constants for the chip and packaged device are quite different. Setting the external resistor to a very high value, transform the circuit to three-terminal.

To enable only three-terminal behavior, comment out the THERMAL_NODE definition in the Verilog-A file:

```
// ‘define THERMAL_NODE
```

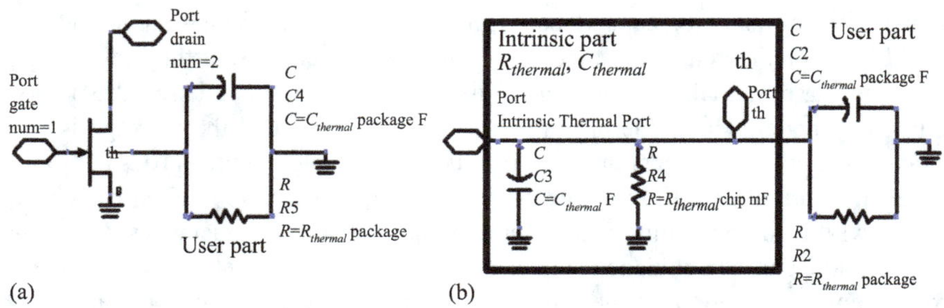

Figure 2.93 Equivalent circuits for thermal nodes arrangement (a) External thermal resistance of the package added to the thermal node and (b) Intrinsic and user part arrangement

The three-terminal version, with user access to the thermal node, will function like the thermal part in the ordinary FET models with internal thermal node, if user leave thermal node open or connect very high resistance (>1 MΩ). The user enters the thermal resistance, thermal capacitance, in the range of μF. When user is having access to the thermal node, very complicated thermal circuits can be connected to the thermal node Figure 2.93(b)). This facility can be beneficial to model thermally coupled devices in dense packed MMIC.

The thermal coefficients for all FET devices are rather typical, in the range of 0.001–0.003; $TcI_{pk} = -0.002$; $TcP_1 = -0.002$ are negative, $TcC_{gs} = +0.002$, $TcC_{gd} = +0.002$ are positive, etc.

Appendix

Practical GaN device parameters/mm.

1. I_{ds} part

N	I_{pk0w}	V_{pks}	DV_{pks}	P_1	P_2	P_3	α_r	α_s	λ	B_1	B_2
Value	0.412	−1.31	0.2	0.91	−0.05	0.17	0.10	0.25	0.012	0.10	2.8

2. Capacitances

N	C_{gs0w}	C_{gspiw}	C_{dsw}	P_{10}	P_{11}	P_{20}	P_{21}
Value	170	40	300	5.3	2.1	0.51	0.03
N	C_{gd0w}	C_{gdpi}	C_{gdpe}	P_{40}	P_{41}	P_{30}	P_{32}
Value	170	40	50	5.3	2.1	0.51	0.03
P_{111}	P_{222}	m					
0.001	0.00	0.20					

3. Thermal

N	R_{thermw}	C_{thermw}	TcP_1	TcI_{pk0}	TcC_{gs0}	TcC_{gd0}	TcR_s	TcR_{therm}
	12.30	0.001	−0.0015	−0.0015	0.002	0.002	0.02	0.003

4. I_{gs} (I_{jw}, P_g, V_{jg}) and parasitics

N	I_{jw}	P_g	V_{jg}	R_{gw}	R_{iw}	R_{sw}	R_{dw}	R_{d2}	R_{gdw}	L_{gw}	L_{dw}	L_{sw}
	0.00020	14.0	0.91	0.17	0.12	0.291	0.35	0.25	0.350	20	20	10

5. Breakdown: I_{ds} (L_{sb0}, V_{tr}, V_{sb2}); Schottky gates: GS, GD (V_{bdgs}, V_{bdgd}, P_{bdgd})

N	L_{sb0}	V_{tr}	V_{sb2}	E_{bd}	V_{bdgs}	V_{bdgd}	P_{bdgd}	K_{bdgate}
	0.002	50	−0.00001	0.4	10	120	0.4	0

6. Delay, dispersion

N	T_{auw}	R_{cminw}	R_{cw}	C_{rfw}	K_{bgate}	R_{cinw}	C_{rfinw}	C_{delw}	R_{delw}
	0.45	400	4,000	200	0.01	100,000	10	1	100

Capacitances (fF); inductances (pH); resistances (Ω); I (A); τ (pS)

Example basic scaling: $N_{fing} = 10$; $W_{fing} = 100$ (μm); W_{tot} (mm) $= N_{fing} \times W_{fing}/1{,}000$

$I_{pk0} = I_{pk0w} \times W_{tot}$; $C_{ds} = 1 + C_{dsw} \times W_{tot}$; $C_{gspi} = 1 + C_{gspiw} \times W_{tot}$; $C_{gs0} = 1 + C_{gs0w} \times W_{tot}$

$C_{gdpi} = 1 + C_{gdpiw} \times W_{tot}$; $C_{gd0} = 1 + C_{gd0w} \times W_{tot}$; $C_{gdpe} = 1 + C_{gdpew} \times W_{tot}$

Linear increase C_{gs} at high V_{ds} and C_{gd} crossing P_{111}: peaking capacitance: m; P_{222}; $P_{222} = 0$ (default off)

$R_g = 0.0001 + R_{gw}/W_{tot}$; $R_i = 0.0001 + R_{iw}/W_{tot}$; $R_s = 0.0001 + R_{sw}/W_{tot}$; $R_d = 0.0001 + R_{sw}/W_{tot}$; $R_{gd} = 0.0001 + R_{sw}/W_{tot}$

$R_{therm} = 0.001 + R_{thermw}/W_{tot}$; $L_g = 1 + L_{gw} \times W_{tot}/N_{fing}$; $L_d = 1 + L_{dw} \times W_{tot}/N_{fing}$; $L_s = 1 + L_{sw} \times W_{tot}/N_{fing}$

$I_j = I_{jw} \times W_{tot}$; $\tau = 0.001 + \tau_w \times W_{tot}$

Breakdown I_{ds} breakdown L_{sb0}, V_{tr}, V_{sb2}; Schottky gates: $K_{bdgate} = 0$ (default off)

$R_{cmin} = 0.001 + R_{cminw}/W_{tot}$; $R_c = 0.001 + R_{cw}/W_{tot}$; $C_{rf} = 1 + C_{rfw} \times W_{tot}$; $C_{del} = C_{delw} \times W_{tot}$; $R_{del} = R_{delw} \times W_{tot}$

Small numbers added for numerical stability.

Acknowledgments

We are thankful to H. Zirath, C. Fager, M. Ferndahl, N. Rorsman, M. Mierzwinski, F. Sischka, S. Maas, W. Curtice, and M. Rudolph, colleagues from Mitsubishi, Agilent for the help, support, and valuable discussions, and GHZ Centre Chalmers, Goteborg, Sweden, SSF for their help.

References

[1] D. Ji, and S. Chowdhury, "A Comprehensive Study of the Design Space to Achieve 1.2 kV GaN-based Vertical JFET with Low Ron," in CSW, 2015, pp. 481, 482.

[2] A.R. Jha, "High-Power GaN-HEMT Devices operating at MM-wave Frequencies," in Sources, Detectors and Receivers Semiconductor, Superconductor and Other, 2005, pp. 640–641.

[3] K. Kikuchi, M. Nishihara, H. Yamamoto, S. Mizuno, F. Yamaki, and T. Yamamoto, "A 65 V Operation High Power X-band GaN HEMT Amplifier," in APMC, 2014, pp. 585–587.

[4] X. Huang, Z. Liu, Q. Li, and F.C. Lee, "Evaluation and Application of 600 V GaN HEMT in Cascode Structure," *IEEE Trans. Power Electron.*, vol. 29, no. 5, pp. 2453–2461, 2014.

[5] P. Feng, K.H. Teo, T. Oishi, M. Nakayama, C. Duan, and J. Zhang, "Design and Simulation of Enhancement-Mode N-Polar GaN Single Channel and Dual-Channel MIS-HEMTs," in Semiconductor Device Research Symposium, 2011, pp. 1–2.

[6] U. Singisetti, M. Hoi Wong, J.S. Speck, and U.K. Mishra, "Enhancement-Mode N-Polar GaN MOS-HFET with 5-nm GaN Channel, 510-mS/mm gm, and 0.66-ORon," *IEEE Electron Device Lett.*, vol. 33, pp. 26–28, 2012.

[7] W. Saito, Y. Takada, M. Kuraguchi, K. Tsuda, and I. Omura, "Recessed-Gate Structure Approach toward Normally off High-Voltage AlGaN/GaN HEMT for Power Electronics Applications," *IEEE Trans. Electron Devices*, vol. 53, no. 2, pp. 356–362, 2006.

[8] X. Liu, C. Zhan, K.W. Chan, *et al.*, "AlGaN/GaN-on-Silicon MOS-HEMTs with Breakdown Voltage of 800 V and On-state Resistance of 3 mOhm/cm2 U," in VLSI Technology, Systems, and Applications (VLSI-TSA), 2012 International Symposium on, 2012.

[9] Z. Dong, S. Tan, Y. Cai, *et al.*, "5.3 A/400V Normally-off AlGaN/GaN-on-Si MOS-HEMT with High Threshold Voltage and Large Gate Swing," *Electron. Lett.*, vol. 49, no. 3, pp. 221–222, 2013.

[10] S. Piotrowicz, E. Morvan, R. Aubry, *et al.*, "Overview of AlGaN/GaN HEMT Technology for L- to Ku-band Applications," *Int. J. Microwave Wireless Technol.*, vol. 2, no. 1, pp. 105–114, 2010.

[11] R. Anholt, "Electrical and Thermal Characterization of MESFETs, HEMTs, and HBTs," Artech House, 1995.

[12] L.D. Nguyen, L. Larson, and U. Mishra, "Ultra-High-Speed MODFET: A Tutorial Review," *Proc. IEEE*, vol. 80, no. 4, pp. 494, 1992.

[13] S. Maas, "Nonlinear Microwave and RF Circuits," Artech House, 2003.

[14] H. Rohdin, and P. Roblin, "A MODFET DC Model with Improved Pinch off and Saturation Characteristics," *IEEE Trans. Electron Devices*, vol. ED-33, no. 5, pp. 664–672, 1986.

[15] R. Johnoson, B. Johnsohn, and A. Bjad, "A Unified Physical DC and AC MESFET Model for Circuit Simulation and Device Modeling," *IEEE Trans. Electron Devices*, vol. ED-34, no. 9, pp. 1965–1971, 1987.

[16] M. Weiss, and D. Pavlidis, "The Influence of Device Physical Parameters on HEMT Large-Signal Characteristics," *IEEE Trans. MTT*, vol. 36, no. 2, pp. 239–244, 1988.

[17] C. Rauscher, and H.A. Willing, "Simulation of Nonlinear Microwave FET Performance Using a Quasi-Static Model," *IEEE Trans. MTT*, vol. MTT-27, no. 10, pp. 834–840, 1979.

[18] W. Curtice, "A MESFET Model for Use in the Design of GaAs Integrated Circuit," *IEEE Trans. MTT*, vol. 28, no. 5, pp. 448–455, 1980.

[19] A. Materka, and T. Kacprzak, "Computer Calculation of Large-Signal GaAs FET Amplifiers Characteristics," *IEEE Trans. MTT*, vol. 33, no. 2, pp. 129– 135, 1985.

[20] T. Brazil, "A universal Large-Signal Equivalent Circuit Model for the GaAs MESFET," in Proc. 21st EuMC, 1991, pp. 921–926.

[21] G. Dambrine, and A. Cappy, "A new Method for Determining the FET Small Signal Equivalent Circuit," *IEEE Trans. MTT*, vol. 36, pp. 1151–1159, 1988.

[22] M. Berroth, and R. Bosch, "High-Frequency Equivalent Circuit of GaAs FETs for Large-Signal Applications," *IEEE Trans. MTT*, vol. 39, no. 2, pp. 224– 229, 1991.

[23] M. Berroth, and R. Bosch, "Broad-Band Determination of the FET Small-Signal Equivalent Circuit," *IEEE Trans. MTT*, vol. 38, no. 7, pp. 891–895, 1990.

[24] M. Ferndahl, C. Fager, K. Andersson, P. Linner, H.-O. Vickes, and H. Zirath, "A General Statistical Equivalent-Circuit-Based De-embedding Procedure for High-Frequency Measurements," *IEEE Trans. MTT*, vol. 56, no. 12, pp. 2962–2700, 2008.

[25] S. Manohar, A. Pham, and N. Evers, "Direct Determination of the Bias-Dependent Series Parasitic Elements in SiC MESFETs," *IEEE Trans. MTT*, vol. 51, no. 2, pp. 597, 2003.

[26] V. Sommer, "A New Method to Determine the Source Resistance of FET from Measured S-Parameters Under Active-Bias Conditions," *IEEE Trans. MTT*, vol. 43, no. 3, p. 504, 1995.

[27] K. Shirakawa, H. Oikawa, T. Shimura, *et al.*, "An Approach to Determining an Equivalent Circuit for HEMT's," *IEEE Trans. Microwave Theory Tech.*, vol. 43, no. 3, pp. 499, 1995.

[28] T. Palacios, S. Rajan, and A. Chakraborty, "Influence of the Dynamic Access Resistance in the gm and fT Linearity of AlGaN/GaN HEMTs," *IEEE Trans. Electron Devices*, vol. 52, no. 10, pp. 2117, 2005.

[29] C.F. Campbell, and S.A. Brown, "An Analytic Method to Determine GaAs FET Parasitic Inductances and Drain Resistance Under Active Bias Conditions," *IEEE Trans. MTT*, vol. 49, no. 7, pp. 1241, 2001.

[30] D. Root, and B. Hughes, "Principles of Nonlinear Active Device Modeling for Circuit Simulation," 32nd ARFTG Conference Digest, Tempe, AZ, USA, 1988, pp. 1–24.

[31] D. Root, S. Fan, and J. Meyer, "Technology-Independent Large-Signal FET Models: A Measurement-Based Approach to Active Device Modeling," 15 ARMMS Conf, Bath, UK, Sep 1991.

[32] D. Root, "Measurement-based Mathematical Active Device Modeling for High Frequency Circuit Simulation," *IEICE Trans. Electron.*, vol. E82-C, no. 6, pp. 924–936, 1999.

[33] D. Root, "Nonlinear Charge Modeling for FET Large-Signal Simulation and Its Importance for IP3 and ACPR in Communication Circuits and Systems," in MWSCAS 2001. *Proc. 4th IEEE Midwest Symp.* vol. 2, 14–17, 2001, pp. 768–772.

[34] R.B. Hallgren, "TOM3 Capacitance Model: Linking Large- and Small-signal MESFET Models in SPICE," *IEEE Trans. MTT*, vol. 47, no. 5, pp. 556, 1999.

[35] R.J. Trew, "MESFET Models for Microwave CAD Applications," *Int. J. Microwave Millimeter Wave CAE*, vol. 1, no. 2, pp. 143–158, 1991.

[36] J.P. Teyssier, J.P. Viaud, and R. Quere, "A New Nonlinear I(V) Model for FET Devices Including Breakdown Effects," *IEEE Microwave Guid Wave Lett.*, vol. 4, no. 4, pp. 104–107, 1994.

[37] K. Kunihiro, and Y. Ohno, "A Large-Signal Equivalent Circuit Model for Substrate-Induced Drain-Lag Phenomena in HJFET's," *IEEE Trans. Electron Devices*, vol. 43, no. 9, pp. 1336–1342, 1996.

[38] J. Conger, A. Peczalski, and M. Shur, "Modeling Frequency Dependence of GaAs MESFET Characteristics," *IEEE J. Solid-State Circuits*, vol. 29, no. 1, pp. 71–76, 1994.

[39] N. Scheinberg, R. Bayruns, and R. Goyal, "A Low-Frequency GaAs MESFET Circuit Model," *IEEE J. Solid-State Circuits*, vol. 23, no. 2, pp. 605–608, 1988.

[40] P. Canfield, S. Lam, and D. Allstot, "Modelling of Frequency and Temperature Effects in GaAs MESFETs," *IEEE J. Solid-State Circuits*, vol. 25, no. 1, pp. 299–306, 1990.

[41] M. Lee, and L. Forbes., "A Self-Back-Gating GaAs MESFET Model for Low-Frequency Anomalies," *IEEE Trans. Electron Devices*, vol. 37, no. 10, pp. 2148–2157.

[42] C. Camacho-Penalosa, and C. Aitchison, "Modeling Frequency Dependence of Output Impedance of a Microwave MESFET at Low Frequencies," *Electron. Lett.*, vol. 21, no. 12, pp. 528–529, 1985.

[43] J. Reynoso-Hernandez, and J. Graffeuil, "Output Conductance Frequency Dispersion and Low-Frequency Noise in HEMT's and MESFET's," *IEEE Trans. MTT*, vol. 37, no. 9, pp. 1478–1481, 1989.

[44] P. Ladbrooke, and S. Blight, "Low-Field Low-Frequency Dispersion of Transconductance in GaAs MESFETs with Implication for Other Rate-Dependent Anomalies," *IEEE Trans. Electron Devices*, vol. 35, no. 3, pp. 257, 1988.

[45] G. Kompa, "Modeling of Dispersive Microwave FET Devices Using a Quasi-Static Approach," *Int. J. Microwave Millimeter Wave Comput. Aided Eng.*, vol. 5, no. 3, pp. 173–194, 1995.

[46] J.P. Teyssier, R. Quere, and S. Augaudy, "A Pulsed S-parameter Measurement Set-Up for the Nonlinear Characterization of FETs and Bipolar Transistors," in Proc. 23rd EuMC, Madrid, 1993, pp. 489.

[47] J.W. Bandler, Q.J. Zhang, S. Ye, and S.H. Chen, "Efficient Large-Signal FET Parameter Extraction Using Harmonics," *IEEE Trans. MTT*, vol. 37, no. 12, pp. 2099–2108, 1989.

[48] I. Angelov, L. Bengtsson, and M. Garcia, "Extensions of the Chalmers Nonlinear HEMT and MESFET model," *IEEE Trans. MTT*, vol. 46, no. 11, pp. 1664–1674, 1996.

[49] I. Angelov, H. Zirath, and N. Rorsman, "Validation of a Nonlinear HEMT Model by Power Spectrum Characteristics," in IEEE MTT-S Digest, 1994, pp. 1571–1574.

[50] I. Angelov, A. Inoue, T. Hirayama, D. Schreurs, and J. Verspecht, "On the Modelling of High Frequency and High Power Limitations of FETs," in INMMIC, Rome, 2004.

[51] I. Angelov, V. Desmaris, K. Dynefors, P.A. Nilsson, N. Rorsman, and H. Zirath ., "On the Large-Signal Modelling of AlGaN/GaN HEMTs and SiC MESFETs," in EGAAS, 2005, pp. 309–312.

[52] I. Angelov, K. Andersson, D. Schreurs, *et al.*, "Large-signal modelling and comparison of AlGaN/GaN HEMTs and SiC MESFETs," 2006 APMC, 2006, pp. 279–282.

[53] Advance Design Station (ADS), Spectre Cadence, Microwave Office (MO) AWR. National Instrument, Ansoft Designer user manuals.

[54] I. Angelov, N. Rorsman, J. Stenarson, M. Garcia, and H. Zirath, "An Empirical Table-Based FET Model," *IEEE Trans. MTT*, vol. 47, no. 12, pp. 2350–2357, 1999.

[55] U. Radhakrishna, S. Lim, P. Choi, T. Palacios, and D. Antoniadis, "GaNFET Compact Model for Linking Device Physics, High Voltage Circuit Design and Technology Optimization," 2015 IEEE International Electron Devices Meeting (IEDM), 2015, pp. 9.6.1–9.6.4

[56] S.E. Gunnarsson, N. Wadefalk, I. Angelov, H. Zirath, and I. Kallfass, "220 GHz (G-Band) Microstrip MMIC Single-Ended Resistive Mixer," *IEEE Microwave Wireless Compon. Lett.*, vol. 18, no. 3, pp. 215–217, 2008.

[57] S.E. Gunnarsson, N. Wadefalk, J. Svedin, *et al.*, "A 220 GHz Single-Chip Receiver MMIC with Integrated Antenna," *IEEE Microwave Wireless Compon. Lett.*, vol. 18, no. 4, pp. 284–286, 2008.

[58] S.E. Gunnarsson, N. Wadefalk, I. Angelov, H. Zirath, I. Kallfass, and A. Leuther, "A G-Band (140–220 GHz) Microstrip MMIC Mixer Operating in Both Resistive and Drain-Pumped Mode," in IEEE MTT-S, 2008, pp. 407–410.

[59] T. Oishi, H. Otsuka, K. Yamanaka, Y. Hirano, and I. Angelov, "Semi-Physical Nonlinear Model for HEMTs with Simple Equations," in INMMIC, Goteborg, 2010.

[60] I. Angelov, M. Thorsel, K. Andersson, A. Inoue, K. Yamanaka, and H. Noto, "On the Large Signal Evaluation and Modeling of GaN FET," *IEICE Trans.*, Jul 2010, pp. 1225–1332

[61] B. Chi, Z. Song, H. Jia, L. Kuang, J. Lin, and Z. Wang, "CMOS Circuit Techniques for mm-Wave Communications," in 2018 IEEE MTT-S International Wireless Symposium (IWS), 2018, pp. 1–3.

[62] H. Jia, B. Chi, L. Kuang, *et al.*, "Research on CMOS mm-Wave Circuits and Systems for Wireless Communications," *China Communications*, vol. 12, no. 5, pp. 1–13, 2015.

[63] P. Reynaert, W. Steyaert, A. Standaert, D. Simic, and G. Kaizhe, "mm-Wave and THz Circuit Design in Standard CMOS Technologies: Challenges and

Opportunities," in 2017 IEEE Asia Pacific Microwave Conference (APMC), 2017, pp. 85–88.

[64] S. Moghadami, and S. Ardalan, "A 205 GHz Amplifier with 10.5 dB Gain and 1.6 dBm Saturated Power Using 90 nm CMOS," *IEEE Microwave Wireless Compon. Lett.*, vol. 26, no. 3, pp. 207–209, 2016.

[65] B. Park, D. Jeong, J. Kim, Y. Cho, K. Moon, and B. Kim, "Highly Linear CMOS Power Amplifier for mm-Wave Applications," in 2016 IEEE MTT-S International Microwave Symposium (IMS), 2016, pp. 1–3.

[66] D. Zhao, and P. Reynaert, "Efficiency Enhancement Techniques for mm-Wave CMOS PAs: A Tutorial," in 2016 IEEE International Symposium on RadioFrequency Integration Technology (RFIT), 2016, pp. 1–3.

[67] D. Simic, and P. Reynaert, "A 14.8 dBm 20.3 dB Power Amplifier for D-Band Applications in 40 nm CMOS," in 2018 IEEE Radio Frequency Integrated Circuits Symposium (RFIC), 2018, pp. 232–235.

[68] P. Candra, V. Jain, and P. Cheng, "A 130 nm SiGe BiCMOS Technology for mm-Wave Applications Featuring HBT with fT/fMAX of 260/ 320 GHz," in 2013 IEEE Radio Frequency Integrated Circuits Symposium (RFIC), 2013, pp. 381–384.

[69] M. Schroter, and A. Pawlak, "SiGe Heterojunction Bipolar Transistor Technology for Sub-mm-Wave Electronics—State-of-the-Art and Future Prospects," 2018 IEEE 18th Topical Meeting on Silicon Monolithic Integrated Circuits in RF Systems (SiRF), 2018, pp. 60–63.

[70] M. Schroter, T. Rosenbaum, and P. Chevalier, "SiGe HBT Technology: Future Trends and TCAD-Based Roadmap," *Proc. IEEE*, vol. 105, no. 6, pp. 1068–1086, 2017.

[71] W.R. Deal, A. Zamora, K. Leong, *et al.*, "A 670 GHz Low Noise Amplifier with <10 dB Packaged Noise Figure," *IEEE Microwave Wireless Compon. Lett.*, vol. 26, no. 10, pp. 837–839, 2016.

[72] W.R. Deal, K. Leong, A. Zamora, *et al.*, "A Low-Power 670-GHz InP HEMT Receiver," *IEEE Trans. Terahertz Sci. Technol.*, vol. 6, no. 6, pp. 862–864, 2016.

[73] M. Thorsell, K. Andersson, M. Fagerlind, M. Sudow, P.A. Nilsson, and N. Rorsman, "Characterization of the temperature dependent access resistances in AlGaN/GaN HEMTs," Integrated Nonlinear Microwave and Millimetre-Wave Circuits 2008. Workshop on INMMIC 2008, pp. 17–20.

[74] R.J. Trew, Yueying Liu, L. Bilbro, Weiwei Kuang, R. Vetury, and J.B. Shealy, "Nonlinear source resistance in high-voltage microwave AlGaN/ GaN HFETs," *IEEE Trans. Microw. Theory Tech.*, vol. 54, no. 5, pp. 2061–2067, 2006.

[75] R.J. Trew, Y. Liu, W. Kuang, *et al.*, "RF Breakdown and Large-Signal Modeling of AlGaN/GaN HFET's," 2006 IEEE MTT-S International Microwave Symposium Digest, San Francisco, CA, 2006, pp. 643–646.

[76] A. Inoue, H. Amasuga, S. Goto, *et al.*, "A Nonlinear Drain Resistance Model for a High Power Millimeter-wave PHEMT," 2006 IEEE MTT-S

International Microwave Symposium Digest, San Francisco, CA, 2006, pp. 647–650.

[77] M. Thorsell, K. Andersson, H. Hjelmgren, and N. Rorsman, "Electrothermal Access Resistance Model for GaN-Based HEMTs," *IEEE Trans. Electron Devices*, vol. 58, no. 2, pp. 466–472, 2011.

[78] D. Schwantuschke, B.J. Godejohann, P. Brückner, A. Tessmann, and R. Quay, "Mm-Wave Operation of AlN/GaN-Devices and MMICs at V- and W-Band," in 2018 22nd International Microwave and Radar Conference (MIKON), 2018, pp. 238–241.

[79] W. Xing, Z. Liu, H. Qiu, *et al.*, "InAlN/GaN HEMTs on Si with High fT of 250 GHz," *IEEE Electron Device Lett.*, vol. 39, no. 1, pp. 75–78, 2018.

[80] B. Romanczyk, M. Guidry, S. Wienecke, *et al.*, "Record 34.2% Efficient mm-Wave N-Polar AlGaN/GaN MISHEMT at 87 GHz," *Electron. Lett.*, vol. 52, no. 21, pp. 1813–1814, 2016.

[81] R. Weber, M. Cwiklinski, S. Wagner, *et al.*, "A Beyond 110 GHz GaN Cascode Low-Noise Amplifier with 20.3 dBm Output Power," in 2018 IEEE/ MTT-S International Microwave Symposium–IMS, 2018, pp. 1499–1502.

[82] W. Heinrich, N. Wolff, O. Bengtsson, *et al.*, "GaN Pushing the Limits of High-Speed Switching," in 2018 22nd International Microwave and Radar Conference (MIKON), 2018, pp. 73–76.

[83] K. Yuk, G.R. Branner, and C. Cui, "Future Directions for GaN in 5G and Satellite Communications," in 2017 IEEE 60th International Midwest Symposium on Circuits and Systems (MWSCAS), 2017, pp. 803–806.

[84] I. Angelov, K. Choumei, and A. Inoue, "An Empirical HBT Large Signal Model for CAD," *Int. J. RF Microwave CAE*, vol. 13, no. 6, pp. 518–533, 2003.

[85] I. Angelov, M. Thorsell, D. Kuylenstierna, *et al.*, "Hybrid measurement-based extraction of consistent large-signal models for microwave FETs," 2013 European Microwave Conference, Nuremberg, 2013, pp. 267–270.

[86] U. Lott, "Measurement of magnitude and phase of harmonics generated in nonlinear microwave two-ports," *IEEE Trans. Microw. Theory Tech.*, vol. 37, no. 10, pp. 1506–1511, 1989.

[87] T. Van den Broeck, and J. Verspecht, "Calibrated vectorial nonlinear-network analyzers," 1994 IEEE MTT-S International Microwave Symposium Digest, San Diego, CA, USA, vol. 2, 1994, pp. 1069–1072.

[88] T. van den Broeck, "Calibrated measurements of nonlinearities in narrow-band amplifiers applied to intermodulation and cross modulation compensation," 1995 IEEE MTT-S International Microwave Symposium Digest, Orlando, FL, vol. 3, 1995, pp. 1155–1158.

[89] P. Blockley, D. Gunyan, and J.B. Scott, "Mixer-based, vector-corrected, vector signal/network analyzer offering 300 kHz–20 GHz bandwidth and traceable phase response," IEEE MTT-S International Microwave Symposium Digest, 2005. pp. 4–7, 2005.

[90] W. Van Moer, and L. Gomme, "NVNA versus LSNA: enemies or friends?," *IEEE Microw. Mag.*, vol. 11, no. 1, pp. 97–103, 2010.

[91] P. Wright, A. Sheikh, C. Roff, P.J. Tasker, and J. Benedikt, "Highly efficient operation modes in GaN power transistors delivering upwards of 81% efficiency and 12W output power," 2008 IEEE MTT-S International Microwave Symposium Digest, 2008, pp. 1147–1150.

[92] P. J. Tasker, "Conventional Transistor Non-Linear Model Extraction/ Verification using Time Domain Microwave Waveform Measurements," 58th ARFTG Conference Digest, San Diego, CA, USA, 2001, pp. 1–6.

[93] B.M. Green, V. Tilak, V.S. Kaper, J.A. Smart, J.R. Shealy, and L.-F. Eastman, "Microwave power limits of AlGaN/GaN HEMTs under pulsed-bias conditions," *IEEE Trans. Microw. Theory Tech.*, vol. 51, no. 2, pp. 618–623, 2003.

[94] M. Thorsell, and K. Andersson, "Fast Multiharmonic Active Load–Pull System With Waveform Measurement Capabilities," *IEEE Trans. Microw. Theory Tech.*, vol. 60, no. 1, pp. 149–157, 2012.

[95] F. H. Raab, "High-efficiency linear amplification by dynamic load modulation," IEEE MTT-S International Microwave Symposium Digest, vol. 3, 2003, pp. 1717–1720.

[96] M. Thorsell, K. Andersson, and C. Fager, "Characterization setup for device level dynamic load modulation measurements," 2009 IEEE MTT-S International Microwave Symposium Digest, 2009, pp. 1197–1200.

[97] J.M. Cusack, S.M. Perlow, and B.S. Perlman, "Automatic Load Contour Mapping for Microwave Power Transistors," *IEEE Trans. Microw. Theory Tech.*, vol. 22, no. 12, pp. 1146–1152, 1974.

[98] Y. Takayama, "A New Load-Pull Characterization Method for Microwave Power Transistors," 1976 IEEE-MTT-S International Microwave Symposium, 1976, pp. 218–220.

[99] G.P. Bava, U. Pisani, and V. Pozzolo, "Active load technique for load-pull characterisation at microwave frequencies," *Electron. Lett.*, vol. 18, no. 4, pp. 178–180, 1982.

[100] K. Andersson, M. Thorsell, G. Paillোncy, and F. Verbeyst, "Large-signal waveform acquisition of pulsed signals," 2011 41st European Microwave Conference, 2011, pp. 910–913.

[101] M. Demmler, P.J. Tasker, M. Schlechtweg, and A. Huilsmann, "Direct extraction of non-linear intrinsic transistor behavior from large signal waveform measurement data," 1996 26th European Microwave Conference, vol. 1, 1996, pp. 256–259.

[102] J. Flucke, F. Schmuckle, W. Heinrich and M. Rudolph, "An accurate package model for 60W GaN power transistors," 2009 European Microwave Integrated Circuits Conference (EuMIC), Rome, 2009, pp. 152–155.

[103] M. Rudolph, "Modeling GaN power transistors," 2010 IEEE 11th Annual Wireless and Microwave Technology Conference (WAMICON), Melbourne, FL, 2010, pp. 1–4.

[104] F. Schnieder, F. Schmückle, W. Heinrich and M. Rudolph, "An analysis of source connections in GaN power transistor packages," 2012 The 7th German Microwave Conference, Ilmenau, 2012, pp. 1–4.
[105] F. Schnieder, O. Bengtsson, F. Schmückle, M. Rudolph and W. Heinrich, "Simulation of RF Power Distribution in a Packaged GaN Power Transistor Using an Electro-Thermal Large-Signal Description," *IEEE Trans. Microw. Theory Tech.*, vol. 61, no. 7, pp. 2603–2609, 2013.
[106] N. Kosaka, S. Fujiwara, A. Okamura, *et al.*, "A high-efficiency and high-gain, plastic packaged GaN HEMT for 3.5-GHz-band LTE base stations," 2016 IEEE International Symposium on Radio-Frequency Integration Technology (RFIT), Taipei, 2016, pp. 1–3.
[107] S. Imai, S. Watanabe, Y. Komatsuzaki, *et al.*, "A 2.6-GHz-band 78-W Doherty power amplifier with GaN HEMT unit-cell structure robust for layout-dependent loop oscillation," 11th European Microwave Integrated Circuits Conference (EuMIC), London, 2016, pp. 512–515.
[108] K. Krishnamurthy, T. Driver, R. Vetury, and J. Martin, "100 W GaN HEMT power amplifier module with > 60% efficiency over 100–1000 MHz bandwidth," 2010 IEEE MTT-S International Microwave Symposium, 2010, pp. 940–943.

Chapter 3

Load pull characterization

Christos Tsironis[1] and Tudor Williams[2]

As an introductory remark I would like to emphasize that our contribution is not the result of extensive study of existing literature and neither has it the ambition of a complete coverage of all historical and theoretical facts. It rather represents our and our colleague's experience and is the result of years in the laboratory to develop products that will help engineers and scientists to understand and improve their products. Our research and findings over the years has been, and is, applied and directly oriented toward final products and practical solutions.

3.1 Definition of load pull

Modern active devices in radio frequency (RF)/microwave power amplifiers can deliver high output-power levels, at times over broad frequency ranges. But this requires establishing optimum source and load conditions for those active devices at as many operating frequencies as possible. Finding those optimum load conditions is possible by characterizing a device under test (DUT) while changing the impedance presented to the load at different frequencies while measuring to find the maximum output power at each frequency. This can be done using load pull "LP" (or source pull, "SP"); this is a test method whereby a potentially nonlinear microwave device (typically a two-port, especially a microwave power transistor) is presented with varying loads while its RF and direct current (DC) behavior is measured and registered. As in all experiment and measurement, only one external parameter (stimulus) is allowed to change, for being able to extract useful information. All other externally imposed parameters are test conditions and must remain controlled and constant. In the case of load pull the said changing parameter is the load (or source) impedance. At this point the first hurdle appears, since, in nonlinear devices the input signals are usually deformed creating undesired harmonic components. This means that, when the load impedance is defined at one harmonic frequency, the impedances at the other harmonic frequencies shall be dealt with as additional external parameters and must remain fixed. Thus, we are

[1]Focus Microwaves, Canada
[2]Mesuro, UK

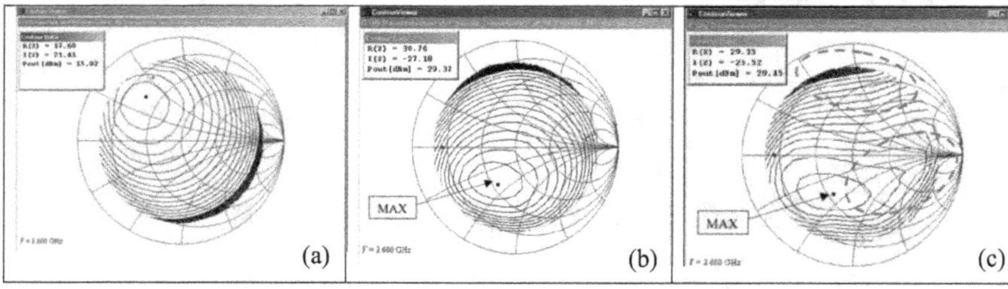

Figure 3.1 Load pull contour: (a) small signal load pull, (b) load pull with fixed harmonic impedances, and (c) load pull without harmonic impedance control

already speaking of "harmonic load pull, HLP," that is, of a test method, whereby the load impedance at one harmonic frequency is changing, all other parameters, environmental, RF and DC and all impedances at all other harmonic frequencies remain constant. This can only be done using harmonic tuners. The result of load/source pull tests are plots, whereby the measured quantity is mapped over the impedance, best represented on a Smith Chart. The measured quantity can be anything, not only output power (P_{out}) or gain (G). It can be DC current (Id), collector efficiency or power-added efficiency (PAE), intermodulation distortion (IMD), adjacent channel power ratio (ACPR), error vector magnitude (EVM) or else, or a combination of those. When the device is driven in "linear mode" (Class A) the gain or power isometric contours are circles around the optimum impedance, corresponding to the conjugate internal impedance of the device (gain circles which can be calculated using S-parameters). When the device is driven in "nonlinear mode" (Class AB, B, C, etc.) the ISO contours resemble potatoes. In this case control of harmonic impedances is crucial because they now become additional test conditions. Load pulling a device operating in nonlinear mode without controlling the harmonic impedances leads to false results. The pictures above, measured at the same frequency on the same transistor show this phenomenon (extracted from [1]). Figure 3.1(a) shows small signal load pull; Figure 3.1(b) shows load pull with fixed harmonic impedances; and Figure 3.1(c) shows load pull without harmonic impedance control. All figures are of the same transistor and frequency; Figure 3.1(b) and (c) is at the same bias and input power. The contour distortion, purely due to dangling harmonic impedances, is obvious.

3.2 Scalar and vector load pull

Depending on the measuring equipment available and the measurement objectives, we distinguish between traditional or "scalar" and "vector" load pull. For scalar load pull we only need two power meters, one to measure (through a coupler) the injected power to the DUT and one to measure the power delivered to the load (Figure 3.2).

A scalar load pull setup allows measuring output power, transducer gain, transducer efficiency, and the spectral quantities (ACPR, EVM) if a spectrum analyzer is used. It does not allow measuring PAE (power added efficiency) and

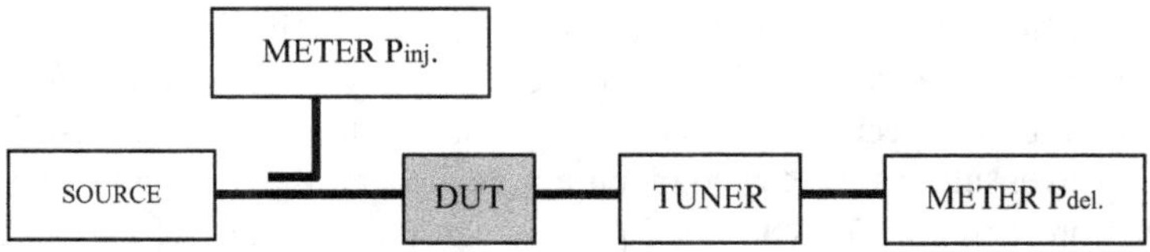

Figure 3.2 Block diagram of a scalar load pull system

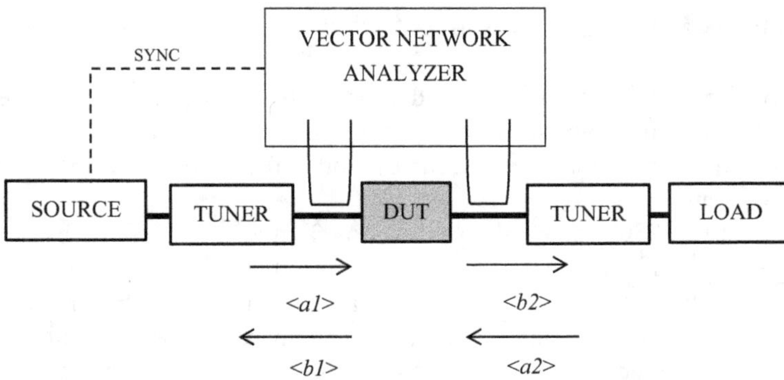

Figure 3.3 Block diagram of a vector load pull system

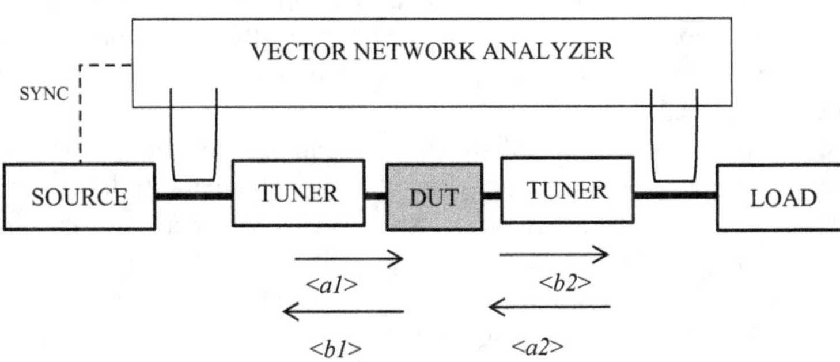

Figure 3.4 Block diagram of a vector load pull system with couplers behind the tuners

input impedance Z_{in} or Γ_{in} of the DUT. Scalar load pull relies 100% on the accuracy/ repeatability of the tuners.

For vector load pull we need directional couplers and a vector network analyzer to measure forward and reverse travelling waves <***a***>, <***b***>, and a phase reference calibrator for fully corrected time domain wave forms (Figure 3.3).

To avoid reducing the tuning range through the insertion loss of the couplers between tuners and DUT one can insert the couplers on the other side of the tuners. This requires higher VNA dynamic range due to signal attenuation by the tuners (Figure 3.4).

Vector load pull allows measuring the large signal input impedance of the DUT, delivered power to the DUT, transducer and power gain, power added efficiency and all other spectral components same as scalar load pull. Measuring <***a***> and <***b***> waves allows vector load pull to calculate the real time tuner load impedance presented to the DUT and does not rely fully on tuner calibration and accuracy. However, tuner calibration is useful to be able to steer into the right area of the Smith chart without long search.

3.3 Why is load pull needed?

The motivation for investing time and money to put together a complex and expensive load pull setup and making time-consuming tests comes from the increasing requirement for better accuracy and efficiency of the many modern amplifier designs.

In the early 1980s load pull was exotic [2], the privilege of few research institutions. I myself, then, designed my first amplifiers using S-parameters taken from the transistor datasheet, and I was not alone. But being 20% or more away from the target specs was, then, tolerable. Not anymore. In particular in the era of mobile phones, whereby the absence of cross-talking of crowded adjacent channels and the talk time on a single charge of a compact battery are key-selling arguments, and whereby one company may sell over 1 million smartphone handsets in a single day, changed everything. Today nobody can afford designing within 20% off target, or not meeting, within weeks of launching a product on the market, the competition's specs. Today all amplifier designs, especially for mobile phones, are based on accurate and extensive load pull data, or nonlinear models, also generated and verified using load pull.

There is a widespread misunderstanding that automatic load pull is about fast and convenient collection of large amounts of exploitable data. In fact, as much as this is helpful, it is not the real reason. The real reason is that, in a noncalibrated manual load pull system (Figures 3.5 and 3.6) it is practically impossible to characterize and, in particular, optimize the load of a DUT itself. In fact, as the tuner moves, its own loss (and phase) change. The user can only observe the performance of the overall network, "DUT plus tuner." If this becomes optimum (in the case of output power this translates to Maximum) this does not mean the DUT power is maximum, or that the user found the optimum load. Depending on the tuner loss and phase the optimum may be elsewhere. It is impossible to identify the location

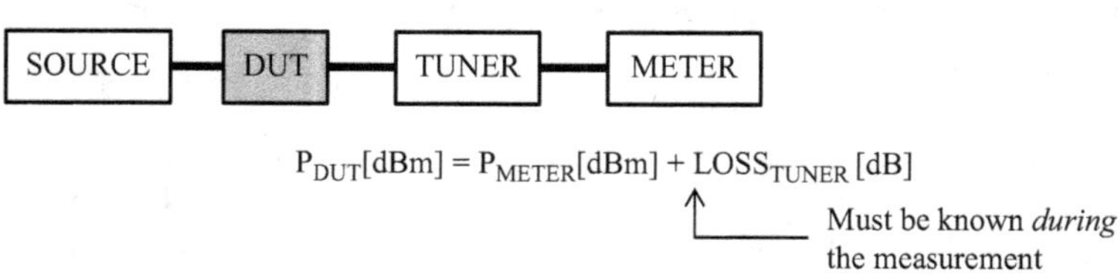

Figure 3.5 Traditional scalar load pull

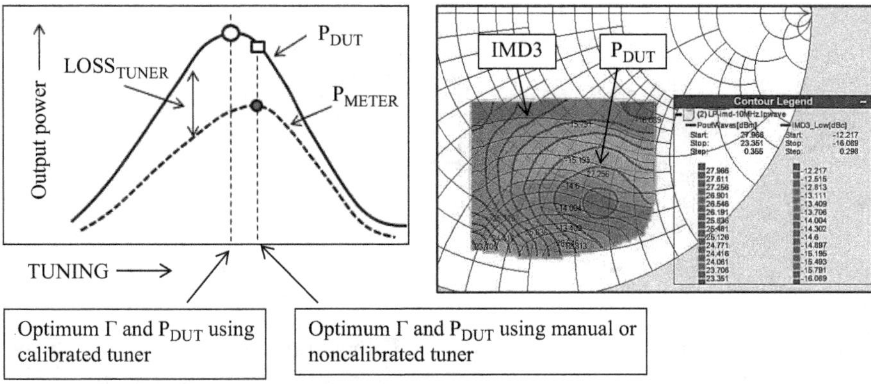

(a) (b)

Figure 3.6 (a) DUT versus system power and (b) ISO power and intermod contours

and the optimum performance of the DUT if the tuner is not precalibrated, that is, if its loss and phase transformation are not instantaneously corrected for, during the measurement. It is like walking with closed eyes.

Whereas still today mostly microstrip prototype amplifier modules are load pull tested, in near future entire wafers, with or without amplifier MMICs, will have to go through this characterization to be qualified, mapped, priced, or rejected.

This means two things:

1. Automated "on wafer" tests and
2. Characterizing chips at high speed (< 1 second/device) at given bias, power, frequency, load, and source impedances are necessary.

A provider of load pull test equipment, which is not at the front line here, will not survive.

3.4 Load pull methods

To do load pull, one has to control the impedance (or reflection factor) presented to the device.

In terms of reflection factor Γ, this is the ratio of reflected to injected power wave into the load at each frequency; assuming a two-port (Figure 3.7), Γ_{load} = <*a*>/<*b*>. <*b*> is created by the two-port (device under test, DUT). By controlling <*a*> we can control Γ.

There are three ways for controlling <*a*>:

1. Reflection on a variable *passive* load;
2. Injection of an *active* signal, coherent to <*b*>; and
3. *Hybrid* combination of 1 and 2.

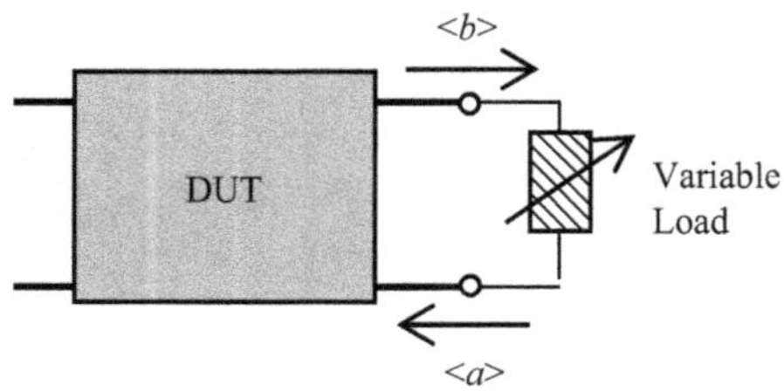

Figure 3.7 Reflection on a variable load

In the case of a passive load (tuner, case 1) the returning signal <*a*> is always smaller than <*b*>, because of losses in the transmission and tuner reflection capacity (tuning range). Therefore $|\Gamma|<1$: all load pull impedance points reside inside the Smith chart.

In the cases of active injection (case 2) and hybrid combination (case 3) the returning signal <*a*> can be larger than <*b*>, leading to the unwanted condition $|\Gamma|>1$: some impedance points would reside outside the Smith chart: ($\text{Re}\{Z_{\text{load}}\} < 0$). Active load pull algorithms should avoid this situation.

The maximum limit of $|\Gamma|$ that can be presented to the DUT is called "tuning range." Tuning range is important, because the majority of microwave power transistors have low internal output impedance R_i, in the range of 1 to 3 Ohms. This corresponds to reflection factors $|\Gamma|$ of 0.96 to 0.89 (VSWR between 50:1 and 17:1; VSWR $= 50/R_i$; $|\Gamma| =$ (VSWR $-$ 1)/(VSWR $+$ 1).

3.5 Reflection on a variable passive load

A variable passive load is an impedance tuner. The established technology for such devices is the "slide screw tuners"; these are made of a slotted, low loss airline, mostly in form of a parallel plate airline (slabline), inside which reflective probes are moved. The probes are made of metal or metallized plastic, are capacitively coupled with the center conductor and create a variable capacitive load and strong field deformation. The probes have a concave bottom which matches the cylindrical center conductor (Figure 3.8).

The wings of the bottom contour of the tuning probes extend below the center of the center conductor to capture a maximum of the electric field, which is concentrated between the center conductor and the side walls. Approaching the probe to the center conductor vertically controls the amplitude of Γ and moving it horizontally controls the phase Φ ($\Gamma = |\Gamma|\exp(j\Phi)$). The overall accuracy of the system relies on mechanical repeatability of positioning the tuning probe. The mechanics of such tuners is ambitious, because sufficient accuracy and mechanical repeatability at high VSWR is possible only if the small gap "S" between probe and center conductor (of the order of 50 μm—one hairwidth) can be kept constant over horizontal probe movement of more than one-half of a wavelength at the lowest frequency of operation ($\lambda = 300$ mm/freq(GHz)), $\lambda/2$(1 GHz) $= 150$ mm; this is required in order to be able to cover 360° of reflection factor phase control; or a free (not regulated) XY movement control tolerance of 50 ± 5 μm/15 cm ≈ 67 ppm.

Figure 3.8 shows a tuning probe in the slabline and Figure 3.9 shows typical tuner response.

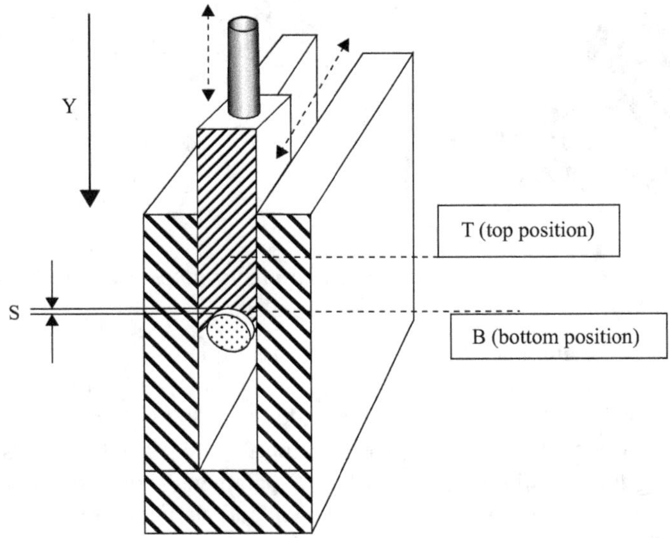

Figure 3.8 Tuning probe

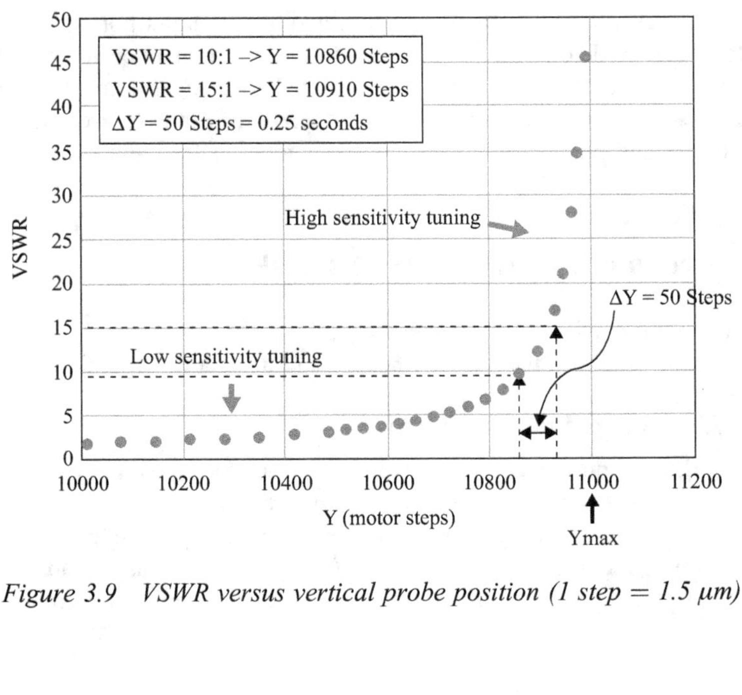

Figure 3.9 VSWR versus vertical probe position (1 step = 1.5 μm)

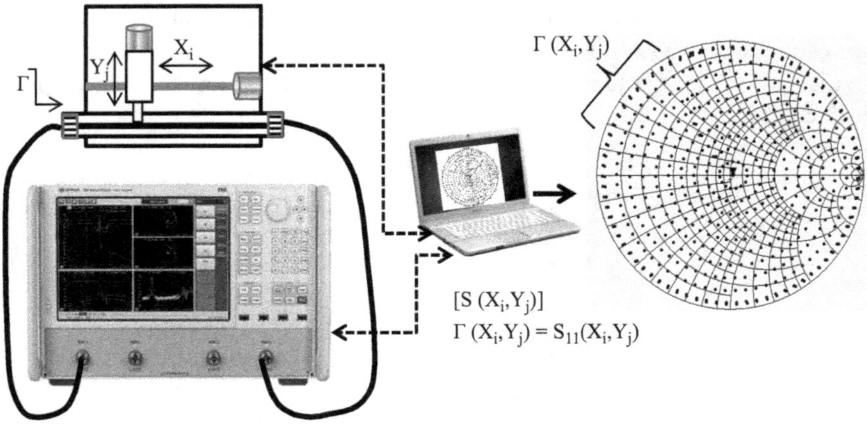

Figure 3.10 Setup and typical results of tuner calibration

The tuners are automated using remotely controlled stepper motors and gear. All tuner operations are described logically using motor steps (X_i horizontal, Y_j vertical). Typical mechanical resolution is 2,000 to 5,000 horizontal positions and 1,000 to 2,000 vertical positions, corresponding to 2 to 10 million possible tuner states. During tuner calibration on a VNA, S-parameters for a fraction of above states is measured and saved. The horizontal and vertical probe positions are selected such as to cover the Smith chart homogenously and are saved in calibration files for each frequency. Each calibration file contains, typically, from 400 to 1400 points. 2D interpolation routines allow synthesizing the several millions of possible tuner states with vector accuracies exceeding 40 dB (or 1%) (Figure 3.10).

During measurement, the tuner is connected to the output port of the test fixture holding the DUT, or, using a cable or a low loss airline extension to the wafer probe, and the output of the tuner is connected to a measurement instrument. Data collected are de-embedded to the DUT reference plane (corrected for tuner and setup loss and phase offset from the tuner to the DUT) and saved in a load pull file.

3.6 Injection of coherent (active) signal

There are a number of basically distinct ways for creating a "virtual" load, that is, creating a return signal at the DUT output without using a real (passive) reflection.

3.6.1 The "split signal" method

This method, first introduced by Takayama [3] consists in using a power splitter after the input signal source and sending part of the signal into the DUT input and part into the output, after amplifying and phase-controlling it (Figure 3.11).

Takayama's method allows $|\Gamma| = |\langle a\rangle/\langle b\rangle| \geq 1$ and has the advantage of using a single source, but requires continuous attenuation and phase control of the

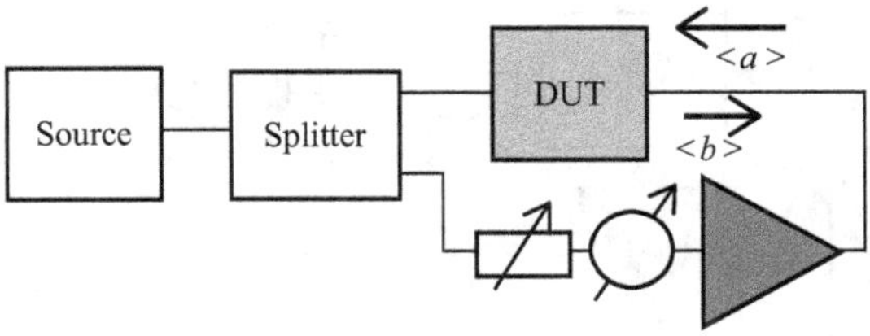

Figure 3.11 "Split signal" method block diagram

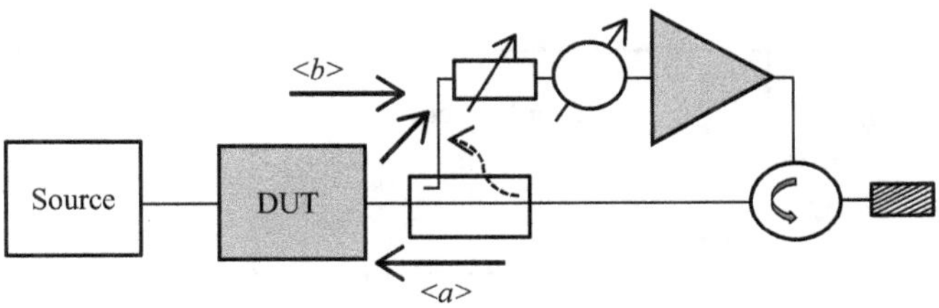

Figure 3.12 "Active closed loop" method block diagram

feedback signal, not available automated. It also needs, if implemented, attenuation and phase control algorithms to keep the load impedance constant when measuring P_{in}/P_{out} saturation plots, since the gain/phase of the feedback amplifier do not track those of the DUT, and by consequence the ratio <a>/<b> does not remain constant. Without instantaneous impedance measurement and fast analog attenuation and phase regulation of the feedback signal, to keep Γ constant during power sweeps, Takayama's method is of no use, but in the 1970s it opened the door to active injection load pull.

3.6.2 The "active load" method

This method is a closed loop configuration, and uses a coupler at the output of the DUT, sampling part of the signal and reinjecting it into the DUT, after amplifying and adjusting its amplitude and phase, same as Takayama (Figure 3.12). This method is relatively simple, but, because of the closed loop in the active load, it bears risks of spontaneous oscillations of the feedback loop through the directivity (leakage) of the coupler. If the feedback amplifier is large enough, saturation plots will not need autocorrection algorithms, as Takayama. This simplifies the implementation.

3.6.3 "Open loop" active injection

This method requires a second signal source, which is synchronized with the primary source. Depending upon the second source being amplitude and phase controlled, an extra variable attenuator and phase shifter may or may not be required in the feedback injection loop (Figure 3.13). Since synchronization of both sources occurs at a base frequency around 10 MHz, we can speak of an "open loop" system.

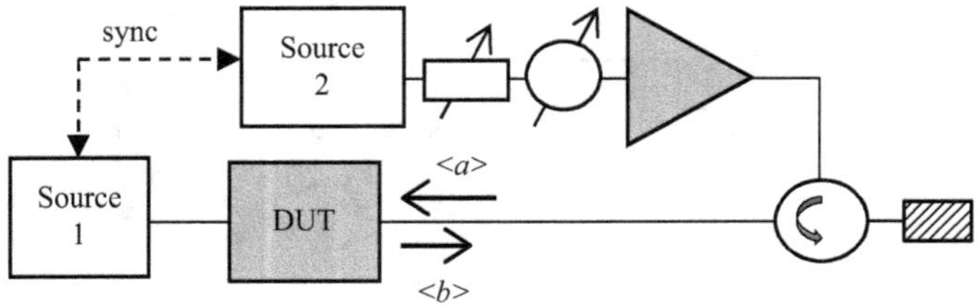

Figure 3.13 "Open loop" method block diagram

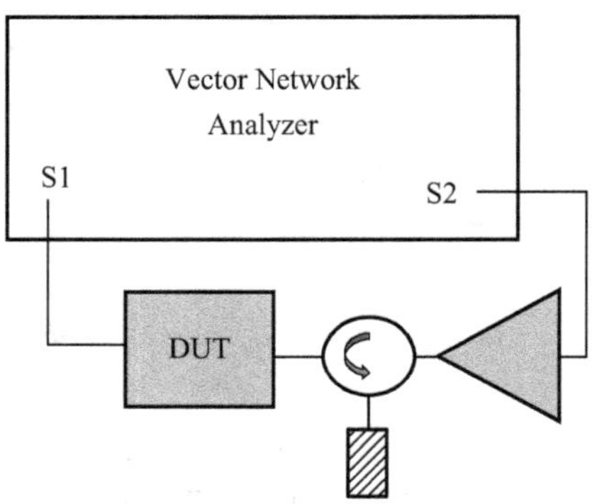

Figure 3.14 "Open loop" method block diagram using a vector source

If the second source is adjustable (vector), like an internal second source of a VNA, the external attenuator and phase shifter can be spared (Figure 3.14). Isolators are used to protect the feedback amplifiers.

3.6.4 "Hybrid" combination

A basic shortcoming of the "pure" active solutions is the requirement for high power feedback amplifiers, for the simple reason that the amplifiers, including their protection isolators, have an internal impedance of 50 Ω; instead the DUT has an internal impedance R_i of 1–3 Ω. We obviously face a strong mismatch situation here, that requires excessive feedback power to overcome (Figure 3.15): $Vo/Vi = (50 + R_i)/R_i$ or $P_o/P_i \approx (50\ \Omega/R_i) + 2$; assuming $R_i = 3\ \Omega$ gives $P_o \approx 18.7\ P_i$ or 13 dB higher. In case of $R_i = 1\ \Omega$ this ratio becomes ≈17 dB. Or, in order to inject 1 Watt into the DUT we would need ≈19 W or 52 W, respectively.

This, obvious, bottleneck, can be alleviated using a "hybrid" configuration.

In this case some form of impedance transformer is needed to prematch the amplifier's internal impedance to the DUTs: (a) employing a prematched test fixture (using either single stage $\lambda/4$ transformer, or, for higher bandwidth, multistage transformers or Klopfenstein wideband ramped transformers) or (b) using adjustable transformers in form of impedance tuners. Fixed transformers are usually made on

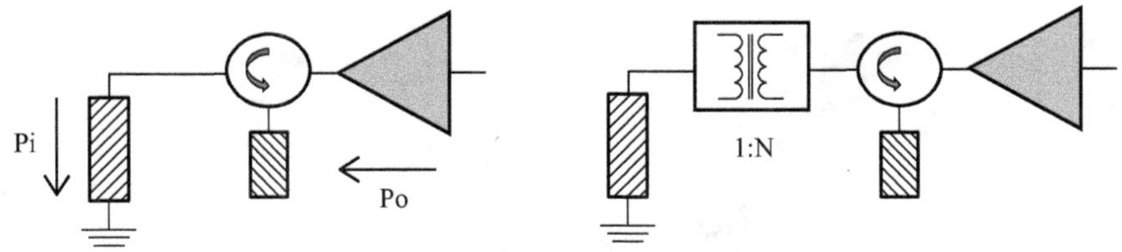

Figure 3.15 Active injection

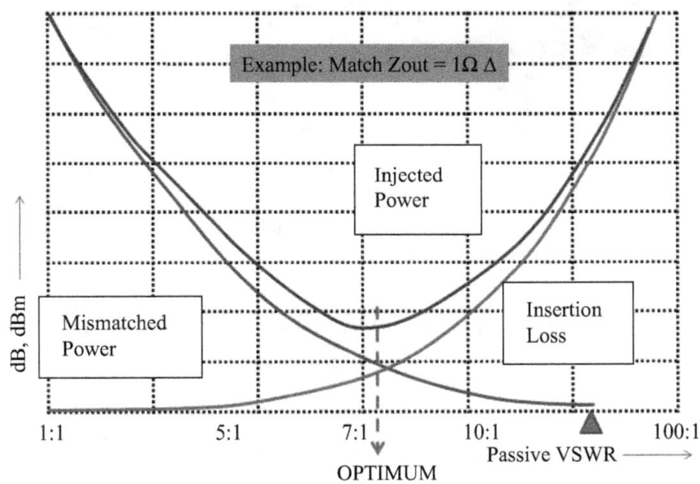

Figure 3.16 Requirement for injected power as a function of insertion loss and tuner prematch

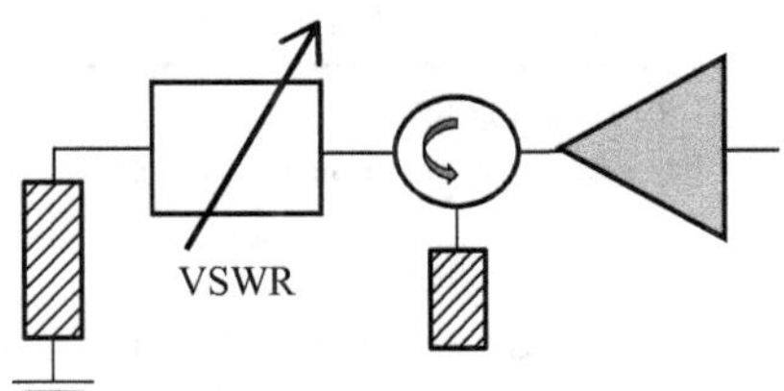

Figure 3.17 "Hybrid-active" method block diagram

microstrip substrates, are cheap and low loss, but un-flexible. They must be re-designed and made for each device and frequency range. Tuners are adjustable and wideband, but have, at high VSWR, considerable loss. When employing tuners and minimizing the feedback power, one has to compromise between transforming ratio (VSWR) and tuner loss (Figures 3.16–3.18) (Table 3.1).

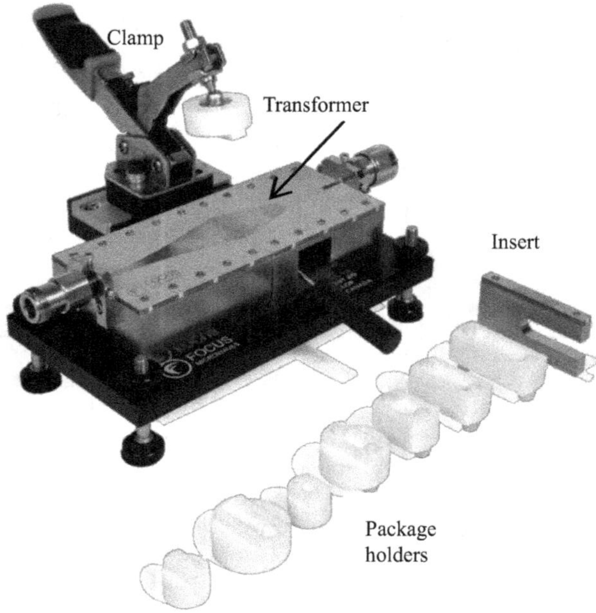

Figure 3.18 Klopfenstein test fixture 10 Ω, 1–10 GHz

Table 3.1 Comparison of tuning methods

Tuning method	Complexity (**)	Cost	Gamma	Speed	Bandwidth	Power	Modulated signals
Passive	Low	Med	Med	Low	High	High	No
Active	High	High*	High	High	Low*	Med*	Yes
Hybrid	High	High	High	High	Low*	High	Yes

*Due to cost and band limitations of power amplifiers.
**Active systems require careful handling to avoid damaging the active component (DUT).

Several experiments with impedance tuners at various frequencies have shown that an optimum VSWR for maximum power transfer between the amplifier and the DUT is approximately 7:1 (Figure 3.16).

In this case the realistic power of the amplifier, needed to match the DUT output power is, considering losses, mismatch, fixtures, etc. approximately double (or ~3 dB higher) than the DUT power. Or, in the case of $R_i = 1\ \Omega$, the impedance mismatch of 50:1 becomes 7:1 × 7:1 and in case of 2 Ω from 25:1 it becomes 5:1 × 5:1.

It is clear that a hybrid system, that can be operated also as simple active or simple passive system***, combines most advantages, but at a price. The best systems, from the user point of view, are systems that can be upgraded, from active or passive to hybrid.

(***) this operation offers easy confirmation of system accuracy: at the same *pure active* or *pure passive* load the measured result must be the same.

Therefore, depending on requirement and budget, the choice of system is easier.

3.7 Impedance tuners

The equipment required to manipulate the impedance presented to the DUT (device under test, mostly transistor) are impedance tuners. Impedance tuners are used to scan a large area of the Smith chart, because power is not the only characteristic of interest. Many other parameters, like efficiency, intermodulation, adjacent channel power, EVM, and many more are meanwhile critical for multichannel communications.

We can distinguish three basic tuner types: (a) passive, (b) active, and (c) hybrid (combination of active and passive tuners).

Each tuner type has its advantages and shortcomings, in particular when the practical needs of industrial users are concerned. In any case the main interest is for automatic or programmable tuners, that is, tuners that can synthesize any specific impedance, on request. Manual, or not programmable tuning, resembles a process of "trial and error." An additional shortcoming of manual tuning lies in the fact that the operation cannot optimize the DUT itself, rather it allows optimizing the assembly "tuner + DUT" without the possibility of actual phase and amplitude correction. In short, manual tuning does not allow finding the real optimum DUT performance (Figure 3.6).

3.7.1 Passive tuners

Historically [1] passive RCA tuners were made of two dielectric $\lambda/4$ long donut-shaped dielectric resonators travelling on the center conductor of a slotted airline (Figure 3.19). Changing the distance between the resonators creates a butterfly kind of trajectory, whereas both resonators moving along the airline change the angle of Γ. The tuner has the advantage of easy manufacturing and alignment. It is bandwidth-limited (1 octave) by nature and 50 Ω is not its natural retreat state, instead 50 Ω is an impedance to be synthesized frequency by frequency. This means test devices are always at risk of oscillation. For these reasons this technology was abandoned.

The present passive tuners are based on the slide screw principle, meaning the reflective (metallic) tuning probe was moved vertically toward the center conductor to control the amplitude and horizontally along the slabline to control the phase of the reflection factor.

Automatic slide screw tuners were initially made using a slotted impedance bridge, where the E-field sensor was replaced by the capacitive metallic probe (slug) (Figures 3.8, 3.20 [4] and 3.21).

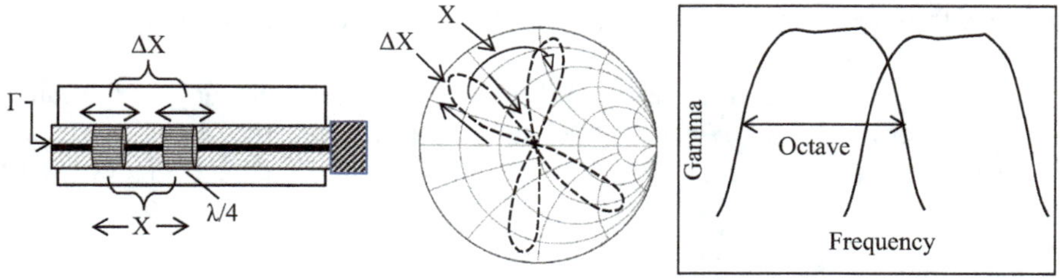

Figure 3.19 The RCA tuner

Figure 3.20 HP 806B universal probe carriages (3–12 GHz)

Figure 3.21 First published slide screw tuner. Courtesy: University of Leeds

The first tuner of this kind was published in 1984 by the University of Leeds in the United Kingdom. The tuner was based on a HP precision slabline bridge model 809B/810B used for measuring standing waves and impedances. Meanwhile this technology has come a long way. From the original 1.8 to 18 GHz tuners of 1987 we have now (2018) 20–110 GHz coaxial tuners (Figures 3.22 and 3.36).

These tuners behave exactly opposite of the RCA two-donut tuners: (a) they are very wideband and (b) their natural retreat state is close to 50 Ω, but they are demanding in manufacturing precision and alignment.

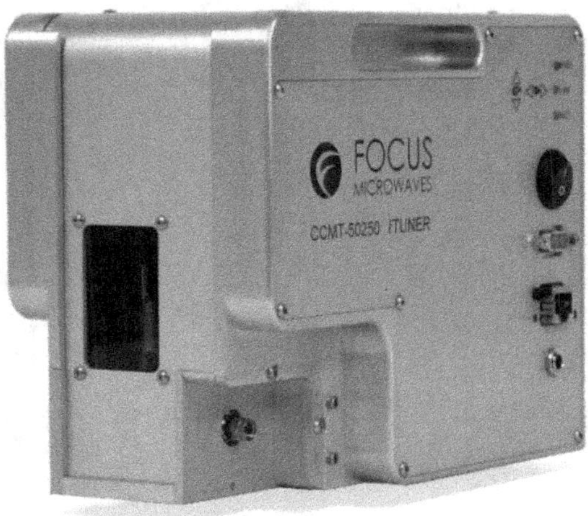

Figure 3.22 Wideband tuner (25–50 GHz, $VSWR_{MAX}$ > 30:1)

3.7.2 Electronic (passive) tuners

A passive tuner, based on fast ON and OFF electronic switching of a series of capacitors (mostly using blocks of PIN diodes followed by fixed capacitors or Varactor diodes) has been introduced in the 1980s by ATN Microwave [6] and used initially for noise figure and noise parameter measurements (Figures 3.23 and 3.24). Those tuners were very fast and repeatable but came out of fashion since around

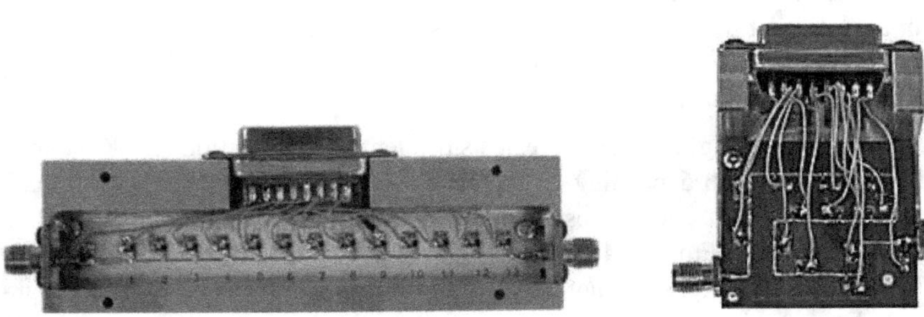

Figure 3.23 Electronic tuner prototypes with 14 diodes mounted at equidistant and nonequidistant spaces along the microstrip transmission line

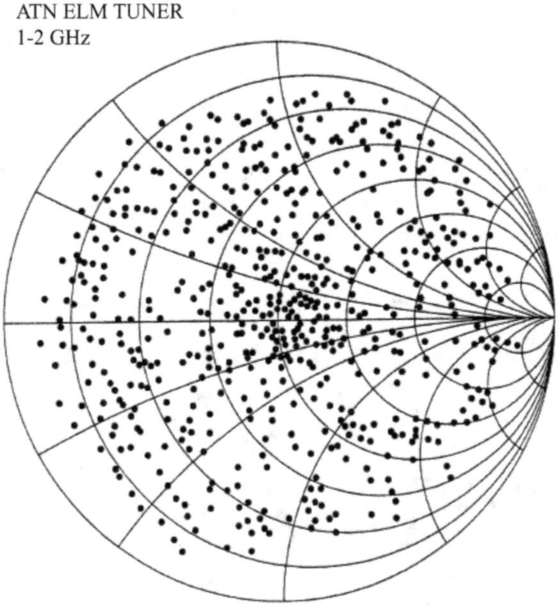

Figure 3.24 Example of tuning pattern of electronic tuner at 1.5 GHz [5]

Table 3.2 Electronic versus electro-mechanical tuners

Passive tuner type	Complexity	Cost	Gamma	Speed	Band-width	Power	Modulated signals
Electronic	High	High	Med	High	Med	Med	No
Electro-mechanical	Low	High	High	Low	High	High	No

2000, because of power, tuning resolution and bandwidth limitations, and the concurrent evolution of mechanical tuner solutions (Table 3.2).

3.7.3 Wideband tuners

Each tuning probe (slug) in a tuner creates sufficient high reflection (>0.85) over a certain frequency band; depending on the detail structure, the manufacturing precision, and the alignment effort invested, a single probe can cover a range of up to 2, even 3 octaves (Fmax:Fmin = 4 or 8).

Tuners combining two or more probes in the same slabline can reach instantaneously Fmax:Fmin ratios of up to 45 or 50 (e.g., three slug tuner covering 0.4 to 18 GHz or 1 to 50 GHz, Figures 3.25 and 3.26). Other examples of wideband tuners operate instantaneously from 2 to 67 GHz, etc. The final limitations of the technology

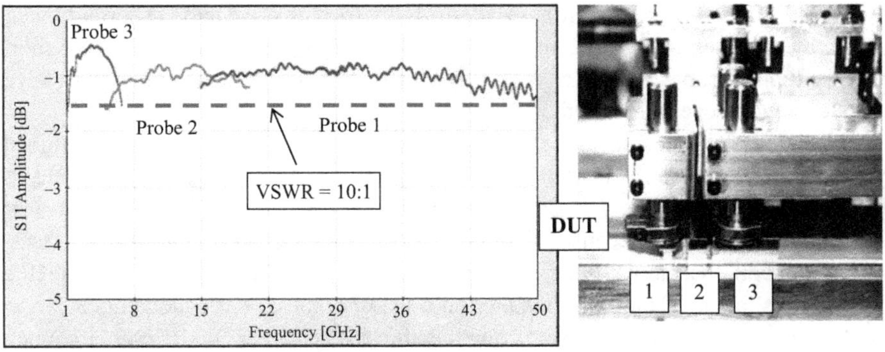

Figure 3.25 Maximum reflection of wideband three-probe tuner and associated mechanics

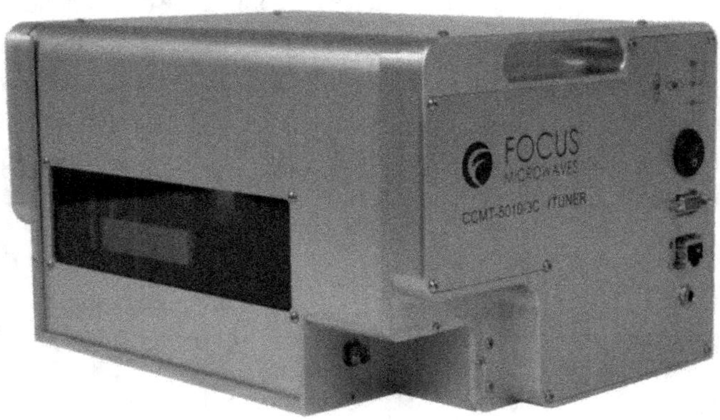

Figure 3.26 Three-probe tuner covering 1 to 50 GHz

are (a) the width of the slug which limits the maximum obtainable capacitance at low frequencies, (b) the cutoff frequency of the transmission line (slabline) which requires smaller structures and center conductors with smaller diameter, thus wider tuning probes, causing an upper frequency limitation, because of self-resonance. Decade-long simulations and trial and error experiments have brought the technology to its limits with what is available today (Figures 3.26 and 3.36).

3.7.4 High power tuners

High power transistors are, usually, made by paralleling several cells. This, obviously, reduces the output impedance, down to 1 Ω or less. It is therefore of outmost importance to be able to characterize (or conjugate match) such devices. As can be seen from Figure 3.9, reaching the high VSWR (50:1) required using slide screw tuners means bringing the tuning probes (slugs) very close (30–50 μm) to the center conductor. At this point the electric field becomes very high, reaching values close to sparking

(Corona discharge, or 3 kV/cm). In addition, the center conductor absorbs power, caused by RF tuner loss and DC resistance, heats up deforms, creating a galvanic short with the tuning probes. This is a typical problem with tuners operated at very high RF and DC power (several hundred of Watts).

Heat can be created on the center conductor itself or at worn out or badly aligned connectors which feed the heat to the center conductor. APC-7 and SMA connectors are to be avoided versus 7/16, or 3.5 mm N type connectors; APC-7 because of sensitive contact, SMA because of generally poor quality of the connecting surfaces.

In general, harmonic tuners are more sensitive to high power because the "slenderness factor" (length-to-diameter ratio L/D), of the center conductors is larger and the mechanical stability lower. Long metallic rods "buckle" under pressure much easier than short ones. The rod stability decreases with $(L/D)^2$ [7].

To avoid such phenomena, it is advisable to use prematching test fixtures comprising the DC bias circuit and operate the tuners at lower VSWR or use special high power tuners. Focus has developed low thermal expansion center conductors, airflow, and mineral oil submersion techniques for cooling the center conductor [8–10].

In general power handling of slide screw tuners decreases with frequency and VSWR for obvious reasons (Figures 3.27 and 3.28):

1. High frequency means thinner center conductor and high RF tuner loss and
2. High VSWR means high tuner loss and small gap between tuning probes and center conductor (Figure 3.8).

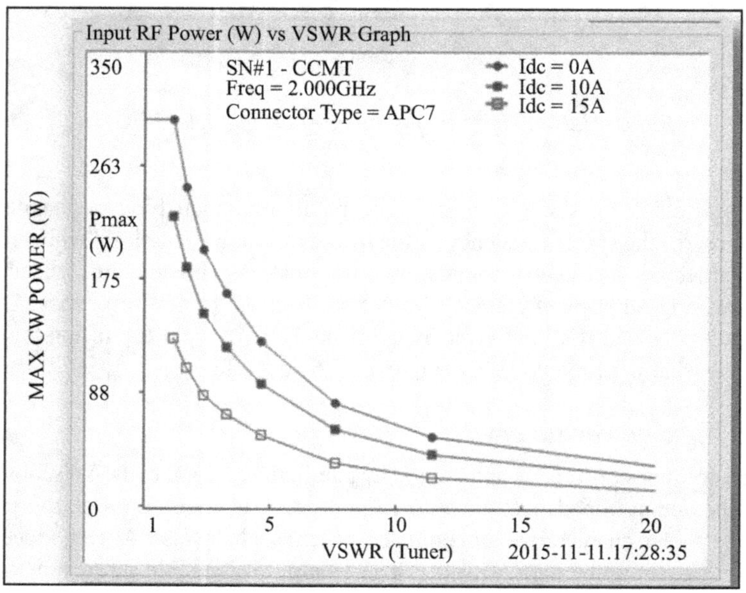

Figure 3.27 Example of tuner power handling

CCMT-303-UHP (0.3-3GHz)

CCMT-975-UHP
(7.5-9.5GHz)

Figure 3.28 Examples of 500 Watt (DC + RF CW) power tuners

3.8 Harmonic load pull

Harmonic load pull is the measurement method whereby the impedances presented to the DUT at harmonic frequencies 2fo, 3fo, etc. of the fundamental injected frequency fo are controlled. To do so one needs either a test fixture with built-in harmonic loads, such as $\lambda/4$ parallel stub resonators (also called harmonic traps, because they stop harmonic energy to propagate and they reflect it back into the DUT) or harmonic frequency tuners. Again, harmonic tuners can be active, passive, or hybrid. In the case of active harmonic tuners at least one harmonic frequency signal must be injected into the output of the DUT. Of course, the combination of fast active tuning with passive power prematching is, conceptually, the optimum solution. And, again the best alternative from the user point of view are "upgradable" systems, from pure passive or active to hybrid. One of the least complex harmonic load pull solutions, combining high speed active tuning with reduced injected power is the combination of passive harmonic tuner with active fundamental power injection (Figure 3.29).

3.8.1 Passive harmonic load pull using di-tri-plexers

Because of the inherent wideband nature of the slide screw tuners it was, up to circa 1998, thought impossible to make harmonic mechanical tuners. Harmonic load pull employed wideband tuners and frequency discriminators (di- or triplexers) (Figure 3.30).

The method has shortcomings: harmonic triplexers are (a) difficult to find, (b) have limited bandwidth, (c) they have in-band insertion loss, and (d) they have high reflection outside the bands; this reduces the tuning range and creates risk of spurious DUT oscillations. The method has been the only solution in the 1990s but is now practically abandoned.

3.8.2 Harmonic rejection tuners

The harmonic rejection tuner (PHT) [11,12] have been introduced in 2000. The concept is simple: several $\lambda/4$ open stub resonators slide along the center

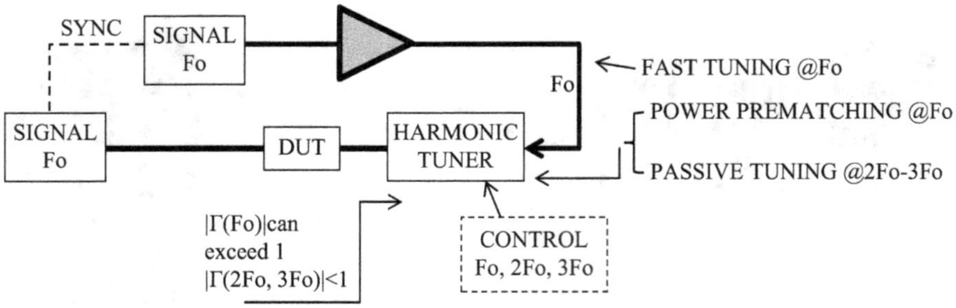

Figure 3.29 Harmonic load pull using focus MPT harmonic tuner

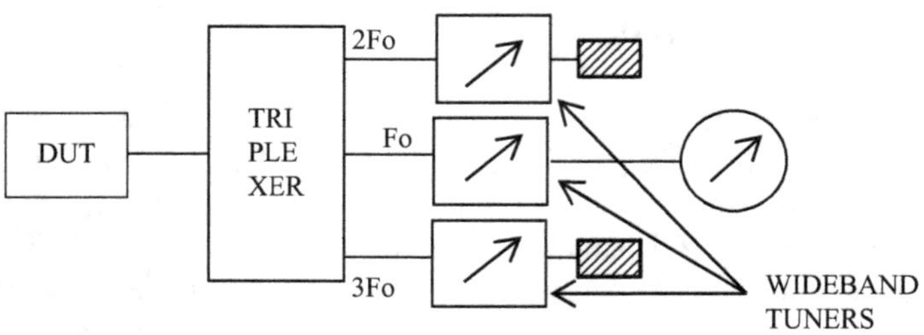

Figure 3.30 Harmonic load pull using di- or triplexers

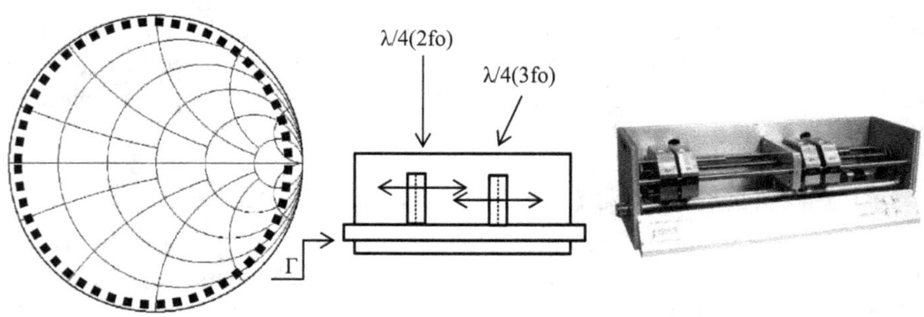

Figure 3.31 Harmonic rejection tuner

conductor of the slabline and create a rotating short at the resonant (harmonic) frequency. That is, we can present to the DUT a short with adjustable phase at a harmonic frequency with limited effect on the fundamental, which is sufficient for most applications. There are two problems: (a) the frequency coverage is only around 5% and (b) tuning at 2fo affects tuning at fo, because of the not high enough Q factor of the used components. The problems were solved using double resonators, fundamental back-tuning, and simple manual exchange of resonators [13] (Figure 3.31).

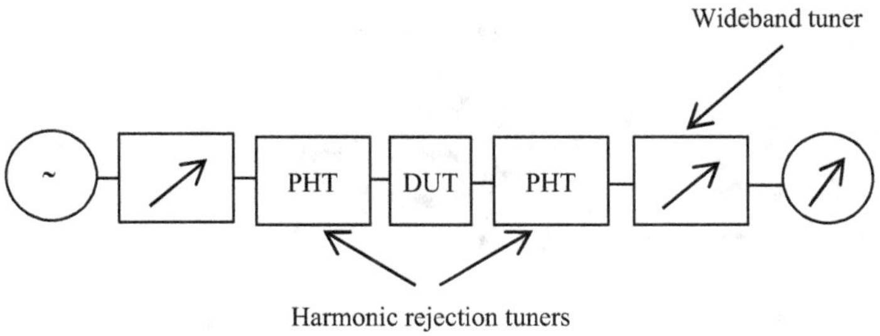

Figure 3.32 Harmonic load pull using Focus PHT harmonic tuners

Harmonic rejection tuners are inserted between DUT and wideband tuner to catch any harmonic power (2fo, 3fo, …) created by the DUT and reflect it back into the DUT, while allowing the fundamental wave (fo) to travers freely (Figure 3.32). They have been popular between 2000 and 2008; several users own dozens of resonators and replace them when switching frequencies. The tuners do not cause spurious oscillations (their low frequency reflection is zero) and reduce the fundamental tuning range only moderately (because of low insertion loss of the airline).

A typical PHT test setup is shown in Figure 3.32.

3.8.3 Wideband multiharmonic tuners

In 2004 the new concept of multipurpose tuner (MPT) [14] was introduced. This tuner uses three independent wideband probes (slugs), as shown in Figures 3.33 and 3.34. It was developed as a mechanically stable unit, to be operable on wafer without breaking the wafer-probes by tilting, caused by horizontal movement of the massive carriages. By placing the three probes 120° apart we can tune by moving the slugs only vertically. Because the slugs are wideband, this device could also operate at different frequencies by re-arranging the horizontal positions. But still, this is all single frequency.

Multiharmonic tuners are possible using wideband nonresonant probes. Based on an idea first introduced by ATN Microwave [15] using an electronic, PIN diode-based tuner, a search algorithm allows independent tuning at two or three harmonics out of millions of possible combinations, that is, finding tuner settings synthesizing simultaneously two or three user defined harmonic impedances. First PIN diode-based devices were crude and targeting impedances were unfocused. Harmonic rejection tuners [12] mentioned above displaced harmonic electronic tuners on the market in the early stages (approx. year 2000).

Then, later in 2004, came the idea of using the trillions of possible tuning states, available because of the permutations of horizontal and vertical resolutions of the mechanics of three independent slugs [16]. Typical maximum numbers of motor steps are 2,000 in vertical and 6,000 in horizontal direction, or 12×10^6 per slug or, for three slugs, $12^3 \times 10^{18} \approx 1.7 \times 10^{21}$ combinations per frequency. It was heuristically assumed that this kind of number of states would include solutions for tuning accurately enough at three frequencies anywhere on the Smith chart.

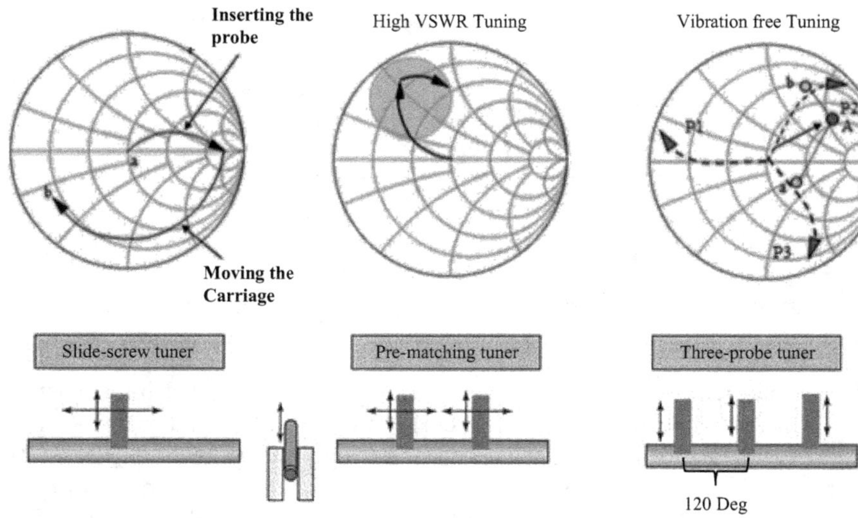

Figure 3.33 Basic operation of single, double, and triple probe tuner

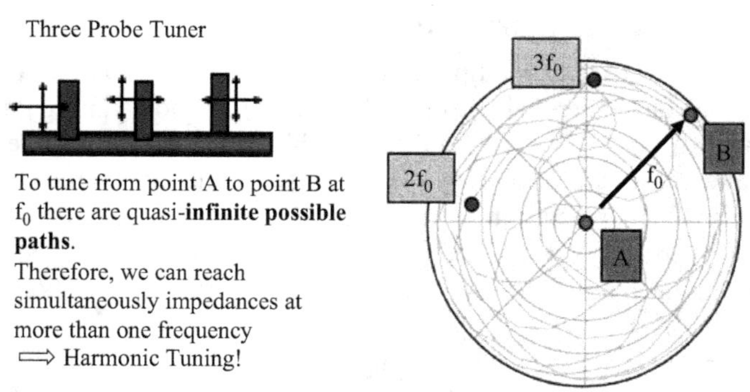

Figure 3.34 Wideband frequency response (S11) of MPT

All experience to date shows this to be true, though we do not have analytical proof of that. Figure 3.34 shows the complex multiple-reflection trace of the multiprobe tuner reflection factor as a function of frequency. The tuning search routine must find the path from A to B, which includes the arbitrarily defined harmonic impedances Γ(2fo) and Γ(3fo).

More than that, it has been experimentally shown that, a two carriage-two wideband probe tuner tunes everywhere on the Smith chart and within the frequency coverage of the wideband probes at two harmonic frequencies and that four

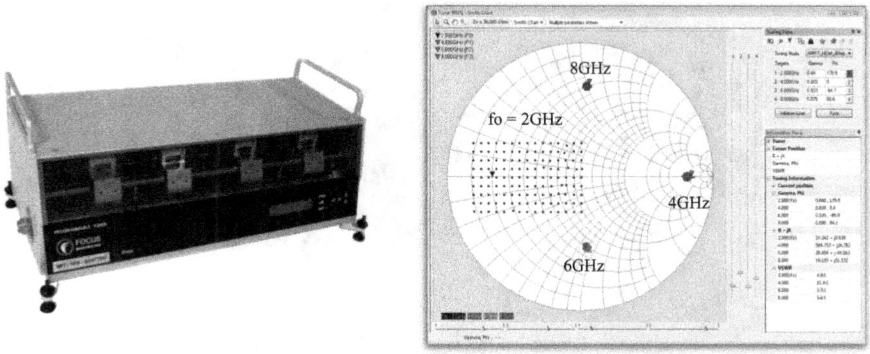

Figure 3.35 Four-probe four-harmonic frequency tuner (1.2–10 GHz)

probes tune impedances at four-harmonic (or not) frequencies (Figure 3.35). In fact, two probes can tune at three frequencies, but leave blank areas on the Smith chart, the same with three probes and four frequencies. Why is that so? We do not know analytically but it is related to the phase difference between reflection factor vectors.

Finding tuner states for harmonic impedances all over the Smith chart, is possible using today's high-speed personal computers within 2 seconds, with a vector accuracy exceeding 40 dB (1%). Because of the necessity to move three probes, real tuning time is longer, up to 10 seconds, faster with increasing frequency, because of shorter tuner movements. Multiple models of such tuners are available for frequencies between fo = 200 MHz (3fo = 600 MHz) and fo $\leq$ 36 GHz (3fo $\leq$ 108 GHz) over large instantaneous bandwidths. These tuners can be operated also as single frequency wideband tuners from (0.2 GHz < fo < 6 GHz) to (20 GHz < fo < 110 GHz) (Figure 3.36) or dual frequency tuners (fo and 2fo), in which case fo can reach 55 GHz. The harmonic tuners can create all harmonic impedances continuously at all frequencies inside the operation band without hardware interventions, other than with PHTs.

A useful application of multiple probe tuners is single frequency high Gamma tuning. If one slug is used as a prematching section, the second slug can tune around it and reach very high reflection factor in a selected area of the Smith chart (Figure 3.33). Reflection factors as close to 1 as 0.99 (VSWR $\approx$ 200:1) at tuner reference plane have been reached at low frequencies.

A critical reader, would wonder, how this, quasi-infinite, number of tuner states are calibrated. This is possible using the de-embedding tuning algorithm [14], in which case for 1,200 calibrated points per tuning probe, instead of $1{,}200^3 \approx 1.7 \times 10^9$ calibration points one only needs to calibrate $3 \times 1200 = 3{,}600$ points.

3.8.4 *Low frequency tuners*

Slide screw tuners are big and heavy. They are at least one-half a wavelength long at the lowest frequency, plus some fringe components (carriages, walls, and

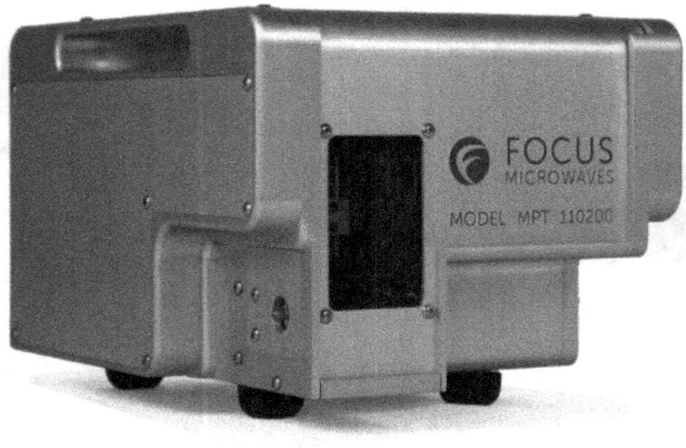

Figure 3.36 Example of three harmonics tuner (20–110 GHz)

connectors). At 100 MHz this is close to 1.6 meters. Multiprobe (harmonic) tuners are twice or three times as long. In view of the required mechanical precision and available laboratory space, this becomes unbearable. Therefore, a new type of tuner, the low-frequency tuners (LFT), was developed [17,18]. This type of tuner uses parallel blade rotary capacitors as adjustable components and sections of semirigid coaxial cable between them. The frequency coverage of such tuners can be optimized to more than one octave. Three or four capacitor stages are necessary. These tuners solve the size and associated mechanical precision, alignment effort and laboratory space requirement problems. Harmonic tuners on the same base are in development. Figure 3.37 shows a four capacitor 60–130 MHz tuner in a 19″ rack, and calibrated points at 20 MHz; also: (a) 5–26 MHz, (b) 10–52 MHz, (c) 95–170 MHz maximum $|\Gamma|$ Smith chart coverage (tuning range) over frequency, and (d) tuned (synthesized) load pull impedances at 12 MHz.

3.8.5 Special tuners

Beyond limited tuning speed, compared to all other solutions, electro-mechanical tuners have superior performance such as high power handling and multioctave bandwidth, but insertion loss limited tuning range. Whereas tuning range of the tuners themselves is sufficient (VSWR $\geq$ 30:1), when integrated in wafer setups the loss of the cables and probes reduces the VSWR values to $\approx$6–7:1, which is often insufficient.

Extended slablines (insert in Figure 3.38) have alleviated the problem, because they are less lossy than cables, but do not represent the best compromise, since the tuning probe (slug), which creates the reflection, is still inside the tuner body far away

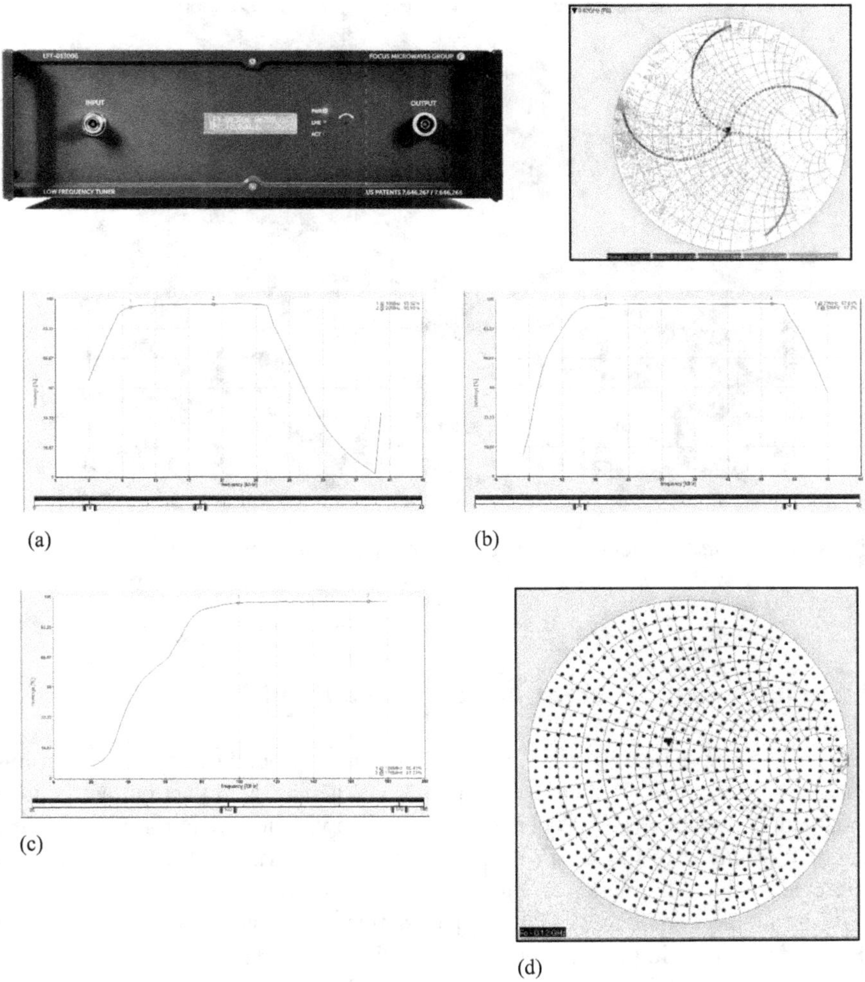

Figure 3.37 Low frequency tuner 60–130 MHz, calibration points, frequency coverage of various models and tuned load pull points at 12 MHz

from the DUT. A new tuner type, the "DELTA" tuner solves this problem: in these tuners, designed mainly for 5G applications (10–110 GHz), the slabline is short and in direct contact with the wafer probe. This allows VSWR at DUT reference plane higher than 10:1, which is sufficient for most applications; assuming, of course, that with increasing frequency the transistors also have higher internal impedance, mostly due to parasitic elements.

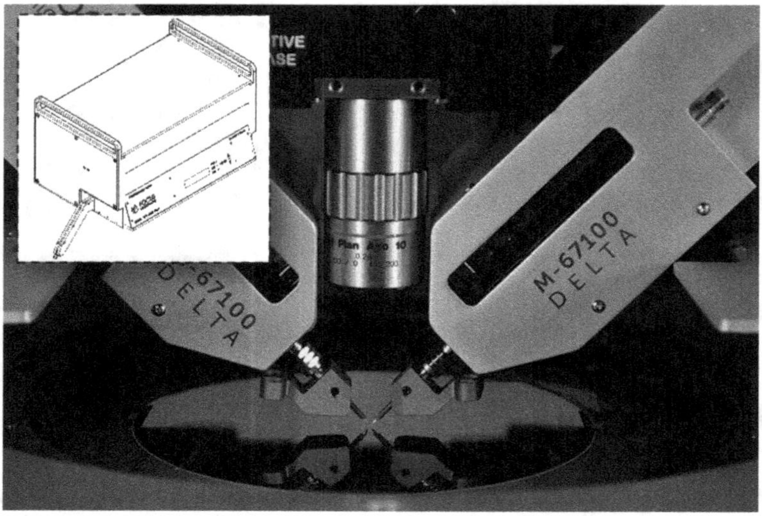

Figure 3.38 *Delta tuners (left: fundamental-wideband; right: three harmonics wideband, 10–67 GHz)*

3.9 Fundamental versus harmonic load pull

Traditional passive tuners, based on the slide-screw principle generate reflections over a large frequency bandwidth (Figure 3.25). Whereas the reflection factor at the fundamental frequency Γ(fo) is calibrated and known, the reflection factors at the harmonic frequencies Γ(2fo), Γ(3fo), etc. may be "known" (if the tuner has been calibrated at those frequencies), but they are not controlled. Since the DUT may create harmonic components of the injected frequency, such control of the load reflection factors is important since they constitute important measurement conditions. This is not obvious, because the effect of "harmonic tuning" depends on the compression level of the DUT, and this, on the other hand, does not only depend on DC bias point and the level of the injected signal, but also on the load impedance at fo as well. It shall not be forgotten that Gain contours, measured in class A using load pull tuners, are the same as gain circles, calculated using small signal S-parameters. It is, quasi, impossible to predict the importance of harmonic tuning. Until the dawn of passive harmonic tuners in 2000 [12], this phenomenon was silently overlooked [19].

Since then, the appearance of true passive harmonic tuners has settled the issue. Control of harmonic impedances is, meanwhile, part of any professional load pull characterization, simply because the effect is not predictable; in short, "if you need load pull, you need harmonic load pull." If you operate in small signal under no compression conditions, then S-parameters are enough (Figures 3.7 and 3.39).

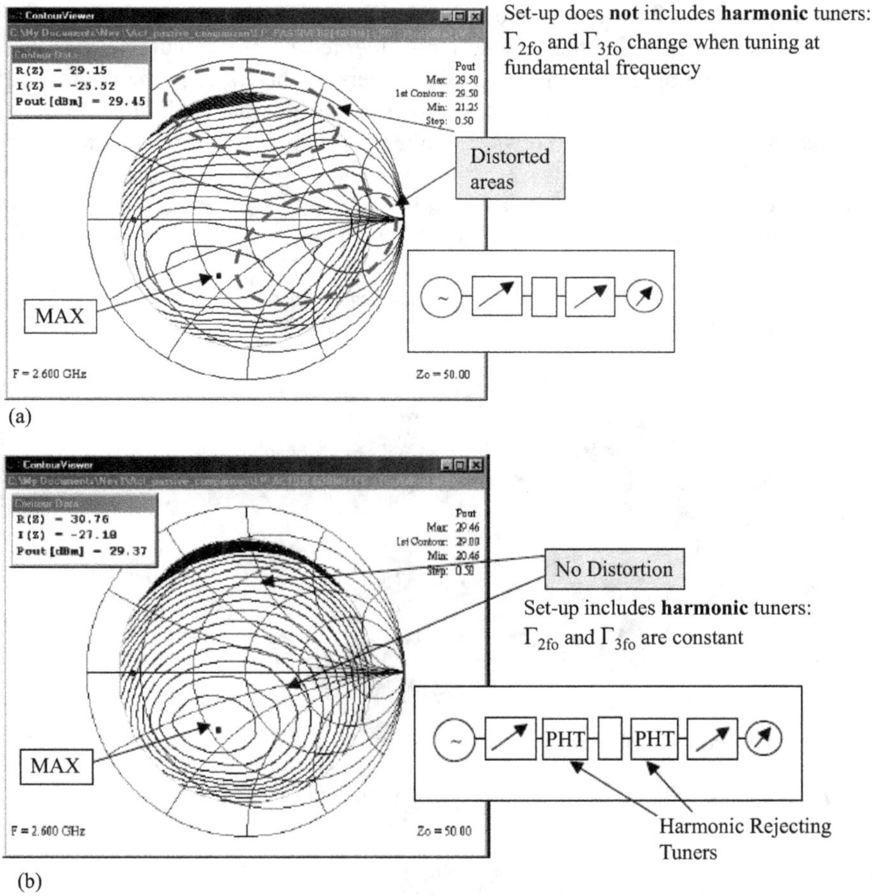

Figure 3.39 Load pull contours of the same transistor, same frequency, and same input power: (a) no harmonic tuning and (b) fixed harmonics load

3.10 On wafer integration

Load pull testing moves inexorably from connectorized RF module testing to on-wafer chip testing. This is the only way to determine the exact, parasitic-free, capacity of the transistors.

As already mentioned, this requires on-wafer high VSWR and high speed testing methods, which is obtainable using active or hybrid tuning and through electronic or active tuning.

Because passive tuning is more stable and easier to trouble-shoot, a number of users prefer this solution. Therefore it is imperative to reduce insertion loss between the wafer probes and the tuning probe inside the tuners (not simply the tuner test port). This is possible using the new DELTA tuner technology (Figure 3.38).

The tuners are placed on three axis positioners on the wafer probe platform. If the tuners are connected with the wafer probes using flexible cable then the tuner body may stay fixed and only the wafer probe itself moves. If the tuners are rigidly connected with the wafer probes (for higher VSWR) then the whole tuner body must be controlled and stabilized. This led to the concept of extended slabline and balance mechanism (Figures 3.40 and 3.41) [20,21].

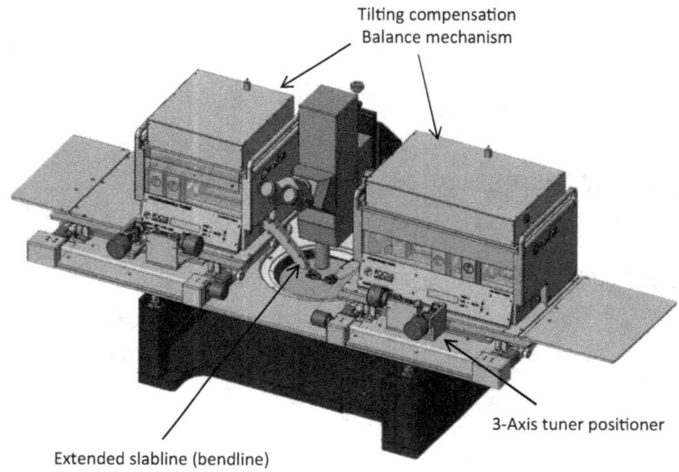

Figure 3.40 Wafer setup using tuners with low loss extended slablines

Figure 3.41 Wafer setup using DELTA technology [patents pending]

3.11 Base-band load pull

The impedance presented to the base-band frequency components affects the overall device measurements. If a, through the down-up mixing behavior of the nonlinear DUT, generated current modulation component at base-band frequency is presented with a high impedance, then it will naturally generate additional control voltage. This generated voltage manifests itself as modulation on the bias lines of the device. Furthermore, as measurements of modulated signals are often conducted with spectrum analyzers that give only scalar information, and such electrical effects usually occur as a device is driven close to compression, they can be misinterpreted as device related memory, when in fact, they are due to the impedance environment presented at the base-band [22,28]. Any modulation on the drain voltage supplied can result in the dynamic characteristic interacting with the device's boundary conditions, causing additional distortion. Introducing low frequency (MHz range) calibrated impedance tuners in the bias lines allows quantifying this phenomenon (Figure 3.42).

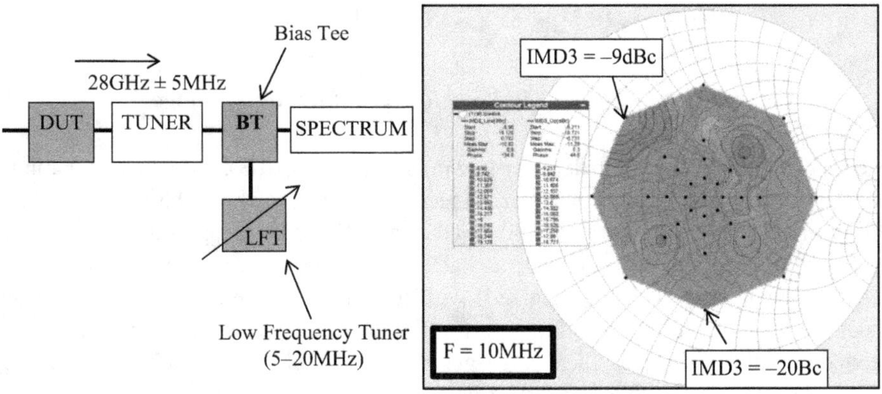

Figure 3.42 Test setup and 28 GHz intermod contours plotted over the 10 MHz Smith chart impedance

3.12 Advanced considerations on active tuning

3.12.1 Introduction

The reflection factors created by the impedance tuners cover, typically, the largest part of the reflection factor plan (Smith chart) (Figure 3.43); appropriate tuning and interpolation algorithms between the calibration points allow creating almost every impedance within the tuning range (shadowed areas) with vector accuracy better than 40 dB; the insertion loss between DUT and tuner reduces the effective tuning range at the DUT ports (small shadowed area, (D) compared with the "tuning range at tuner test port" (C); therefore the actual tuning capacity of such a "passive" tuner system is shown as "tuning range at DUT port"). Many power transistors (DUT)

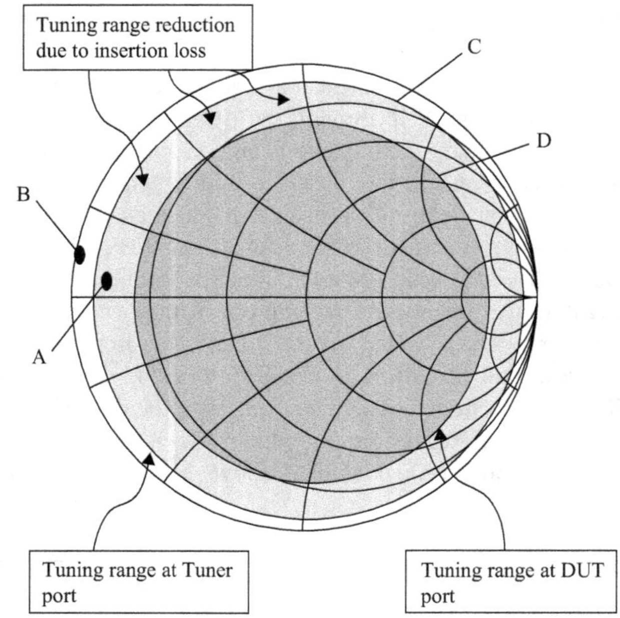

Figure 3.43 The insertion loss effect on tuning range

need to be matched at impedances shown as dots (A) or (B). In case (A) a minimization of the test fixture and interconnection loss might allow the tuner to reach this point. But in many other cases, like in the case (B) this is simply impossible using a passive tuner system. In this case an "active" solution is necessary. In this case a second signal, coherent (synchronized) with the signal delivered to the DUT at its input port, is injected into the output port of the DUT; the phase and amplitude of the second signal is adjusted, such as to generate a virtual load, which can create a reflection factor larger than the reflection factor created by the passive tuner alone.

Traditionally, load-pull or "emulation" of load impedance has been achieved using passive techniques, as discussed earlier; these systems having become vastly more capable, with faster tuning times, increased frequency coverage and harmonic capabilities, still dominate the market. More recently, active systems have begun to penetrate the market along with a clear increase in customer demand; so, what factors drive this shift toward active tuning?

In the past the key advantage of active systems over passive systems was considered to be the ability to overcome the loss between the device under test and the tuner, which meant that any impedance could be emulated. In reality, however, in the vast majority of cases the precision engineered tuners can reach the desired VSWR; in addition, methods such as impedance transformers, using for example Klopfenstein lines (Figure 3.18) to change the system impedance can successfully shift the emulated VSWR circles to allow characterization of even the lowest impedance devices. There are of course still some cases where loss can be an issue, for example at higher frequencies (particularly at harmonics) or where couplers are placed between the test device and the tuner (for measurement of distorted waveforms), Figure 3.3, and here active offers an advantage, but in reality the surge in interest is triggered by two key benefits namely (a) speed and (b) the ability to control impedance over a modulated bandwidth, that are, both, alien to passive electro-mechanical tuner systems.

In terms of speed, advances in test equipment and general computing power along with the development of modular instrumentation platforms such as PXI, allowing for enhanced data transfer speeds, mean that by far the overriding factor in test speed over a passive load-pull setup is the time taken for the mechanical movement of the probe within the tuner, thus moving to an active approach, where the injected signal can be manipulated at a very fast rate, can have a significant impact on the overall test time. This becomes critical at a time with increasing desire for full wafer test.

In terms of bandwidth, the advancement of communications standards has, and continues to be driven by a range of bandwidth-hungry applications, not least the once humble mobile phone, which has been transformed into a personal communications device, providing multimedia wireless solutions including voice, video, graphics, audio, and broad-band Internet access. This advancement has led directly to the creation of several complex wider-band, multicarrier modulation schemes employing, amongst other things, a combination of phase and amplitude modulation, that strive to make the most efficient use of the spectrum to meet demand. As we move closer to the era of 5G and even wider-band signals, the validity of designing for a wide-band end application using only CW or pulsed drive signals is questioned, and users need to move the test scenarios much closer to the end application, or the tuner should be able to emulate the future matching network's response for the entire bandwidth of the modulated signal.

3.12.2 Closed loop (active load)

The first closed-loop active load-pull architecture was presented in 1982 [23]. In this closed-loop architecture, depicted in Figure 3.44, a direct RF feedback loop is employed.

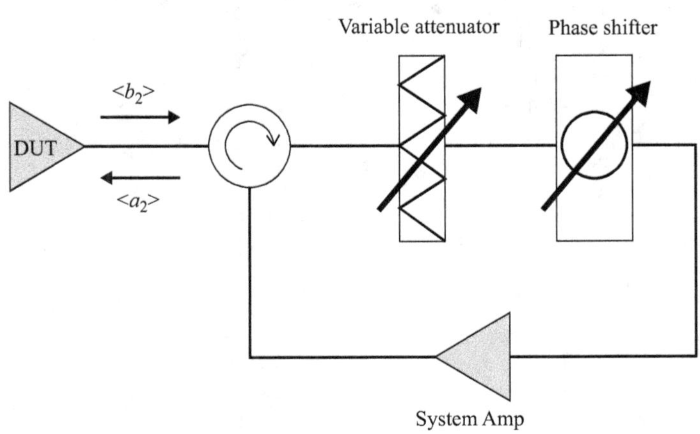

Figure 3.44 Block diagram of a closed loop load pull system

The output of the device is passed through a circulator and is then attenuated and phase-shifted to emulate the desired impedance, before being amplified to overcome the losses in the system. The load-pull (injected) signal is a directly related, modified version of the DUT output signal. This ensures the emulated load impedance to be independent of drive level and any changes in phase of the device output. The use of such systems in industrial environments has however uncovered a major problem with the architecture in that it is prone to instability, due to the closed RF loop employed. Such oscillations are particularly dangerous as they not only destroy the device but could also destroy any measurement equipment in the test set due to maximum input powers of the instruments being exceeded. An attempt to solve the oscillation problem involved a tunable YIG filter, incorporated into the active closed loop, to limit the possibility of oscillation, but the phase change at the resonance frequency is large and creates strong skewing for two-tone or modulated signals.

In terms of modulated capabilities, passive tuners and closed loop architectures have a severe limitation, as delay around the loop over a modulated bandwidth cannot be compensated, leading to reflection factor phase spreading or impedance "skewing" over a modulated bandwidth (Figure 3.45), following the formula $\Delta\Phi^\circ = 0.024 \times L_{el}(\text{cm}) \times \Delta f(\text{MHz})$, whereby L_{el} is the length of the loop; in fact the impedance spread is even greater that that created by passive tuners, with measurements on a commercial setup suggesting that this spread can be as high as

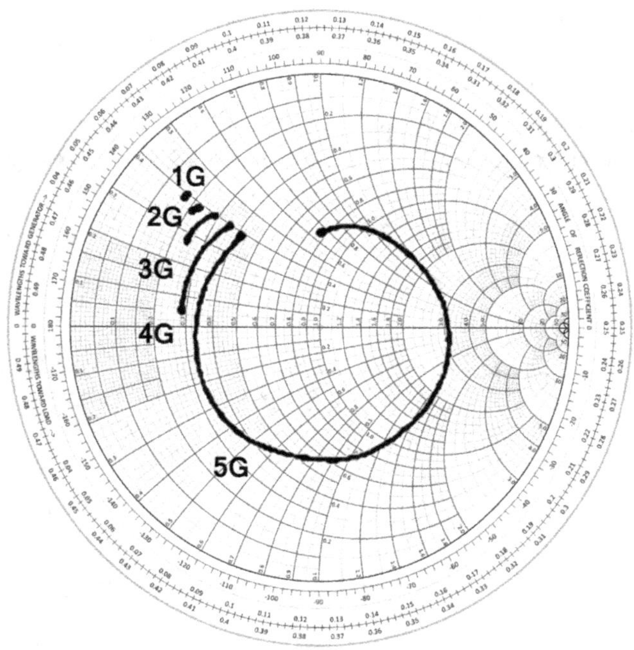

Figure 3.45 Possible "skewing" for the various modulation schemes

30°/MHz in the impedance space, compared to around 3°/MHz for a typical passive setup causing misleading values for important figures of merit such as output power, efficiency, and adjacent channel power ratio (ACPR). Figure 3.45 depicts "skewing" for the various modulation schemes.

3.12.3 Open loop—split signal

Due to the potential oscillation problems with the closed-loop architecture discussed above, active load-pull systems in open-loop configuration have been developed. Open-loop architectures do not rely on feedback of the device output signal, instead they work by generating a new signal to "inject" into the output of the test device to emulate load reflection. The concept was first demonstrated by Takayama [1] (Figure 3.46); it uses a split signal technique whereby the signal is split with one-half used to drive the test device, the other half is modified in terms of magnitude and phase and injected into the output of the test device.

By its nature, this method avoids the stability problems associated with closed-loop feedback architectures, as there is no longer a direct RF feedback path. There is however no longer any relationship between the device output and input-derived load-pull signals. Any change in the reflected power wave changes the output state of the device, thus changing the achieved load impedance. This leads to an iterative process where a number of magnitude and phase adjustments of the injected signal are required to maintain the target reflection coefficient.

The open-loop architecture was improved using an I-Q vector modulator, thus providing a solution that was both easier to automate. A further extension came with the introduction of microwave signal generators with phase lock-capabilities (Figure 3.47).

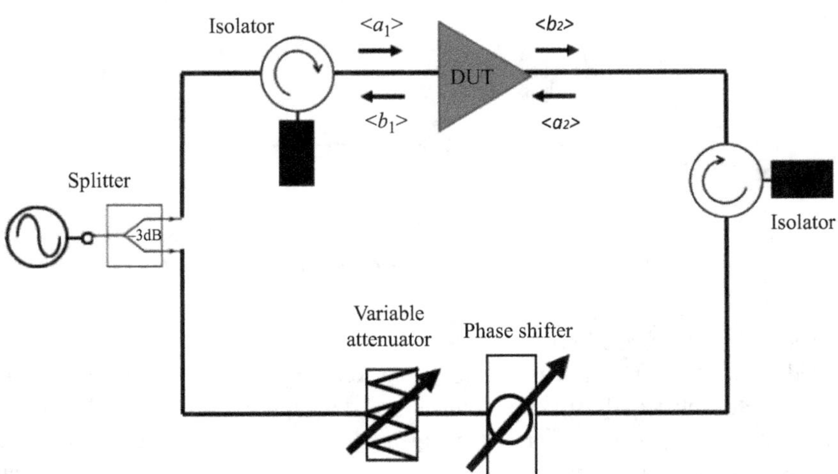

Figure 3.46 "Open loop" method block diagram based on split-signal architecture

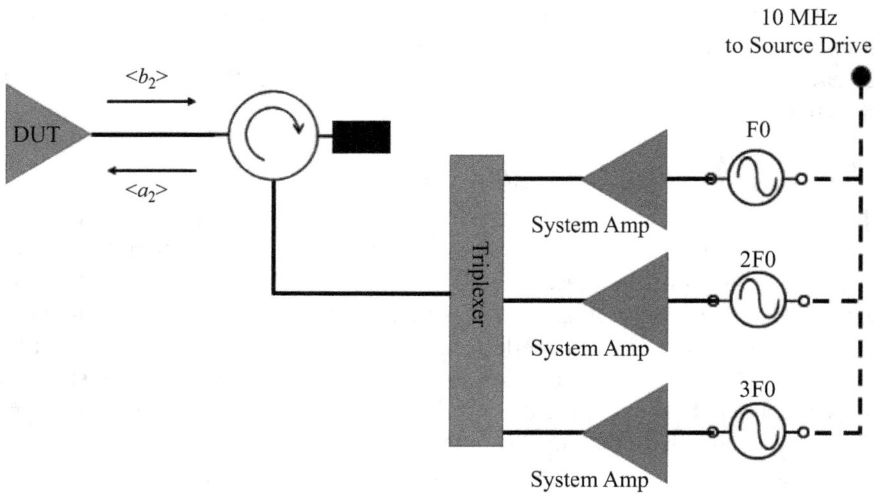

Figure 3.47 "Open loop" method with independent phase locked signal generators

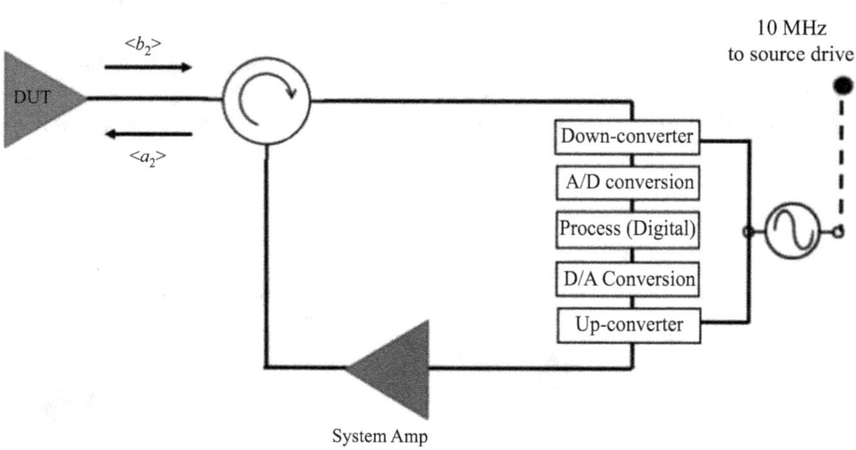

Figure 3.48 Quasi-closed-loop active load pull

The ability to phase-lock signal generators using a 10 MHz reference signal allowed multiple signal generators to be locked together which meant relative magnitude and phase adjustments could be made (Figure 3.48). This allowed the injected signal (a_2) to be generated independently, using signal generator(s) that are phase locked to the input signal generator. The output of the device is again terminated into a passive 50 Ω load using a circulator. This extension allowed the use of more than one signal generator to control the impedance of harmonics, as well as the fundamental frequency component, and allowed for full automation of the active load pull approach as the desired impedance could be controlled with a software algorithm defined to deal with the iteration required to set the load. Such systems have become more prevalent as VNA architectures have developed to include multiple internal signal generators. The introduction of internal signal generators based on direct digital synthesis (DDS) allows sources with absolute phase coherence making them ideal for application in open loop active load-pull systems; for example, the ZVA-67 from Rhode and Schwarz has four DDS-based internal signal generators providing all necessary receivers and signal generators in a single instrument, making it an ideal base instrument for harmonic active load-pull [24]. Open-loop solutions do however suffer from the fact that measurements are rather time consuming as the injected signal has become decoupled from the device output. The problem is further compounded when drive conditions are varied; with no link between the device output and injected signal, any change in the input signal leads to a further round of iteration to reestablish the required load. Clearly as we move to modulated signals, we can no longer use the established approach of using a new signal generator for injection at each generated signal frequency; while the method was practical for CW load-pull with harmonics, clearly this will not scale to complex modulations with thousands of generated frequency components both in and out of band. Marchetti [25] published a modulated system based around the open loop architecture. Here PXI-based arbitrary waveform generators (AWGs) and swept receivers are used to allow the capture and control of complex modulated waveforms. They also introduce the concept of time segmented RF injection waves to dramatically increase the speed of CW and pulsed load-pull measurements; however, for complex modulated signal, due to the coupled effect of in-band and harmonic components of the signal, the control for iterative setting of the load becomes very complex with an average of 100 iterations required to set a single load or source impedance.

3.12.4 Quasi-closed-loop load pull

The benefits of the closed loop architecture without the RF stability issues offer Mesuro's RAPID system [26]; in this system (Figure 3.49), the device output is fed into a feedback loop but this time it is down-converted, then processed digitally before being up converted to form the desired injected signal. This architecture retains the real time load tracking of the closed loop architecture but the risk of uncontrolled oscillations is removed as the RF loop is now discontinued. The RAPID system is based around a PXI architecture, and integrates measurement of RF parameters such as power added efficiency, EVM and ACPR; the PXI-based architecture, combined with the digital loop, leads to very fast load-pull extraction, including impedance synthesis iterations and output power measurement. The architecture also lends itself to load-pull of wide-band modulated signals and emulation of matching networks. The digital processing also allows for delay compensation of the loop allowing for a true wide-band load to be presented across a modulated bandwidth.

The open loop architecture and FPGA algorithms allow for quasi real-time tuning. The speed of a CW load pull is approximately 5 ms per load synthesized at

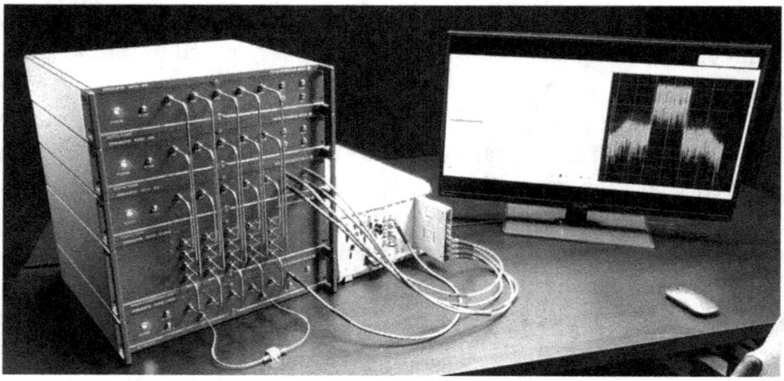

Figure 3.49 Harmonic RAPID quasi-closed-loop active tuner

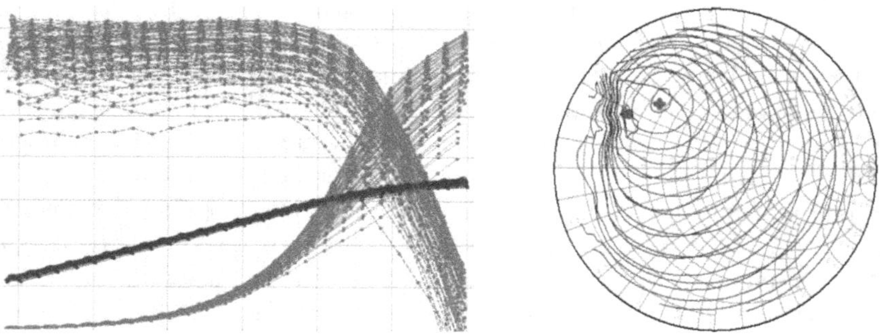

Figure 3.50 Measurement results using quasi-closed-loop load pull system

1 KHz IFBW (200 points/second). This includes impedance synthesis and output power measurement (Figure 3.50)

Combined with a calibrated phase reference allows accurate time-domain waveforms (Figure 3.51) for ultra-fast behavioral model extraction.

3.13 Data transfer into CAD and nonlinear models

There are two main methods of importing load-pull data in to the simulation environment, namely direct data import or the creation of a model of the test device. For simple data analysis computer-aided design (CAD) vendors provide the ability to load and display data from load-pull systems, these scripts have evolved along with the load-pull setups to deal with complex swept load-pull data sets including nested harmonic load and source-pull with power. This offers the user the possibility to interpolate data, view results, and even design matching networks; this approach is limited however by a number of factors, the ability of the simulator to correctly predict performance is highly dependent upon the density of the

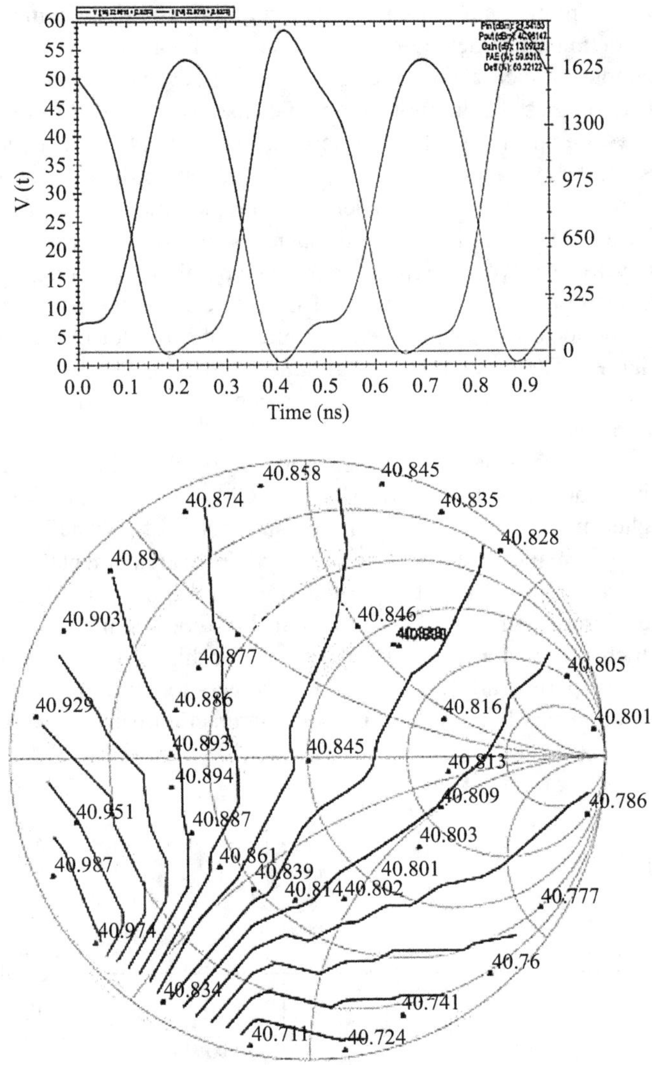

Figure 3.51 Time domain waveforms

original dataset; and this approach is also limited to linear simulator types so cannot be used for example with nonlinear simulators such as harmonic balance, and this is a problem particularly where more than one active device is used as would be the case in a multistage or parallel PA design.

Traditionally, modeling has involved the extraction of a physical description of a device usually achieved by fitting measured pulsed DC, S-parameter, and load-pull data to a physical model of the resistors, capacitors, inductors, and current sources that make up the semiconductor device. This is clearly a time consuming process and requires a good understanding of the underlying device physics.

More recently nonlinear behavioral models aim in providing a more seamless process of modeling a test device for use in the simulator.

One example of such a nonlinear behavioral model is X-Parameters™ from Keysight Technologies™; this poly harmonic distortion (PHD) formulation first introduced by Verspecht [27] is described as a mathematical superset of s-parameters and has the same use case as S-parameters in the simulator but in the nonlinear domain. It allows for models that predict the magnitude and phase response of a test device at harmonic frequencies, these models can also be cascaded to allow for example system or multistage/parallel simulation. When combined with a load-pull setup, a number of models can be extracted to produce a model using load look up. A simplified measurement setup for model extraction is shown in Figure 3.52; here small perturbation ("tingle") tones at the input and output close to fundamental and harmonic frequencies are used to map the input and output frequencies to one another.

Another nonlinear behavioral model proposed by Woodington *et al.* [28] and now commonly known as the Cardiff Model+ (CM+) extends this formulation to allow for higher order polynomial fit of the data; this has the benefit of being able to model the performance of the device over a range of fundamental and harmonic impedances in a single model, the formulation shown in (3.1), with equivalent Fourier series description in (3.2). The model does not assume the superposition principle, which means that the full mixing relationship of the harmonics can be uncovered; this is particularly important for designers looking to model devices in modes of operation involving strong harmonic interaction, such as Class F:

$$B_{n,h} = P_1^h f\left(V_{1,0}, A_{1,1}, V_{2,0}, \frac{A_{2,1}}{P_1}\right) \tag{3.1}$$

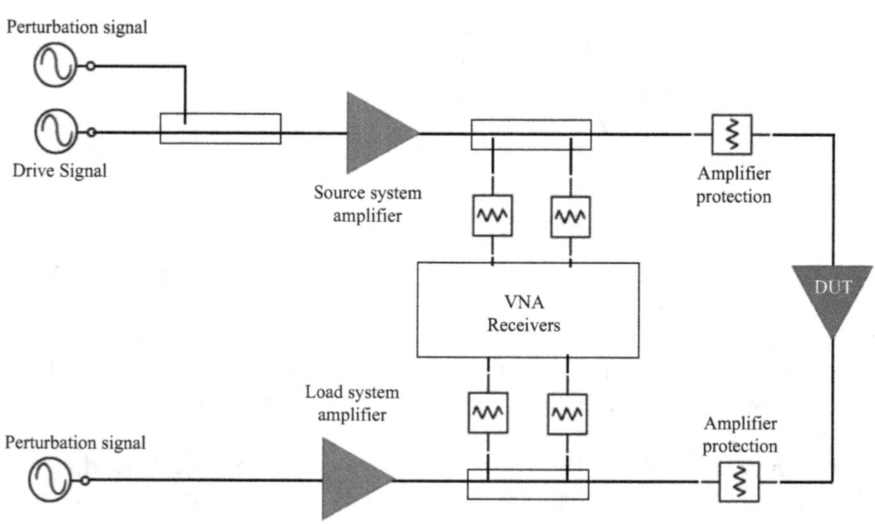

Figure 3.52 Block diagram of a simplified measurement setup for model extraction

where h is the harmonic index, n is the port index, and P_1 is $< A_{11}$ and Q_1 is $< A_{21}$:

$$B_{n,h} = P_1^h \sum_{-(\omega-1)/2}^{+(\omega+1)/2} K_{n,h,m} \left(\frac{Q_1}{P_1} \right)^m \tag{3.2}$$

where m is the phase index.

The extraction process is also simplified: rather than using small perturbation signals the Cardiff Model+ can be extracted directly from standard vector load-pull measurements providing a very simple and quick way of producing a nonlinear model of a test device [29].

While easy to extract and less time consuming to create, compared to a physical device model, behavioral models should be used with care, they are formulations to describe the measured behavior of a test device, so while interpolation is accurate, extrapolation should only be used with care to avoid erroneous results.

Acknowledgments

The authors are of the opinion that the contributions to this work presented here are not theirs alone, but a collective effort. Over 30 years, engineers, scientists, and technicians have carried the burden of trying and improving many of the ideas. They are summarily recognized and appreciated, in particular, Randeep Saini (ex. Mesuro), Simon Woodington (Mesuro), Bryan Hosein (Focus), Roman Meierer (ex. Focus), Dominique Dubouil (ex. Focus), Baoli Tian (Focus), Raymond Jallad (Focus), Neven Misljenovic (ex. Focus), Paul Tasker (University of Cardiff), and several others.

References

[1] Focus Microwaves Inc., "Harmonic Effects in Load Pull using Wideband Tuners," Application Note 56, August 2003, https://focus-microwaves.com/application-notes/

[2] Cusack J.M., Perlow S.M., and Perlman B.S., "Automatic load contour mapping for microwave power transistors," *IEEE Trans. Microwave Theory Tech.,* vol. MTT-22, pp. 1146–1152, 1974.

[3] Takayama Y., "A New Load-pull Characterization Method for Microwave Power Transistors," [online], 1976 IEEE Microwave Theory & Techniques Society (MTT-S) International Microwave Symposium, pp. 218–220. Retrieved from http://ieeexplore.ieee.org/document/1123701/ [Accessed on April 06, 2017].

[4] Hewlett Packard Company, "809B/8146 Universal Probe Carriages," 1962 Short Form Catalog, p. 11.

[5] Woodin C.E., and Wandrei D.L., US Patent 5,276,411, "*High Power Solid State Programmable Load*," January 1994.

[6] Adamian V.A., and Phillips P.V., US Patent 5,034,708, "*Programmable Broadband Electronic Tuner*," July 1991.

[7] "Buckling," Wikipedia, https://en.wikipedia.org/wiki/Buckling#Buckling_under_tensile_dead_loading

[8] Focus Microwaves Inc., "Kilo-Watt Range Tuners," Product Note 90, July 2017, https://focus-microwaves.com/articles-2/

[9] Tsironis C., US Patent 8,975,988, "*An Impedance Tuner Using Dielectrically Filled Airline*," March 2015.

[10] Tsironis C., US Patent pending, "*High Power Tuners and Method.*"

[11] Tsironis C., US Patent 6,297,649, "*Harmonic Rejection Load Tuner*," October 2001.

[12] Focus Microwaves Inc., "Programmable Harmonic Tuner, PHT," Product Note 44, November 1997, https://focus-microwaves.com/articles-2/

[13] Focus Microwaves Inc., "Harmonic Tuning Isolation in Load Pull Setups using PHT," Application Note 33, May 1999, https://focus-microwaves.com/application-notes/

[14] Tsironis C., US Patent 7,135,941, "*Triple Probe Automatic Slide Screw Load Pull Tuner and Method*," November 2006.

[15] ATN Microwave Inc., "A Load Pull System with Harmonic Tuning," *Microwave Journal*, Product Feature, pp. 128–132, 1996.

[16] Focus Microwaves Inc., "MPT, a Universal Multi-Purpose Tuner," Product Note 79, October 2004, https://focus-microwaves.com/articles-2/

[17] Tsironis C., US Patent 8,212,629, "*Wideband Low Frequency Impedance Tuner*," July 2012.

[18] Tsironis C., US Patent 7,646,267, "*Low Frequency Electro-mechanical Impedance Tuner*," January 2010.

[19] Focus Microwaves Inc., "Harmonic Effects in Load Pull using Wideband Tuners," Application Note 56, August 2003, https://focus-microwaves.com/application-notes/

[20] Tsironis C., US Patent 7,102,457, "*Mechanically Balanced Microwave Load Pull Tuner*," September 2006.

[21] Tsironis C., US Patent 9,431,999, "*Intelligent Mechanical Balancing Apparatus for Slide Screw Tuners*," August 2016.

[22] Tsironis C., US Patent 9,310,410, "*Load and Source Pull Test System for RF and Baseband Frequencies*," April 2016.

[23] Bava G.P., Pisani U., and Pozzolo V., "Active Load Technique for Load-Pull Characterisation at Microwave Frequencies," *Electronics Letters*, Volume 18, Issue 4, February 18, 1982, pp.178 – 180.

[24] Rohde & Schwarz, "R&S®ZVA Vector Network Analyzers," Datasheet, https://www.rohde-schwarz.com/ca/product/zva-productstartpage_63493-9660.html.

[25] Marchetti M., Pelk M.J., and Nicolaas De Vreede L.C., US Patent 8,456,175, "*Open Loop Load Pull Arrangement with Determination of Injections Signals*," June 2013.

[26] Tasker P.J., and Benedikt J., US Patent 6,639,393, "*Methods and Apparatus for Time-domain Measurement with a High Frequency Circuit Analyzer,*" October 2003.

[27] Verspecht J., and Teyssier J.-P., US Patent 7,282,926, "*Method and an Apparatus for Characterizing a High-frequency Device-under-test in a Large Signal Impedance Tuning Environment,*" October 2007.

[28] Woodington S., Williams T., Qi H., *et al.*, "A Novel Measurement Based Method Enabling Rapid Extraction of a RF Waveform Look-Up Table Based Behavioural Model," 2008 IEEE MTT-S International Microwave Symposium Digest, 2008, pp. 1453–1456.

[29] Tsironis C., US Patent 10,281,510, "Load Pull Method for Transistors driven by modulated signal," May 2019.

Chapter 4

Matching networks: automated Darlington synthesis of immittance functions

Binboga Siddik Yarman[1]

Real frequency matching network design techniques require the "Darlington synthesis" of positive real immittance functions. Therefore, this chapter is devoted to "automated Darlington synthesis" to construct lossless matching networks with lumped and distributed elements.

In this chapter, several high precision lowpass/bandpass/high-pass LC (Inductor L, Capacitor C) ladder synthesis algorithms are presented. Novel algorithms directly work on the driving-point input immittance or bounded-real reflection functions which describe any kind of cascade-connection LC ladders, perhaps with Brune and C-sections in resistive termination.

The crux of the idea is that, at each step of the proposed synthesis methods, a simple pole at DC, or at infinity, or at a finite frequency is removed; then, the remaining immittance function is corrected using what we call is the parametric method. At this point, parametric method warrants the exact LC ladder nature of the remaining immittance function. Thus, at the end of synthesis process, an LC ladder is obtained with high numerical precision.

Examples are presented to exhibit the implementation of the synthesis algorithms. For example, a randomly generated driving-point input immittance is synthesized with 19 elements yielding a relative error less than 10^{-6}. This is an outstanding result introduced in the literature. Furthermore, numerical robustness of the novel synthesis methods is tested. Based on the tests, we can confidently state that proposed synthesis algorithms can safely extract more 40 elements from the original immittance function with relative error less than 10^{-2}. This is a remarkable result. Newly developed synthesis algorithms are coded on MATLAB® environment and it is successfully combined with the "real frequency-direct computational or simplified real frequency techniques (SRFT)" to construct practical impedance matching networks.

Employing the RFT together with new synthesis algorithms, we completed the design examples for various purposes such as to design broadband-RF and microwave amplifiers, antenna-matching networks to be used modern communication systems.

[1]Department of Electrical and Electronics Engineering, Istanbul University-Cerrahpasa, Istanbul, Turkey

4.1 High-precision lowpass ladder synthesis via parametric approach

Real frequency techniques (RFT) are known as the best semi-analytic methods to provide solutions to broadband matching problems [1–4]. Success of these techniques stems from the fact that they directly operate on driving-point network functions such as impedance, admittance or reflectance functions to optimize the electrical performance of the matched systems. No matter which form is used, eventually, driving-point input immittance $F(p) = a(p)/b(p)$ of the matching network must be synthesized as a lossless two-port in resistive termination. This is known as Darlington synthesis [5–7]. Hence, accurate synthesis of immittance functions becomes inevitable.

Unfortunately, over last the three decades, we have suffered due to lack of numerically stable synthesis package which works on a general-purpose computing platform. As far as physical realization of matching networks is concerned, construction of lowpass LC ladders is in high demand. Therefore, in this section, we introduce a robust computer algorithm to synthesize immittance functions as lowpass LC ladders [8]. At each step of the algorithm, a pole at infinity is extracted from the properly specified immittance function, then the remaining function is corrected using the parametric approach. Parametric method is known to generate realizable network functions with outstanding precision [9–11].

Hence, we end up with a robust synthesis algorithm, which can synthesize lowpass LC ladders with more than 40 elements. This is the first of its kind in the literature.

In the following, first, we introduce mathematical description of immittance functions, which can be synthesized as lowpass LC ladders. Then, details of the parametric method to generate immittance functions are presented (Section 4.1.2). In Section 4.1.3, high precision lowpass LC ladder synthesis algorithm is developed step-by-step by means of live examples. For example, in Example 4.5, newly proposed synthesis algorithm is successfully combined with the real frequency direct computational technique (RF-DCT) to construct practical matching networks. An actual impedance-transforming filter is synthesized using the optimized immittance function obtained as the output of the RF-DCT design package. In Example 4.6, numerical robustness of the proposed algorithm is tested. In this regard, we can confidently state that our high-precision lowpass ladder synthesis algorithm can handle synthesis problems with more than 40 elements. In our recollection, this is a remarkable result [8].

Computer codes of the proposed synthesis methods are developed on MATLAB [12] and placed at the end of this chapter. Synthesis packages introduced in this book are unique and made available to the readers of this book.

4.1.1 Lowpass LC ladder form

Due to their robust sensitivity behavior of element values, in many communication engineering practices, lowpass LC ladders are preferred to construct matching networks for various applications. Therefore, driving-point input immittance $F(p)$ of the matching networks is selected to yield a lowpass LC ladder topology. In this case, synthesis becomes a simple long division process as described by the following equations:

$$F(p) = \frac{a(p)}{b(p)} = \frac{a_{0,1}p^n + a_{0,2}p^{n-1} + \ldots + a_{0,n}p + a_{0,(n+1)}}{0p^n + b_{0,2}p^{n-1} + \ldots + b_{0,n}p + b_{0,n+1}} \tag{4.1a}$$

$$= x_1 p + \cfrac{1}{x_2 p + \cfrac{1}{x_3 p + \cfrac{1}{\ddots \cdots + \cdots \cfrac{1}{x_n p + x_{n+1}}}}} \tag{4.1b}$$

Corresponding ladder networks are depicted in Figure 4.1(a) and (b).

The above long division can be performed step-by-step such that at step i, $x_i p$ is removed (a pole at infinity) yielding:

$$F_i(p) = \frac{a_i(p)}{b_i(p)} = x_i p + \frac{R_i(p)}{b_i(p)} \tag{4.2a}$$

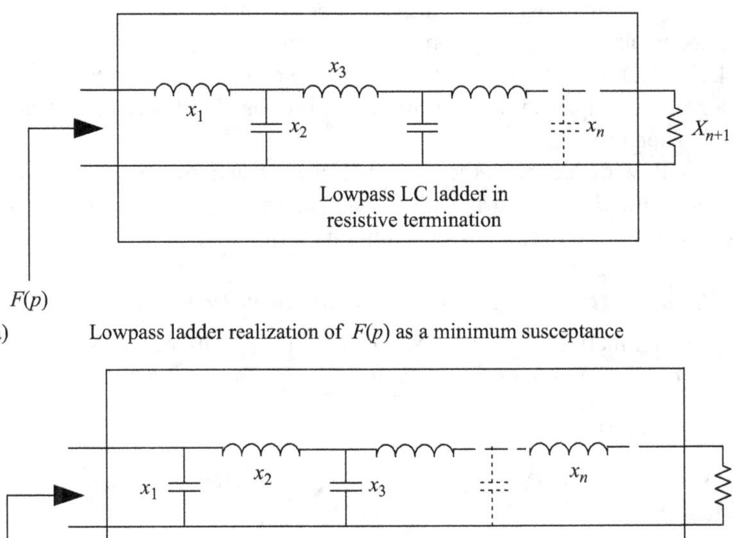

(a) Lowpass ladder realization of $F(p)$ as a minimum susceptance

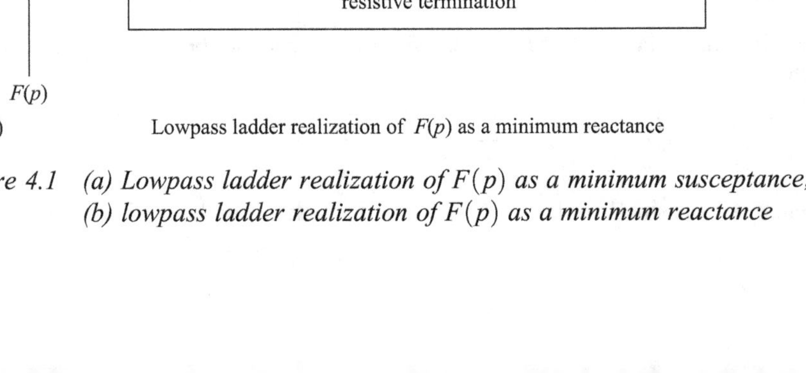

(b) Lowpass ladder realization of $F(p)$ as a minimum reactance

Figure 4.1 (a) Lowpass ladder realization of $F(p)$ as a minimum susceptance; (b) lowpass ladder realization of $F(p)$ as a minimum reactance

or

$$F_i(p) = x_i p + \frac{1}{F_{i+1}(p)} \tag{4.2b}$$

where

$$F_{(i+1)}(p) = \frac{b_i(p)}{R_i(p)} = \frac{a_{(i+1)}(p)}{b_{(i+1)}(p)} \tag{4.2c}$$

In (4.2a)–(4.2c), quotient x_i and the remainder $R_i(p)$ are given by

$$x_i = \frac{a_{i,1}}{b_{i,2}} \tag{4.3a}$$

$$R_i(p) = a_i(p) - (x_i p)b_i(p) \tag{4.3b}$$

$$= R_{i,1}p^{(n-i)} + R_{i,2}p^{(n-i-1)} + \ldots + R_{i,(n-i)}p + R_{i,(n-i+1)} \tag{4.3c}$$

It should be noted that due to the nature of division process, leading coefficient $R_{i,1}$ is zero $(\text{i.e.}, R_{i,1} = 0)$. Furthermore, if the original immittance function $F(p)$ is specified to yield an exact lowpass LC ladder, then the second coefficient $R_{i,2}$ must also be zero $(\text{i.e.}, R_{i,2} = 0)$ to end up with degree reduction in $F_{(i+1)}(p) = (a_{(i+1)}(p))/(b_{(i+1)}(p))$. However, as synthetic division progresses on the computer, mild or severe numerical errors are accumulated yielding a nonzero $R_{i,2}$. Thus, $F_{(i+1)}(p)$ no longer belongs to a lowpass LC ladder, or may not even be a positive real function. Therefore, one must develop a meaningful algorithm to warrant the lowpass LC ladder nature along with synthetic division process.

In the following section, the parametric approach to generate positive real functions is outlined [9,10]. It is shown that this method warrants the lowpass LC ladder realization of the reprocessed immittance function.

4.1.2 Parametric representation of an immittance function

A positive real function $F(p)$ can be expressed in terms of the summation of a Foster function $F_F(p)$ and a minimum function $F_M(p)$ such that

$$F(p) = F_F(p) + F_M(p) \tag{4.4}$$

For the case under consideration, Foster function $F_F(p)$ may have a simple pole at infinity such that

$$F_F(p) = k_f p; k_f > 0$$

The minimum function $F_M(p)$ is free of $j\omega$ poles and can be written in parametric form as

$$F_M(p) = \frac{a_M(p)}{b_M(p)} = \sum_{j=1}^{n} \frac{k_j}{p - p_j} \tag{4.5}$$

where p_j are the simple poles of $F_M(p)$ which are placed in the left half plane (LHP).

Even part of $F_M(p)$ must be as same as the even part of $F(p)$, and it is given by

$$R(p) = \frac{1}{2}[F(p) + F(-p)] = \frac{1}{2}[F_M(p) + F_M(-p)] \tag{4.6}$$

Employing (4.5) and (4.6), $R(p)$ can be given in parametric form as

$$R(p) = \frac{A(p)}{B(p)} = \sum_{j=1}^{n} \frac{k_j p_j}{p_j^2 - p^2} \tag{4.7a}$$

where even polynomials $A(p)$ and $B(p)$ are described as

$$\begin{aligned} A(p) &= \frac{1}{2}[a_M(p)b_M(-p) + a_M(-p)b_M(p)] \\ &= A_1 p^{2n} + A_2 p^{2(n-1)} + \cdots + A_n p^2 + A_{n+1} \end{aligned} \tag{4.7b}$$

and

$$\begin{aligned} B(p) &= b_M(p)b_M(-p) \\ &= B_1 p^{2n} + B_2 p^{2(n-1)} + \cdots + B_n p^2 + B_{n+1} \end{aligned} \tag{4.7c}$$

or

$$B(p) = B_1 \prod_{j=1}^{n} \left(p_j^2 - p^2\right) \tag{4.7d}$$

If $R(p)$ is known then, the residues k_j of (4.5) and (4.7a)–(4.7d) is generated as

$$k_j = \frac{\left(p_j^2 - p^2\right) R(p)}{p_j} \Big|_{p=p_j} \tag{4.8a}$$

or

$$k_j = \frac{A(p_j)}{B_1 p_j \prod_{\substack{i=1 \\ i \neq j}}^{n} \left(p_j^2 - p_i^2\right)} \tag{4.8b}$$

On the other hand, since $F_M(p)$ corresponds to a lowpass LC ladder network as shown in Figure 4.1, then $R(p)$ must possess the following form:

$$R(p) = \frac{A(p)}{B(p)} = \frac{0.p^{2n} + 0.p^{2(n-1)} + \cdots + 0.p^2 + A_{n+1}}{B(p)} \tag{4.9a}$$

or in short,

$$R(p) = \frac{A_{n+1}}{b(p)b(-p)}; A_{n+1} > 0 \tag{4.9b}$$

In this case, once the polynomials $a_M(p)$ and $b_M(p)$ are specified, one must verify that polynomial $A(p)$ has all zero coefficients except $A_{n+1} > 0$. As far as numerical implementation of (4.9a) and (4.9b) is concerned, perhaps using MATLAB, the machine zero could be set to a small positive threshold, which is designated by $\varepsilon_{zero} = 10^{-m}$ with $m \geq 0$. In this case, if polynomial $A(p)$ is represented by a MATLAB vector $A = [A(1)A(2)\cdots A(n)A(n+1)]$, entrees of A in absolute values must be less than ε_{zero} except A_{n+1}.

On the other hand, if we start from the even part $R(p)$ as specified by (4.9a) and (4.9b), the minimum function $F_M(p)$ can be uniquely determined by means of the residues k_j and the poles p_j of $R(p)$ employing (4.8a) and (4.8b) and (4.5) to yield an LC lowpass ladder network when it is synthesized.

Let us investigate this situation by means of an example.

Example 4.1

Let a minimum impedance function $Z_M(p)$ be

$$Z(p) = \frac{0p^3 + 0.2755p^2 + 0.1051p + 0.1182}{1.0000p^3 + 0.3816p^2 + 0.7390p + 0.1182} \tag{4.10}$$

Determine its even part and comment if it describes a lowpass ladder structure as specified by (4.9a) and (4.9b).

Solution

For this problem, we developed a MATLAB function called $[A, B] = even_part(a, b)$ (see Appendix: Program List 4.8).

This function generates the even part $R(p)$ as

$$R(p) = \frac{0p^6 + (-1.34) \times 10^{-4}p^4 + (-7.06) \times 10^{-4}p^2 + 0.0140}{-1p^6 - 1.3328p^4 - 0.4561p^2 + 0.0140} \tag{4.11}$$

For the example under consideration, (4.11) reveals that selection of $\varepsilon_{zero} > 10^{-3}$ results in ladder form of (4.9a) and (4.9b). Thus, the exact ladder form of $R(p)$ is approximated as

$$R(p) \cong \frac{0.0140}{-1p^6 - 1.3328p^4 - 0.4561p^2 + 0.0140} \tag{4.12}$$

All the above computations are gathered under a MATLAB program called "Main_Example 4_1.m" and it is given as an attachment to this book.

In the following example, we generate a minimum function from its even part.

Example 4.2

Employing (4.5–4.8) generate the minimum function $Z_1(p) = a_1(p)/b_1(p)$ from the given exact lowpass ladder form of the even part specified by (4.12) and compare your result with that of (4.10).

Solution

For this purpose, (4.8a) and (4.8b) is programmed under a MATLAB function called *function* $[p, k]$ = residues_evenpart(A, B) to generate the LHP poles p_j of $R(p)$ and corresponding residues k_j as shown in Table 4.1.

Table 4.1 For the given even part $R(p)$ of Example 4.2, computation of LHP poles and corresponding residues employing function $[p, k] = residues\_evenpart(A, B)$

p_j	k_j
−0.1066−0.8318*I*	−0.0600+0.0234*I*
−0.1066+0.8318*I*	−0.0600−0.0234*I*
−0.1682	−0.1558

By straightforward algebraic manipulations, rational form of $Z(p) = a(p)/b(p)$ is generated. These algebraic manipulations are gathered under a MATLAB function called *function* $[a, b] = RtoZ(A, B)$. Execution of this function yields the desired rational form $Z_1(p) = a_1(p)/b_1(p)$ such that

$$a1 = [0\ \ 0.27582199\ \ 0.10528040\ \ 0.11832159]$$

$$b1 = [1.0.38180859\ \ 0.73908890\ \ 0.11832159]$$

whereas in Example 4.1, $Z(p) = a(p)/b(p)$ is given by

$$a = [0\ \ 0.2755\ \ 0.1051\ \ 0.1182]$$

$$b = [1.0000\ \ 0.3816\ \ 0.7390\ \ 0.1182]$$

As it can be observed from the above vectors, pairs $(a1, a)$ and $(b1, b)$ are close to each other within an error of measurement in the order of $\sim 10^{-4}$. More specifically, norms of the errors occurred during computations are determined as

$$\varepsilon_a = norm(a - a1) = 3.26 \times 10^{-4}$$

$$\varepsilon_b = norm(b - b1) = 5.59 \times 10^{-4}$$

All the above computations are combined under a MATLAB program called "Main$_{\text{Example2m}}$," and it is made available as an attachment to this book.

4.1.3 Warranted ladder network synthesis via parametric synthetic division

Even though ladder network synthesis seems to be a straightforward synthetic division as detailed by (4.2a)–(4.2c), it is very tricky due to accumulation of numerical errors occur at each step. As we go along with division process, numerical precision is lost, and remaining positive real functions $F_{i+1}(p) = (a_{i+1}(p))/(b_{i+1}(p))$ may no longer describe an exact lowpass ladder structure as specified by (4.9a) and (4.9b). However, this problem can be overcome by regenerating $F_{i+1}(p)$ from its real part using parametric approach, as described in the previous section. In this section, high precision lowpass ladder network synthesis is outlined as step-by-step process as given next.

4.1.4 Lowpass LC ladder network synthesis

A driving-point input immittance function $F(p) = a(p)/b(p) = Even(p) + Odd(p)$ describes an "Exact Lowpass LC Ladder (ELCL)" if and only if it is even part $R(p) = Even(p)$ is an "All-Pole Even-Rational (APER) Function" in complex variable $p = \sigma + j\omega$ such that $R(p) = A_{n+1}/B(p)$ as specified by (4.9a) and (4.9b). In practice, however, corresponding immittance function $F(p) = a(p)/b(p)$ may not yield an APER $R(p)$. In this case, one can purposely regenerate $F(p) = a(p)/b(p)$ from the APER form of $R(p) = (A_{n+1})/(B(p))$. Thus, new $F(p)$ belongs to an exact Lowpass LC ladder in resistive termination.

Now, let us construct a minimum impedance function $F(p)$ which yields an APER form for $R(p)$.

Let the APER form of $R(p)$ be

$$R(p) = \frac{0.0140}{-1p^6 - 1.3324p^4 - 0.4559p^2 + 0.0140}$$

Generate a minimum function $F(p)$ via parametric approach.

From the recalculated $F(p)$ generate $R(p)$ and comment on the result.

Solution

1. Employing MATLAB function $[a, b] = R\,\mathrm{to}\,Z(A, B)$ with $A = [0, 0, 0, 0.0140]$; $B = [-1 - 1.3324 - 0.4559\ 0.0140]$, we obtained polynomials $a(p)$ and $b(p)$ as in Table 4.2. At this point, we know for sure that $Z(p)$ belongs to an ELCL in resistive termination.
2. Even part $R(p)$of$F(p)$ is regenerated using MATLAB function $[A1, B1] =$ even_part$(a1, b1)$ as in Table 4.3.

Close examination of Table 4.3 indicates that $R(p) = A_1(p)/B_1(p)$ describes almost exact lowpass LC ladder network within 1.667×10^{-16} accuracy. In other words, in generating $F_1(p)$ as the driving-point immittance of an ELCL via parametric approach, accumulated numerical error is better than 10^{-15}. This is a significant improvement compared to that of 10^{-4} of Example 4.2, in generating exact lowpass LC ladder structures via parametric approach.

All the above computations are combined under a MATLAB main program called "YarmanKilinc_Example3.m" and made available as an attachment to this book.

Table 4.2 Generation of minimum function $F_1(p) = a_1(p)/b_1(p)$ from $R(p) = A(p)/B(p)$

MATLAB polynomial $a(p)$	MATLAB polynomial $b_1(p)$
0	1.000000000000000
0.275685028606870	0.381808594742783
0.105258913364013	0.739088901509729
0.118321595661992	0.118321595661992

Table 4.3 Regeneration of even part $R(p) = A_1(p)/B_1(p)$ from $F(p) = a_1(p)/b_1(p)$

MATLAB polynomial $a_1(p)$	**MATLAB polynomial $b_1(p)$**
0	−1.000000000000000
$1.665334536937735 \times 10^{-16}$	−1.332800000000000
$-1.4571677198205 \times 10^{-16}$	−0.456100000000000
0.014000000000000	0.014000000000000

Thus, based on the above discussions, the following algorithm is proposed to synthesize lowpass LC ladder from a given immittance function $F(p)$.

4.1.5 Algorithm: guaranteed synthesis of a lowpass LC ladder from a given minimum driving-point immittance function $F(p) = a(p)/b(p)$ using MATLAB

Inputs

eps_zero: Threshold for the "algorithmic zero."

It should be noted that for many applications, it is sufficient to choose eps_sq $= 10^{-8}$.

Driving-point input immittance $F(p) = a(p)/b(p)$ where $a(p)$ and $b(p)$ are described as MATLAB polynomials:

$$a = [0\,a(2)\cdots a(n)a(n+1)]$$
$$b = [b(1)b(2)\cdots b(n)b(n+1)]$$

In the above presentation, integer n designates the degree of the denominator polynomial $b(p)$ and obviously, degree of $a(p)$ is $(n-1)$ since $F(p)$ is a minimum immittance function.

Computational steps:

Step 1

Check if $F(p)$ is a proper function to be synthesized as a lowpass LC ladder network:

For this purpose, a MATLAB function called *function* $[a, b, ndc] =$ Check_immitance(a, b) is developed to check if the given $F(p)$ is a minimum function describing a lowpass LC ladder. During computations, first, the even part of $F(p)$ is generated and then, its zeros are computed. As explained in the previous sections, for a lowpass LC ladder network, numerator $A(p)$ of $R(p)$ must be a positive real constant. Literally speaking, $A(p)$ must be free of zeros. For many design problems, rational form of $R(p)$ is expressed as $R(p) = (-1)^{ndc}A_0p^{2(ndc)}/B(p)$. Therefore, for lowpass *ndc* must be zero. Hence, when function $[a, b, ndc] =$ Check_immitance(a, b) is executed, it must yield

$ndc = 0$; otherwise, function $F(p)$ cannot be synthesized as a lowpass LC ladder. Details are skipped here. Nevertheless, interested readers are referred to [3,8].
Step 2
Flip over $F(p)$ and remove the first pole at infinity:
Since $F(p)$ is a minimum function, it is flipped over and the first pole of $1/F(p)$ is removed at infinity by setting

$$F_1(p) = \frac{1}{F(p)} = \frac{b(p)}{a(p)} = (q) \times p + \frac{R(p)}{a(p)}$$

where the term $(q) \times p$ is the quotient and the positive real number q is the residue of the first pole at infinity. Furthermore, the reminder polynomial $R(p)$ is given by

$$R(p) = b(p) - (q) \times p \times a(p)$$

Here, it should be noted that an ELCL function $F(p)$ yields a reminder polynomial $R(p)$ such that the first two leading coefficients must be zero so that degree reduction occurs in $R(p)/a(p)$.
Step 3
Perform the degree reduction operation in $R(p)/a(p)$ by properly shifting the entrees of $R(p)$ and $a(p)$ and define a new minimum function

$$F_1(p) = \frac{a_1(p)}{b_1(p)}$$

where $a_1(p) = R(p)$ of degree $(n-2)$ and $b_1(p) = a(p)$ of degree $(n-1)$.
At this step, result of synthetic long division operation may summarized as

$$\frac{b(p)}{a(p)} = (q) \times p + \frac{a_1(p)}{b_1(p)}$$

Step 4
Employing parametric approach, regenerate $F_1(p) = a_1(p)/b_1(p)$ as an ELCL function for further pole extraction.
The above algorithm is programmed under a MATLAB function called $[q, a1, b1, ndc] = \mathit{ExtractTrZero\_infinity}(a, b)$ which yields the positive residues q_i and the ELCL function $F_1(p) = a_1(p)/b_1(p)$. Successive usage of function $[q, a1, b1, ndc] = \mathit{ExtractTrZero\_infinity}(a, b)$ guarantees the lowpass ladder realization of a given minimum function $F(p)$.

Now, let us use the above algorithm to synthesize an exact lowpass ladder function $F(p)$ generated by Example 4.4.

Example 4.3

Let $F(p) = a(p)/b(p)$ be a minimum impedance function as specified by Table 4.2.

1. Show that it belongs to a lowpass ladder by evaluating its even part.
2. Synthesize $F(p)$ via parametric approach.

3. Regenerate $F(p) = a_1(p)/b_1(p)$ using the element values of the LC ladder network and compute the norm of the accumulated errors occurred in generating $a_1(p)$ and $b_1(p)$. In other words, determine

$$\varepsilon_a = norm(a - a_1)$$
$$\varepsilon_b = norm(b - b_1)$$

Solution

For the example under consideration, we have developed a MATLAB program called "$\text{Main}_{\text{Example4}_3}$.m."

1. This program removes the poles of $F(p)$ at infinity as detailed previously.

Using function $[A, B] = even_part(a, b)$, we generate the even part $R(p) = A(p)/B(p)$. Hence, numerator polynomial $A(p)$ is computed as

$$R = \left[0 - 1.38 \times 10^{-17}\, 0\ \ 0.014\right].$$

As it is observed from above, threshold for algorithmic zero ε_{zero} is less than $= 10^{-16}$ which practically yields an exact lowpass LC ladder.

2. Employing our MATLAB function $[Q, a1, b1, ndc] = ExtractTrZero\_infinity(eps\_zero, aa, bb)$.

Residues $q(i)$ are sequentially generated within the main program developed for the example under consideration.

Thus, synthetic division via parametric approach yields the element values of the ladder as shown in Table 4.4 and Figure 4.2.

3. Using the above element values, rational form of $F(p)$ is generated by means of a MATLAB function called $[a1, b1] = ExactLowpassLadder(q)$. In this function, long-division representation of $F(p)$ is converted into rational a form $F(p) = a_1(p)/b_1(p)$ in a sequential manner. Thus, accumulated errors are found as

$$\varepsilon_a = norm(a - a_1) = 4.16 \times 10^{-17}$$
$$\varepsilon_b = norm(b - b_1) = 2.48 \times 10^{-16}$$

Table 4.4 Synthesis of $F(p)$ as an impedance function via long division using parametric approach

Residues of the poles at infinity	Element values of lowpass ladder
3.627443143211162	Shunt Capacitor *C*1
0.888601178998401	Series Inductor *L*2
2.624671916010501	Shunt Capacitor *C*3
1.000000000000001	Resistive termination

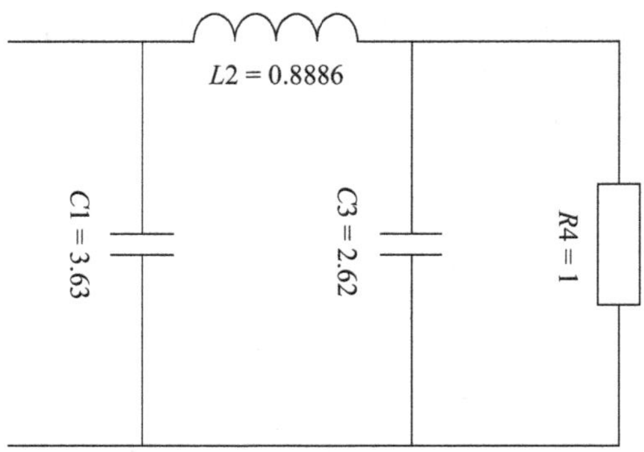

Figure 4.2 Lowpass ladder synthesis of $F(p)$

Now, let us run an example which cannot be handled with a straight forward long division process.

Example 4.4

Let $F(p)$ be a minimum impedance function, which is specified, in rational form as shown by Table 4.5.

By "Parametric Long Division Algorithm," synthesize $F(p)$ and compute the accumulated errors as in Example 4.4 such that

$$\varepsilon_a = norm(a - a_1)$$
$$\varepsilon_b = norm(b - b_1)$$

Solution

In this example , MATLAB program developed for Example 4.3 is revised for the new driving-point input impedance function. The revised program is called *Main$_E$xample4_4.m*. Hence, execution of *YarmanKlinic$_E$xample5.m* results in 19-element ladder in resistive termination as shown in Table 4.5 and Figure 4.3.

Finally, by recalculating the driving-point input impedance $F(p) = a_1(p)/b_1(p)$ form the element values, accumulated numerical error is obtained as

$$\varepsilon_a = norm(a - a_1) = 1.35 \times 10^{-6}$$
$$\varepsilon_b = norm(b - b_1) = 3.92 \times 10^{-6}$$

This is a very significant result. We should note that, for the problem under consideration, we are not able to carry out the synthesis process using a simple long division algorithm.

It should be mentioned that synthesis part of the main program Main_Example_4.m is gathered under a MATLAB function called $q = LowpassLadder\_Yarman(a, b)$

Table 4.5 Rational for $F(p) = a(p)/b(p)$

$a(p)$ $1.0 \times 10^3 * [.]$	$B(p)$ $1.0 \times 10^4 * [.]$
0	0.000100000000000
0.000243570982302	0.002439695870041
0.005942391195838	0.019914726154049
0.048441163667878	0.092525662301846
0.223771800301324	0.294017502526219
0.703166932392809	0.700877094955499
1.647563920080328	1.323370333733525
3.038406878703025	2.047310742070293
4.561480281120445	2.650311854216540
5.691843698340120	2.907173323430787
5.973794366521135	2.719656869729844
5.301481077973682	2.173275287659293
3.977617930885949	1.479042415540437
2.509345559121446	0.850450799503025
1.315779401276350	0.407297891509801
0.562150827674645	0.158774880986098
0.189558067899127	0.048562164265494
0.048032011167581	0.010905748935087
0.008431529662470	0.001542112526954
0.000827704427793	0.000082770442779

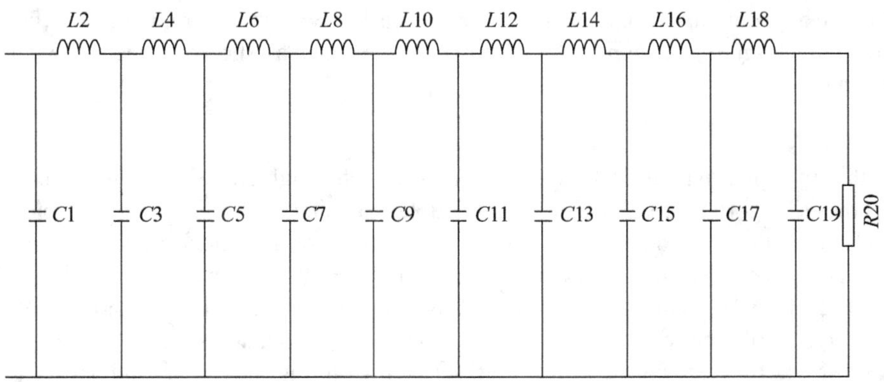

Figure 4.3 Synthesis of $F(p)$ as specified by Table 4.6

Table 4.6 *Synthesis of* $F(p)$ *as an impedance function via long division with parametric approach*

Residues of the poles at infinity	Type of the elements
4.105579369714495	Shunt capacitor $C1$
0.908103724557547	Series inductor $L2$
3.894982368200534	Shunt capacitor $C3$
1.074937488509994	Series inductor $L4$
3.160131787019010	Shunt capacitor $C5$
1.547460187794941	Series inductor $L6$
1.903836653032907	Shunt capacitor $C7$
2.119819628827039	Series inductor $L8$
1.951649082349190	Shunt capacitor $C9$
1.732250507192499	Series inductor $L10$
1.510739692650735	Shunt capacitor $C11$
1.290751815708178	Series inductor $L12$
1.077714680523259	Shunt capacitor $C13$
0.873806438230660	Series inductor $L14$
0.677786692385867	Shunt capacitor $C15$
0.488188311392944	Series inductor $L16$
0.307789188067298	Shunt capacitor $C17$
0.151325075403820	Series inductor $L18$
0.040988715531210	Shunt capacitor $C19$
0.999999999996369	Terminating conductor or $R20$

so that any minimum function $F(p)$ can be easily synthesized from the specified polynomials $a(p)$ and $b(p)$ in a very practical manner as shown in the following example:

Example 4.5

In this example, "RF-DCT" Package is combined with "high precision lowpass ladder synthesis" algorithm to design practical double matching networks. For the problem under consideration, output of a power amplifier, which is simply modeled with a resistance $R_G = 12\,\Omega$, drives a resistive load $R_L = 50\,\Omega$, over $f_{c1} = 850$ MHz to $f_{c2} = 2,100$ MHz. Thus, we face to design a doubly terminated filter between R_G and R_L. Furthermore, it is desired to suppress second and third harmonics of 2,100 MHz by means of two trap circuits as shown in Figure 4.4. Second harmonic $f_G = 2f_{c2} = 4,200$ MHz is suppressed on the generator side by means of a parallel resonance circuit ($L_G = 0.947$ nH, $C_G = 1.515$ pF) which is placed in series with R_G. Similarly, third harmonic $f_L = 3f_{c2} = 6,300$ MHz is suppressed at the load-end using a parallel resonance circuit ($L_L = 0.410$ nH, $C_L = 1.515$ pF) in series with R_L.

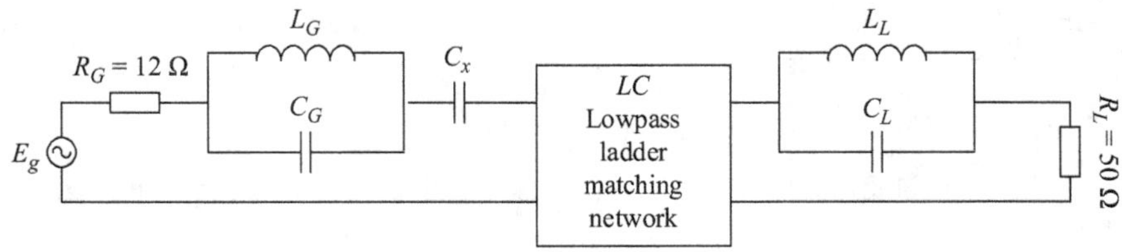

Figure 4.4 Construction of an impedance transforming filter with second and third harmonic suppression

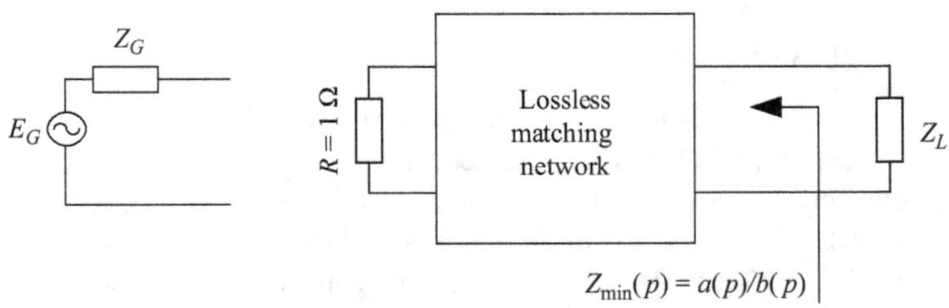

Figure 4.5 Description of a lossless matching network in terms of Darlington's input impedance

Furthermore, we introduce a series capacitor $C_x = 7.622$ pF on the generator-end to force the gain to be zero at DC.

Hence, as it stands, Figure 4.4 describes a double matching problem from a complex generator to a complex load [2,10]. We can solve this problem employing the RF-DCT [13]. Details of RF-DCT is presented in the next chapter. Therefore, RF-DCT is skipped here. However, for the problem under consideration, RF-DCT package is perfectly combined with our high precision lowpass LC ladder synthesis algorithm to construct an actual equalizer. Results are summarized next.

Referring to Figure 4.4, RF-DCT optimizes the transducer power gain (TPG) over the frequency band of operation. In this regard, TPG is expressed as a function of generator and load immittances, namely, $F_G(j\omega)$ and $F_L(j\omega)$, respectively, as well as the driving-point input immittance $F(j\omega) = a(j\omega)/b(j\omega)$ of the lossless matching network such that

$$TPG = f(F_G, F, F_L) \tag{4.13}$$

For the problem under consideration, $F(p)$ describes the matching network from the back-end as shown in Figure 4.5.

In the above formulation, we can either work with impedances or admittances depending on the success of optimization. In RF-DCT package, this fact can be control by means of a flag which is designated by an integer *KFlag*. If *KFlag* = 1, TPG is expressed in terms of impedances. If it is set to zero (*KFlag* = 0), TPG is generated as a function of admittances.

In (4.4), if we skip the Foster portion, $F(p)$ becomes a minimum function and on the real frequency axis it is written as

$$F(j\omega) = F_M(j\omega = R(\omega) + jX_M(\omega)$$

where the realizable ladder form of $R(\omega)$ is described by means of an auxiliary polynomial:

$$c(\omega) = c_1\omega^{2n} + c_2\omega^{2(n-1)} + \cdots + c_n + 1 \tag{4.14}$$

such that, for ladder structures at large, real part $R(\omega)$ is expressed as

$$R(\omega) = \frac{W(\omega)(\omega^2)^{ndc}(a_0)^2}{\frac{1}{2}[c^2(\omega) + c^2(-\omega)]} \geq 0 \tag{4.15}$$

In (4.15), $W(\omega)$ includes all the finite $j\omega$ zeros of the real part $R(\omega)$. Integer *ndc* refers to total count of zeros at DC. Once W, a_0 are set and the coefficients c_i are initialized, we can uniquely determine polynomial $a(p)$ and $b(p)$ of $F_M(p) = a(p)/b(p)$ as described in Section 4.1.3. However, for simple lowpass LC ladder structures, there is no finite $j\omega$ zero. Therefore, in (4.15), $W(\omega)$ must be skipped and *ndc* must be set to zero.

As a matter of fact, once $W(\omega)$, a_0 and *ndc* are selected and coefficients $c_i s$ are initialized, our MATLAB function

$$[a, b] = Minimum\_Function(ndc, W, a0, c)$$

immediately generates polynomials $a(p)$ and $b(p)$.

In RF-DCT, real coefficients of the auxiliary polynomial $c(p)$ is the unknowns of the matching problem and they are determined by optimizing the TPG.

For this problem, we set $KFlag = 1$. That means, we have chosen the liberty to work with impedance functions to optimize the TPG. Furthermore, we select $W = [0]$ and $ndc = 0$ to end up with a simple lowpass LC ladder structure as

$$R(\omega) = \frac{(a_0)^2}{(1/2)[c^2(\omega) + c^2(-\omega)]} \geq 0 \tag{4.16}$$

Close examination of (4.16) reveals that at $\omega = 0, R(0) = a_0^2$. This is the termination resistor of the ladder. For this example, we wish to end up with unity normalized termination. Therefore, we set $a_0 = 1$.

Referring to Figure 4.5, $F(p) = Z_{in}(p)$ is generated using our MATLAB function $[a, b] = Minimum_Function(ndc, W, a0, c)$ and polynomial $c(p)$ is determined via optimization of TPG.

Upon completion of the TPG optimization, numerator polynomial $a(p)$ and denominator polynomial $b(p)$ are determined using the optimized coefficients of the auxiliary polynomial $c(p)$. Thus, RF-DCT yields $c(p)$ as

$$c = \begin{bmatrix} -14.3684189015323 & -18.0423604030418 & 19.017623094473 \\ 26.41841021342986 & -6.13233082734834 & -8.2097202383532 \end{bmatrix}$$

which in turn results in $a(p)$ and $b(p)$ as in Table 4.7.

Table 4.7 Minimum reactance function $Z_{\min}(p) = a(p)/b(p)$ of Example 4.6

a(p)	*b(p)*
0	1.
0.5317255405252561	1.612713484895965
0.8575209494686762	1.83560770765771
0.7103323870755622	2.213214814579984
0.748313398653478	0.8784885664075210
0.1814517460254957	0.6239712155615226
0.06959708001646280	0.06959708001646278

Using our lowpass ladder synthesis algorithm, normalized element values are computed by

$$[q] = LowpassLadder\_Yarman(a, b)$$

where normalized element value vector $[q]$ starts with a shunt capacitor $q_1 = C_1$, ends with a series inductor $q_6 = L_6$ and finally LC ladder is terminated in $q_7 = R_7 = 1$ as detailed below.

$$[q] = \begin{bmatrix} 1.880669487894412 \\ 1.064074072964803 \\ 3.603530743693160 \\ 0.923027679356771 \\ 3.481279576098693 \\ 0.620072944987193 \\ 1 \end{bmatrix}$$

Then, the actual elements are obtained by de-normalization with respect to $R0 = 50\,\Omega$ and $f0 = 2,100$ MHz such that capacitors are replaced by

$$C_{i\text{-}actual} = \frac{C_i}{(2\pi f_0 \times R_0)}$$

and inductors are replaced by

$$L_{i\text{-}actual} = L_i \times R_0/(2\pi f_0)$$

Thus, actual elements are obtained as

$$Cx = 3.4\,\text{pF}, C1 = 2.8\,\text{pF}, L4 = 3.49\,\text{nH}, C5 = 5.27\,\text{pF}, L2 = 4\,\text{nH}, C3 = 5.46\,\text{pF},$$
$$L6 = 2.34\,\text{pH}, R7 = 50\,\Omega$$

Resulting lowpass LC ladder network is depicted in Figure 4.6.

In order to end up with the complete design, Figure 4.6 is flipped over, $R7$ is removed and the remaining ladder is connected between the generator and the load impedances as shown in Figure 4.7.

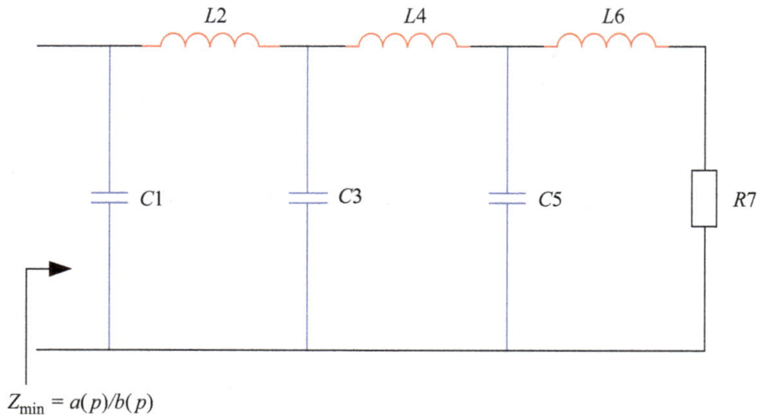

Figure 4.6 Synthesis of $Z_{min}(p)$ of Example 4.6

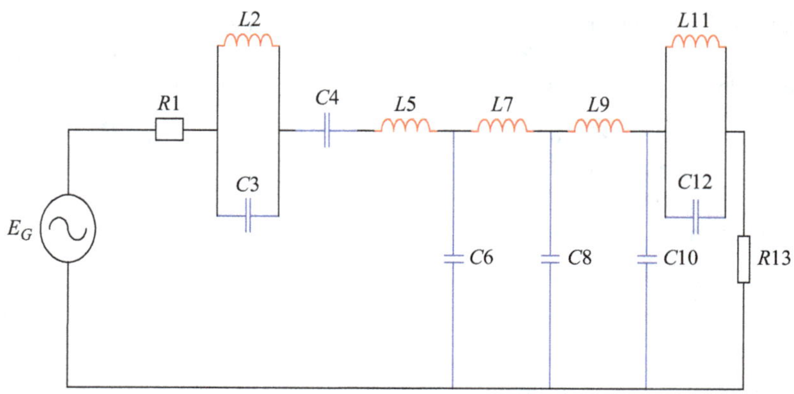

Figure 4.7 Impedance transforming filter with actual elements of Example 4.6 with actual elements: R1 = 12 Ω, L2 = 947.35 pH, C3 = 1.515 pF, C4 = 7.62 pF, L5 = 2.349 nH, C6 = 5.27 pF, L7 = 3.49 nH, C8 = 5.46 pF, L9 = 4.03 nH, C10 = 2.85 pF, L11 = 421 pH, C12 = 1.15 pF, R13 = 50 Ω

For the sake of completeness, in Figure 4.8, TPG of the impedance-transforming filter is depicted. Close examination of this figure reveals that minimum of the passband is less than -0.8 ($T_{min} < -0.8$) at $f_{min} = 1,288$ MHz. Furthermore, gain between second and third harmonic is less than -50 dB which is an excellent solution for practical usage.

All the above computations are gathered under a MATLAB program "Main_Example4_5.m." This program simply operates on RF-DCT package. It optimizes the TPG and generates the driving-point input impedance $Z_{in}(p) = a(p)/b(p)$. Finally, plots TPG and the complete matched system as shown in Figure 4.8.

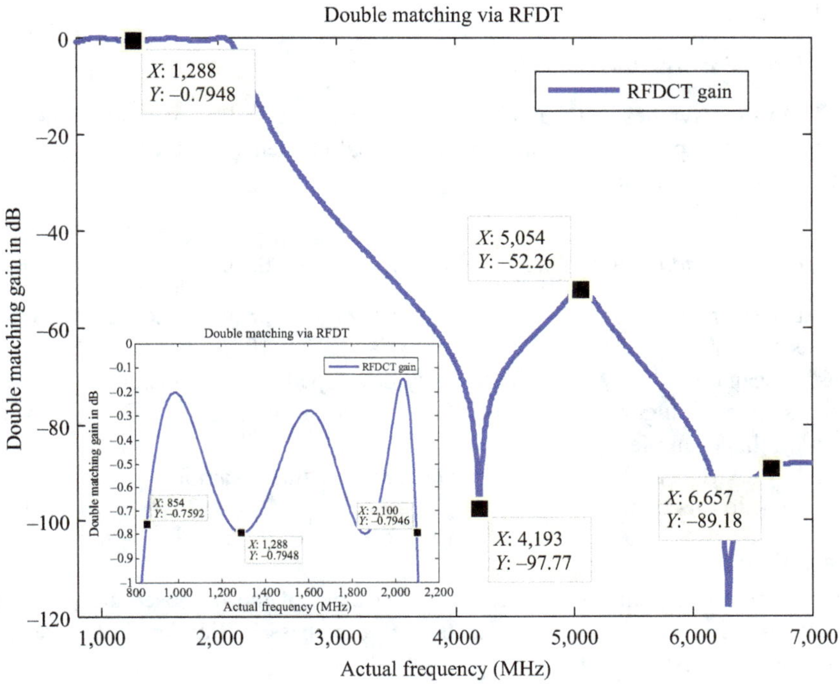

Figure 4.8 Transducer power gain of the impedance transforming filter

Main program calls several MATLAB functions as follows:
Function

$$[c] = nonlinear\_optimization(c0 \cdots)$$

Takes the initial value of the polynomial coefficients $c0$ as a MATLAB vector and determines the coefficients of the auxiliary polynomial $c(p)$ to optimize TPG.
Function

$$[a, b] = Minimum\_Function(ndc, W, a0, c)$$

Computes the driving-point minimum immittance $F(p) = a(p)/b(p)$ from the auxiliary polynomial $c(p)$.
Function

$$[T\_Double] = Gain\_DoubleMatching(\cdots)$$

Computes the double matching gain over the frequencies point by point.
Function

$$[CT, CV] = CircuitPlot\_Yarman(KFlag, R0, f0, a, b, ndc)$$

Synthesizes $F(p) = a(p)/b(p)$ and puts the result in MATLAB vectors CT and CV. CT includes the circuit codes of the network elements. CV contains the element values. Details are skipped here.

Function

$$Plot\_Circuitv1(CT, CV)$$

Draws the synthesized matching network as an LC ladder network alone.

Finally, complete circuit diagram with actual elements together with generator and load networks is drawn using

Function

$$DrawLCLadder\_withGenLoad(KFlag, ndc, a, b, R0, f0).$$

At last, main program *YarmanKilinc_Example6.m* generates the absolute synthesis error as $\varepsilon_a \cong 21.39 \times 10^{-14}$ and $\varepsilon_b \cong 1.95 \times 10^{-15}$.
Thus, we can confidently state that recently developed "high precision LC-lowpass ladder synthesis" algorithm is successfully integrated with the RF-DCT as exhibited with this example.

All the MATLAB programs developed for this example is given as an attachment to this book.

Example 4.6

This example is organized to assess the robustness of the proposed algorithm by increasing the total number of elements in the lowpass LC ladder. In this regard, choosing

$$KFlag = 1, ndc = 0, W = 0, a0 = 1,$$

we generate the coefficients of auxiliary polynomial $c(p)$ using the random number generator command on *MATLAB* as

$$c = 10 \times rand(1, n)$$

The above command produces a random vector $c = [.]$ with n entrees. By trial and error, we tried to determine maximum number reactive elements in the lowpass ladder, which can be safely synthesized numerically. After generating $c(p)$ as above, minimum function $F(p) = a(p)/b(p)$ is computed using our *MATLAB* function

$$[a, b] = Minimum\_Function(ndc, W, a0, c)$$

and it is synthesized employing

$$q = LowpassLadder\_Yarman(a, b)$$

Finally, LC ladder is drawn by means of our *MATLAB* function pairs

$$[CT, CV] = CircuitPlot\_Yarman(KFlag, R0, f0, a, b, ndc);$$

$$Plot\_Circuitv1(CT, CV)$$

To end up with normalized element values, we select

$$R0 = 1; f0 = 1/2/pi$$

We found that we can carry the synthesis of an LC Lowpass ladder with 40 elements with relative errors

$$\varepsilon_{ar} = \frac{norm(a_1 - a)}{norm(a)}$$

$$\varepsilon_{br} = \frac{norm(b_1 - b)}{norm(b)}$$

less than 2×10^{-2} as shown in Figure 4.9.

Element values of Figure 4.9 is given as

L1 = 7.20009 F	L12 = 1.74775 H	C23 = 1.19489 F	L34 = 427.607 mH
L2 = 3.08353 H	C13 = 1.70491 F	L24 = 1.13137 H	C35 = 355.629 mF
C3 = 2.41606 F	L14 = 1.66382 H	C25 = 1.06608 F	L36 = 285 mH
L4 = 2.20857 H	C15 = 1.62203 F	L26 = 999.152 mH	C37 = 216.497 mF
C5 = 2.1204 F	L16 = 1.57799 H	C27 = 930.749 mH	L38 = 150.909 mH
L6 = 2.06687 H	C17 = 1.53105 F	L28 = 861.04 mH	C39 = 88.6464 mF
C7 = 2.01552 F	L18 = 1.48119 H	C29 = 1790.217 mF	C40 = 29.1779 mH
L8 = 1.96063 H	C19 = 1.4266 F	L30 = 718.488 mH	R41 = 1 Ω
C9 = 1.90342 F	L20 = 1.37349 H	C31 = 646.082 mF	
L10 = 1.84712 H	C21 = 1.31608 F	L32 = 573.255 mH	
C11 = 1.7948 F	L22 = 1.31608 F	C33 = 500.308 mF	

All the above computations are gathered under the MATLAB main program called $Main_{Example4_6}.m$, and it is given as an attachment to this book.

Up to this point, a parametric method is presented to synthesize a driving-point input immittance functions as a lowpass LC ladder with high numerical precision. In the process of numerical synthesis, after each element extraction, remaining immittance function is regenerated using parametric approach to warrant lowpass ladder realization. It is shown that difficult synthesis problems such as a "20-element lowpass LC ladder" can be constructed using the high precision synthesis algorithm with accumulated numerical error better than 10^{-6}, whereas classical long division algorithm can never complete the synthesis due to severe error accumulations. Furthermore,

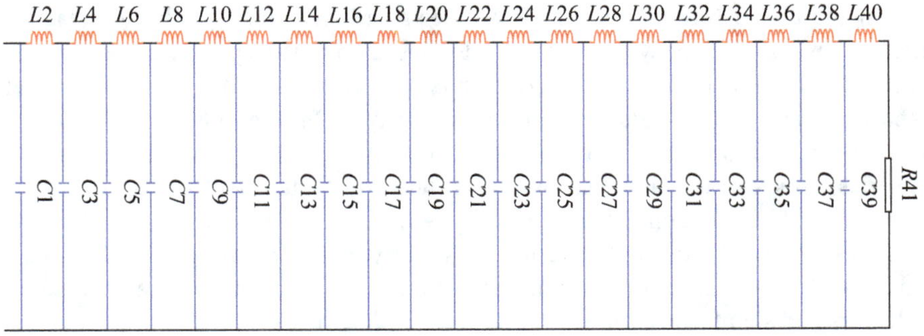

Figure 4.9 Synthesis of driving-point impedance of degree 40

robustness of the proposed synthesis algorithm is tested by synthesizing a randomly generated minimum immittance function with 40 elements.

In the next section, the above-presented synthesis algorithm is extended to synthesize bandpass LC ladders [14].

4.2 LC ladder forms of bandpass structures

A minimum immittance function $F(p)$ is specified as a positive real function in rational form as in (4.1a) and (4.1b). In other words:

$$F(p) = \frac{a(p)}{b(p)} = \frac{a_1p^n + a_2p^{n-1} + a_3p^{n-2} + \cdots + a_np + a_{n+1}}{b_1p^n + b_2p^{n-1} + b_3p^{n-2} + \cdots + b_np + b_{n+1}} \tag{4.17}$$

where the denominator polynomial $b(p)$ is degree of n and it is free of $j\omega$ poles. For many engineering applications, it is desired to set $a_1 = 0$ which makes $F(p)$ zero as frequency approaches to infinity (a zero of transmission at infinity).

Transmission zeros of a minimum immittance function is specified by its zeros and poles as well as the zeros of its even part $R(p)$ which is defined as

$$\begin{aligned} R_{even}(p) &= \frac{1}{2}[F(p) + F(-p)] = \frac{A(p)}{B(p)} \\ &= \frac{A_1p^{2n} + A_2p^{2(n-1)} + \cdots + A_np^2 + A_{n+1}}{B_1p^{2n} + B_2p^{2(n-1)} + \cdots + B_np^2 + B_{n+1}} \end{aligned} \tag{4.18}$$

If $R(p)$ includes transmission zeros only at DC (meaning at $p = 0$) and infinity (meaning $p = \infty$), $A(p)$ is simplified as

$$A(p) = A_0p^{2(ndc)}; n - ndc \geq 0; (-1)^{ndc}A_0 > 0 \tag{4.19a}$$

or

$$R_{even}(p) = \frac{A_0p^{2(ndc)}}{B_np^{2n} + B_2p^{2(n-1)} + \cdots + B_np^2 + B_{n+1}} \geq 0 \tag{4.19b}$$

where $A_0 = A_{n-ndc+1}$ and all the other terms A_k of (4.2a)–(4.2c) are zero.

In the above formulation, integer n designates the total number of reactive elements when $F(p)$ is synthesized as a lossless two-port [*N*] in resistive termination. Integer *ndc* is the count of DC transmission zeros which are realized as series capacitors and shunt inductors in two-port [*N*]. On the other hand, integer $n_\infty = n - ndc > 0$ is the count of transmission zeros at infinity, which are realized as series inductors and shunt capacitors as shown in Figure 4.10.

It should be mentioned that $ndc = 0$ case results in a lowpass LC ladder with all series inductors and shunt capacitors. Similarly, $ndc = n$ case yields a high-pass ladder with all series capacitor and shunt inductors. Therefore, we call the series inductors and shunt capacitor as lowpass reactive elements or in short lowpass elements (LE). Similarly, series capacitors and shunt inductors are called high-pass reactive elements or in short high-pass elements (HE).

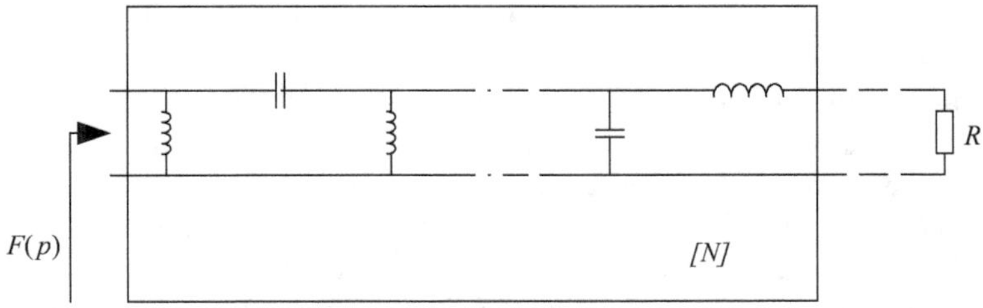

Figure 4.10 Darlington synthesis of $F(p)$ which includes transmission zeros at DC and infinity

If $ndc > 0$, the last term a_{n+1} of $a(p)$ must be zero (i.e., $a_{n+1} = 0$) introducing zeros of transmission at zero.

Synthesis of $F(p)$ can be carried out in a sequential manner by extracting transmission zeros step-by-step.

In cascade synthesis of network functions, at a given pole location $p = p_i$, a transmission zero is realized by removing that pole from the given immittance function. Obviously, in this section, we are only concerned with transmission zeros at $p = 0$ and $p = \infty$. For example, assuming $ndc > 0$, first, transmission zeros at DC can be removed one by one. In this case, a_{n+1} must be zero. Therefore, at step 1, first $F(p)$ is flipped over as $H(p) = 1/F(p) = b(p)/a(p)$ to introduce a pole in $H(p)$. Then, a transmission zero at DC is extracted by removing the pole of $H(p)$ at $p = 0$. At this step:

$$H(p) = \frac{b(p)}{a(p)} = \frac{k_1}{p} + F_r(p) \tag{4.20}$$

where

$$F_r(p) = \frac{R(p)}{a_r(p)}$$

The remainder $R(p)$ and the denominator polynomial $a_r(p)$ are obtained after proper degree cancellations. At the end of this process, $F_r(p)$ may not be a minimum function. Continuing with the pole extraction process with care, after ndc step, remaining immittance function $F_r(p)$ must include transmission zeros only at infinity. In this case, $F(p)$ is expressed as

$$H(p) = \frac{k_1}{p} + \cfrac{1}{(k_2/p) + \cfrac{1}{\cdots + \cdots + \cfrac{1}{(k_{ndc}/p) + F_r(p)}}} \tag{4.21}$$

From the programming point of view, after each step, remaining positive real function $F_r(p)$ can be renamed or initialized as $F_r = a(p)/b(p)$ by setting

$a(p) = R(p)$ and $b(p) = a_r(p)$ so that the same algorithm is employed to remove the poles at $p = 0$ from the inverse function $H_r(p) = b(p)/a(p)$.

As far as actual realization of $H(p)$ is concerned, if the synthesis begins with a minimum reactance impedance $F(p)$, $H(p) = 1/F(p)$ is an admittance function. Therefore, the first element of the ladder must be a shunt inductor and it is specified by $L_1 = 1/k_1$. In this case, the second element will be a series capacitor with $C_2 = 1/k_2$ and so on. As a rule of thumb, we can say that, if the initial minimum function $F(p)$ is an impedance function then, the odd index terms of $H(p)$ of (4.21) will be shunt inductors and even index terms will be series capacitors. Obviously, if *ndc* is an even integer, the front high-pass section will end with a series capacitor, or if it is an odd integer, the high-pass section will end with a shunt inductor as shown in Figure 4.11(a) and (b), respectively.

On the other hand, if $F(p)$ is a minimum susceptance function (or equivalently minimum admittance), its inverse $H(p) = 1/F(p)$ will be an impedance function. Therefore, in this case, actual synthesis starts with a series capacitor $C_1 = 1/k_1$ and continues with a shunt inductor $L_2 = 1/k_2$ as shown in Figure 4.12(a) and (b).

Remaining lowpass section can be synthesized by extracting transmission zeros at infinity from the immittance function $F_r(p) = a(p)/b(p)$. As mentioned before, $F_r(p) = a(p)/b(p)$ may be a minimum function with $a_1 = 0$ and $b_1 \neq 0$ or it may not be minimum. In this case, $a_1 \neq 0$ and $b_1 = 0$. If $F_r(p)$ is a non-minimum function, lowpass synthesis directly starts by expressing $F_r(p)$ as

$$\frac{k_{ndc}}{p} + F_r(p) \tag{4.22a}$$

(last step of poles extractions at $p = 0$)

$$F_r(p) = \frac{a(p)}{b(p)} = q_1 p + \frac{R(p)}{b_r(p)} \tag{4.22b}$$

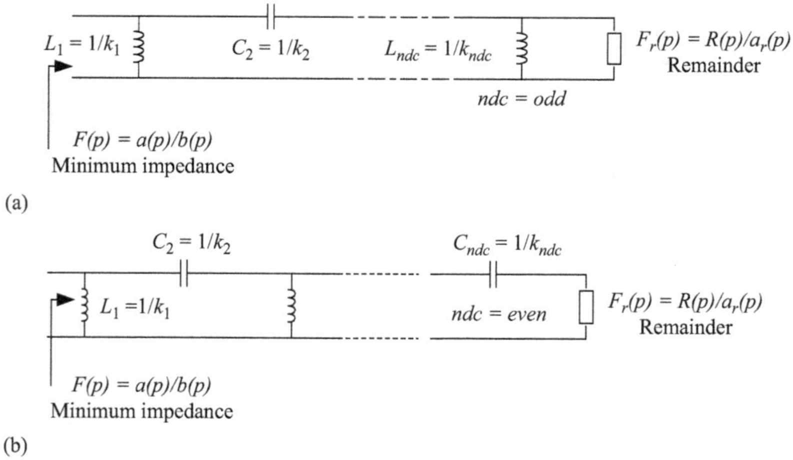

Figure 4.11 (a) Darlington synthesis of $F(p)$ for ndc = odd case, (b) Darlington synthesis of $F(p)$ for ndc = even case

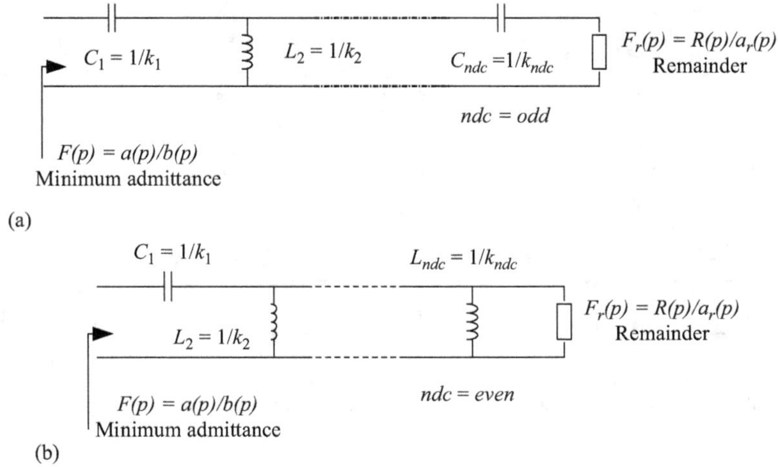

Figure 4.12 *(a) Synthesis of the high-pass section of a minimum admittance $F(p)$ for ndc = odd case, (b) Synthesis of the high-pass section of a minimum admittance $F(p)$ for ndc = even case*

or the last step is expressed as

$$\frac{k_{ndc}}{p} + q_1 p + \frac{R(p)}{b_r(p)} \tag{4.22c}$$

In this case, at the very beginning of the synthesis process, if the starting function $F(p) = a(p)/b(p)$ is a non-minimum reactance function, and if $ndc =$ even then, the lowpass section starts with a series inductor $L_{(ndc+1)} = q_1$, and it ends with a shunt capacitor $C_n = q_{n\infty}$ in parallel with a terminating conductance G if $n_\infty = n - ndc$ is an even integer as shown in Figure 4.13(a).

On the other hand, if $F_r(p)$ is a minimum function, it must be flipped over to be able to extract a pole at infinity from the inverse function $H_r(p) = b(p)/a(p)$ such that

$$\frac{k_{ndc}}{p} + F_r(p); \text{(last step of pole extractions at } p = 0) \tag{4.23a}$$

or

$$\frac{k_{ndc}}{p} + \frac{1}{H_r(p)} = \frac{k_{ndc}}{p} + \frac{1}{q_1 p + (R(p)/a_r(p))} \tag{4.23b}$$

In this case, series capacitor C_{ndc} is followed by a shunt capacitor $C_{(ndc+1)}$ as shown in Figure 4.13(b), and so on for $ndc = even$ and $F_r(p) = R(p)/a_r(p)$ minimum impedance case; $n_\infty = n - ndc = odd$.

Just for the sake of completion, let us mention that sequential cascaded connection of series capacitors and shunt inductors constitute a high-pass ladder network. Therefore, these elements are called "HE" (i.e., shunt inductors, series capacitors).

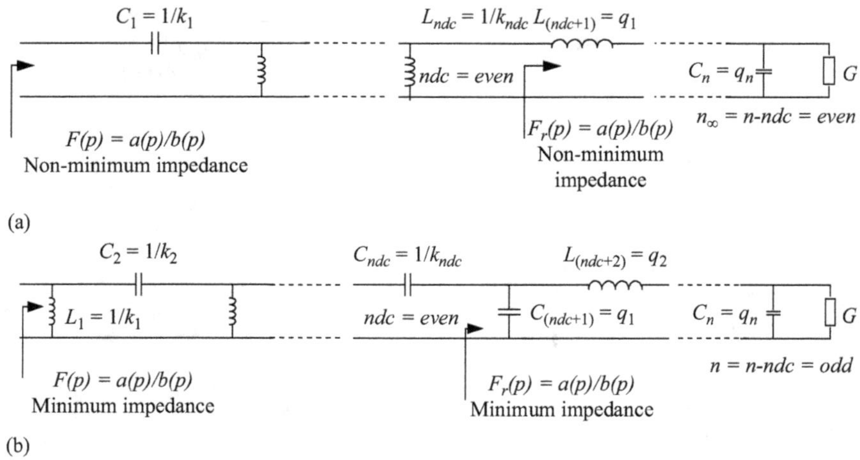

Figure 4.13 *(a) Synthesis of the lowpass section of a non-minimum admittance $F(p)$ for $n_\infty = n - ndc = $ even case, (b) synthesis of the lowpass section of a non-minimum admittance function $F(p)$*

Similarly, connection of series inductors and shunt capacitors form a lowpass ladder. Therefore, these elements are called "LE."

Even though extraction process looks straight forward throughout the numerical computations, one may end up with severe accumulation errors which destroys the ladder form of [N] by introducing nonzero terms in $A(p)$ beyond $A_0p^{2(ndc)}$. Unfortunately, this is a classical problem of network synthesis.

In our previously introduced high-precision lowpass LC ladder synthesis algorithm, for $ndc = 0$ case, we have employed the "parametric method" to overcome severe accumulation errors by enforcing the analytic ladder form of (4.9a) and (4.9b) and regenerating $F_r(p)$ from its even part at each step. In this approach, we start with a minimum function $F(p) = a(p)/b(p)$ with $a_1 = 0$. In this case, even part of $F(p)$ is given by

$$R_{even}(p) = \frac{A_0}{B_n p^{2n} + B_2 p^{2(n-1)} + \cdots + B_n p^2 + B_{n+1}} \tag{4.24}$$

The first transmission zero at infinity is completely removed from the inverse function $H(p) = 1/F(p) = b(p)/a(p)$ such that

$$H(p) = q_r p + F_r(p) \tag{4.25a}$$

$$F_r(p) = \frac{R(p)}{a_r(p)} \tag{4.25b}$$

The remaining function $F_r(p)$ must be minimum and possess an even part as in (4.24) of degree $(n-1)$. However, due to accumulation errors, numerator polynomial $A(p)$ of the even part $R_{even}(p) = [1/2][F_r(p) + F_r(-p)]$ may have very small coefficients but not exactly zero beyond $A_{n+1} = A_0$. In this case, by defining an algorithmic zero such as $\varepsilon_{zero} = 10^{-m}; m > 1$, we can set all nonzero terms to zero if they are smaller than ε_{zero}. In this way, we force the even part $R_{even}(p) = [1/2][F_r(p) + F_r(-p)]$ to be lowpass ladder form, regenerate $F_r(p)$ by means of parametric approach from the corrected even part as specified by (4.8a) and (4.8b). This process continues until all the transmission zeros are extracted.

In the previous section, it was shown that one can synthesize a lowpass ladder up to 40 elements from a given driving-point function $F(p)$ with a computational resolution better than 10^{-1} which may be acceptable for many practical problems.

Unfortunately, extractions of DC transmission zeros are tricky and requires special care since $F_r(p)$ of (4.17) may not be a minimum function. Therefore, in this section, "parametric method" is modified to carry out the high precision synthesis of non-minimum immittance functions which includes transmission zeros at DC and infinity.

From the presentation point of view, it is easy to understand new synthesis procedure step-by-step by means of the case studies as we develop the algorithm. In the following sections, first we introduce a practical procedure to generate a minimum realizable immittance function which has transmission zeros at DC and infinity, then high precision synthesis is detailed.

4.2.1 Generation of a minimum function via parametric approach for a bandpass LC ladder network

A minimum function $F(p) = a(p)/b(p)$ can be generated from its even part using parametric approach.

On the real frequency axis, even part is given by

$$R_{even}(j\omega) = \frac{(a_0^2)W(j\omega)\omega^{2ndc}}{b(j\omega)b(j\omega)} \tag{4.26}$$

where $W(p)$ includes all the finite transmission zeros of $F(p)$ beyond $p = 0$. For this section, we only use transmission zeros at $p = 0$ and $p = \infty$. Therefore, the term $W(j\omega)$ of (4.26) is omitted (i.e., finite transmission zeros which are included in a MATLAB vector $W(j\omega)$ is set to zero in the programs).

Denominator polynomial $C(\omega^2) = b(j\omega)b(j\omega)$ must be a positive even polynomial. It may be generated by means of an auxiliary polynomial $c(\omega) = c_1\omega^n + c_2\omega^{n-1} + \cdots + c_n\omega + 1$ such that

$$b(j\omega)b(j\omega) = C(\omega^2) > 0 \tag{4.27a}$$

where

$$C(\omega^2) = \frac{1}{2}\left[c^2(\omega) + c^2(-\omega)\right] = C_1\omega^{2n} + C_2\omega^{2(n-1)} + \cdots + C_n\omega^2 + 1 \tag{4.27b}$$

Replacing ω^2 by $-p^2$, a realizable ladder form of the even part is given by

$$R_{even}(p^2) = \frac{A(p^2)}{B(p^2)} \tag{4.28a}$$

$$A(p)^2 = (-1)^{ndc} a_0^2 p^{2ndc} = A_0 p^{2ndc}; \text{ with } A_0 = (-1)a_0^2 \tag{4.28b}$$

$$\begin{aligned} B(p)^2 &= (-1)^n C_1 p^{2n} + (-1)^{n-1} C_2 p^{n-1} + \cdots - C_1 p + 1 \\ &= B_1 p^n + B_2 p^{2(n-1)} + \cdots + B_1 p^2 + 1 \end{aligned} \tag{4.28c}$$

where

$$B_i = (-1)^{(n-i+1)} C_i; \; i = 1, 2, \cdots, n \tag{4.28d}$$

It should be emphasized that once *ndc* is selected and the coefficients $\{a_0, c_1, c_2, \cdots, c_n\}$ are initialized, coefficients of $B(p^2)$ can easily be generated on MATLAB which in turn yields a realizable ladder form for $R_{even}(p^2)$.

In parametric approach, a positive real minimum immittance function is represented in terms of its poles p_i and its corresponding residues k_i as

$$F(p) = F_0 + \sum_{i=1}^{n} \frac{k_i}{p - p_i} \tag{4.29}$$

Using (4.26), $R_{even}(p^2)$ is expressed in terms of the same residues and poles as

$$R_{even}(p^2) = \frac{1}{2}[F(p) + F(-p)] = F_0 + \sum_{i=1}^{n} \frac{k_i p_i}{p^2 - p_i^2} \tag{4.30}$$

where the residues k_i are computed from the realizable form of $R_{even}(p^2)$ as specified by the following equation:

$$k_i = (-1)^n \frac{(-1)^{ndc} a_0^2}{p_i B_1 \prod_{\substack{j=1 \\ j \neq i}}^{n} p_i^2 - p_j^2} \tag{4.31}$$

For the section under consideration if $ndc < n$ then, $F_0 = A_1/(-1)^n C_1 = 0$ and for $n = ndc$ $F_0 = (-1)^{n-ndc}/C_1$ (notice that $C_1 = c_1^2 > 0$).

Hence, positive real minimum immittance function production starts by selecting *ndc* and initializing the coefficients $\{a_0, c_1, c_2, \cdots, c_n\}$. Then, the even polynomial $C(\omega^2) = (1/2)[c^2(\omega) + c^2(-\omega)] = C_1\omega^{2n} + C_2\omega^{2(n-1)} + \cdots + C_n p^2 + 1$ computed and replacing $\omega^2$by $-p^2$ $B(p^2)$ is obtained. Thereafter, poles p_i are computed and k_i are determined by means of (4.31). Finally, $F(p)$ is generated as a minimum rational function $F(p) = a(p)/b(p) = ((a_1 p^n + a_2 p^{n-1} + \cdots + a_n p + a_{n+1})/(b_1 p^n + b_2 p^{n-1} + \cdots + b_n p + b_{n+1}))$.

The above formulation is programmed as a MATLAB function

$$[a, b] = Minimum\_Function(ndc, W, a0, c).$$

In the following section, we introduce a procedure to extract a transmission zero at DC with high resolution from a given immittance function.

4.2.2 Extraction of a transmission zero at DC

In the previous section, we generated a minimum function which has built-in DC transmission zeros. These zeros can be extracted by removing the poles at $p = 0$ from the inverse function as follows.

Initially, the full form of a minimum immittance function $F(p)$ is specified as

$$F(p) = \frac{a(p)}{b(p)} = \frac{a_1p^n + a_2p^{n-1} + \cdots + a_np + a_{n+1}}{b_1p^n + b_2p^{n-1} + \cdots + b_np + b_{n+1}} \tag{4.32}$$

with $a_1 = 0$ (indication of a transmission zero at infinity) and

$a_{n+1} = 0$ (indication of at least one DC transmission zero)

Inverse of this function is given by

$$H(p) = \frac{1}{F(p)} = \frac{b_1p^n + b_2p^{n-1} + \cdots + b_np + b_{n+1}}{a_2p^{n-1} + \cdots + a_np} \tag{4.33}$$

Thus, a DC transmission zero is extracted by removing the pole $p = 0$ as follows:

$$H(p) = \frac{b(p)}{a(p)} = \frac{k_r}{p} + F_r(p) \tag{4.34a}$$

with

$$F_r(p) = \frac{R}{a_r(p)} \tag{4.34b}$$

where

$$k_r = lim_{p\to 0}\mathrm{p}\mathrm{H}(p) = \frac{b_{n+1}}{a_n} \tag{4.35a}$$

$$a_r(p) = \frac{1}{p}a(p) = a_2p^{n-2} + a_3p^{n-3} + \cdots + a_{n-1}p + a_n \tag{4.35b}$$

$$R(p) = \frac{1}{p}[b(p) - k_ra_r(p)] = R_1p^{n-1} + R_2p^{n-2} + \cdots + R_{n-1}p + R_n \tag{4.35c}$$

Here, it is important to note the following points:

If $ndc > 1$, after we remove the first transmission zero at DC, the last term R_n of $R(p)$ must be zero to introduce the left-over transmission zeros at DC.

Due to cancellation of the common term p which appears in $R(p)$ and already built-in $a(p)$, degree of $a_r(p)$ is $(n - 2)$.

Degree of $R(p)$ is $(n - 1)$ since the leading term of $R(p)$ is $(1/p)b_1p^n = b_1p^{n-1}$.

$F_r(p) = R(p)/a_r(p)$ is not a minimum function since $\lim_{p\to\infty} F_r(p) \to (R_1/a_2)$ $p \to \infty$ indicating that $F_r(p)$ has a pole at ∞.

Even part of $F_r(p)$ must yield a ladder form as in (4.26) with total of $(ndc - 1)$ transmission zeros at DC. However, due to error accumulations, this may not be the case.

Above introduced essential points are programmed under the following MATLAB functions as presented at the end of this chapter.

Function $[A, B] = even\_part(a, b)$ generates the even part of a positive real function $F(p) = a(p)/b(p)$.

Function $[kr, R, ar] = Highpass\_Remainder(a, b)$ extracts a pole at DC ($p = 0$) with residue k_r as detailed by (4.35a). Once extraction is completed, for the next step, remainder $R(p)$ is set to $a(p)$ and $a_r(p)$ is set to $b(p)$ to continue for the follow-up steps of the synthesis.

In the following section, we present a fine process to remove the pole at infinity from a given proper function $F(p)$.

4.2.3 Extraction of a pole at infinity

A non-minimum function $F(p)$ with a pole at infinity, is described as $F(p) = a(p)/b(p)$ such that

$$a(p) = a_1 p^n + a_2 p^{n-1} + \cdots + a_n p + a_{n+1}$$

and

$$b(p) = 0p^n + b_2 p^{n-1} + \cdots + b_n p + b_{n+1}$$

It should be noted that a pole at infinity is introduced when the coefficient b_1 is zero (i.e., $b_1 = 0$). In this case, $F(p)$ is expressed as

$$F(p) = k_\infty p + F_{\min}(p)$$

where the minimum function is described as

$$F_{\min}(p) = \frac{a_{\min}(p)}{b_{\min}(p)}$$

The residue k_∞ is given by

$$\begin{aligned} k_\infty &= \lim_{p\to\infty} \frac{1}{p}\frac{a(p)}{b(p)} \\ &= \lim_{p\to\infty} \frac{a_1 p^n + a_3 p^{n-2} + \cdots + a_n p + a_{n+1}}{b_2 p^{n-1} + \cdots + b_n p + b_{n+1}} = \frac{a_1}{b_2} \end{aligned} \tag{4.36}$$

Once a pole at infinity is extracted, the remaining function is expressed as

$$F(p) = \frac{a(p)}{b(p)} = k_\infty p + \frac{R(p)}{b_r(p)}$$

where the reminder $R(p)$ is specified as

$$R(p) = a(p) - [k_\infty p][b(p)] \tag{4.37}$$

Degree of $b_r(p)$ is $(n - 1)$.

In order to end up with degree reduction, first two terms of $R(p)$ must be zero introducing leftover transmission zeros at infinity.

In order to execute above computations, we developed the following MATLAB functions.

Function

$$[kinf, R, br] = removepole\_atinfinity(anew, bnew)$$

removes a pole at infinity as described by (4.36) and (4.37). Then, $F_r(p) = R(p)/b_r(p)$ becomes a minimum positive real function.

$$Function[a1, b1, ndc] = check\_immittance(a\_min, b\_min)$$

Checks the immittance function $F_{\min}(p) = a_{\min}(p)/b_{\min}(p)$ if it satisfies an LC bandpass ladder form. This function first generates the even part $R(p) = A(p)/B(p)$ then compares all the coefficients of $A(p)$ with a selected small number $\varepsilon_{zero} = 10^{-m}$. Then decides whether it is a ladder or not. If it is a ladder within a precision of 10^{-m}, from the corrected real part, $F_{\min}(p)$ is regenerated as an exact ladder using parametric approach.

Function

$[a\_G, b\_G, ndc] = kinfFmin\_ToF(kinf, a\_min, b\_min)$ generates the corrected non-minimum function from the given pole at infinity described by the residue k_∞ and corrected minimum function $F_{\min}(p) = a_{\min}(p)/b_{\min}(p) = R(p)/b_r(p)$.
In the following case study, we introduce the complete synthesis of the above minimum immittance function step-by-step, employing the MATLAB functions developed for the section under consideration. Eventually, complete synthesis procedure is gathered under an algorithm implemented on the MATLAB environment.

4.2.4 Bandpass LC ladder synthesis algorithm by means of case studies

4.2.4.1 Case Study 1

In this section, let us synthesize the function $F(p)$ which is specified by Table 4.8 as an impedance function step-by-step.

Table 4.8 Minimum reactance function of case study 1 $F(p) = Z(p) = a(p)/b(p)$

a(p)	*b(p)*
0	1.000000000000000
0.513976968750998	4.198433713455139
1.157898233543669	7.913422823138356
1.213142248104650	9.268979838320304
0.694401166177243	7.284067053656422
0.188516938299314	3.683494822490302
0	1.000000000000000

Step 1a: First, we must check the impedance function if it yields an LC ladder form.

The leading coefficient of $a(p)$ is $a(1) = 0$. $F(p)$ has a chance to be a minimum reactance function (or equivalently a minimum impedance). Furthermore, $F(p)$ must include at least one transmission zero at DC since $a(7) = 0$. However, concrete decision can be made after we drive the even part of $F(p)$ and check poles and zeros if they are placed in the closed left half plane. We can find the zeros and the poles of $F(p)$ using the MATLAB statement $pa = roots(a)$ and $pb = roots(b)$, respectively.

The above MATLAB statements result in Table 4.9.

Close examination of Table 4.9 indicates that numerator polynomial has a simple zero at $p = 0$ and the other zeros are placed in the closed LHP and the poles of $F(p)$ are all also in the closed LHP. Therefore, $F(p)$ must be a minimum function (minimum reactance).

Step 1b: Let us now investigate if $F(p)$ can be realized as an LC ladder network. For this purpose, we can employ our MATLAB function *even_part* to generate the even part of $F(p)$. Execution of $[A, B] = even\_part(a, b)$ yields

$$R_{even-part}(p^2) = A(p^2)/B(p^2)$$

with

$$A = [0, -1, -7.993 \times 10^{-15}, -1.95 \times 10^{-14} \times 10^{-14}, -3.88 \times 10^{-15}, 0]$$

Ignoring the terms with absolute values less than 10^{-13}, we can say that $F(p)$ describes an LC ladder network with $ndc = 5$ transmission zeros at DC and $n_\infty = n - ndc = 6 - 5 = 1$ transmission zero at infinity. Hence, algorithmic zero $\varepsilon_{zero} = 10^{-13}$ would be sufficient to consider $F(p)$ as a proper impedance function which can be synthesized with five high-pass and one LE.

Step 1c: We can always refine the coefficients of numerator and denominator of $F(p)$ via parametric approach using our MATLAB function *Check_immittance*(a, b). Now, we can initiate the synthesis process.

Step 2: Start synthesis by extracting HE.

It is reminded that a transmission zero at $p = p_i$ is extracted by removing a pole at that location from the impedance or admittance function. For the case

Table 4.9 Minimum reactance function of case study 1

Zeros of the numerator polynomial *a*(*p*)	**Zeros of the denominator polynomial *a*(*p*)**
0	−0.1986+0.8602*I*
−0.3897+0.6586*I*	−0.1986−0.8602*I*
−0.3897−0.6586*I*	−1.2019
−0.7367+0.2890*I*	−1.3161
−0.7367−0.2890*I*	−0.6415+0.6321*I*
	−0.6415−0.6321*I*

under consideration, our transmission zeros are placed at $p = 0$ and $p = \infty$. In this work, we use the freedom to extract all the transmission zeros at $p = 0$(i.e., at DC) at the first glance. Then, transmission zeros at infinity are removed.

Step 2a: Extract the first DC transmission zero as a pole from the inverse function

$$H(p) = \frac{b(p)}{a(p)} = \frac{k_1}{p} + F_r(p) \text{ such that } F_r(p) = \frac{R(p)}{a_r(p)}$$

At this step, we use our MATLAB function $[kr, R, ar] = Highpass_{Remainder(a,b)}$. Thus, the remainder polynomial $R(p)$ is given by

$$R = \begin{bmatrix}1 & 4.198 & 5.186 & 3.126 & 0.848 - 1.1 \times 10^{-14}\end{bmatrix}$$

The denominator polynomial is found as

$$a_r = [0 \; 0.5139 \; 1.1578 \; 1.213 \; 0.6944 \; 0.188]$$

To be able to preserve the ladder form, the last term of $R(p)$ must be zero to keep the left-over transmission zeros at DC. Notice that $R(6) = -1.1 \times 10^{-14}$ which is very small but not exactly. So, we can confidently set $R(6) = 0$. In other words, previously set algorithmic zero $\varepsilon_{zero} = 10^{-13}$ successfully works to make $R(6) = 0$. Once we set $R(6) = 0$, we may refine the polynomial coefficients of $R(p)$ and $a_r(p)$ of $F_r(p) = R(p)/a_r(p)$ via parametric approach. Before we start the operation, we express $F_r(p) = a(p)/b(p)$ as a new function by storing $R(p)$ and $a_r(p)$ in new polynomials $a(p) = R(p)$ and $b(p) = a_r(p)$, respectively. We observe that the single transmission zero at infinity has been preserved by having $a_r(1) = 0$ which is the leading coefficient of $a_r(p)$. Therefore, $F_r(p)$ has a pole at infinity. In this case, first we can remove the pole at infinity by expressing

$$F_r(p) = \frac{a(p)}{b(p)} = k_\infty p + F_{\min}(p)$$

where $F_{\min}(p)$ is the minimum function and it is generically specified by $F_{\min}(p) = a_{\min}(p)/b_{\min}(p)$

The above from is obtained employing our MATLAB function $[kinf, R, br] = removepole\_atinfinity(a, b)$. In this case, the new residual function $F_r(p) = R(p)/b_r(p)$ is defined as $F_{\min}(p) = a_{\min}(p)/b_{\min}(p) = R(p)/b_r(p)$. At this point, coefficients of $a_{\min} = R(p)$ and $b_{\min}(p) = b_r(p)$ can be refined by means of parametric approach to yield an LC ladder structure using our MATLAB function $[a\_min, b\_min, ndc] = Check\_immittance(a\_min, b\_min)$. Then we can regenerate original full coefficient form of $F_r(p) = a(p)/b(p)$ with high numerical precision. These numerical operations are gathered within our MATLAB function

$$[a, b] = Laddercorrection\_onF(eps\_zero, a, b)$$

This function refines the coefficients of a given immittance function $F(p) = a(p)/b(p)$ whether it is a minimum function or not. Function laddercorrection is given at the end of this chapter. Execution of laddercorrection yields the high precision ladder form with absolute error less than 10^{-13}.

Step 2b: Now we are ready to extract the second DC transmission zero from the new

$$F(p) = \frac{a(p)}{b(p)}$$

where

$$a = [1.9458.16810.0916.0831.6510]$$

$$b = [01\ \ 2.252\ \ 2.360\ \ 1.351\ \ 0.3667]$$

It must be noted that if both $a(p)$ and $b(p)$ are divided by $= a(1) =$ 1.94561247059392, we end up with the above form of $R(p)$ and $a_r(p)$.

Using our MATLAB function $[kr, R, ar] = Highpass\_Remainder(a, b)$, we can extract the pole at DC from the new inverse function $H(p) = b(p)/a(p)$. Thus, we have,

$$\begin{aligned} k_r &= k_2 = 0.222 \\ R &= \left[0\ \ 0.567\ \ 0.4387\ \ 0.11911 - 1.33 \times 10^{-15}\right] \\ a_r &= [1.945\ \ 8.168\ \ 10.0918\ \ 6.0836\ \ 1.6515] \\ R(5) &= -1.33 \times 10^{-15} \end{aligned}$$

So, as in Step 2a, we can set $R(5) = 0$ and define the new function $F(p) = F_r(p) = a(p)/b(p)$ such that $a(p) = R(p)$ and $b(p) = a_r(p)$. Then, we refine the coefficients of $a(p)$ and $b(p)$ employing out MATLAB function laddercorrection to be ready for the next step.

Step 2c: At this step, third transmission zero is extracted using our MATLAB function

$$[kr, R, ar] = Highpass\_Remainder(a, b)$$

and the above steps are repeated.

Thus, we end up with the following results.

$$\begin{aligned} kr &= k_3 = 13.865 \\ R &= [1\ \ 4.198\ \ 1.1397\ \ 0.000000000000004] \\ a_r &= [0\ \ 0.2918\ \ 0.2255\ \ 0.06122] \\ a &= [3.425\ \ 14.383\ \ 3.904\ \ 0] \\ b &= [0\ \ 1\ \ 0.7725\ \ 0.2097] \end{aligned}$$

Notice that, at this step, $b(1) = 0$ and new $F(p) = a(p)/b(p)$ has a pole at infinity as in Step 2a. Furthermore, $a(4) = 0$ which is the indication of left-over transmission zeros at DC.

Let us run Step 2d to extract the fourth transmission zero.

Step 2d: As in Step 2b, at this step, we end up with a minimum function $F_r(p) = R(p)/a_r(p) = F(p) = a(p)/b(p)$. Results are summarized as

$$kr = k_4 = 0.053714976706065$$
$$R = [0 \;\; 0.8159809749122820]$$
$$ar = [3.425 \;\; 14.383 \;\; 3.9047]$$
$$a = [0 \;\; 0.238 \;\; 0]$$
$$b = [1 \;\; 4.1984 \;\; 1.1397]$$

Last transmission zero at DC is extracted at Step 5.

Step 2e: This is the final step to extract the last DC transmission zero. The final residue at $p = 0$ is

$$kr = k_5 = 4.785359148234052$$

In this step, the last term of $R(p)$ should not be zero since the remaining function must be free of DC transmission zeros. In fact

$$R = [14.198433713455144]$$

On the other hand,

$$ar = [0\,0.238184062974532]$$

Thus, at this step, resulting remainder function $F_r(p) = R(p)/a_r(p)$ is not a minimum function since $ar(1) = 0$ indicating a transmission zero at infinity.

Step 3: In Step 2, all the DC transmission zeros are extracted by removing the poles at $p = 0$ from the inverse function $H(p) = 1/F(p) = b(p)/a(p)$. Hence, high-pass section of the original impedance function $F(p)$ is completely removed. Now, we are ready to synthesize the remaining function $F_r(p) = R(p)/a_r(p)$ as a lowpass ladder.

At this step, we can use our MATLAB function

$$q = LowpassLadder\_Yarman(a, b)$$

Function *LowpassLadder_Yarman* works on minimum functions. However, currently computed remainder function $F_r(p) = R(p)/a_r(p)$ is not a minimum function but its inverse is minimum function. Therefore, at this step, we can flip over $F_r(p)$ and then call "LowpassLadder_Yarman" as $q = LowpassLadder\_Yarman(b, a)$.

Execution of $q = LowpassLadder\_Yarman(b, a)$ reveals that

$$q = [4.198417.6268]$$

Thus, we end up with the complete synthesis as shown in Figure 4.14. Minimum impedance function starts with a shunt inductor $L_1 = 1/k_1 = 5.3046$.

The second element is a series capacitor $C_2 = 1/k_2 = 0.2221$, the third element is a shunt inductor $L_3 = 1/k_3 = 13.8651$, the fourth element is a series capacitor $C_4 = 1/k_4 = 0.0537$ and the last high-pass element is a shunt

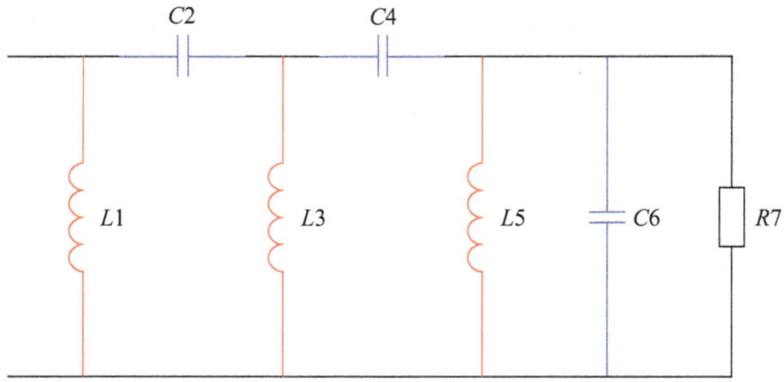

Figure 4.14 Darlington synthesis of $F(p)$ as specified by Case Study 1 with normalized element values: L1 = 188.51 mH, 4.5 F, L3 = 72.12 mH, C4 = 18.61, L5 = 208.97mH, C6 = 4.19 F, R7 = 56.73 mΩ

inductor $L_5 = 1/k_5 = 4.7854$. Once the high-pass section synthesis is completed, the remaining non-minimum admittance function includes only one transmission zero at infinity. In this case, $F_r(p) = q_1 p + q_2$, where q_1 is the value of a shunt capacitor $C_6 = q_1 = R(1)/ar(2) = a(1)/b(2) = 4.1984$ and the terminating constant is a conductance which is computed as $q_2 = G = R(2)/ar(2) = a(2)/b(2) = 17.6268$ or equivalently terminating shunt resistance is $R_7 = 1/q_2 = 0.0567$.

The above major steps constitute an algorithmic structure for the high precision synthesis of a minimum function as an *LC* ladder terminated in a positive constant $q(n+1)$. For the sake of clear understanding, we gathered all the above computational steps under a *MATLAB* program called *CaseStudy*1*.m*.

Based on the above discussions, we can state the following rule of thumbs for the synthesis of a minimum reactance function. These rules constitute the outline of the high precision band pass *LC* – *Ladder* synthesis algorithm.

4.2.5 General rules for bandpass LC ladder synthesis

For the sake of easy explanation, here, we assume that driving-point input immittance $F(p) = a(p)/b(p)$ is a minimum reactance (impedance) function without loss of generality.

If the synthesis process starts with the extractions of HE, the first element must be a shunt inductor since $F(p)$ is a minimum reactance function expressed by means of its inverse with a pole at $p = 0$ as

$$H(p) = \frac{1}{F(p)} = \frac{k_1}{p} + F_r(p) = \frac{1}{L_1 p} + F_r(p)$$

Obviously, the second high-pass element is a series capacitor if $ndc > 1$.

In general, we can state that, if the synthesis of a minimum reactance function starts with the extraction of HE, the odd indexed terms are shunt inductors and the even-indexed terms are series capacitors.

During the extraction process of odd-indexed elements, the remaining function $F_r(p) = R(p)/a_r(p)$ is a non-minimum admittance function. This is useful information to refine coefficients of numerator and denominator polynomial $F_r(p)$ to yield a high precision LC ladders employing the MATLAB function laddercorrection.

During the extraction process of even indexed elements, the remaining function $F_r(p) = R(p)/a_r(p)$ is a minimum reactance function. This is also useful information to refine coefficients of numerator and denominator polynomial $F_r(p)$ to yield a high-precision LC ladders using the MATLAB function check_immittance.

After extracting all transmission zeros at $p = 0$ as shunt inductors and series capacitor within *ndc* steps, we can start extraction of LE from the remaining function $F_r(p)$ as series inductors and shunt capacitors in $n_\infty = n - ndc$ steps.

If $ndc =$ even, synthesis of the lowpass section starts with an impedance function.

If $ndc =$ odd, synthesis of lowpass section starts with an admittance function.

No matter what *ndc* is, for both odd and even-indexed components, high-pass circuit elements of the LC ladder network can be extracted employing our MATLAB function

$$[kr, R, ar] = Highpass\_Remainder(a, b)$$

where the immittance function to be synthesized is described as

$$F(p) = \frac{a(p)}{b(p)} = \frac{k_r}{p} + F_r(p)$$

such that

$$F_r(p) = \frac{R(p)}{a_r(p)}$$

Whether it is minimum or not, the remaining function $F_r(p) = R(p)/a_r(p)$ is redefined as $F(p) = a(p)/b(p)$ by setting $a(p) = R(p)$ and $b(p) = a_r(p)$. Then all the HE of the LC ladder network can be extracted one by one within a loop which runs on index i up to ndc.

Once high-pass section is constructed, resulting residual function $F_r(p) = R(p)/a_r(p)$ includes only transmission zeros at infinity.

If $F_r(p) = R(p)/a_r(p) = F(p) = a(p)/b(p)$ is a minimum function, our MATLAB function

$$q = LowpassLadder\_Yarman(a, b)$$

completely removes the transmission zeros at infinity as series inductors and shunt capacitors. On the other hand, if $F(p) = a(p)/b(p)$ is not a minimum function, its inverse $H(p) = b(p)/a(p)$ must be minimum. In this case,

$$q = LowpassLadder\_Yarman(b, a)$$

completely removes transmission zeros at infinity.

All the above steps are combined under a MATLAB function

$$[k, q, Highpass\_Elements, Lowpass\_Elements] = GeneralSynthesis\_Yarman(a, b)$$

which completely removes all the transmission zeros at DC and infinity with high-precision. Function GeneralSynthesis_Yarman is given as an attachment to this book.

If the starting function is a minimum susceptance (minimum admittance) function then, all the above statements are true with the following provisions: In the high-pass section of the LC ladder, shunt inductors and series capacitors are replaced by their counterparts of series capacitors and shunt inductors, respectively. Similarly, in the lowpass section, shunt capacitors and series inductors are replaced by their counter parts of series inductors and shunt capacitors, respectively.

In immittance synthesis, resulting lossless two-port topology is not unique. Essential issue is that, during the synthesis process, the zeros of the even-part function $R_{even}(p)$ appear as the transmission zeros of the lossless two-port. These zeros are realized as the poles of the immittance function $F_r(p) = a(p)/b(p)$ or $H_r(p) = 1/F_r(p)$ at each step. In the course of synthesis, we are free to extract transmission zeros in any order, the way we wish to extract. In other words, for the case under consideration, synthesis can start by extracting either a pole at DC (k_1/p) or a pole at infinity $(q_1 p)$. This statement is true at each step of the synthesis. For the sake of simplicity, we have chosen the freedom to start with the extraction of DC transmission zeros first, then continued with the extraction of transmission zeros at infinity. However, we could have done the other way around or we could have extracted transmission zeros at infinity and DC in ad-hoc manner. Hence, we end up with different combinations of circuit topologies such that they yield a variety of element values. This may be a desirable situation from the practical implementation point of view.

Now, let us work on the synthesis of the impedance function $F(p)$ as specified by Table 4.8 with a different circuit layout.

4.2.5.1 Case Study 2: An alternative circuit topology

Referring to the minimum impedance function $F(p) = a(p)/b(p)$ as specified by Table 4.8, we can start the synthesis process by removing the single transmission zero at infinity and then continue with the extraction of remaining transmission zeros at DC as follows.

Since $F(p) = Z(p) = a(p)/b(p)$ is a minimum impedance function, its single transmission zero at infinity is removed as the pole of the admittance function

$$H(p) = \frac{1}{F(p)} = \frac{b(p)}{a(p)} = k_\infty p + \frac{R(p)}{b_{r(p)}}$$

This operation is completed using our *MATLAB* function

$$[kinf, R, br] = removepole\_atinfinity(b, a).$$

The resulting residue and the reminder function $H_r(p) = R(p)/br(p)$ are given by

$$k_\infty = 1.9456$$
$$R = [1.9456\ \ 5.5531\ \ 7.9179\ \ 6.9173\ \ 3.6835\ \ 1.0000]$$
$$br = [0.5140\ \ 1.1579\ \ 1.2131\ \ 0.6944\ \ \ 0.18850]$$

As we see from above, the remainder function $H_r(p) = R(p)/b_r(p)$ has a pole at DC since $b_r(6) = 0$. This pole can be removed using our MATLAB function $[kr, R, ar] = Highpass\_Remainder(a, b)$ where $a(p)$ and $b(p)$ are the numerator and the denominator polynomials of the impedance function $F_r(p) = a(p)/b(p) = b_r(p)/R(p)$. Rest of the transmission zeros at DC can be extracted in a similar manner. Hence, the original admittance function is expressed as

$$H(p) = \frac{b(p)}{a(p)}$$
$$= q_1 p + \frac{k_1}{p} + \frac{1}{(k_2/p) + (1/((k_3/p) + (1/((k_4/p) + (1/((k_5/p) + G))))))}$$

In the synthesis process, $q_1 = k_\infty$ corresponds to a shunt capacitor $C_1 = q_1$. Similarly, k_1 represents a shunt inductor $L_2 = 1/k_1$. The rest of the element values are given by

$$C_3 = \frac{1}{k_2}, L_4 = \frac{1}{k_3}; C_5 = \frac{1}{k_4}; L_6 = \frac{1}{k_5}; \text{Terminating Conductance } G = k_6$$

Resulting synthesis is depicted in Figure 4.15.

Comparing Figures 4.14 and 4.15, we see that the same impedance function yields two different circuit topologies with different element values. Obviously, we can generate many other circuit topologies by changing the extraction order of the poles at infinity and zero.

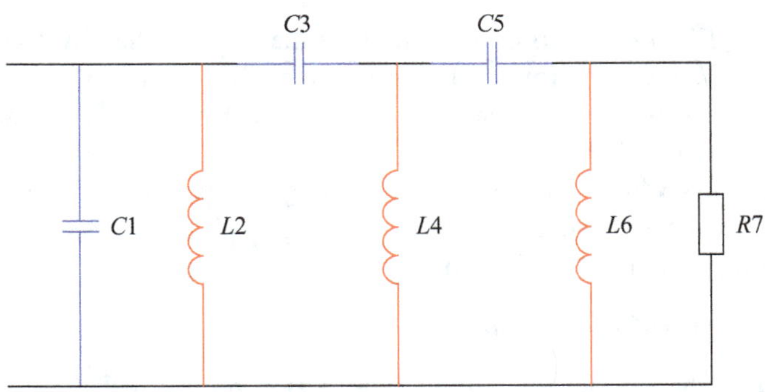

Figure 4.15 Darlington synthesis of $F(p)$ as in Case Study 2 with normalized element values: C1 = 1.94 F, L2 = 188.51 mH, C3 = 2.55 F, L4 = 223.61 mH, C5 = 4.89 F, L6 = 973.07 mH, R7 = 264.17 mΩ

Table 4.10 Execution of MATLAB Program Case Study2.m

Residues		**Component values**	
q_1	1.9456	C_1	1.9456
k_1	5.3046	L_2	0.1885
k_2	0.3910	C_3	2.5573
k_3	4.4720	L_4	0.2236
k_4	0.2041	C_5	4.8996
k_5	1.0277	L_6	0.9731
$G = R/ar$	3.7854	$R = 1/G$	0.2642

All the above computations are carried out using our MATLAB program Case$_S$tudy2.m which is given on our webpage [15]. Execution of Case$_S$tudy2.m is summarized in Table 4.10.

In the following, we present the major *MATLAB* function developed for high precision LC ladder synthesis.

4.2.6 *A general synthesis function on* MATLAB

As indicated above, high-precision synthesis process is gathered under a MATLAB function called

$$[k, q, Highpass\_Elements, Lowpass\_Elements] = GeneralSynthesis\_Yarman(a, b)$$

This function synthesizes a minimum function $F(p) = a(p)/b(p)$ as a Darlington lossless two-port in resistive termination for which all the transmission zeros are specified at DC and infinity. Outputs of this function are given as follows:

k : Residues of the poles at DC

q : Residues of the poles at infinity

Highpass_Elements: Normalized values of the HE specified by $1/k(i)$.

Lowpass_Elements: Normalized values of the LE specified by $q(i)$.

It should be noted that the last entry of array q which is $q(n_\infty + 1)$ specifies the terminating constant of the lossless ladder.

This MATLAB function consists of three major steps. In the first step, $F(p)$ is checked if it can be realized as a lossless LC ladder with transmission zeros only at DC and infinity. At this step, function

$$[a1, b1, ndc] = Check\_immittance(a, b)$$

is employed. In the second step, all the transmission zeros at DC are extracted one by one employing our function

$$[kr, R, ar] = Highpass\_Remainder(a, b)$$

and coefficients of the remaining function $F_r(p) = R(p)/a_r(p) = a(p)/b(p)$ are refined using MATLAB function

$$[a, b] = Laddercorrection\_onF(eps\_zero, a, b)$$

In the third step, all transmission zeros at infinity is removed using our MATLAB function

$$q = LowpassLadder\_Yarman(a, b)$$

Finally, circuit topology is drawn as detailed in the following case study.

4.2.6.1 Case Study 3: Generation of complete circuit topology

Let the driving-point minimum function be $F_{inB}(p) = a(p)/b(p)$ subject to synthesize. Then, it is synthesized, and its circuit diagram is sketched using our MATLAB function.

$$[CT, CV] = CircuitPlot\_Yarman(KFlag, R0, f0, a, b, ndc)$$

In this function, input variable *KFlag* describes the minimum function $F_{inB}(p)$ whether it is minimum reactance or minimum susceptance. If $Kflag = 1$, $F_{\min B}(p) = a(p)/b(p)$ is a minimum reactance impedance. If $Kflag = 0$, $F_{inB}(p) = a(p)/b(p)$ is a minimum susceptance admittance. In this regard, ladder synthesis is handled accordingly. Input variables $R0$ and $f0$ are the normalization numbers to compute the actual element values of the ladder network, which are stored in the output vector CV. As explained previously, MATLAB vectors a and b are the polynomial coefficients of the input immittance $F_{inB} = a(p)/b(p)$. *ndc* is the total number of HE in the ladder. Size $n1 = n + 1$ of vector b specifies the total number of elements in the ladder including the final resistive termination. The output vector CT refers to codes of the circuit elements. For example, $CT(i) = 2$ refers to a series capacitor, $CT(i) = 7$ is a shunt (or parallel) inductor, $CT(i) = 8$ is a shunt (or parallel) capacitor, $CT(i) = 1$ is a series inductor and $CT(i) = 9$ is a resistor which must be the termination for the ladder with $i = n1$.

Function CircuitPlot_Yarman includes two major MATLAB functions, namely,

$$SynthesisMinimumReactance\_Yarman(R0, f0, a, b, ndc)$$

$$SynthesisMinimumSusceptance\_Yarman(R0, f0, a, b, ndc)$$

Function $SynthesisMinimumReactance\_Yarman(R0, f0, a, b, ndc)$ synthesizes a minimum reactance impedance function $F_{inB}(p) = a(p)/b(p)$ as an LC ladder network with actual elements for a specified normalization resistance $R0$ and for a specified normalization frequency $f0$. In many applications, $R0$ is selected as 50 Ω. Usually, $f0$ is selected as the cutoff frequency of the passband where the lossless ladder network is going to be utilized.

Similarly, $SynthesisMinimumSusceptance\_Yarman(R0, f0, a, b, ndc)$ synthesizes a minimum susceptance admittance function $F_{inB}(p) = a(p)/b(p)$.

Both functions work with the synthesis package called

$$[A, CVal] = LadderSynthesis\_BSYarman(KFlag, R0, f0, ndc, a, b)$$

LadderSynthesis_BSYarman either synthesizes a minimum reactance function ($KFlag = 1$ Case) or a minimum susceptance function ($KFlag = 0$ case). In the course of synthesis, corresponding codes (CT) for the element values are generated.

Actual element values of the LC Ladder are stored in the matrix designated by $CVal(i,j)$. First column of this matrix $[Cval(:,1)]$ contains series capacitors of the ladder which are designated by CHP (meaning capacitors as high-pass circuit elements). Second column $[CVal(:,2)]$ contains shunt or parallel inductors which are designated by LHP (meaning inductors L as high-pass circuit elements). Third column of $CVal(:,3)$ is filled by parallel capacitors which are designated by CLP (meaning capacitors as lowpass circuit elements). Fourth column of $CVal(:,4)$ is filled by the series inductors which are designated by LLP (meaning Inductors L as lowpass circuit elements).

In the course of the synthesis, MATLAB function

$$[A, CVal] = LadderSynthesis\_BSYarman(KFlag, R0, f0, ndc, a, b)$$

first extracts the transmission zeros at DC; in other words, the high-pass circuits elements (namely, CHP and LHP) are extracted from the given immittance function $F_{inB}(p) = a(p)/b(p)$. Then, LE are extracted (namely, CLP and LLP). Finally, lossless ladder is terminated in a resistor R. It is mentioned that straight forward synthesis of $F(p)$ yields normalized element values of the LC Ladder which are obtained by setting $R_0 = 1$ and $\omega_0 = 2\pi f_0 = 1$ or corresponding $f_0 = 1/(2\pi)$.

For example, entries of the component value matrix *CVal* of the LC-ladder network shown in Figure 4.16 is given by Table 4.11 with

$$a = \begin{bmatrix} 0 & 2.775 \times 10^{-2} 0.17465 & 0.43146 & 0.53803 & 0.33366 & 5.7976 \times 10^{-2} 0 \end{bmatrix}$$
$$b = \begin{bmatrix} 1 & 6.2926 & 16.298 & 24.121 & 23.464 & 15.061 & 5.7552 & 1 \end{bmatrix}$$

We start reading the *CVal* matrix from the left-hand side. CHP and LHP columns (first and the second columns) refer to high-pass section of the ladder. First nonzero element of the first row is the first element of the ladder which must be either a series capacitor ($CT(1) = 2$) or a shunt inductor ($CT(1) = 7$). For the example under consideration, $CVal(1,1) = 0$ and $CVal(1,2) = 0.05.797$ meaning that first element of the ladder is a shunt inductor $L_1 = 0.05797$. Then, we start to read the second row of column 1 and column 2. $CVal(2,1) = 99.712$ and $CVal(2,2) = 0$ meaning that the second component of the ladder is a series capacitor ($CT(2) = 7$) with $C_2 = 99.712$. Rest of the entrees of the first two columns is zero meaning that there is no more HE in the ladder.

Component values of the lowpass section are stored in the third and the fourth column. We see that third element of the ladder is a shunt capacitor (CLP) with circuit code $CT(3) = 8$ and $CVal(1,3) = C_3 = 56.41$. Fourth element is a series inductor (LLP) with circuit code $CT(4) = 1$ and $CVal(2,4) = L_4 = 0.04134$. Fifth element is a shunt capacitor with circuit code $CT(5) = 1$ and $CVal(3,3) = C_5 = 30.894$. Sixth

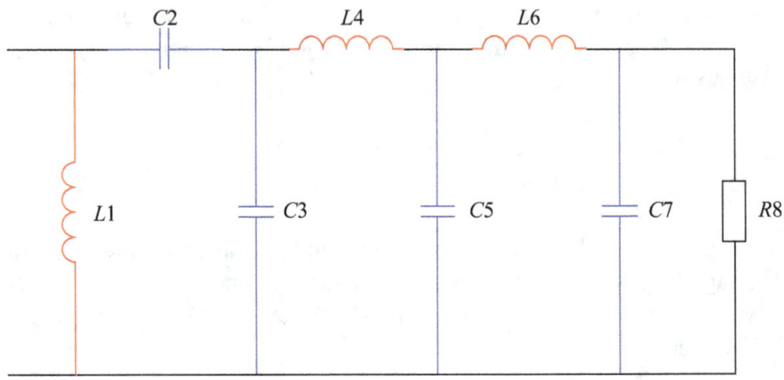

Figure 4.16 Darlington synthesis of F(p) as in Case Study 3 with normalized element values: L1 = 57.98 mH, C2 = 99.729 F, C3 = 56.42 F, L4 = 41.33 mH, C5 = 30.89 F, L6 = 15.10 mH, C7 = 5.31 F, R8 = 29.91 mΩ

Table 4.11 Entrees of the component value matrix CVal of Figure 4.16

[*CVal*] =	CHP	LHP	CLP	LLP	*Resistor R*
	0	0.0579	56.41	0	0.0299
	99.7	0	0	0.04134	0
	0	0	30.89	0	0
	0	0	0	0.0151	0
	0	0	5.3108	0	0
Circuit codes	2	7	8	1	9

element is a series inductor with circuit code $CT(6) = 1$ and $l(6,4) = 0.0151$. Seventh element is a shunt capacitor with circuit code $CT(7) = 8$ and $CVal(3,7) = C_7 = 5.3108$, and finally the resistive termination (with circuit code $CT(8) = 9$) is given by $CVal(1,5) = R_8 = 0.0299$ as depicted in Figure 4.16.

Finally, output vector A prints the alpha numeric text array to describe the columns of the actual component values matrix *CVal*. In other words, *A* is given by $A = [SCaps(CHP)PInds(LHP)PCaps(CLP)SInds(LLP)ResistorR]$ as in Table 4.11.

All the above computations are collected under MATLAB program called "CaseStudy3.m"

4.2.7 Assessment of the numerical error accumulated due to numerical computations

In order to assess the numerical stability of the synthesis package first, we generate the input immittance $F_{inA}(j\omega)$ of the LC ladder in terms of the component values point by point as a function of angular frequency ω. Then, it is compared with the rational form of the original input immittance $F_{inB}(p)|_{p=j\omega} = a(p)/b(p)|_{p=j\omega}$ using the relative norm error given by

$$\varepsilon_r(\omega) = \left| \frac{F_{inA}(j\omega) - F_{inB}(j\omega)}{F_{inB}(j\omega)} \right| \tag{4.38}$$

which is computed over the selected frequency band. In this regard, let N be the total number of sampling points over the angular frequency band $\Delta B = \omega_2 - \omega_1$. Starting from ω_1, we can sweep the passband up to ω_2 with step size $\Delta\omega = \Delta B/(N-1)$. For each i, relative error

$$\varepsilon_r(\omega_i) = \frac{F_{inA}(j\omega) - F_{inB}(j\omega)}{F_{inB}(j\omega)}$$

is computed at the angular frequency $\omega_i = \omega_{i-1} + \Delta\omega$. Then, we end up with the relative error vector

$$[\varepsilon_r] = [\varepsilon_r(\omega_1)\varepsilon_r(\omega_2)\cdots\varepsilon_r(\omega_N)]$$

Eventually, the relative norm error is calculated as

$$\varepsilon_r = \sqrt{\varepsilon_r{}^2(\omega_1) + \varepsilon_r{}^2(\omega_2) + \cdots + \varepsilon_r{}^2(\omega_N)} \tag{4.39}$$

We can also define another error measure to assess the numerical stability of the high precision LC ladder synthesis procedure. For example, let $F_{inC}(p) = \hat{a}(p)/\hat{b}(p)$ be the rational form of the input impedance directly calculated from the component values of the LC ladder. In this case, as function of component values, $\hat{a}(p)$ and $\hat{b}(p)$ are expressed as

$$\hat{a}(p) = \widehat{a_1}p^n + \widehat{a_2}p^{n-1} + \cdots + \widehat{a_n}p + \widehat{a_{n+1}}; \widehat{a_i} > 0 \quad \text{for all } i \tag{4.40a}$$

and

$$\hat{b}(p) = \widehat{b_1}p^n + \widehat{b_2}p^{n-1} + \cdots + \widehat{b_n}p + \widehat{b_{n+1}}; \widehat{b_i} > 0 \quad \text{for all } i \tag{4.40b}$$

At $p = 1$, $F_{inB}(1)$ and $F_{inC}(1)$ are given as

$$\begin{aligned} F_{inB}(1) &= \frac{a_1 + a_2 + \cdots + a_n + a_{n+1}}{b_1 + b_2 + \cdots + b_n + b_{n+1}} = \frac{\sum_{i=1}^{n+1} a_i}{\sum_{i=1}^{n+1} b_i} > 0 \\ F_{inC}(1) &= \frac{\widehat{a_1} + \widehat{a_2} + \cdots + \widehat{a_n} + \widehat{a_{n+1}}}{\widehat{b_1} + \widehat{b_2} + \cdots + \widehat{b_n} + \widehat{b_{n+1}}} = \frac{\sum_{i=1}^{n+1} \hat{a}_i}{\sum_{i=1}^{n+1} \hat{b}_i} > 0 \end{aligned} \tag{4.41}$$

Then, we can define a relative error ε_{RC} such that at $p = 1$

$$\varepsilon_{RC} = \frac{norm\left[\left(\left(\sum_{i=1}^{n+1} a_i\right)/\left(\sum_{i=1}^{n+1} b_i\right)\right) - \left(\left(\sum_{i=1}^{n+1} \hat{a}_i\right)/\left(\sum_{i=1}^{n+1} \hat{b}_i\right)\right)\right]}{norm\left[\left(\sum_{i=1}^{n+1} a_i\right)/\left(\sum_{i=1}^{n+1} b_i\right)\right]} \tag{4.42}$$

Obviously, ε_{RC} gives an excellent idea about the fit between the rational forms of $F_{inB}(p)$ and $F_{inC}(p)$. Of course, one should expect that $\varepsilon_{RC} < \varepsilon_r$.

Quality of the fit between the rational forms of $F_{inB}(p)$ and $F_{inC}(p)$ depends on the variation of the coefficients of the numerator and the denominator polynomials as well as the LC ladder components.

Regarding the computer implementation of the relative error, first we generate the component value-based input immittance of the ladder using our MATLAB function

$$[Zin, Yin] = InputimmittanceLadder\_viacodes(CT, CV, ndc, w)$$

This function takes the outputs of

$$[CT, CV] = CircuitPlot\_Yarman(KFlag, R0, f0, a, b, ndc)$$

Then, for specified circuit codes CT and element values CV, the input impedance and the input admittance of the LC ladder are computed at a specified angular frequency ω. Let us investigate the numerical accuracy of the high precision synthesis algorithm by means of the following case study.

4.2.7.1 Case Study 4: Evaluation of relative errors accumulated in synthesis process

In this case study, first, we generate a random impedance $F(p) = a(p)/b(p)$ of degree $n = 25$ out of which we have $ndc = 8$ transmission zeros at DC. In this case, we use the following data to generate the minimum function.

$$KFlag = 1; \quad R0 = 1; \quad f0 = 1/2/pi;$$

$$W = 0, \quad a0 = 1,$$

$$c = [1.0\ 0.2 - 0.3\ 0.4\ 0.5\ 0.6 - 0.7\ 0.8\ 0.91.0\ 0.2 - 0.3\ 0.4\ 0.5\ 0.8\ 0.6$$
$$- 0.6\ 0.7\ 0.35 - 0.45\ 0.75\ 0.651.0 - 1.01.0]$$

$$ndc = 8$$

Using function $[a, b] = Minimum\_Function(ndc, W, a0, c)$, we generate the polynomials $a(p)$ and $b(p)$ as listed in Table 4.12.

Employing our MATLAB function $[CT, CV] = CircuitPlot\_Yarman$ $(KFlag, R0, f0, a, b, ndc)$ synthesis is completed as shown in Figure 4.17 and its component values are given as

L1 = 4.4128 mH	C8 = 423.436 kF	C15 = 63.1398 kF	L22 = 7.95912 μH
C2 = 284.603 F	C9 = 88.2257 kF	L16 = 20.153 μ H	C23 = 16.4918 kF
L3 = 310.081 μH	L10 = 29.2548 μH	C17 = 52.8569 kF	L24 = 3.44258 μH
C4 = 5.8885 kF	C11 = 80.5046 kF	L18 = 16.397 μH	C25 = 3.31676 kF
L5 = 77.9769 μH	L12 = 26.5123 μH	C19 = 41.51.05 μH	R26 = 18.6178 μΩ
C6 = 46.1522 kF	C13 = 72.2868 kF	L20 = 12.3058 μH	
L7 =68.3219 μH	L14 = 23.5183 μH	C21 = 29.3037 kF	

All the above computations were completed using the MATLAB program *CaseStudy4.m*.

This program first synthesizes $F_{inB}(p) = a(p)/b(p)$ where polynomials $a(p)$ and $b(p)$ are specified as in Table 4.12 and then draws the LC ladder circuit

Table 4.12 Generation of 25-degree minimum immittance function using function [a, b] = Minimum_Function (ndc, W, a0, c) with KFlag = 1;R0 = 1; $f0 = \frac{\frac{1}{2}}{pi}$*; W = 0; a0 = 1; ndc = 8; c = [1 0.2 −0.3 0.4 0.5 0.6 −0.7 0.8 0.9 1 0.2 −0.3 0.4 0.5 0.8 0.6 −0.6 0.7 .35 −0.45 0.75 0.65 1 −1 1]*

a(p)	**b(p)**
0	1.000000000000000e+000
3.718850024315442e−003	1.619411702763076e+001
6.022349250197179e−002	1.314047131523003e+002
4.853813550806636e−001	7.097111650457864e+002
2.585981093342458e+000	2.860171343248478e+003
1.020687356734017e+001	9.142732263644626e+003
3.171271091008718e+001	2.406410019908477e+004
8.046983987287547e+001	5.345535619089486e+004
1.708015072973604e+002	1.019368551431189e+005
3.081816790475702e+002	1.688845047643391e+005
4.778640187996920e+002	2.451418966174715e+005
6.413107222734160e+002	3.135405564427417e+005
7.480077599976447e+002	3.546109608864103e+005
7.595430203296277e+002	3.552298737080938e+005
6.710738797336259e+002	3.151545104929310e+005
5.144754093761967e+002	2.471686316643337e+005
3.404942744406864e+002	1.707266807229726e+005
1.930061965193032e+002	1.032492038271484e+005
9.263283692557589e+001	5.420292444096737e+004
3.702934006809269e+001	2.440314307833654e+004
1.203623936421057e+001	9.262455804295103e+003
3.067068301401550e+000	2.891673083324884e+003
5.769688042588983e−001	7.154072888131586e+002
7.156486466112633e−002	1.320049320237501e+002
4.412795848466732e−003	1.621757885898816e+001
0	9.999999999999976e−001

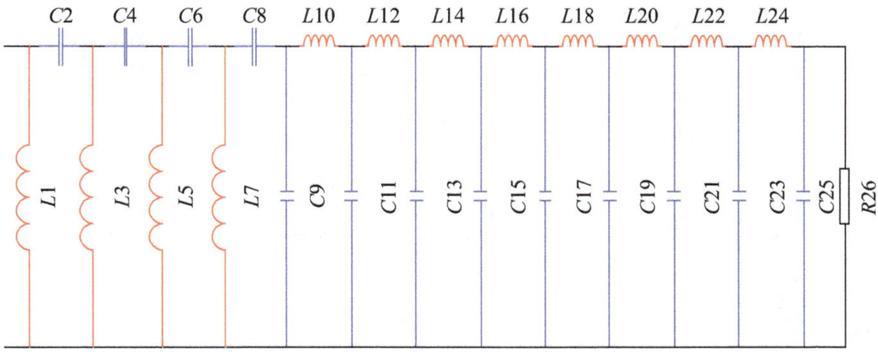

Figure 4.17 Darlington synthesis of $F(p)$ as specified in Case Study 4

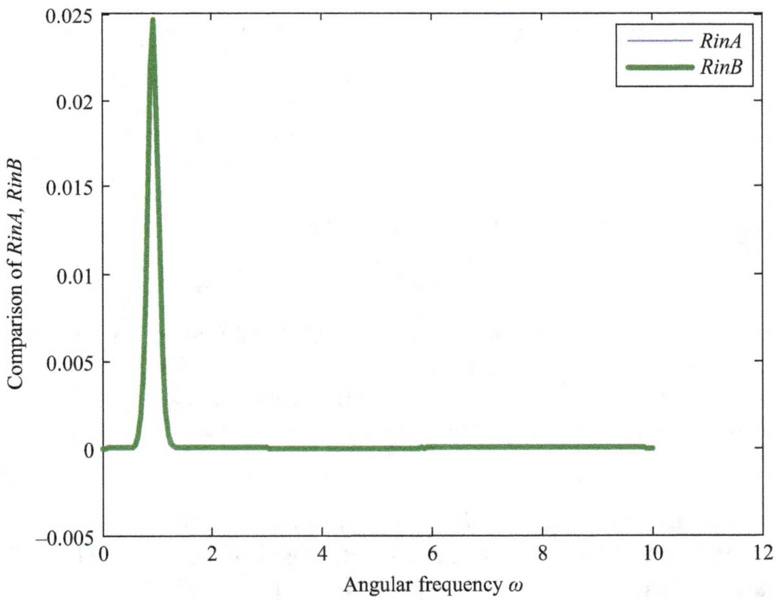

Figure 4.18 Real parts of $F_{inA}(j\omega)$ and $F_{inB}(j\omega)$

diagram as shown in Figure 4.17. Afterwards, input impedance $F_{inA}(p)$ of the LC ladder is generated over the frequency band $(\omega 2 - \omega 1)$ where $\omega 1 = 0$ and $\omega 2 = 10$ is selected. Similarly, $F_{inB}(j\omega) = a(j\omega)/b(j\omega)$ is generated over the same frequency band. Plots of the real and imaginary parts of these functions are depicted in Figures 4.18 and 4.19, respectively.

Interested reader can verify the above results using the main program CaseStudy4.m which is given as an attachment to this book.

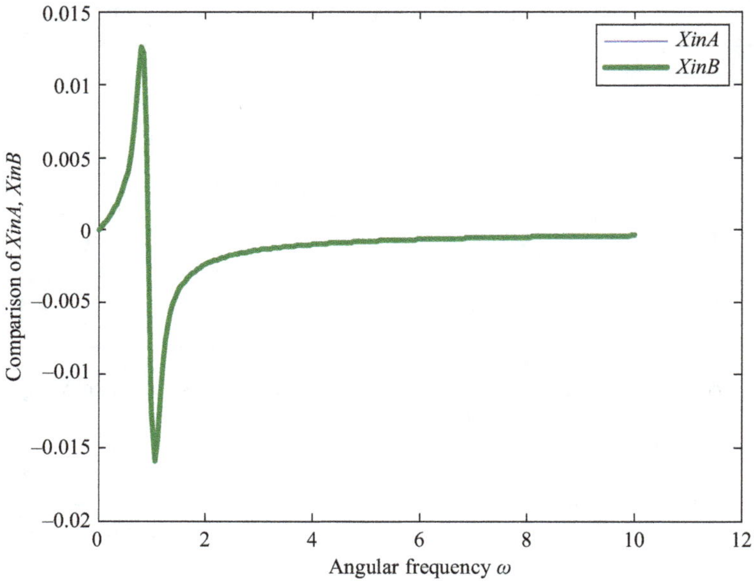

Figure 4.19 Imaginary parts of $F_{inA}(j\omega)$ and $F_{inB}(j\omega)$

Finally, relative errors ε_r of (4.39) and ε_{RC} of (4.42) are computed as

$\varepsilon_r = 7.64 \times 10^{-7}$ and $\varepsilon_{RC} = 9.5 \times 10^{-7}$

Thus above result verifies that numerical error achieved in high precision synthesis algorithm presented in this section is less than 10×10^{-7} which is outstanding. On the other hand, it was impossible to carry out the synthesis of $F_{inB}(p)$ of Table 4.12 without using high precision synthesis algorithm.

4.3 Computer-aided Darlington synthesis of an immittance functions with transmission zeros at DC and infinity, at finite frequencies and in RHP

However, some applications may demand such matching networks with transmission zeros at finite frequencies as well as at DC and infinity. Hence, realization of finite transmission zeros becomes inevitable. Therefore, in this work, previously introduced high-precision lumped element synthesis techniques are extended to include transmission zeros at finite frequencies in Laplace domain. Finite transmission zeros are realized as Darlington's type C-sections using our newly introduced synthesis algorithms [15–17].

In this section, we presume that the driving-point impedance is minimum reactance. This fact does not affect the generality of the synthesis approach since the $j\omega$ poles can be removed from the given impedance as a Foster function remaining a positive real impedance which is a minimum reactance (i.e., an impedance which is free of $j\omega$ poles) [15,17].

Newly proposed high precision synthesis algorithms consist of three steps. In the first step, finite frequency and right half plane (RHP) transmission zeros are extracted as Brune and Type C-sections, respectively, from the given impedance one by one in a sequential manner. Each Brune/type-C section includes a coupled coil (with inductors L_1, L_2 and the coupling coefficients M) connected to ground via a capacitor C. Then, transmission zeros at DC are removed as series capacitors and shunt inductors. Finally, transmission zeros at infinity are realized as series inductors and shunt capacitors. Finite frequencies and RHP transmission zeros are realized using modified zero shifting method. In this section, we introduce two different Brune/type-C section extraction methods [6,16,18]. The first one directly works on the given impedance function. Therefore, it is called "Impedance Based Brune/type-C section Extraction." The second method utilizes the chain parameters. Hence, it is called "chain-parameters-based Brune/type-C section extraction." After each transmission zero extraction, remaining impedance function is corrected employing the parametric approach to yield an exact LC ladder. In this process, leftover transmission zeros are imbedded into the remaining impedance function as the zeros of the even part. The parametric approach is summarized in Section 4.3.4. Moreover, comprehensive coverage of the method can be found in [9,10].

In the following sections, first we introduce Brune/type-C section-extraction algorithms based on impedance and chain parameters approach. Then, examples are presented. It is shown that newly proposed high precision computer-aided Darlington's synthesis algorithms yield successful synthesis of positive real functions with forty reactive elements yielding accumulated—relative numerical error in the size of 10^{-1}.

4.3.1 Brune section extraction using impedance-based approach

In this section, we deal with the extraction of a finite frequency transmission zero from the given minimum reactance impedance [16]. In this regard, let:

$$Z(p) = \frac{a(p)}{b(p)} = \frac{a_1p^n + a_2p^{n-1} + \cdots + a_np + a_{n+1}}{b_1p^n + b_2p^{n-1} + \cdots + b_np + b_{n+1}} \tag{4.43}$$

be a minimum reactance impedance with a real frequency finite transmission zero at ω_a and DC transmission zeros of order ndc to be synthesized in Darlington sense. In this case, even part of $Z(p)$ is given by

$$R(p) = \frac{F(p)}{[b(p)b(-p)]} \tag{4.44a}$$

Zeros of the even polynomial $F(p)$ are called the transmission zeros of $Z(p)$ including the ones at DC. For the case under consideration, $F(p)$ is given by

$$F(p) = a_0^2\left(p^2 + \omega_a^2\right)^2(-1)^{ndc}p^{2ndc} = \mp f^2(p) \tag{4.44b}$$

where

$$F(p) = a_0(p^2 + \omega_a^2)p^{ndc} \tag{4.44c}$$

At the finite frequency transmission zeros $p = \mp j\omega_a$

$$Z(j\omega_a) = R(j\omega_a) + jX_a(\omega_a) \tag{4.45a}$$

where

$$R(j\omega_a) = 0 \tag{4.45b}$$

Referring to (4.45a) and (4.45b), we may extract an inductor L_a from $Z(p)$ such that

$$X_a = \omega_a L_a \tag{4.46}$$

In (4.46), the inductor L_a could be positive or negative without disturbing the positive real feature of $Z(p)$.

Based on (4.46), we can express $Z(p)$ as follows and partially synthesize it as shown in Figure 4.20:

$$Z(p) = pL_a + Z_1(p) \tag{4.47a}$$

where

$$Z_1(p) = \frac{a_1(p)}{b_1(p)} \tag{4.47b}$$

and

$$b_1(p) \equiv b(p) \tag{4.47c}$$

$$a_1(p) \equiv a(p) - (pL_a)b(p) \tag{4.47d}$$

In (4.47d), $a_1(p)$ is a degree of $(n + 1)$ polynomial.

Obviously, as it is introduced above, $Z_1(p) = Z(p) - L_a p$ is zero when $p = \mp j\omega_a$. In other words:

$$Z_1(j\omega_a) = Z(j\omega_a) - j\omega_a L_a = 0 \tag{4.47e}$$

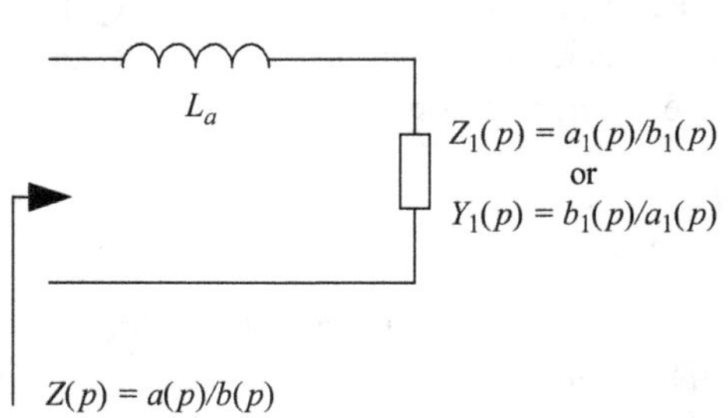

Figure 4.20 Extraction of an inductor L_a from $Z(p)$

Therefore, the numerator polynomial $a_1(p)$ must include the term $(p^2 + \omega_a^2)$ such that

$$a_1(p) = (p^2 + \omega_a^2)a_2(p) \tag{4.48}$$

or the admittance function $Y_1(p)$ has poles at $p = \mp j\omega_a$. That is,

$$Y_1(p) = \frac{b_1(p)}{a_1(p)} = \frac{b_1(p)}{(p^2 + \omega_a^2)a_2(p)} \tag{4.49}$$

By extracting the poles at $p = \mp j\omega_a$, $Y_1(p)$ can be written as

$$Y_1(p) = \frac{k_b p}{p^2 + \omega_a^2} + Y_2(p) \tag{4.50a}$$

where

$$Y_2(p) = \frac{b_2(p)}{a_2(p)} \tag{4.50b}$$

In the above formulation, $a_2(p)$ and $b_2(p)$ are found as

$$a_2(p) = \frac{a_1(p)}{p^2 + \omega_a^2} \tag{4.51a}$$

$$b_2(p) = \left[\frac{1}{p^2 + \omega_a^2}\right][b(p) - (k_b p)a_2(p)] \tag{4.51b}$$

Partial synthesis of $Y_2(p)$ is depicted in Figure 4.21.

Finally, we set:

$$Z_2(p) = \frac{a_2(p)}{b_2(p)} = L_c p + Z_3(p) \tag{4.52a}$$

where

$$Z_3(p) = \frac{a_3(p)}{b_3(p)} \tag{4.52b}$$

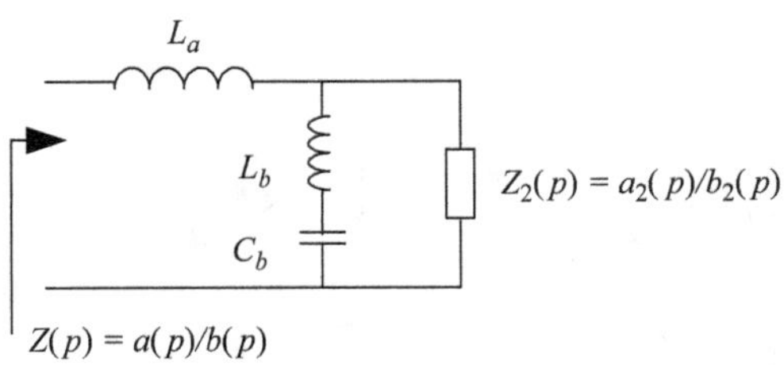

Figure 4.21 Extraction of the poles at $p = \mp j\omega_a$ from the admittance function $Y_1(p)$

$$L_c = \frac{a_{21}}{b_{21}} \tag{4.52c}$$

and

$$a_3(p) = a_2(p) - (L_c p) b_3(p) \tag{4.52d}$$

$$b_3(p) = b_2(p) \tag{4.52e}$$

In (4.52c), a_{21} and b_{21} or denote the leading coefficients of the polynomial polynomials $a_2(p)$ and $b_2(p)$, respectively.

Thus, partial synthesis of (4.43) is shown by Figure 4.22.

The above realization process of the finite transmission zero ω_a is called the "Brune section extraction." In this form, one of the inductors L_a or L_c may have a negative value. However, it may be eliminated by introducing a coupled coil with mutual inductance $M > 0$ as depicted in Figure 4.23.

At the first glance, it is straight forward to show that the way inductors L_a, L_b and L_c are derived must satisfy the following equation [9]:

$$L_a \cdot L_b + L_a \cdot L_c + L_b \cdot L_c = 0 \tag{4.53}$$

In this regard, the negative inductor is removed employing (4.53) and realization of it is depicted in Figure 4.23:

$$L_1 = L_a + L_b > 0 \tag{4.54a}$$

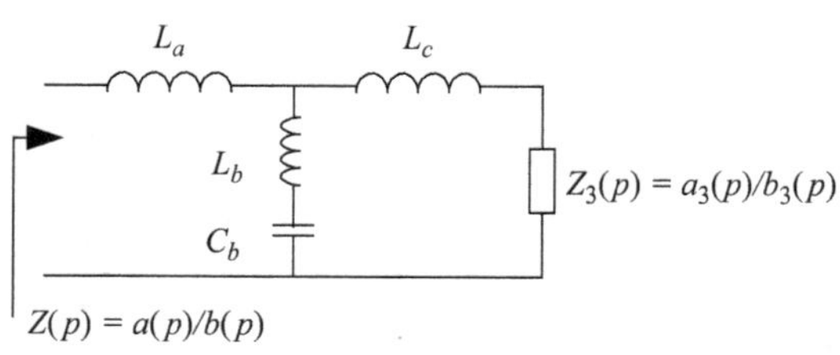

Figure 4.22 Extraction of the poles at ∞ from the impedance function $Z_2(p)$

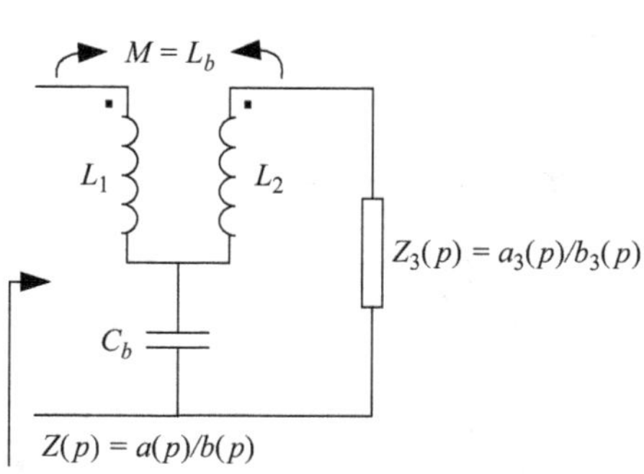

Figure 4.23 Brune section: elimination of the negative inductor using coupled coils with a positive mutual inductance M

$$L_2 = L_b + L_c > 0 \tag{4.54b}$$

$$M = L_b > 0 \tag{4.54c}$$

In the following section, we summarize the upgraded version of our high precision synthesis algorithms introduced in [8,14] to include the extraction of finite frequency transmission zeros a Brune sections.

4.3.2 MATLAB implementation of the new synthesis algorithm

In cascade synthesis [8,16,17] transmission zeros are realized as the poles of the immittance function at each step. In this regard, DC transmission zeros are realized either as series capacitors which are the poles of impedance functions at $p = 0$ or as shunt inductors which are the poles of admittance functions at $p = 0$. In a similar manner, in Brune synthesis, a finite frequency transmission zero at ω_a is realized by introducing a pole at that frequency into the admittance function in the second step of the synthesis. In this regard, synthesis algorithm can be implemented within three steps. In step 1, at a given finite frequency transmission zero ω_a, an inductance L_a is extracted from the given impedance function $Z(p) = a(p)/b(p)$ to introduce a zero in to the remaining impedance function $Z_1(p) = a_1(p)/b_1(p)$ which is the pole of the admittance function $Y_1(p) = b_1(p)/a_1(p)$. This fact is described by (4.50–4.52). Thus, in MATLAB, first we generate:

$$Z(j\omega_a) = R_a + jX_a \tag{4.55}$$

In this step, R_a must be zero since the even part of the given impedance is zero at $p = \mp j\omega_a$ as specified by (4.45b).

However, due to numerical computational errors, R_a will not be exactly zero, rather it will be a small number. In this regard, we define an algorithmic zero such that $\varepsilon_{zero} = 10^{-m}; m > 0$. If $R_a \leq \varepsilon_{zero}$ then, we can go ahead with the synthesis. Otherwise, synthesis algorithm must be stopped meaning that the given impedance does not include a finite transmission zero at ω_a. In this case, if $R_a \leq \varepsilon_{zero}$, then by (4.45b) we set:

$$L_a = \frac{X_a}{\omega_a} \tag{4.56}$$

In (4.56), value of L_a may be positive or negative.

In the second step, we generate the numerator and denominator polynomials of $Z_1(p)$ as in the following equation:

$$a_1(p) = a(p) - pL_a b(p) \tag{4.57a}$$

$$b_1(p) = b(p) \tag{4.57b}$$

At this point, we should mention that degree of $a_1(p)$ is increased by one. Then, the numerator polynomial of $Z_2(p) = a_2(p)/b_2(p)$ is determined as in (4.57a) and (4.57b):

$$a_2(p) = \frac{a_1(p)}{p^2 + \omega_a^2} \tag{4.58}$$

Employing (4.51b), residue k_b is found as

$$k_b = \frac{b(j\omega_a)}{(j\omega_a)a_2(j\omega_a)} \tag{4.59}$$

k_b must be a real positive number. At this point, we completed the extraction of the finite transmission zero ω_a as a series resonance circuit in shunt configuration as shown in Figure 4.22 with element values:

$$L_b = \frac{1}{k_b} > 0 \tag{4.60}$$

and

$$C_b = \frac{1}{\omega_a^2 L_b} > 0 \tag{4.61}$$

The last computation line of this step is to determine the denominator polynomial $b_2(p)$ of $Z_2 = a_2(p)/b_2(p)$ such that

$$b_2(p) = \left[\frac{1}{p^2 + \omega_a^2}\right][b(p) - (k_b p)a_2(p)] \tag{4.62a}$$

In MATLAB environment, division by $(p^2 + \omega_a^2)$ is performed using the function:

$$[q, r] = deconv(u, v) \tag{4.62b}$$

deconv performs the polynomial division operation $u(p)/v(p)$ resulting in the quotient polynomial $q(p)$ and the remainder $r(p)$. Obviously, in computing $a_2(p)$ and $b_2(p)$ using (4.58) and (4.62a) and (4.62b), remainders must be zero. However, due to accumulated numerical errors for both operations, remainders may be small numbers but not exactly zeros. In this case, we can compare the norm of remainders with the algorithmic zero if they are less than ε_{zero}. If so, then we can go ahead with Step 3; otherwise, the algorithm is stopped indicating that extraction of the given finite transmission zero is not possible. At this step, degree of polynomial $a_2(p)$ is $n-1$ and degree of $b_2(p)$ is $n-2$. In this case, the impedance function $Z_2(p) = a_2(p)/b_2(p)$ must include a pole at infinity.

In Step 3, the remaining pole at infinity of $Z_2(p) = a_2(p)/b_2(p)$ is removed as an inductor:

$$L_c = \frac{a_2(1)}{b_2(1)} \tag{4.63a}$$

It should be noted that, in MATLAB environment, a polynomial $P(x) = p_1x^n + p_2x^{n-1} + \cdots + p_nx + p_{n+1}$ of degree n is described by means of a vector P which includes all the coefficients $\{p_1, p_2, \cdots, p_n, p_{n+1}\}$ such that

$$P = [p(1)p(2)\cdots p(n)p(n+1)] \tag{4.63b}$$

Furthermore, norm of a vector P is defined as

$$norm(P) = \sqrt{p^2(1) + p^2(2) + \cdots + p^2(n) + p^2(n+1)} \tag{4.63c}$$

Based on the above MATLAB notation, $a_2(1)$ and $b_2(1)$ are the leading coefficients of polynomials $a_2(p)$ and $b_2(p)$, respectively.

If L_a is negative, L_c must be positive. Otherwise, L_c may take a negative value. Upon completion of this process, we end up with the remaining positive real impedance $Z_3(p) = a_3(p)/b_3(p)$. In this case,

$$a_3(p) = a_2(p) - [L_c p][b_3(p)] \tag{4.64a}$$

$$b_3(p) = b_2(p) \tag{4.64b}$$

In the above equation set, degree of $a_3(p)$ must be $n_{a3} = n - 3$ or $n_{a3} = n - 2$ and degree of $b_3(p)$ must be $n_{b3} = n - 2$.

It should be mentioned that the above Brune section extraction process is also known as zero shifting and it is programmed in MATLAB under the functions called

$$[Even\_Part, La] = Zeroshifting\_Step1(a, b, wa, eps\_zero)$$

$$[Lb, Cb, kb, a2, b2, r\_norm] = Zeroshifting\_Step2(wa, La, a, b, eps\_zero)$$

and

$$[Lc, a3, b3, L1, L2, M] = Zeroshifting\_Step3(La, Lb, a2, b2)$$

In Step 4, we plot the result of zero shifting synthesis using our general purpose MATLAB plot function

$$Plot\_Circuitv1(CT, CV)$$

where CT designates the type of component to be drawn and CV is the value of that component.

Based on our nomenclature, $CT(i) = 1$ describes a series inductor L_i. Components of a series resonance circuit $(pL + 1/pC)$ in shunt configuration is described by $CT(i) = 10$ and $CT(i+1) = 11$ referring to inductor L and capacitor C, respectively. Terminating resistor R_T is designated by $CT(i) = 9$ with the component value $CV(i) = R$. For our Brune section, we use the following MATLAB codes:

$$CT(1) = 1, CV(1) = L_a; CT(2) = 10, CV(2) = L_b \tag{4.65a}$$

$$CT(3) = 11, CV(3) = C_b; CT(4) = 1, CV(4) = L_c \tag{4.65b}$$

For $n = 2$, terminating resistor R_T is given by

$$CT(5) = 9, CV(5) = R_T \tag{4.66}$$

If $n > 2$, then in step 5, the remaining impedance $Z_3(p) = a_3(p)/b_3(p)$ is synthesized using our high precision LC ladder synthesis algorithm, and the final circuit schematic is plotted.

It is noted that if the driving-point immittance function $F(p) = a(p)/b(p)$ is specified as an admittance, then we should flip it over to make it impedance $Z(p) = b(p)/a(p)$ to be able to apply zero shifting synthesis algorithm. In this case, $Z(p)$ may have a pole at infinity and a pole at DC. If it is so, then poles of $Z(p)$ are extracted as a Foster function as follows:

$$Z(p) = \frac{b(p)}{a(p)} = L_x p + \frac{1}{C_x p} + Z_1(p) \tag{4.67}$$

All the above steps are gathered under the major MATLAB function called

$$[CT, CV, L1, L2, M] = SynthesisbyTranszeros(KFlag, W, ndc, a, b, eps\_zero)$$

This function synthesizes the immittance function $F(p) = a(p)/b(p)$ as described above. If the input variable $KFlag = 1$ selected, $F(p)$ is an impedance, if $KFlag = 0$, then $F(p)$ is an admittance. $F(p)$ may include poles at $p = 0$ and/or at $p = \infty$. At the beginning of the synthesis, these poles are extracted as in (25) leaving a minimum reactance function. Then, total number of *nz* finite transmission zeros are extracted in a sequential manner, as they are provided by the input vector *W* of size *nz*. Thereafter, total number of *ndc* transmission zeros at DC are removed and, finally, remaining transmission zeros at infinity are extracted using our high-precision synthesis algorithms introduced in [8,14,15,17].

Remarks

It should be emphasized that, in the above synthesis process, after each pole extraction, remaining impedance is corrected using parametric approach. Therefore, the algorithm introduced above may be called "direct synthesis with impedance correction" or in short "DS-with ImC."

It is experienced that "DS-with ImC" is able to extract 10 Brune/type-C sections with accumulated numerical error less than 10^{-1} as outlined in Example 4.1. However, straightforward impedance synthesis without correction fails due to over/under flows after three or four Brune section extractions.

In the following section, we propose an alternative method to synthesize Brune sections using the impedance-based chain parameters in a similar manner to that described by Cameron *et al.* [19].

4.3.3 Synthesis via chain matrix method

Referring to Figure 4.22 and (4.43), a lossless two-port terminated in a unit resistance can be described by means of its driving-point impedance-based chain parameter matrix $T(p)$such that

$$T = \frac{1}{f_T}\begin{pmatrix} A_T & B_T \\ C_T & D_T \end{pmatrix} \tag{4.68a}$$

where

$$A_T(p) = \{\text{either even or odd part of } a(p)\} \tag{4.68b}$$
$$B_T(p) = \{\text{either odd or even part of } a(p)\} \tag{4.68c}$$
$$C_T(p) = \{\text{either odd or even part of } b(p)\} \tag{4.68d}$$
$$D_T(p) = \{\text{either even or odd part of } b(p)\} \tag{4.68e}$$

and the polynomial $f_T(p)$ includes all the finite transmission zeros of $Z(p)$ and may be expressed as in (4.45a) and (4.45b) such that

$$f_T(p) = a_0 p^{ndc} \prod_{i=1}^{nz} (p^2 + \omega_i^2) \tag{4.68f}$$

In (4.68f), ω_i is a finite transmission zero of $Z(p)$ which is realized as a Brune section.

Referring to Figure 4.24(a), chain matrix of a Brune section with a transmission zero $\omega_i = \omega_a$ is given as

$$T_1 = \frac{1}{f_1}\begin{pmatrix} A_1 & B_1 \\ C_1 & D_1 \end{pmatrix} \tag{4.69a}$$

where

$$A_1(p) = RC_b(L_a + L_b)p^2 + R \tag{4.69b}$$
$$B_1(p) = (L_a + L_c)p \tag{4.69c}$$
$$C_1(p) = RCR_b p \tag{4.69d}$$
$$D_1(p) = C_b(L_b + L_c)p^2 + 1 \tag{4.69e}$$
$$f_1(p) = a_{01}(p^2 + \omega_a^2) = L_b C_b(p^2 + \omega_a^2) \tag{4.69f}$$
$$a_{01} = L_b C_b \tag{4.69g}$$

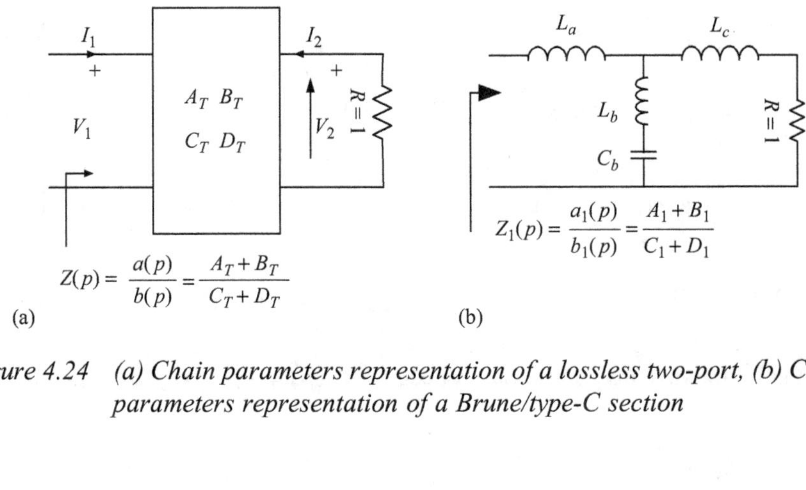

Figure 4.24 (a) Chain parameters representation of a lossless two-port, (b) Chain parameters representation of a Brune/type-C section

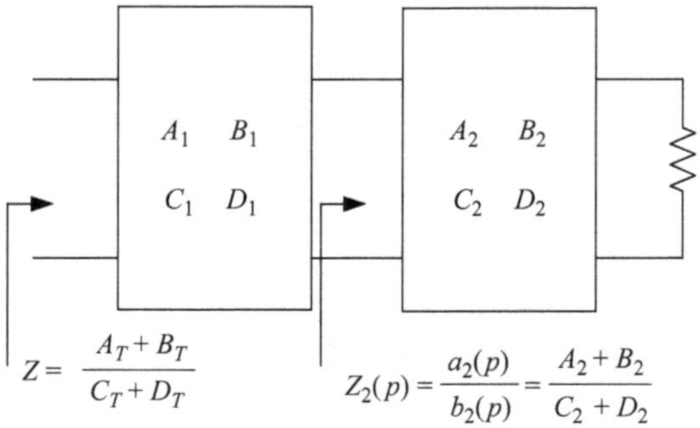

Figure 4.25 Extraction of a type-C section from $T = T_1T_2$

Referring to (4.69a), Figures 4.24(a) and 4.25, synthesis of $Z(p)$ may be initiated by extracting the first finite transmission zero ω_1 as a Brune section as described by the chain matrix T_1 of (4.69a)–(4.69g).

In this case,

$$T = \frac{1}{f_T}\begin{bmatrix} A_T & B_T \\ C_T & D_T \end{bmatrix} = T_1T_2 = \left\{\frac{1}{f_1}\begin{bmatrix} A_1 & B_1 \\ C_1 & D_1 \end{bmatrix}\right\}\left\{\frac{1}{f_2}\begin{bmatrix} A_2 & B_2 \\ C_2 & D_2 \end{bmatrix}\right\} \tag{4.70}$$

where T_2 is the chain matrix of the remaining lossless two-port after the extraction of T_1 from the given chain matrix T and it is given by

$$T_2 = \frac{1}{f_2}\begin{pmatrix} A_2 & B_2 \\ C_2 & D_2 \end{pmatrix} = T_1^{-1}T \tag{4.71a}$$

where

$$A_2 = \frac{1}{f_1^2}(A_TD_1 - C_TB_1) \tag{4.71b}$$

$$B_2 = \frac{1}{f_1^2}(B_TD_1 - D_TB_1) \tag{4.71c}$$

$$C_2 = \frac{1}{f_1^2}(C_TA_1 - A_TC_1) \tag{4.71d}$$

$$D_2 = \frac{1}{f_1^2}(D_TA_1 - B_TC_1) \tag{4.71e}$$

Eventually, the driving-point input impedance $Z_2(p) = a_2(p)/b_2(p)$ is given by

$$Z_2 = \frac{A_2 + B_2}{C_2 + D_2} = \frac{a_2(p)}{b_2(p)} \tag{4.72a}$$

where

$$a_2(p) = \frac{1}{f_1^2}[a(p)D_1(p) - b(p)B_1] \tag{4.72b}$$

and

$$b_2(p) = \frac{1}{f_1^2}[b(p)A_1(p) - a(p)C_1] \tag{4.72c}$$

Thus, the alternative method of zero shifting may be implemented as described in the following algorithm.

4.3.4 Algorithm: impedance synthesis via chain matrix approach

Inputs

Polynomials $a(p)$ and $b(p)$ of degree n of a given driving-point impedance $Z(p) = a(p)/b(p)$ with nz finite transmission zeros of $W = \{\omega_1, \omega_2, \cdots, \omega_{nz}\}$ and ndc transmission zeros at DC. It should be noted that $n \geq ndc + 2nz$.

Computational steps:

Step 1: Compute the chain parameters $\{A_1, B_1, C_1, D_1\}$ of T_1 as in (4.70).

Step 2: Compute the chain parameters $\{A_2, B_2, C_2, D_2\}$ of T_2 as in (4.71a)–(4.71e).

Step 3: Compute the driving-point impedance $Z_2(p) = a_2(p)/b_2(p)$ employing (4.72a)–(4.72c).

Step 4a: Determine the component values of the extracted Brune section as in (4.51a) and (4.51b), (4.55), (4.56), (4.60) and (4.61). Store all the component codes and values.

Step 4b: Correct the impedance $Z_2(p) = a_2(p)/b_2(p)$ by inserting the remaining transmission zeros in its even part employing the parametric approach.

Step 5: Repeat the steps 1–4 nz times until all the finite transmission zeros are extracted. Then, store the remaining impedance as $Z_2(p) = Z_r(p) = a_r(p)/b_r(p)$ which will only have transmission zeros at DC and infinity.

Step 6: Finally, synthesize $Z_r(p) = a_r(p)/b_r(p)$ using the high precision synthesis method introduced in [14].

The above algorithm is programed under a MATLAB function called "Zeroshifting_viaChainMatrixImpedanceCorrection."

4.3.5 Real and complex transmission zeros

For the sake of completeness, let us mention that a driving-point input impedance may as well include real and complex transmission zeros.

A real transmission zero $\sigma_a > 0$ appears as the square of a second order even polynomial $(p^2 - \sigma_a^2)^2$, and the complex conjugate mirror image paired transmission zeros appear as the square of a fourth order even polynomial $(p^4 + dp^2 + e)^2$ in the numerator polynomial $F(p)$ of the even part $R(p)$. Thus:

$$R(p) = F(p)/[b(p)b(-p)] \tag{4.73a}$$

$$F(p) = a_0^2(p^2 - \sigma_a^2)^2(p^4 + dp^2 + e)^2\hat{F}(p) \tag{4.73b}$$

$$\hat{F}(p) = (-1)^{ndc}p^{2ndc}\prod_{i=1}^{nz}(p^2 + \omega_i^2)^2 \tag{4.73c}$$

A positive–real transmission zero σ_a can be extracted as a Darlington type-C section like a transmission zero located at a finite frequency ω_a. Thus, one can employ the Brune section extraction algorithm replacing ω_a^2 by $-\sigma_a^2$.

In this case, even part $R(p)$ of $Z(p)$ will be zero at $p = \mp\sigma_a$ such that

$$Z(p) = R(p) + odd(p) \tag{4.74a}$$

with

$$R(\sigma_a) = 0 \tag{4.74b}$$

At this point, we presume that $odd(p)$ consist of a single positive inductor L_a such that

$$odd(p) = pL_a \tag{4.75a}$$

or

$$L_a = \frac{odd(\sigma_a)}{\sigma_a} = Z(\sigma_a) = \frac{a(\sigma_a)}{b(\sigma_a)} \tag{4.75b}$$

or

$$L_a = \frac{a(\sigma_a)b(-\sigma_a) - a(-\sigma_a)b(\sigma_a)}{b(\sigma_a)b(-\sigma_a)} \tag{4.75c}$$

After extracting pL_a from $Z(p)$, the remaining impedance, $Z_1(p) = Z(p) - pL_a$ must have a zero at $p = \mp\sigma_a$ or equivalently, the admittance $Y_1(p) = a_1(p)/b_1(p)$ has a pole at $p = \mp\sigma_a$. Then, (4.75a)–(4.75c) will have the following form:

$$Y_1(p) = \frac{b_1(p)}{a_1(p)} = \frac{k_b p}{p^2 - \sigma_a^2} + Y_2(p) \tag{4.76a}$$

where

$$\begin{aligned} a_1(p) &= a(p) - pL_1b(p) \\ b_1(p) &= b(p) \end{aligned} \tag{4.76b}$$

and

$$Y_2(p) = \frac{b_2(p)}{a_2(p)} \tag{4.76c}$$

$$a_2(p) = \frac{a_1(p)}{p^2 - \sigma_a^2} \tag{4.76d}$$

$$k_b = \frac{b(\sigma_a)}{\sigma_a a_2(\sigma_a)} > 0 \tag{4.76e}$$

Then, we define the negative inductor L_b as

$$L_b = -\frac{1}{k_b} < 0 \tag{4.76f}$$

and the series capacitor C_b is given by

$$C_b = \frac{k_b}{\sigma_a^2} > 0 \tag{4.76g}$$

The rest of the elements of Darlington type-C section is given as in (4.75a)–(4.76g).

As we did before, the negative inductor L_b is removed using a perfectly coupled coil as specified by (4.77) with a negative coupling coefficient:

$$M = -\frac{1}{k_b} = L_b \tag{4.77}$$

The negative coupling coefficient M is realized by reversely winding $L_1 = L_a + L_b > 0$ and $L_2 = L_c + L_b > 0$ of Figure 4.24(b).

Complex transmission zeros in quadruplet symmetry are realized as a Darlington type-D sections. However, usage of type-D section does not have much practical importance. Therefore, its synthesis and realization details are skipped in this section. Interested readers are referred to [6,18,20,21].

In the following sections, impedance correction via parametric approach is summarized and then we present examples.

4.3.6 Impedance correction via parametric approach

In this section, we presume that a minimum reactance impedance $Z_k(p) = a_k(p)/b_k(p)$ of degree of nk is obtained with a random numerical error as the result of Brune/type-C section extractions and/or any pole extractions at DC and infinity. However, its remaining transmission zeros are precisely known in advance and they are specified under a MATLAB vector $W = [\omega_1 \omega_2 \cdots \omega_{nz}]$ and as $\mathrm{p}^{2\mathrm{ndc}}$. Due to the random numerical errors, the computed even part of $Z_k(p)$ which is given by

$$\begin{aligned} R(p) &= \frac{1}{2}[Z(p) + Z(-p)] = \frac{a_k(p)b_k(-p) + a_k(-p)b_k(p)}{b_k(p)b_k(-p)} \\ &= \frac{A_1 p^{2m} + A_2 p^{2(nk-1)} + \cdots + A_m p^2 + A_{m+1}}{b_k(p)b_k(-p)}; m \geq nk \end{aligned} \tag{4.78a}$$

may not precisely yield the predefined transmission zeros at finite frequencies $\{\omega_1\omega_2\cdots\omega_{nz}\}$ and/or at DC of order $2ndc$. In this case, keeping the same denominator $b(p)$, we can generate a corrected even part function such that

$$R_{ck}(p) = \frac{a_0^2(-1)^{ndc}p^{2ndc}\prod_{i=1}^{nz}\left(p^2+\omega_i^2\right)^2}{b_k(p)b_k(-p)} \tag{4.78b}$$

where a_0 is determined from the leading coefficient of the even polynomial $F(p) = a_k(p)b_k(-p) + a_k(-p)b_k(p)$ such that

$$a_0 = \sqrt{|A_1|} \tag{4.78c}$$

Hence, using (4.78a)–(4.78c), even part of $Z_k(p)$ is regenerated precisely to include exact transmission zeros. In this case, corrected minimum reactance impedance $Z_{kc}(p)$ may be determined by means of parametric approach [9]. In this method, first, roots p_{ik} of $b_k(p) = b_{1k}p^{nk} + \cdots + b_{nk}p + 1$ are computed. Then, the corrected impedance is expressed as

$$Z_{kc} = Z_{0k} + \sum_{i=1}^{nk}\frac{K_i}{p-p_{ik}} = \frac{a_{kc}(p)}{b_{kc}(p)} \tag{4.79a}$$

where the residues K_i are given by

$$K_i = (-1)^{nk}\frac{(-1)^{ndc}a_0^2}{p_{ik}B_{1k}\prod_{\substack{j=1\\ j\neq i}}^{n} p_{ik}^2 - p_{jk}^2} \tag{4.79b}$$

and

$$B_1 = (-1)^{nk}b_{1k}^2 \tag{4.79c}$$

$$Z_{ok} = \frac{a_{1k}}{b_{1k}} \tag{4.79d}$$

Thus, the given impedance $Z_k(p)$ is corrected to warrant the desired network topology with prescribed transmission zeros.

4.3.7 Assessment of the synthesis error

We can device several methods to assess the accumulated numerical error that occurs during immittance synthesis. For example, as described in Section 4.3.4, we can randomly generate a positive real impedance function $Z(p) = a(p)/b(p)$ with transmission zeros located in the RHP as well as on the finite frequencies using parametric approach. Then, synthesis is carried out, which in turn yields the lossless Darlington two-port in resistive termination. In this regard, impedance-based accumulated numerical error may be derived by generating the actual driving-point impedance $Z_T(p) = a_T(p)/b_T(p)$ from the element values of the synthesized network. In this case, relative error may be described by

$$Error_R = \frac{norm(a - a_T)}{C_{\max}} + \frac{norm(b - b_T)}{C_{\max}} \tag{4.80a}$$

where the pair vectors $\{a, a_T\}$,$\{b$ and $b_T\}$ include the coefficients of the numerator and denominator polynomials of $Z(p)$ and $Z_T(p)$, respectively, and

$$C_{\max} = \max(a, a_T, b, b_T) \tag{4.80b}$$

Similarly, we can define an absolute error as

$$Error_A = norm(a - a_T) + norm(b - b_T) \tag{4.80c}$$

On the other hand, for practical problems, one may as well define a TPG-based synthesis error. For example, if $Z(p)$ describes a doubly terminated filter, such as an elliptic filter with several finite frequency transmission zeros, then we can generate TPG from $Z(p)$, per say $T_Z(\omega)$, and from the synthesized Darlington lossless two-port, per say $T_s(\omega)$, over the operational frequency band, per say $\omega_1 \leq \omega \leq \omega_2$. Then, TPG-based synthesis error may be expressed as

$$Error_{TPG} = norm(T_Z - T_S); \omega_1 \leq \omega \leq \omega_2 \tag{4.80d}$$

In (4.80d) $T_Z(\omega)$ and $T_s(\omega)$ generated point by point over the passband. Therefore, they define vectors T_Z and T_S in MATLAB. Hence, the norm of the error vector $\varepsilon_{TPG} = T_Z - T_S$ results in the total accumulated error derived based on the TPG within the prescribed frequency band of operation.

4.3.8 Examples

In this section, several examples are presented to exhibit the utilization of the impedance synthesis algorithms introduced in this section. In the first one, we test the numerical robustness of the newly proposed synthesis algorithms on the randomly generated minimum reactance function. In the second example, an elliptic filter is designed with $nz = 6$ finite transmission zeros. In the third example, the proposed impedance synthesis algorithm is integrated with the RFT to design a lossless matching network for an actual monopole antenna.

Example 4.7

In this example, general form of a driving—point input impedance $Z(p) = a(p)/b(p)$ is synthesized in Darlington sense employing the algorithms developed in this section. For this purpose, $Z(p)$ is generated from its even part $R(p^2)$ employing the parametric approach of Section 4.3.4 as a minimum reactance. $R(p^2)$ is constructed with $nr = 3$ positive real—RHP zeros, $nz = 3$ finite frequency zeros and $ndc = 3$ zeros at DC. Thus, as in (4.78a)–(4.78c), its numerator polynomial $F(p^2)$ is given by

$$Z(p) = R(p^2) + Odd\{Z(p)\} \tag{4.81a}$$

$$R(p^2) = \frac{F(p^2)}{B(p^2)} \tag{4.81b}$$

where

$$F(p) = a_0^2(-1)^{ndc}p^{2ndc}\prod_{i=1}^{nr}\left(p^2 - \sigma_i^2\right)^2\prod_{j=1}^{nz}\left(p^2 + \omega_i^2\right)^2 \tag{4.81c}$$

Furthermore, its denominator polynomial $B(p^2) = b(p)b(-p)$ is expressed by means of an auxiliary polynomial $c(\omega)$ such that

$$B(p^2) = b(p)b(-p) = \frac{1}{2}\left[c^2(\omega) + c^2(-\omega)\right]\big|_{\omega^2=-p^2} > 0 \tag{4.82}$$

where $c(\omega)$ is selected as an 18th degree polynomial (i.e., $n = 18$) with real arbitrary coefficients c_i such that

$$c(\omega) = \left[\sum_{i=1}^{n=18} c_i\omega^i\right] + 1 \tag{4.83a}$$

Then, $Z(p)$ is generated as described in Section VI, using our MATLAB function

$$[a, b] = ZeroShiftingMinimum\_FunctionRHP(ndc, W, S, a0, c)$$

In the above function, input vectors $W^T = [\omega_1\ \omega_2\ \infty\ \omega_{nz}]$ includes all the selected finite frequency transmission zeros, $S^T = [\sigma_1\ \sigma_2\ \infty\ \sigma_{nr}]$ contains all the selected positive real—RHP zeros of $R(p^2)$ and $c^T = [c_1\ c_2\ \infty\ c_n]$ includes the real coefficients of the auxiliary polynomial $c(\omega)$. Input variable a_0 is the leading coefficient of (4.81c) and *ndc* is the total number of transmission zeros located at DC. In this regard, degree n is given by

$$n = 2(nr + nz) + ndc + n_L \tag{4.83b}$$

where n_L designates the total number of transmission zeros at infinity. For the example under consideration, n_L is determined as

$$n_L = 18 - 2 \times (3 + 3) + 3 = 3$$

Once the input variables are set, *ZeroShiftingMinimum_FunctionRHP generates* $Z(p) = a(p)/b(p)$ as MATLAB vectors a and b such that

$$a(p) = \sum_{i=1}^{n+1=19} a_ip^i \tag{4.84a}$$

and

$$b(p) = \sum_{i=1}^{n+1=19} b_ip^i \tag{4.84b}$$

All the input variables are listed in Table 4.13a.

Table 4.13a Generation of Z(p) = a(p)/b(p) with transmission zeros located in RHP [S], on the finite frequencies [W], at DC (n_{DC}) and infinity (n_L)

$\varepsilon_{zero} = 10^{-8}$	$a_0 = 500$	$ndc = 3$	$n_L = 3$	$KFlag = 1$
$c(\omega) = c_1\omega^n + c_2\omega^{n-1} + \cdots + c_n\omega + 1$				
$nr = 3; [S]$	$nz = 3; [W]$	$c_1 \cdots c_6$	$c_7 \cdots c_{12}$	$c_{13} \cdots c_{18}$
$\sigma_1 = 0.1$	$\omega_1 = 0.2$	1.6294	0.5570	1.9143
$\sigma_2 = 0.3$	$\omega_2 = 1.5$	1.8116	1.0938	0.9708
$\sigma_3 = 1.7$	$\omega_3 = 2.0$	0.2540	1.9150	1.6006
		1.8268	1.9298	0.2838
		1.2647	0.3152	0.8435
		0.1951	1.9412	1.8315

Table 4.13b Z(p) = a(p)/b(p) such that n = 18, ndc = 3, n_L = 3 nr = 3 with S^T = [0.1 0.3 1.7] and nz = 3 with W^T = [0.2 1.5 2]

a(p)	***b(p)***
0	1.629400000000002e + 00
7.640386374191374e + 04	2.021199578409565e + 01
9.477565802382027e + 05	1.240994058599176e + 02
5.748126748609327e + 06	4.997625503065354e + 02
2.255362397132513e + 07	1.476782670622500e + 03
6.393411837380380e + 07	3.396224286336301e + 03
1.384849904981637e + 08	6.293526108606870e + 03
2.364596946610408e + 08	9.601210825123586e + 03
3.238068035051984e + 08	1.221637655172800e + 04
3.583006791852569e + 08	1.305433366323806e + 04
3.204420015909864e + 08	1.173859256942452e + 04
2.304831538762949e + 08	8.858425592653462e + 03
1.319330848881640e + 08	5.567059212761518e + 03
5.898551922123835e + 07	2.873327778659241e + 03
1.993957497958470e + 07	1.190802369500236e + 03
4.811178659928725e + 06	3.821715429382096e + 02
7.402976617981475e + 05	8.935564490704259e + 01
5.461218267980077e + 04	1.355554064078911e + 01
0	1.000000000000000e + 00

Z(*p*) is synthesized employing the newly proposed algorithms, namely,

$$[CTRHP, CVRHP, LL1, LL2, MM, aR, bR]$$

$$= SynthesisbyRHPTranszeros(KFlag, W, S, ndc, a, b, eps\_zero)$$

and

$$[CT, CV, L1, L2, M] = SynthesisbyTranszeros(KFlag, W, ndc, aR, bR, eps\_zero)$$

Coefficients of *a*(*p*) *and b*(*p*) are listed in Table 4.13b.

During the execution of the above statements, first, $nr = 3$ positive real RHP zeros $\sigma_1 = 0.1$, $\sigma_2 = 0.3$ and $\sigma_3 = 1.7$ are extracted from $Z(p) = a(p)/b(p)$ as type-C sections using "SynthesisbyRHPTranszeros." In this function, the remaining impedance $Z_R(p) = a_R(p)/b_R(p)$ is returned as MATLAB vectors *aR* and *bR*. Then, $Z_R(p) = a_R(p)/b_R(p)$ is synthesized employing the MATLAB function "SynthesisbyTranszeros." In this regards, finite frequency transmission zeros $\omega_1 = 0.2$, $\omega_2 = 1.5$ and $\omega_3 = 2.0$ are extracted as Brune sections. Then $ndc = 3$ transmission zeros at DC and, finally, $nL = 3$ transmission zeros at infinity are removed in a sequential manner. Hence, we end up with the complete synthesis as depicted in Figure 4.26. Element values of the final circuit is listed in Table 4.13c.

At the end of the synthesis, accumulated numerical error is assessed by regenerating the driving-point input impedance $Z_T(p) = a_T(p)/b_T(p)$ from the synthesized circuit. Eventually, the accumulated numerical error is computed

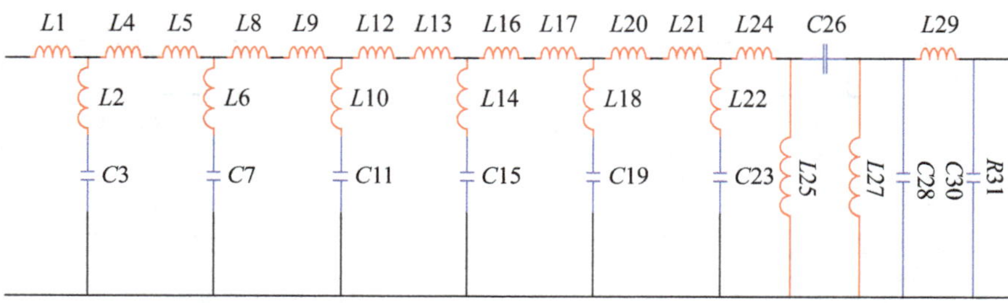

Figure 4.26 Synthesis of Z(p) as specified by Tables 4.13a and b

Table 4.13c Element values of the synthesized network

	Type-C sections		Brune sections	High-pass sections
σ_1	$L1 = 53.9342$ H $L2 = -668.814$ $C3 = 149.518$ F $L4 = 677.212$ H	ω_1	$L13 = 883$ H $L14 = 168.818$ μH $C15 = 148.089$ kF $L16 = -141.744$ μH	$L25 = 23.7736$ μH $C26 = 521.685$ kF $L27 = 107.029$ μH
σ_1	$L5 = 663.21$ mH $L6 = -86.7987$ mH $C7 = 128.01$ F $L8 = 99.8692$ mH	ω_2	$L17 = -1.04075$ μH $L18 = 15.9144$ μH $C19 = 27.9272$ kF $L20 = 1.11358$ μH	Lowpass sections
σ_3	$L9 = 481.815$ μH $L10 = -387.053$ μH $C11 = 893.989$ F $L12 = 1.96795$ mH	ω_3	$L21 = 3.92545$ μH $L22 = 34.2182$ μH $C23 = 7.30606$ kF $L24 = -3.52147$ μH	$C28 = 49.7319$ kF $L29 = 1.9386$ μH $C30 = 10.0286$ kF $R31 = 8.03857$ μΩ
			error accumulated in the course of synthesis	
error_a			***error_b***	***Cmax***
1.64×10^{-9}			2.54×10^{-14}	3.58×10^{8}

using (39). It is found that $error_a = norm(a - a_T)/C_{\max} = 1.64 \times 10^{-9}$ and $error_b = norm(b - b_T)/C_{\max} = 2.54 \times 10^{-14}$ or total relative error is found as $error_R = error_a + error_b = 7.45 \times 10^{-9}$.
All the above computations are gathered under a MATLAB program called Zeroshifting_ExampleGeneralFormZ.m and it is provided as an open source code for the interested readers in [10].

Remarks

It should be mentioned that the size of the accumulated numerical error occurs in the course of synthesis, depends on the total number and the value of the real and finite transmission zeros as well as the values of the selected coefficients of the auxiliary polynomial $c(\omega)$.

To assess the robustness of the newly developed high-precision zero shifting extraction methods, we run many tests. In this regard, finite transmission zeros are selected in ascending order and total number of $n = 2 \times nz$ coefficients of the auxiliary polynomial $c(\omega)$ are initialized using the random number generator of MATLAB. For example, when we run the program with $nz = 9$ finite transmission zeros as given in the MATLAB vector $W = [0.1\ 0.3\ 1\ 1.2\ 1.4\ 1.6\ 1.8\ 2\ 2.1]$, at the end of the synthesis, we end up with 36 reactive elements and various size of relative errors ranging between 10^{-10} and 10^{-4}.

It is experienced that, we can employ "impedance based- zero shifting" algorithm with impedance correction, to extract 13 or beyond Brune sections, whereas the "chain-matrix-based-finite transmission zero extraction with impedance correction" fails after 10 type-C and/or Brune section extractions. On the other hand, it was not possible to utilize impedance or chain-matrix-based extraction algorithms without impedance correction beyond $nr = nz = 3$ or 4 type-C/Brune section extractions. Furthermore, if we increase the number of finite transmission zeros up to $n = 2 \times nz = 20$, then the accumulated relative error increases up to 10^{-1}. In this case, synthesis includes total number of 40 reactive elements. The above results can be reproduced by using our MATLAB program called Zeroshifting_Example1b.m which is provided as an attachment to this book.

Example 4.8

In this example, an $n = 12$ degree lowpass elliptic filter transfer scattering parameter $S_{21}(p)$ is generated using the MATLAB function $[z, p, k] = ellipap(n, eps\_LP, eps\_SB)$ with passband ripple $eps\_LP = 0.3$ dB and stop band ripple $eps\_SB = 40$ dB. This function returns with vectors z, p and the gain factor k of $S_{21}(p)$ such that

$$S_{21}(p) = \frac{f(p)}{g(p)} \tag{4.85a}$$

where

$$f(p) = k\prod_{i=1}^{n}(p - z_i) = k\prod_{i=1}^{nz}(p^2 + \omega_i^2); z_i = \mp j\omega_i \tag{4.85b}$$

and

$$g(p) = \prod_{i=1}^{n} (p - p_i) \tag{4.85c}$$

In other words, vector z includes the zeros and vector p includes the poles of $S_{21}(p)$.

Using Belevitch notation [21], input reflection coefficient $S_{11}(p)$ of the filter is given by

$$S_{11}(p) = \frac{h(p)}{g(p)} \tag{4.86a}$$

where $h(p)$ is constructed on the proper factorization of the even polynomial $H(p^2) = g(p)g(-p) - f(p)f(-p)$. Eventually, driving-point input impedance $Z(p)$ of the filter is generated from $S_{11}(p)$ such that

$$Z(p) = \frac{1 + S_{11}}{1 - S_{11}} = \frac{g(p) - h(p)}{g(p) + h(p)} = \frac{a(p)}{b(p)} \tag{4.86b}$$

All the above derivations are programmed under our MATLAB function $[h, F, f, W, g, a, b] = Elliptic\_hFg(k, z, p)$. Output of this function returns with vectors a and b describing the polynomials $a(p)$ and $b(p)$, respectively. Output vector W includes all the finite frequency transmission zeros ω_i of the filter placed within the stop band. Execution of the above functions is summarized in Table 4.14a.

Table 4.14a Driving-point input impedance $Z(p) = a(p)/b(p)$ of an elliptic filter such that $n = 12$, $ndc = 0$, $n_L = 0$, $nz = 6$ with $W^T = [3.0782\ 1.3057\ 1.0721\ 1.0192\ 1.0059\ 1.0027]$

***a(p)* [10⁵]**	***b(p)* [10⁵]**
0.231380769468048	0.000005784808481
0.195486899350027	0.149480117115023
1.176460252139375	0.126440002829854
0.849911357228007	0.692684296597020
2.401171297629883	0.493531486698597
1.455468782321504	1.255403929991658
2.490343619370675	0.737863109667367
1.223696817536921	1.109648064306197
1.356033460243625	0.518553274426196
0.503026410303298	0.476488920447394
0.351618906714228	0.165052862026282
0.080373634674760	0.079021741736659
0.029844958899418	0.017264737135098

For this example, input impedance $Z(p)$ is synthesized using both "direct synthesis" and "chain matrix" with impedance correction methods employing our MATLAB functions

$$[CT,CV,LL1,LL2,MM] = SynthesisbyTranszeros(KFlag,W,ndc,a,b,epszero)$$

and

$$[CTF,CVF] = Zeroshifting\_viaChainMatrixImpedanceCorrection(W,ndc,a,b,epszero).$$

Result of synthesis is depicted in Figure 4.27(a).

Element values of the synthesized elliptic filter are given by Table 4.14b.

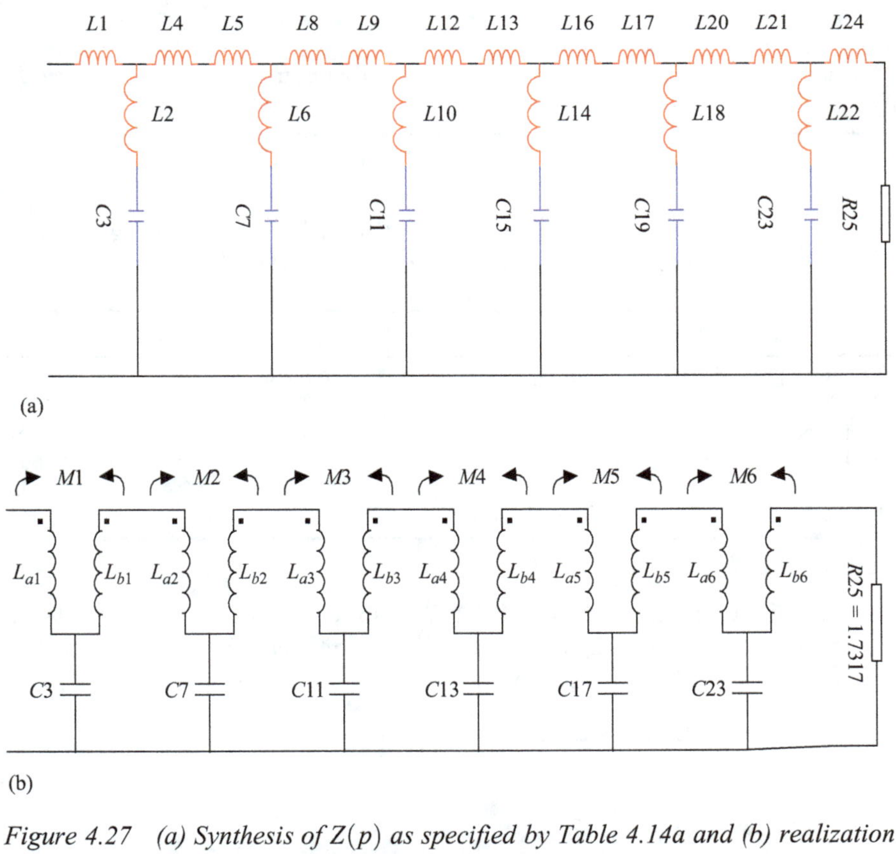

Figure 4.27 (a) Synthesis of $Z(p)$ as specified by Table 4.14a and (b) realization of the elliptic filter with coupled coils

Table 4.14b Element values and the accumulated numerical errors occur in the course of elliptic filter synthesis

ω_1	$L1 = 1.4724$ $L2 = 0.0777$ $C3 = 1.3592$ $L4 = -0.0738$	ω_4	$L13 = -4.3198$ $L14 = 77.2605$ $C15 = 0.0125$ $L16 = 4.5757$
ω_2	$L5 = 2.3355$ $L6 = 0.5749$ $C7 = 1.0203$ $L8 = -0.4614$	ω_5	$L13 = 1.3995$ $L14 = 14.4642$ $C15 = 0.0683$ $L16 = -1.2761$
ω_3	$L9 = 2.7441$ $L10 = 5.2501$ $C11 = 0.1657$ $L12 = -1.8022$	ω_6	$L21 = -1.6235$ $L22 = 35.2613$ $C23 = 0.0282$ $L24 = 1.7018$ $R25 = 1.7317$
$\boldsymbol{error_a}$ (direct syn.)	$\boldsymbol{error_b}$ (direct syn.)	$\boldsymbol{error_{a-chain}}$ (chain matrix)	$\boldsymbol{error_{b-chain}}$ (chain matrix)
0.0107	0.0053	0.0363	0.0175

Table 4.14c Parameters of Brune sections with coupled coils replacing the negative valued inductive T-sections of Figure 4.27(b)

	$\boldsymbol{L_a}$	$\boldsymbol{L_b}$	$\boldsymbol{M}$
SECTION 1	1.5501	0.0039	0.0777
SECTION 2	2.9104	0.1135	0.5749
SECTION 3	7.9942	3.4479	5.2501
SECTION 4	72.9407	72.6848	77.2605
SECTION 5	15.8637	13.1881	14.4642
SECTION 6	33.6378	33.5595	35.2613

As indicated by the last row of Table 4.14b, direct synthesis by impedance correction method yields slightly better relative error than that the chain matrix synthesis method. More precisely, we compare the following error figures.

$$error_a : 0.0107 \text{ versus } 0.063$$

and

$$error_b : 0.0053 \text{ versus } 0.0175$$

Negative inductors of Figure 4.27(a) can be eliminated by replacing tee-inductors using coupled coils as in (4.12). Thus, we end up with Figure 4.27(b).

Element values of Figure 4.27(b) is given in Table 4.14c.

Transducer power gain of the elliptic filter is depicted in Figure 4.28.

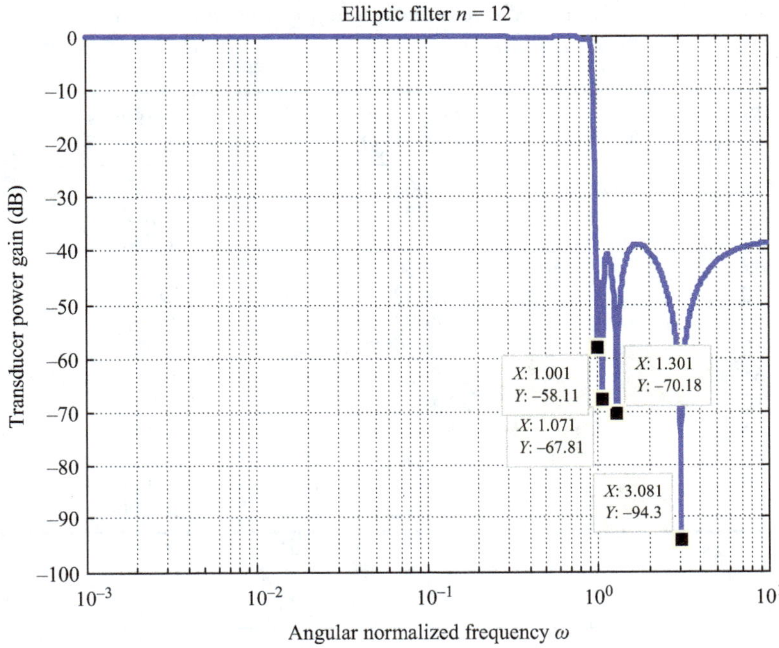

Figure 4.28 Synthesis of $Z(p)$ as specified by Table 4.14a

Example 4.9

In this example, we design a wideband monopole antenna matching network over 40–85 MHz employing the RFT via parametric approach [16]. Measured antenna impedance data is given in Table 4.15a, and its real and imaginary parts are depicted in Figures 4.29 and 4.30, respectively. In the design, we use the real part form of the driving-point impedance as expressed by (4.44). In this regard, we insert two transmission zero at the normalized angular frequencies $\omega_{z1} = 0.2$ and $\omega_{z2} = 1.0$ (meaning that $f_{z1} = 20$ MHz) and $f_{z2} = 100$ MHz).

These transmission zeros help to concentrate the power delivered to the antenna within the desired band of operation. On the other hand, we wish to terminate the lossless equalizer in $R_0 = 50\ \Omega$ which corresponds to a normalized unit termination. In this case, we fixed the coefficient of the numerator as $a_0 = 1/\left(\omega_{z1}^2 \omega_{z2}^2\right) = 25$.

Eventually, using the RFT-parametric method, TPG of the matched antenna is optimized yielding the normalized input impedance $Z_B(p) = a(p)/b(p)$ of the matching network as listed in Table 4.15b. During the optimization of TPG, impedances are normalized with respect to $R_0 = 50\ \Omega$. For the frequency normalization, we use $f_0 = 100$ MHz.

Synthesis of the normalized impedance $Z(p) = a(p)/b(p)$ is depicted in Figure 4.31(a). In this figure, normalized element values are denormalized by replacing inductors L_i by $L_{iA} = L_i \times R_0/\omega_0$ and capacitors C_i by $C_{iA} = C_i/(R_0\omega_0)$.

Table 4.15a Measured impedance data for monopole antenna over 20–100 MHz

Frequency (MHz)	Real part $RLA(f)$ (Ω)	Imaginary part $XLA(f)$ (Ω)
20	0.6000	0.6000
30	0.8000	0.8000
40	0.8000	0.8000
45	1.0000	1.0000
50	2.0000	2.0000
55	3.4000	3.4000
60	7.0000	7.0000
65	15.0000	15.0000
70	22.4000	22.4000
75	11.0000	11.0000
80	5.0000	5.0000
90	1.6000	1.6000
100	1.0000	1.0000

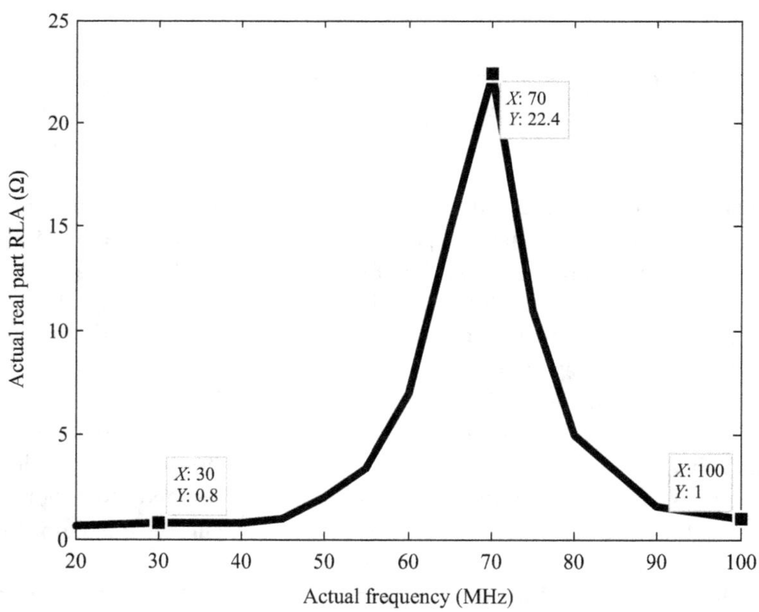

Figure 4.29 Measured antenna impedance data: real part RLA over 20–100 MHz

Furthermore, inductive T-structures ($L1$, $L2$, $L4$) and ($L5$, $L6$, $L8$) are replaced by coupled coils which eliminates the negative inductors as reviewed in Section 4.3.4 by (4.57a)–(4.76g). Thus, we obtained the matching network with actual elements as depicted in Figure 4.31(b). Corresponding TPG versus normalized angular

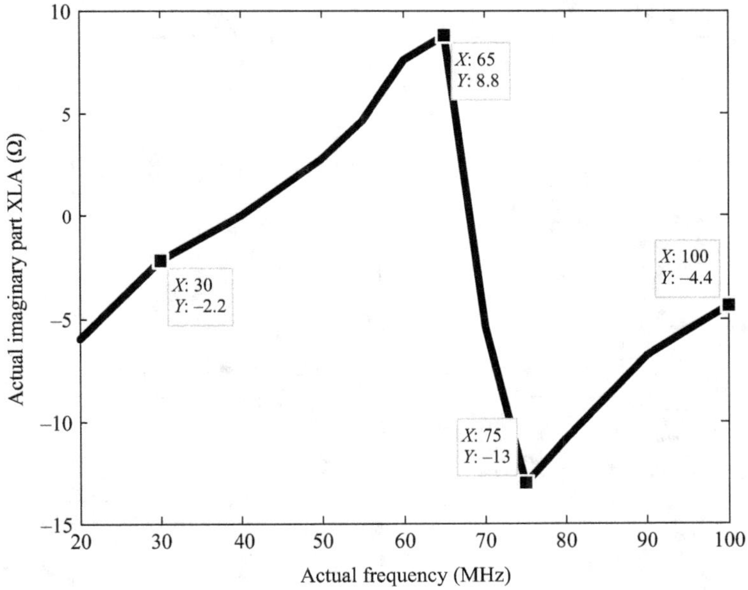

Figure 4.30 Measured antenna impedance data: imaginary part XLA over 20–100 MHz

Table 4.15b Driving-point input impedance $Z_B(p) = a(p)/b(p)$ of the matching network designed for Example 4.10

a(p)	**b(p)**
0	1.0000
0.1285	36.7818
4.7280	46.9881
5.9732	62.7998
5.6160	44.0998
3.7764	26.1939
0.8576	5.0048
0.2657	0.2657

frequency is shown in Figure 4.32. In this figure, normalized angular frequency $\omega = 1$ corresponds to the actual frequency. Actual element values of the matching network are given by Table 4.15c.

Close examination of Figure 4.32 reveals that at $\omega = 0.2$ and $\omega = 1.0$ TPG goes down to minus infinity as forced by means of finite transmission zeros located at $\omega_1 = 0.2$ and $\omega_2 = 1$.

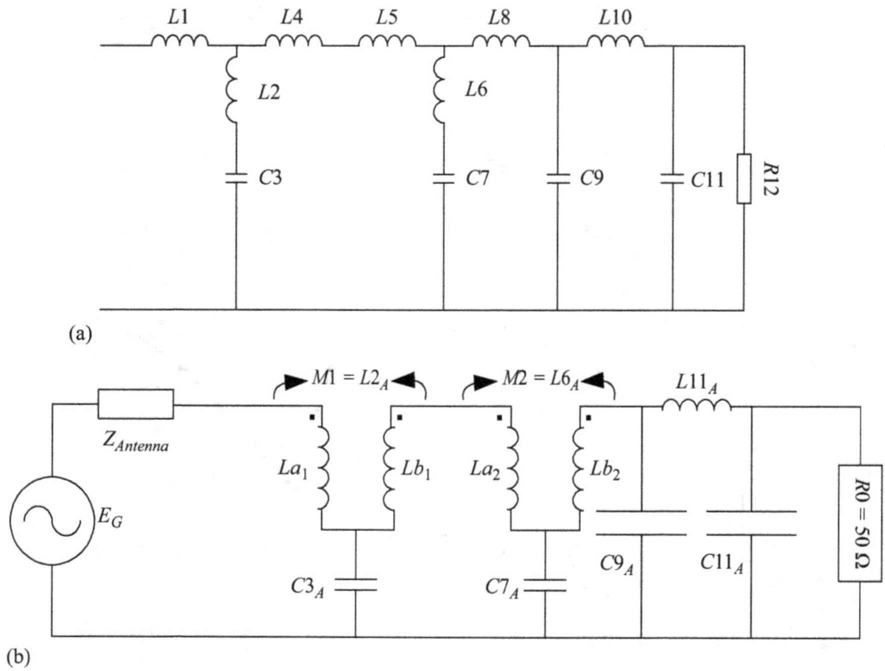

Figure 4.31 *(a) Synthesis of the matching network for the monopole antenna over 40–85 MHz, (b) matching network of Table 4.15d with coupled coils*

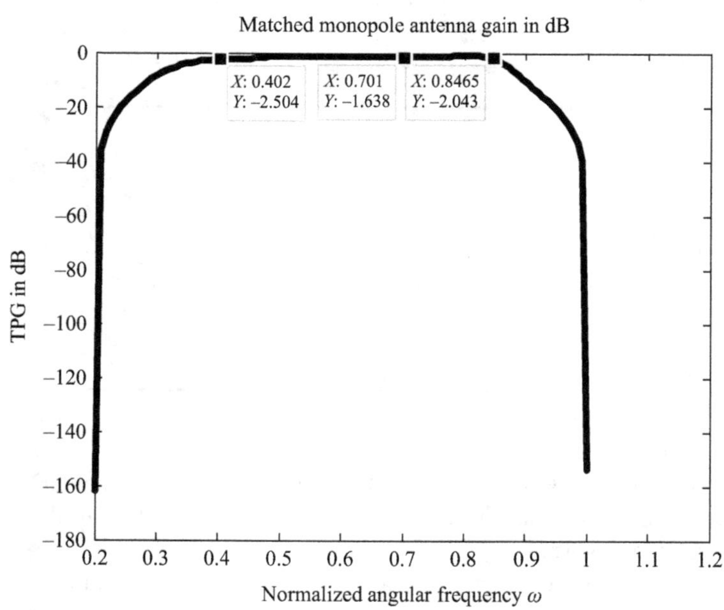

Figure 4.32 *Large view of TPG over $\omega = 0.2$ and $\omega = 1.0$*

Table 4.15c Actual element values of the matching network

First Brune Section			
$L1(mH)$	$L2(H)$	$L4(H)$	$C3(F)$
−936.584	1.47569	2.5637	16.9412
Second Brune Section			
$L5(H)$	$L6(H)$	$L8(H)$	$C7(mF)$
−1.52892	4.49579	2.31683	222.43
Lowpass Section with resistive termination R			
$C9(F)$	$L10(mH)$	$C11(mF)$	$R12\ (\Omega)$
1.64324	812.306	27.1874	1

Table 4.15d Antenna matching network with coupled coils

***La*1 (nH)**	***Lb*1 (nH)**	***M*1 (nH)**	***C3A* (pF)**
42.90	321.44	117.43	539.2
$La2$(nH)	$Lb2$(nH)	$M2$(nH)	$C7A$(pF)
236.1	542.13	357.76	7.08
$C9A$(pF)	$L10A$(nH)	$C11A$(pF)	$R12(\Omega)$
52.306	64.64	0.865	50

4.4 Reflectance-based impedance generation and its synthesis

In the previous sections, we introduced high-precision, numerically robust computer algorithms to synthesize minimum immittance functions which describe LC ladders with real transmission zeros in the left half plane, at DC, at infinity and on finite frequencies on the $j\omega$ axis. Those algorithms were implemented on MATLAB environment. Eventually, the major MATLAB function called

$$[CT, CV] = CircuitPlot\_Yarman(KFlag, R0, f0, a, b, ndc)$$

to synthesize the minimum immittance function as an LC ladder network in resistive termination. In the above MATLAB function, input arguments $a = [a_1 a_2 \cdots a_n a_{n+1}]$ and $b = [b_1 b_2 \cdots b_n b_{n+1}]$ are the MATLAB vectors which include the coefficients of the numerator *a(p)* and denominator *b(p)* polynomials of the normalized driving-point minimum input immittance

$$F(p) = \frac{a(p)}{b(p)} = \frac{a_1 p^n + a_2 p^{n-1} + \cdots a_n p + a_{n+1}}{b_1 p^n + b_2 p^{n-1} + \cdots b_n p + 1} \quad \text{with } a_1 = 0$$

For an LC ladder network, even part of the driving-point input immittance $F(p)$ must yield

$$R(p^2) = \frac{1}{2}[F(p) + F(-p)] = \frac{(-1)^{ndc}(a_0)^2 p^{2ndc}}{B_1 p^{2n} + B_2 p^{2(n-1)} + \cdots + B_n p^2 + 1}$$

In the above expression, *ndc* is total count of even part zeros or equivalently, transmission zeros at DC such that $ndc \leq n$.

Integer *n* refers to total number of reactive elements in the LC ladder.

Kflag is a user-defined control flag which describes the type of immittance. If $F(p)$ is a minimum reactance then, we set $KFlag = 1$. If $F(p)$ is a minimum susceptance then, we select $KFlag = 0$.

Obviously, synthesis of $F(p)$ yields normalized element values of the LC ladder in resistive termination. However, actual element values are computed using the normalization numbers *R*0 and *f*0; where "*R*0" is the normalization resistance and "*f*0" is the normalization frequency, respectively.

If the normalized capacitors are designated by C_{Ni}, actual capacitors are given by $C_{Ai} = C_{Ni}/(2\pi f_0 R_0)$. Similarly, if the normalized inductors are designated by L_{Ni}, actual inductors are specified by $L_{Ai} = R_0/(2\pi f_0)$.

Execution of function CircuitPlot_Yarman results in the output vectors CT and CV where vector CT includes the topological codes of the circuit elements of the LC ladder in sequential order. Size of vector CT is "$n + 1$."

CT(*i*) = 2 designates a series capacitor, *CT*(*i*) = 7 refers to a shunt inductor, *CT* (*i*) = 1 designates a series inductor, *CT*(*i*) = 8 designates a shunt capacitor and finally termination resistor R_{n+1} is designated by *CT*(*n* + 1) = 9. Vector *CV* contains the actual element values of the LC ladder in order.

Employing the MATLAB tools developed in previous sections, in this section, a driving-point input impedance is generated from a given bounded real-reflectance function which also describes an LC ladder. However, the resulting impedance $z_{in}(p) = F(p)$ may be a minimum function or not. From the computer implementation point of view, we can generate an impedance from a bounded real reflectance under six major types referred to as synthesis cases. Synthesis of each case requires special care. In the following sections, first we generate a realizable bounded real reflectance using SRFT [3] to yield an LC ladder network with transmission zeros only at DC and infinity (Section 4.4.1). Then, the driving-point input impedance is generated from the realizable reflectance (Section 4.4.2).

In Section 4.4.3, possible impedance structures are classified under cases. In Section 4.4.4, we present synthesis examples for each synthesis case.

4.4.1 Simplified real frequency technique

SRFT constructs lossless two-ports for designing various matching networks [10–13]. Referring to Figure 4.33, in SRFT, real normalized driving-point reflectance $S_{in} = h(p)/g(p)$ is generated to yield a lossless two-port in resistive termination R.

On MATLAB environment, numerator polynomial $h(p)$ and denominator polynomial $g(p)$ are described as follows:

$$\begin{aligned} h(p) &= h_1 p^n + h_2 p^{n-1} + \cdots + h_n p + h_{n+1} \\ g(p) &= g_1 p^n + g_2 p^{n-1} + \cdots + g_n p + g_{n+1} \end{aligned} \tag{4.87}$$

For broadband matching problems, coefficients of the numerator polynomial $h(p)$ is determined to optimize the electrical performance of the system under consideration. For many practical problems, $g(p)$ is determined in such a way that, when $S_{in} = h(p)/g(p)$ is synthesized, it yields an LC ladder satisfying the following losslessness equation:

$$g(p)g(-p) = h(p)h(-p) + (-1)^{ndc} p^{2ndc} \tag{4.88}$$

In (4.88), $h(p)$ is a user-selected polynomial with arbitrary real coefficients or it is determined to optimize the electrical performance of a matched system under consideration; $g(p)$ is generated as a strictly real Hurwitz polynomial with preset *ndc* which is the total number of transmission zeros placed at DC by the designer.

It should be mentioned that there is no restriction imposed on the real coefficients of $h(p)$ except that $|h(j\omega)|^2 + \omega^{2ndc}$ must be strictly positive over the entire angular real frequency axis $j\omega$ which in turn requires $h(0) \neq 0$. This is a trivial condition to impose on $h(p)$.

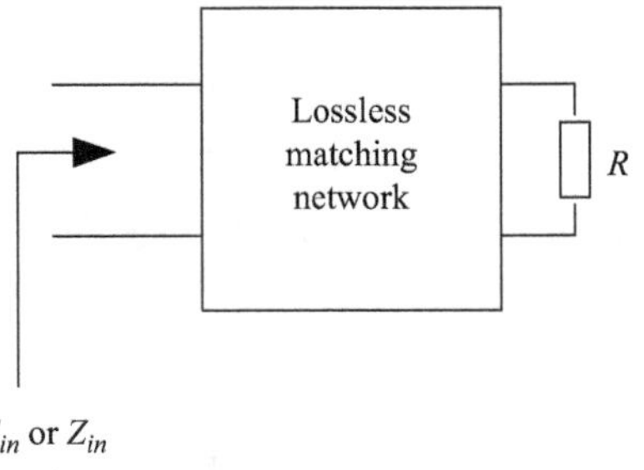

Figure 4.33 Synthesis of bounded real reflection coefficient $S_{\text{in}}(p) = (Z_{\text{in}} - R)/(Z_{\text{in}} + R)$

4.4.2 Generation of driving-point input impedance from a realizable reflectance

$S_{in}(p)$ is described in terms of the actual driving-point input impedance $Z_{in}(p)$ of the ladders as

$$S_{in}(p) = \frac{h(p)}{g(p)} = \frac{Z_{in}(p) - R}{Z_{in}(p) + R} = \frac{z_{in}(p) - 1}{z_{in}(p) + 1} \tag{4.89}$$

where R is known as the port normalization resistance or in short-port termination. It is usually selected as $R = 50\ \Omega$.

Furthermore:

$$z_{in}(p) = \frac{Z_{in}(p)}{R} \tag{4.90}$$

is the normalized input impedance of the LC ladder and it is determined by

$$z_{in}(p) = \frac{1 + S_{in}(p)}{1 - S_{in}(p)} = \frac{g(p) + h(p)}{g(p) - h(p)} = \frac{a(p)}{b(p)} \tag{4.91}$$

where

$$\begin{aligned} a(p) &= g(p) + h(p) \\ b(p) &= g(p) - h(p) \end{aligned} \tag{4.92}$$

In (6), numerator and denominator polynomials are given as follows:

$$a(p) = a_1 p^n + a_2 p^{n-1} + \cdots + a_n p + a_{n+1}$$

and (4.93)

$$b(p) = b_1 p^n + b_2 p^{n-1} + \cdots + b_n p + b_{n+1}$$

For the case where $0 \leq ndc \leq n$, we must have the following identities among the coefficients of $h(p)$ and $g(p)$:

$$g_1 = \mp h_1; \quad 0 < ndc < n$$

or

$$g_1 = \sqrt{h_1^2 + 1} > 1; \quad ndc = n$$

and

$$g_{(n+1-ndc)} = \sqrt{h^2_{(n+1-ndc)} + 1}; \quad 0 < ndc < n \tag{4.94}$$

$$g_{n+1} = \mp h_{n+1}; \quad 0 < ndc < n$$

or

$$g_{n+1} = \sqrt{h_{n+1}^2 + 1}; \quad ndc = 0$$

Depending on the signs of h_1 and h_{n+1} and the value of ndc, coefficients a_1, a_{n+1},b_1 and b_{n+1} are determined to lead the synthesis of $z_{in}(p)$. We found useful to make the following classifications to carry out the synthesis of $z_{in}(p)$.

4.4.3 Synthesis of driving-point impedance $z_{in}(p) = a(p)/b(p)$

4.4.3.1 Case 1

1. When $ndc = 0$ or in general, $ndc < n$ and h_1 is negative real number then, $g_1 = -h_1$ and thus, $a_1 = g_1 + h_1 = 0$ and $b_1 > 0$.
2. When $ndc = 0$ and $h_{n+1} = 0$, then $g_{n+1} = 1$ and, thus, $a_{n+1} = 1$ and $b_{n+1} > 0$.
3. When $ndc < n$ and h_{n+1} is negative, $g_{n+1} = -h_{n+1}$ and thus $a_{n+1} = g_{n+1} + h_{n+1} = 0$ and $b_{n+1} > 0$.

Above conditions indicate that denominator polynomial $b(p)$ is regular on the $j\omega$ axis. Therefore, $z_{in}(p) = a(p)/b(p)$ is a minimum reactance function. Hence, setting $KFlag = 1$, it can be synthesized employing our MATLAB function

$$[CT, CV] = CircuitPlot\_Yarman(KFlag, R0, f0, a, b, ndc)$$

4.4.3.2 Case 2

1. When $0 < ndc < n$ and h_1 is a positive real number, then $g_1 = h_1$ and thus $a_1 > 0$ and $b_1 = g_1 - h_1 = 0$. This situation indicates that degree of polynomial $a(p)$ is higher than that of $b(p)$. Therefore, $F(p)$ has a pole at infinity.
2. When $0 < ndc < n$ and h_{n+1} is a positive number, $g_{n+1} = h_{n+1}$ and, thus, $a_{n+1} > 0$ and $b_{n+1} = g_{n+1} - h_{n+1} = 0$. This situation indicates that $b(p)$ has a zero at $p = 0$.

Above conditions describe a minimum susceptance $H(p) = y_{in}(p) = 1/z_{in}(p) = b(p)/a(p)$. Therefore, by setting $KFlag = 0$, we can directly go ahead with the synthesis employing the MATLAB function

$$[CT, CV] = CircuitPlot\_Yarman(KFlag, R0, f0, a, b, ndc)$$

4.4.3.3 Case 3

1. When $0 < ndc < n$ and h_1 is a positive number, $g_1 = h_1$ and thus $a_1 > 0$ and $b_1 = 0$ which means that $z_{in}(p) = a(p)/b(p)$ has a pole at infinity.
2. When $0 < ndc < n$ and h_{n+1} is a negative number, $g_{n+1} = -h_{n+1}$ and, thus, $a_{n+1} = g_{n+1} + h_{n+1} = 0$ and $b_{n+1} > 0$.

In this case, $z_{in}(p) = a(p)/b(p)$ is not a minimum reactance since has a pole at infinity ($p = \infty$). Therefore, we remove the pole at infinity using the MATLAB function.

$$[L0, a1, b1] = removepole\_atinfinity(a, b)$$

The residue $L0$ at infinity yields the normalized value of the series inductor which is extracted from the impedance function $z_{in}(p) = a(p)/b(p)$. Then, the actual value of this inductor is computed as follows:

$$L0\_Act = L0 \times R0/2/pi/f0; KFlag = 1$$

Eventually, synthesis is completed using the function *CircuitPlot_Yarman*.

$$[CT, CV] = CircuitPlot\_Yarman(KFlag, R0, f0, a1, b1, ndc)$$
$$CT = [1CT]; \quad CV = [L0\_Act\ CV]$$

4.4.3.4 Case 4

1. When $0 < ndc < n$ and h_1 is a negative number, $g_1 = -h_1$ and thus $a_1 = g_1 + h_1 = 0$ and $b_1 > 0$ which means that $F(p)$ has a zero at $p = \infty$.
2. When $0 < ndc < n$ and h_{n+1} is a positive number, $g_{n+1} = h_{n+1}$ and, thus, $a_{n+1} = g_{n+1} + h_{n+1} = 0$ and $b_{n+1} > 0$ which means that $z_{in}(p) = a(p)/b(p)$ has a pole at $p = 0$. In this case, we must remove this pole employing the MATLAB function

$$[k0, a1, b1] = removepole\_atzero(a, b)$$

Here, $k0$ is the residue at $p = 0$, and it yields the normalized value of the seriesly extracted capacitor $C0 = 1/k0$.

Its actual value is determined as follows:

$$C0 = 1/k0; \quad C0\_Act = C0/R0/2/pi/f0; \quad KFlag = 1$$

Hence, we can initiate the complete synthesis as

$$ndc1 = ndc - 1$$
$$[CT, CV] = CircuitPlot\_Yarman(KFlag, R0, f0, a1, b1, ndc1)$$
$$CT = [7CT]; \quad CV = [C0\_Act\ CV]$$

4.4.3.5 Case 5

It does not matter what the value of h_1 is, for $ndc = n$ and if h_{n+1} is a positive number then $g_{n+1} = h_{n+1}$ and $a_1 > 0$, $a_{n+1} > 0$ and $b_1 > 0$, $b_{n+1} = g_{n+1} - h_{n+1} = 0$. This situation describes a minimum susceptance driving-point input admittance function $H(p) = 1/z_{in}(p)$ which has all its transmission zeros at $p = 0$.

Thus, setting $KFlag = 0$, we can complete the synthesis using

$$[CT, CV] = CircuitPlot\_Yarman(KFlag, R0, f0, b, a, ndc)$$

4.4.3.6 Case 6

As in Case 5, it does not matter what the value of h_1 is, for $ndc = n$ and if h_{n+1} is a negative number then, $g_{n+1} = -h_{n+1}$ and $a_1 > 0$, $a_{n+1} = g_{n+1} + h_{n+1=}0$ and $b_1 > 0$, $b_{n+1} > 0$. This situation describes a minimum reactance driving-point input impedance function $z_{in}(p)$ which has all its transmission zeros at $p = 0$.

Thus, complete synthesis is obtained as follows. Thus, setting $KFlag = 1$ g, we can complete the synthesis using

$$[CT, CV] = CircuitPlot\_Yarman(KFlag, R0, f0, b, a, ndc)$$

All the above conditions are summarized in Table 4.16. In this table, actual negative, positive and zero values of the coefficients are designated by logical numbers

Table 4.16 Conditions imposed on the coefficients to generate possible synthesis cases

$F(p) = a(p)/b(p)$: **Impedance function**	*ndc*	h_1	g_1	h_{n+1}	g_{n+1}	a_1	a_{n+1}	b_1	b_{n+1}
Case 1: Minimum reactance	$ndc < n$	$-1/+1$	$+1$	-1	$+1$	0	$0/+1$	$+1$	$+1$
Case 2: Minimum susceptance	$ndc < n$	$+1$	$+1$	$+1$	$+1$	$+1$	$+1$	0	$0/+1$
Case 3: Extract pole at $p = \infty$	$ndc < n$	$+1$	$+1$	-1	$+1$	$+1$	0	0	$+1$
Case 4: Extract pole at $p = 0$	$ndc < n$	-1	$+1$	$+1$	$+1$	0	$+1$	$+1$	0
Case 5: High-pass minimum susceptance	$ndc = n$	∓ 1	$+1$	$+1$	$+1$	$+1$	$+1$	$+1$	0
Case 6: High-pass minimum reactance	$ndc = n$	∓ 1	$+1$	-1	$+1$	$+1$	0	$+1$	$+1$

"−1," "+1" and "0," respectively. For example, if we set the logical values of h_1 and h_{n+1} as -1 (a negative real number) and $+1$ (a positive number), corresponding logical values of $g_1 = \mp h_1$ and $g_{n+1} = \mp h_{n+1}$ are designated by "+1."

Similarly, depending on the choice of $ndc<n$, corresponding logical values of $a_1 = g_1 + h_1 =$ "0" for $h_1 =$ "−1" and $a_1 = g_1 + h_1 =$ "+1" for $h_1 =$ "+1," etc.

All the above cases are combined under

function [*CT CV*] = *Synthesis_ImpedanceBased*(*a*, *b*, *ndc*, *R0*, *f0*)

MATLAB function *Synthesis_ImpedanceBased* synthesizes the given impedance function $z_{in}(p) = a(p)/b(p)$ with actual elements for specified polynomials $a(p)$, $b(p)$ *ndc*, and normalization resistance R_0 and normalization frequency f_0.

It should be noted that integer *ndc* is the total count for DC zeros of the transfer scattering parameter $S_{21}(p)$ which is expressed by means of the following set [10]:

$$S_{21}(p) = \frac{p^{ndc}}{g(p)} \tag{4.95a}$$

such that

$$S_{21}(p)S_{21}(-p) = 1 - S_{in}(p)S_{in}(-p) = \frac{(-1)^{ndc}p^{2ndc}}{g(p)g(-p)} \tag{4.95b}$$

where

$$S_{in}(p) = \frac{h(p)}{g(p)} \tag{4.95c}$$

and

$$g(p)g(-p) = h(p)h(-p) + (-1)^{ndc}p^{2ndc} \tag{4.95d}$$

It can be shown that DC zeros of $S_{21}(p)S_{21}(-p)$ overlaps with those zeros of the even function described by

$$R(p^2) = \frac{1}{2}[z_{in}(p) + z_{in}(-p)] \tag{4.96a}$$

or

$$R(p^2) = \frac{1}{2}\left[\frac{a(p)b(-p) + a(-p)b(p)}{b(p)b(-p)}\right] = \frac{A(p^2)}{B(p^2)} \tag{4.96b}$$

with

$$\begin{aligned} A(p^2) &= a(p)b(-p) + a(-p)b(p) \\ &= A_1p^{2n} + A_2p^{2(n-1)} + .. + A_np^2 + A_{n+1} = A_{(n+1-ndc)}p^{2ndc} \end{aligned} \tag{4.96c}$$

where all the coefficients are zero except $A_{(n+1-ndc)}$. In this case, we can express the even polynomial $A(p^2)$ as a MATLAB vector A such that

$$A = \left[0\,0\cdots A_{(n+1-ndc)}\,0\,0\cdots 0\right] \tag{4.97}$$

Hence, *ndc* can also be determined automatically by a simple sorting algorithm from the given polynomials *a(p)* and *b(p)* by generating the MATLAB vector *A*.

Our MATLAB function `[ndc] = DCZeros_Evenpart(a,b)` performs the above mentioned sorting process accurately to determine *ndc* provided that (2) is satisfied and *a(p)* and *b(p)* are generated as in (4.95).

Now, let us give examples to show the implementation process.

4.4.4 Examples

In this section, we present six examples to show the synthesis of different cases.

Example 4.10

Let $h(p) = [-0.2\; 0.8 - 1\; 0.2 - 1 - 1]$ meaning that $h(p) = -0.2p^5 + 0.8p^4 - 1p^3 + 0.2p^2 - 1p - 1$ and $ndc = 3$.

1. Find the realizable strictly Hurwitz polynomial $g(p)$.
2. Find the polynomials $a(p)$ and $b(p)$ for $z_{in}(p) = a(p)/b(p)$.
3. Identify the logical values of the coefficients $h_1, g_1, a_1, b_1, h_{n+1}, g_{n+1}, a_{n+1}, b_{n+1}$ as in Table I and determine the possible synthesis case.
4. Synthesize $z_{in}(p) = a(p)/b(p)$ and draw the lossless LC-ladder network in resistive termination.

Solution

1. For this purpose, we developed a *MATLAB* program called *Main_SParBased Synthesis.m*. This program takes the standard *MATLAB* polynomial $h(p)$ and *ndc* as inputs and then generates the realizable strictly Hurwitz polynomial $g(p)$. Thus,

execution of *Main_SParBasedSynthesis.m* yields the strictly Hurwitz polynomial $g(p)$ as

$$g = [0.2000\ \ 1.1946\ \ 2.9677\ \ 3.2886\ \ 2.8244\ \ 1.0000]$$

2. In the same program, we have calculated *MATLAB* polynomials $a(p)$ and $b(p)$ for $z_{in}(p) = a(p)/b(p)$. Hence, we have

$$a = [0\ \ 1.9946\ \ 1.9677\ \ 3.4886\ \ 1.8244\ \ 0.0000]$$
$$b = [0.4000\ \ 0.3946\ \ 3.9677\ \ 3.0886\ \ 3.8244\ \ 2.0000]$$

3. Actual value of h_1 $=-0.2$ is a negative number. Therefore, its designated logical level is $h_1 =$ "-1." Similarly, actual value of $h_{n+1} =-1$. It is also a negative number. Hence, its designated logical level is $h_{n+1} =$ "-1." Actual and logical values of a_1 and a_{n+1} are zero (i.e., $a_1 = 0$ and $a_{n+1} = 0$). Finally, logical values of g_1, g_{n+1}, b_{n+1} are equal to "$+1$" since their actual values are always positive real numbers. Thus, all the above indicates that we are dealing with a minimum reactance input impedance of Case 1 as shown in Table 4.17.
4. *Main_SParBasedSynthesis.m* calls

```
function [ CT CV ] = Synthesis_ImpedanceBased( a,b,ndc,R0,f0 )
```

to synthesizes $Z_{in}(p) = a(p)/b(p)$ as shown in Figure 4.34.

Table 4.17 Case study for Example 4.11

$Z_{in}(p) = a(p)/b(p)$	ndc	h_1	g_1	h_{n+1}	g_{n+1}	a_1	a_{n+1}	b_1	b_{n+1}
Case 1: Minimum reactance	$ndc < n$	−1	+1	−1	+1	0	0	+1	+1

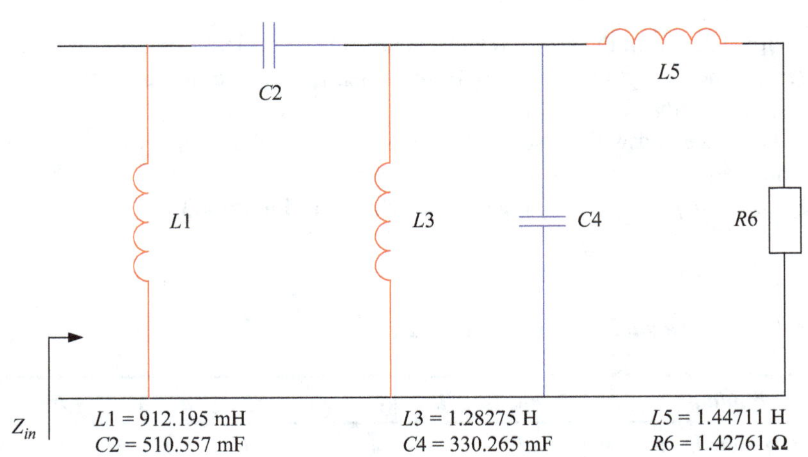

Figure 4.34 Synthesis of $Z_{in}(p) = (0p^5 + 1.9946p^4 + 1.9677p^3 + 3.4886p^2 + 1.8244p + 0)/(0.4p^5 + 0.3946p^4 + 3.9677p^3 + 3.0886p^2 + 3.8244p + 2)$

It is noted that, in *function* [*CT CV*] = *Synthesis_ImpedanceBased*(*a, b, ndc, R*0*, f*0), *ndc* is determined using $[ndc] = DCZeros\_Evenpart(a, b)$. Furthermore, normalization numbers are selected as $R0 = 1$ and $f0 = 1/2/pi$ to end up with normalized angular frequency $w0 = 2 \times pi \times f0 = 1$.

Example 4.11

Let $h = [1\ 0.2\ 0.3\ 0.4\ 0.5\ 0.61]$ and $ndc = 4$

1. Find the realizable strictly Hurwitz polynomial $g(p)$.
2. Find the polynomials $a(p)$ and $b(p)$ for $z_{in}(p) = a(p)/b(p)$.
3. Identify the logical values of the coefficients $h_1, g_1, a_1, b_1, h_{n+1}, g_{n+1}, a_{n+1}, b_{n+1}$ as in Table I and determine the possible synthesis case.
4. Synthesize $z_{in}(p) = a(p)/b(p)$ and draw the lossless LC-ladder network in resistive termination.

Solution

Employing the *MATLAB* program *Main_SParBasedSynthesis.m* with the specified inputs such that

$$h = [1\ 0.2\ 0.3\ 0.4\ 0.5\ 0.61] \quad \text{and} \quad ndc = 4$$

1. Denominator *MATLAB* polynomial $g(p)$ is found as

$$g = [1\ \ 3.6822\ \ 7.0592\ \ 8.3568\ \ 6.8201\ \ 3.6056\ \ 1\ ]$$

2. Using the same program, we generated the polynomials $a(p)$ and $b(p)$ as follows:

$$a = [2\ \ 3.8822\ \ 7.3592\ \ 8.7568\ \ 7.3201\ \ 4.2056\ \ 2]$$

$$b = [0\ \ 3.4822\ \ 6.7592\ \ 7.9568\ \ 6.3201\ \ 3.0056\ \ 0]$$

3. As it is seen from the above results, actual and logical values of b_1 and b_{n+1} are zero, other coefficients $h_1, g_1, a_1, h_{n+1}, g_{n+1}, a_{n+1}$ are all positive numbers yielding logical value "+1."

   Thus, we end with Case 2 which has a minimum susceptance driving-point input admittance as shown in Table 4.18.
4. $Y_{in}(p) = b(p)/a(p)$ is synthesized as shown in Figure 4.35.

Table 4.18 Case study for Example 4.12

$Y_{in}(p) = b(p)/a(p)$	*ndc*	h_1	g_1	h_{n+1}	g_{n+1}	a_1	a_{n+1}	b_1	b_{n+1}
Case 2: minimum susceptance	$ndc < n$	+1	+1	+1	+1	+1	+1	0	0

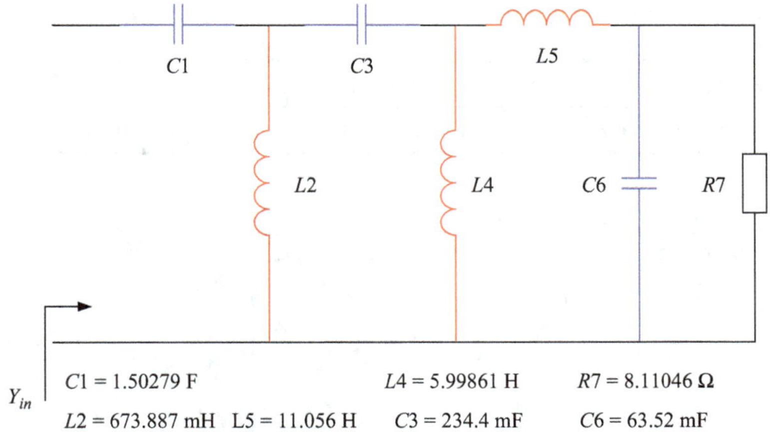

Figure 4.35 Synthesis of $Y_{\rm in}(p) = b(p)/a(p)$ *of Example 4.12*

Example 4.12

Let $h = [0.91 - 1 - 23 - 4]$ and $ndc = 2$.

1. Find the realizable strictly Hurwitz polynomial $g(p)$.
2. Find the polynomials $a(p)$ and $b(p)$ for $z_{in}(p) = a(p)/b(p)$.
3. Identify the logical values of the coefficients $h_1, g_1, a_1, b_1, h_{n+1}, g_{n+1}, a_{n+1}, b_{n+1}$ as in Table 4.19 and determine the possible synthesis case.

Synthesize $z_{in}(p) = a(p)/b(p)$ and draw the lossless LC-ladder network in resistive termination.

Solution

1. We run the same program with the specified input data and then $g(p)$ is obtained as

$$g = [0.9000\ \ 4.7882\ \ 11.1816\ \ 13.8876\ \ 10.2030\ \ 4.0000]$$

2. Polynomials $a(p)$ and $b(p)$ of $z_{in}(p) = a(p)/b(p)$ are determined as the output of *MATLAB* program *Main_SParBasedSynthesis.m*. Hence, we have,

$$a = [1.8000\ \ 5.7882\ \ 10.1816\ \ 11.8876\ \ 13.2030\ \ 0.0000]$$
$$b = [03.7882\ \ 12.1816\ \ 15.8876\ \ 7.2030\ \ 8.0000\ ]$$

as it is seen that $b_1 = 0$ and $b_{n+1} = b_5 > 0$. Therefore, it is concluded that $Z_{in}(p)$ has a pole at infinity.

3. In short, as in the previous examples, logical values for h_1, g_1 are equal to "1" since their actual values are positive real numbers. Logical value for $h_{n+1} = h_5 = -1$ since its actual value is negative (i.e., $h_5 = -4$). Rest of the coefficients is positive. Thus, we are facing Case 3 as indicated by Table 4.19. Therefore, synthesis starts with the extraction of the pole placed at infinity.

Table 4.19 A case study for Example 4.13

$Z_{in}(p) = a(p)/b(p)$	*ndc*	h_1	g_1	h_{n+1}	g_{n+1}	a_1	a_{n+1}	b_1	b_{n+1}
Case 3: extract pole at $p = \infty$	$ndc < n$	+1	+1	−1	+1	+1	0	0	+1

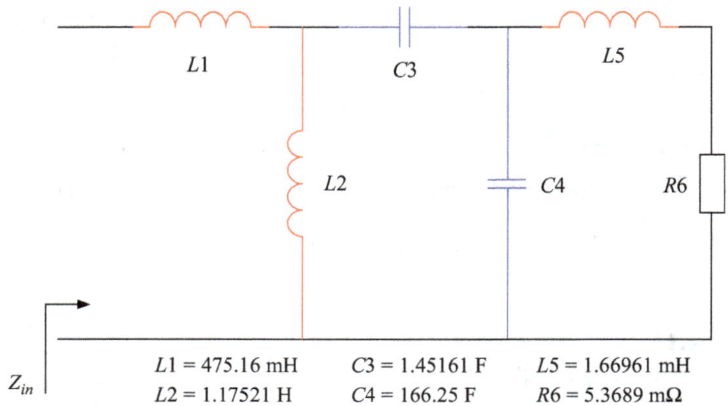

Figure 4.36 Synthesis of $Z_{in}(p)$ of Example 4.13

MATLAB program *Main_SParBasedSynthesis.m* reveals the synthesis as shown in Figure 4.36.

Example 4.13

Let $h = [-1\ 2\ 3\ 4\ 5\ 6\ 7\ 8]$ with $ndc = 5$.

1. Find the realizable strictly Hurwitz polynomial $g(p)$.
2. Find the polynomials $a(p)$ and $b(p)$ for $z_{in}(p) = a(p)/b(p)$.
3. Identify the logical values of the coefficients $h_1, g_1, a_1, b_1, h_{n+1}, g_{n+1}, a_{n+1}, b_{n+1}$ as in Table 4.16 and determine the possible synthesis case.
4. Synthesize $z_{in}(p) = a(p)/b(p)$ and draw the lossless LC-ladder network in resistive termination.

Solution

1. Using *Main_SparBasedSynthesis.m*, we have

$$g = [1.0000\ 7.4318\ 22.6162\ 42.8033\ 54.3621\ 47.3724\ 26.6638\ 8.0000]$$

2. Polynomials $a(p)$ and $b(p)$ are found as

$$a = [09.4318\ 25.6162\ 46.8033\ 59.3621\ 53.3724\ 33.6638\ 16.0000]$$

$$b = [2.0000\ 5.4318\ 19.6162\ 38.8033\ 49.3621\ 41.3724\ 19.6638\ 0.0000]$$

As we see from above, $Z_{in}(p)$ has a pole at $p = 0$ since $b_{n+1} = b_8 = 0$.

Table 4.20 A case study for Example 4.13

$Z_{in}(p) = a(p)/b(p)$	ndc	h_1	g_1	h_{n+1}	g_{n+1}	a_1	a_{n+1}	b_1	b_{n+1}
Case 4: extract pole at $p = 0$	$ndc < n$	-1	$+1$	$+1$	$+1$	0	$+1$	$+1$	0

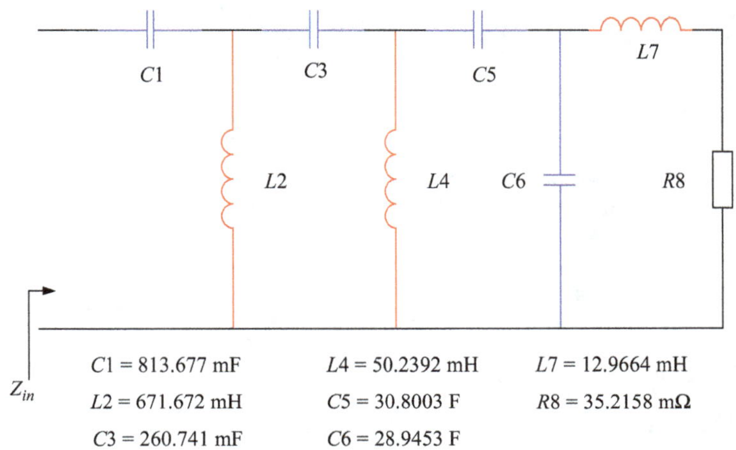

Figure 4.37 Synthesis of $Z_{in}(p)$ as described by Example 4.13

3. In short, we can see that logical values of the coefficients yield Table 4.20. Hence, $Z_{in}(p) = a(p)/b(p)$ satisfies all the conditions of Case 4. Therefore, synthesis starts with a pole extraction at DC. Then, the minimum reactance impedance synthesis follows (Table 4.20).
4. As explained above, synthesis of $Z_{in}(p) = a(p)/b(p)$ starts with the extraction of a series capacitor $C1$. Then, the minimum reactance driving-point impedance synthesis follows. Complete synthesis is shown in Figure 4.37.

Example 4.14

Let $h = [-1\ 2\ 3\ 4\ 5\ 6\ 7\ 8\ 9\ ]$ and $ndc = 8$.

1. Find the realizable strictly Hurwitz polynomial $g(p)$.
2. Find the polynomials $a(p)$ and $b(p)$ for $z_{in}(p) = a(p)/b(p)$.
3. Identify the logical values of the coefficients $h_1, g_1, a_1, b_1, h_{n+1}, g_{n+1}, a_{n+1}, b_{n+1}$ as in Table 4.16 and determine the possible synthesis case.
4. Synthesize $z_{in}(p) = a(p)/b(p)$ and draw the lossless LC-ladder network in resistive termination.

Solution

Using program *Main_SParBasedSynthesis.m* we found that

1. $g(p) = [1.4142\ 9.9024\ 31.1331\ 62.3907\ 88.1645\ 91.6986\ 68.9437\ 34.3364\ 9.0000]$

Table 4.21 A case study for Example 4.14

$Z_{in}(p) = a(p)/b(p)$	*ndc*	h_1	g_1	h_{n+1}	g_{n+1}	a_1	a_{n+1}	b_1	b_{n+1}
Case 5: High-pass minimum susceptance	$ndc = n$	−1	+1	+1	+1	+1	+1	+1	0

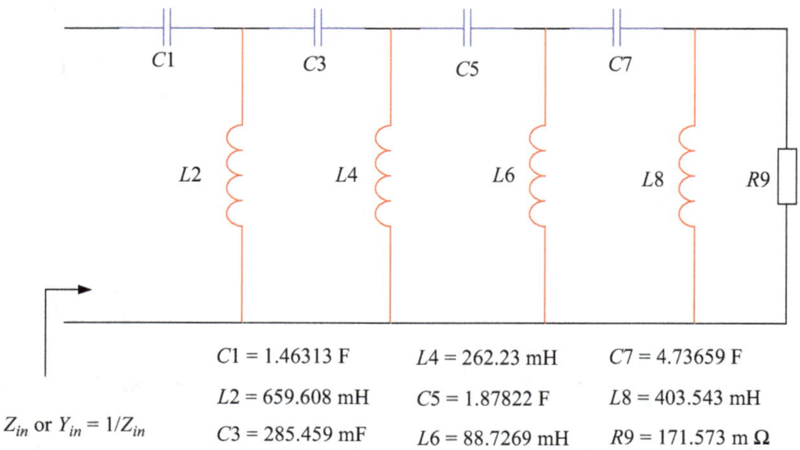

Figure 4.38 Synthesis of $Z_{in}(p) = a(p)/b(p)$ *for Example 4.15*

2. $a =$ [0.4142 11.9024 34.1331 66.3907 93.1645 97.6986 75.9437 42.3364 18.0000]
3. $b =$ [2.4142 7.9024 28.1331 58.3907 83.1645 85.6986 61.9437 26.3364 0.0000]
4. Close examination above polynomials yields Table 4.21.

Thus, Example 4.15 falls into Case 5 where $ndc = n = 8$
$Z_{in}(p) = a(p)/b(p)$ is synthesized as shown in Figure 4.38.

Example 4.15

Let $h =$ [0.9 0.8 0.7 0.6 0.5 0.411 − 11 − 2] and $ndc = 10$

1. Find the realizable strictly Hurwitz polynomial $g(p)$.
2. Find the polynomials $a(p)$ and $b(p)$ for $z_{in}(p) = a(p)/b(p)$.
3. Identify the logical values of the coefficients $h_1, g_1, a_1, b_1, h_{n+1}, g_{n+1}, a_{n+1}, b_{n+1}$ as in Table 4.16 and determine the possible synthesis case.
4. Synthesize $z_{in}(p) = a(p)/b(p)$ and draw the lossless LC-ladder network in resistive termination.

Table 4.22 A case study for Example 4.15

$Z_{in}(p) = a(p)/b(p)$	ndc	h_1	g_1	h_{n+1}	g_{n+1}	a_1	a_{n+1}	b_1	b_{n+1}
Case 6: High-pass minimum reactance	$ndc = n$	+1	+1	−1	+1	+1	0	+1	+1

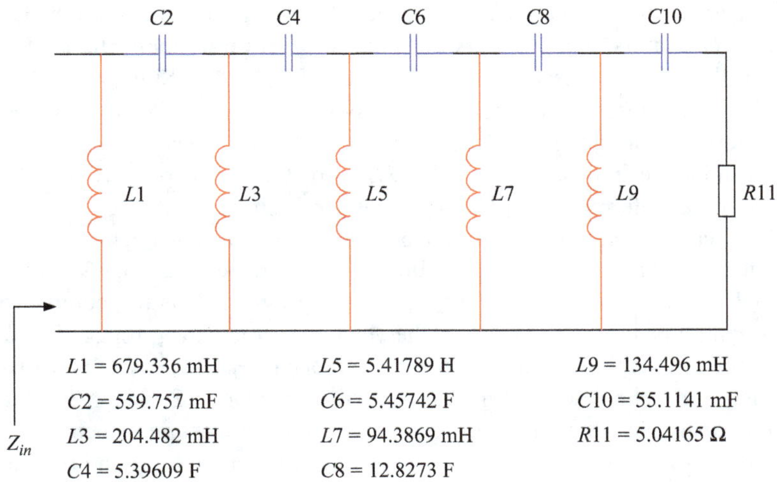

Figure 4.39 Synthesis of $Z_{\text{in}}(p) = a(p)/b(p)$ for Example 4.15

Solution

Using program *Main_SParBasedSynthesis.m*, we have

1. $g =$ [1.3454 8.5438 27.3594 57.8068 89.0732 104.4529 94.3943 65.2328 33.2362 11.3993 2.0000]
2. $a =$ [2.2454 9.3438 28.0594 58.4068 89.5732 104.8529 95.3943 66.2328 32.2362 12.3993 0.0000]
3. $b =$ [0.4454 7.7438 26.6594 57.2068 88.5732 104.0529 93.3943 64.2328 34.2362 10.3993 4.0000]
4. The above polynomial coefficients yield Table 4.22.

Thus, we end up with Case 6. Synthesis of $Z_{in}(p) = a(p)/b(p)$ is depicted in Figure 4.39.

4.5 High precision synthesis of a Richards immittance via parametric approach

4.5.1 Description of lossless two-ports in terms of Richards variable

A Richards driving-point immittance function $F(\lambda) = a(\lambda)/b(\lambda)$ is a positive real rational function expressed in classical complex frequency or Laplace–domain variable $p = \sigma + j\omega$. It possesses all the mathematical properties of a positive real function $F(p) = a(p)/b(p)$ described in complex $p - Plane$. *Complex Richards–Plane* $\lambda = +j\Omega$ is a transformed domain of $p = \sigma + j\omega$ which is obtained under a tangent hyperbolic mapping $\lambda = tanh\,(p\tau)$. In circuit theory, it is well established that any driving-point positive real immittance function $F(p) = a(p)/b(p)$ can be synthesized as a lossless two-port in resistive termination in $p - domain$ using lumped circuit elements such as $p - domain$ inductive impedances $Z_L(p) = pL_L$, capacitive admittances $Y_C = pC_L$ and resistor R where L_L, C_L and R are the lumped inductors, capacitors and resistors, respectively. Several publications describe recent developments in passive network synthesis [3,8,14,15]. Similarly, a Richards driving-point immittance function can be synthesized as a lossless two-port in resistive termination using Richards unit-elements (UE) in cascade configuration, Richards inductive impedances $Z_L(\lambda) = \lambda L_\lambda$ and Richards capacitive admittances $Y_C(\lambda) = \lambda C_\lambda$ as referred to in [22–24]and shown in Figure 4.40.

UE in cascade configuration can be realized as an equal length transverse electromagnetic (TEM) transmission lines with characteristic impedance Z_i and constant fixed delay length $\tau = l_s/v_p$. In this expression, l_s is the commensurate physical length of the lines, and v_p is the velocity of the wave propagation within the transmission medium. Richards inductive impedance $Z_L(\lambda) = \lambda L_\lambda$ is realized as a commensurate short stub with characteristic impedance $Z_i = L_\lambda$. Similarly, a Richards capacitive admittance is realized as a commensurate open stub with characteristic impedance $Z_i = 1/C_\lambda$ as depicted in Figure 4.40(a) and (b), respectively. An equal physical length l_s or corresponding delay length τ is a design parameter, and it is selected properly at an appropriate frequency f_s such that $\omega_s\tau$ is fixed as a fraction of π (i.e., $\omega_s\tau = 2\pi f_s\tau = (\pi/m)\ or\ \tau = (1/2mf_s); m > 0$).

In designing communication systems, use of lossless matching networks is essential. High-quality factor lumped circuit components may be preferred to

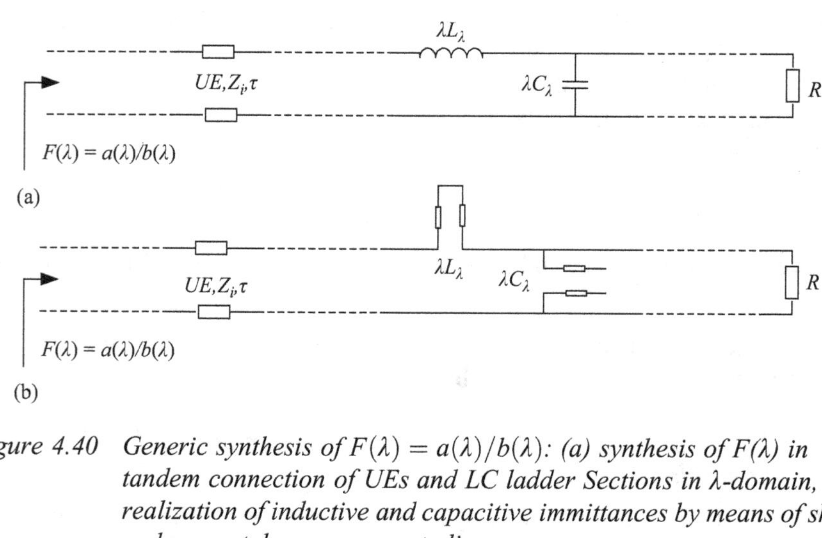

Figure 4.40 Generic synthesis of $F(\lambda) = a(\lambda)/b(\lambda)$: (a) synthesis of $F(\lambda)$ in tandem connection of UEs and LC ladder Sections in λ-domain, (b) realization of inductive and capacitive immittances by means of short and open stub commensurate lines

design matching networks. However, if the operating frequencies go beyond a few GHz, we are forced to utilize distributed elements such as commensurate transmission lines. Details of the microwave engineering applications of transmission lines (e.g., coplanar microstrip and suspended lines) can be found in widely used texts such as [25–29].

The RFT is known to be the best design methods when constructing lossless matching networks for communication systems [30,31]. They work on the network functions either in complex Laplace variable $p = \sigma + j\omega$ or in Richards variable $\lambda = \Sigma + j\Omega$. Therefore, accurate synthesis of the network functions is in high demand.

In our previous sections, we introduced high-precision synthesis of immittance functions generated in complex $p - Domain$. In this section, previously introduced synthesis techniques are extended to synthesize network functions in *Richards – Domain* [24].

In the following sections, we first introduce the parametric method to generate a realizable positive immittance function $F(\lambda) = a(\lambda)/b(\lambda)$ in complex Richards variable λ. Then, the new method of Richards Synthesis is presented. Various examples are given to illustrate the utilization of the new synthesis method.

4.5.2 Generation of a Richards immittance via parametric method

In the classical network theory, the Richards variable λ is the transformed variable of $p = \sigma + j\omega$ which is expressed as

$$\lambda = \Sigma + j\Omega = \tanh(p\tau) \tag{4.98}$$

It should be noted on the real frequency axis ω (for the case where $\sigma = 0, p = j\omega$) the Richards frequency Ω is given by

$$\Omega = \tan(\omega\tau) \tag{4.99}$$

Many practical lossless matching networks, which are designed with commensurate transmission lines, demand minimum driving-point input immittance function. We define a minimum function $F(\lambda) = a(\lambda)/b(\lambda)$ as one which is strictly analytic in the closed RHP. If the minimum function refers to impedance, it is called minimum reactance. Similarly, if it is admittance, it is called minimum susceptance. For the sake of completeness, let us state the following properties of the positive real functions specified in Richards domain.

4.5.3 Properties of a Richards immittance function

1. A positive real rational function $F(\lambda)$ must have all its zeros and poles in the open LHP.
2. $F(\lambda)$ may have poles and zeros on the Richards frequency axis $\lambda = j\Omega$, but these poles and zeros must be simple (i.e., order of 1) with positive residues.
3. If $F(\lambda)$ is a minimum function, then it must be free of poles on the $\lambda = j\Omega$ axis. However, it may have simple zeros on the imaginary axis $\lambda = j\Omega$.
4. Any positive real function $F(\lambda)$ can be expressed as

$$F(\lambda) = F_F(\lambda) + F_M(\lambda) \tag{4.100}$$

where $F_F(\lambda)$ is a Foster reactance function [32] which is purely imaginary for $\lambda = j\Omega$ such that

$$F_F(j\Omega) = jX_F(\Omega); \frac{dX_F}{d\Omega} \geq 0 \text{ for all } \Omega \tag{4.101}$$

Furthermore, the Foster function can be expressed as

$$F_F(\lambda) = k_\infty \lambda + \frac{k_0}{\lambda} + \sum_{i=1}^{N_F} \frac{k_i \lambda}{\lambda^2 + \Omega_i^2} \tag{4.102}$$

1. The minimum function $F_M(\lambda)$ can be expressed in terms of its even and odd parts such that

$$F_M(\lambda) = R(\lambda) + F_{M-odd}(\lambda) \tag{4.103}$$

   On the Ω axis,

$$F_M(j\Omega) = R(\Omega) + jX_M(\Omega) \tag{4.104}$$

   where

$$X_M(\Omega) = \frac{F_{M-odd}(j\Omega)}{j} \tag{4.105}$$

2. From (4.100)–(4.105), we can deduce that $F(j\Omega)$ and $F_M(j\Omega)$ must possess the same real part $R(\Omega)$.
3. Let $F(j\Omega) = R(\Omega) + jX(\Omega)$. Then,

$$X(\Omega) = X_F(\Omega) + X_M(\Omega) \tag{4.106}$$

4. In designing matching networks employing RFT, the real part $R(\Omega)$ of (4.104) is specified as

$$R(\Omega^2) = \left[a_0^2\right] \left[\frac{\Omega^{2q}\left(1 + \Omega^2\right)^k}{(1/2)[c^2(\Omega) + c^2(-\Omega)]} \right] = \frac{A(\Omega^2)}{B(\Omega^2)} \geq 0 \quad \text{for all } \Omega \tag{4.107}$$

where

$$A(\Omega) = A_1 \Omega^{2(q+k)} + A_2 \Omega^{2(q+k-1)} + \cdots + A_{(q+k)} \Omega^2 + A_{q+k+1} \geq 0; \\ A_{q+k+1} = a_0^2 \geq 0 \tag{4.108a}$$

and

$$B(\Omega^2) = B_1 \Omega^{2n} + B_2 \Omega^{2(n-1)} + \cdots + B_n \Omega^2 + 1 \tag{4.108b}$$

In (4.108a) and (4.108b), $B(\Omega^2)$ is a strictly positive polynomial. The integer n must satisfy the inequality of $n \geq k + q$. $c(\Omega)$ is an auxiliary real polynomial of degree n to generate strictly positive denominator polynomial $B(\Omega)$ such that

$$c(\Omega) = c_1\Omega^n + c_2\Omega^{n-1} + \cdots + c_n\Omega + 1 \tag{4.109}$$

Note that by replacing Ω^2 with $-\lambda^2$, we can obtain the Richards domain even function $R(\lambda^2)$ as

$$R(\lambda^2) = \left[a_0^2\right]\frac{(-1)^q\lambda^{2q}(1-\lambda^2)^k}{B(\lambda^2)} = \frac{A(\lambda^2)}{B(\lambda^2)} \tag{4.110}$$

In the above representation, zeros of $R(\lambda^2)$ are known as "zeros of transmission" of the lossless two-port constructed as the result of the synthesis of the immittance function $F(\lambda)$. These zeros are located at $\lambda = \mp 1$ of multiplicity of k, $\lambda = 0$ of multiplicity of $2q$, and perhaps at infinity with multiplicity of $2n_\infty = 2(n - q - k)$ if $n > q + k$.

When developing broadband matching networks via RFTs, the designer specifies the transmission zeros of the lossless two-port under consideration by fixing the integers k, q and n. Then, the unknowns of the matching problem are chosen as the real coefficients $\{c_i; i = 1, 2, \cdots, n\}$ of the auxiliary polynomial $c(\Omega) = c_1\Omega^n + c_2\Omega^{n-1} + \cdots + c_n\Omega + 1$ and the real coefficient a_0 of the numerator.

In RFTs, $F(\lambda)$ is assumed to be a minimum function, then it is generated from its real part $R(\Omega)$ using the parametric approach as described below.

It should be mentioned that the selected form of $R(\lambda^2)$ specified by (4.110) is not unique. It only specifies a practical network structure which can be easily manufactured using coplanar, microstrip or suspended transmission lines by building UE, open or short stubs in series and shunt configurations as described in [24]. However, if desired, much more complicated forms for the even part function $R(\lambda^2)$ can be selected by introducing proper complex transmission zeros beyond DC and infinity.

4.5.4 Parametric approach in Richards domain

There are several versions of the RFT such as real frequency-line segment technique [1], RFT-parametric approach [10] and SRFT [33,34]. It is shown that RFT-parametric approach provides better numerical stability over the other methods [30]. Therefore, in this section, we prefer to employ RFT-parametric approach to generate minimum immittance functions in Richards domain.

A positive real minimum function $F(\lambda)$ can be generated from its nonnegative even-real part $R(\Omega^2)$ which is specified on the transformed frequency axis $\Omega = \tan(\omega\tau)$ as follows.

Once a_0, q, k and arbitrary real coefficients $\{c_1, c_2, \cdots, c_n\}$ are initialized such that $n \geq q + k$, then minimum positive real function $F(\lambda)$ can easily be generated using the parametric approach as follows:

In the parametric approach, minimum function $F(\lambda)$ is expressed in partial fraction expansion form as in (4.111):

$$F(\lambda) = R_0 + \sum_{j=1}^{n} \frac{k_j}{\lambda - \lambda_j} = \frac{a(\lambda)}{b(\lambda)} \tag{4.111}$$

where λ_j are the closed left half plane roots of the denominator $B(\lambda) = B_1 \prod_{j=1}^{n} (\lambda^2 - \lambda_j^2)$ and the residues k_j are given by

$$k_j = \frac{1}{\lambda_j} \frac{A(\lambda_j)}{B_1 \prod_{\substack{i=1 \\ j \neq i}}^{n} \left(\lambda_j^2 - \lambda_i^2\right)} \tag{4.112}$$

and

$$R_0 = lim_{\lambda \to \infty} R(\lambda^2) = \left\{ \begin{array}{ll} 0; & n > (q+k) \\ \frac{a_1}{b_1} > 0; & n = (q+k) \end{array} \right\} \tag{4.113}$$

In parametric approach, once the even part of the immittance function is specified as in (4.110), first the roots λ_i of $B(\lambda^2)$ is computed. Then, selecting the left half plane roots, residues k_j are determined as in (4.112). Finally, open rational form of $F(\lambda) = a(\lambda)/b(\lambda)$ is derived using (4.110)–(4.113). Eventually, the immittance function $F(\lambda) = a(\lambda)/b(\lambda)$ is synthesized using a proper method.

4.5.5 Cascade connection of k-unit elements

Referring to the classical section of P. I. Richards [22] and Figure 4.41, let us consider the driving input impedance $Z_{in}(\lambda) = a(\lambda)/b(\lambda)$ of a cascade connection of k-unit elements terminated in a Richards impedance $Z_{LC}(\lambda)$ such that

$$\begin{aligned} a(\lambda) &= a_1\lambda^n + a_2\lambda^{n-1} + \cdots + a_n\lambda + a_{n+1} \\ b(\lambda) &= b_1\lambda^n + b_2\lambda^{n-1} + \cdots + b_n\lambda + b_{n+1} \end{aligned} \tag{4.114a}$$

Here, we presume that $Z_{LC}(\lambda)$ includes only q transmission zeros at DC and n_L transmission zeros at infinity yielding $n = k + q + n_L$. $Z_{in}(\lambda)$ may be synthesized in two parts. In the first part of the synthesis, unit elements are extracted step-by-step by calculating the characteristic impedance Z_i of each section employing the Richards technique [22,24]. Once all the unit elements are extracted, then in the second part of the synthesis, the remaining driving-point input impedance $Z_{LC}(\lambda)$ is synthesized using our high-precision LC-ladder synthesis method as described in [8,14] in Richards domain-λ.

To keep track of the synthesis steps by means of proper indexing, at the first step, let us initialize the driving-point impedance $Z_{in}(\lambda)$ as

$$Z_{in}(\lambda) = Z_{in1}(\lambda) = \frac{a_1(\lambda)}{b_1(\lambda)} \tag{4.114b}$$

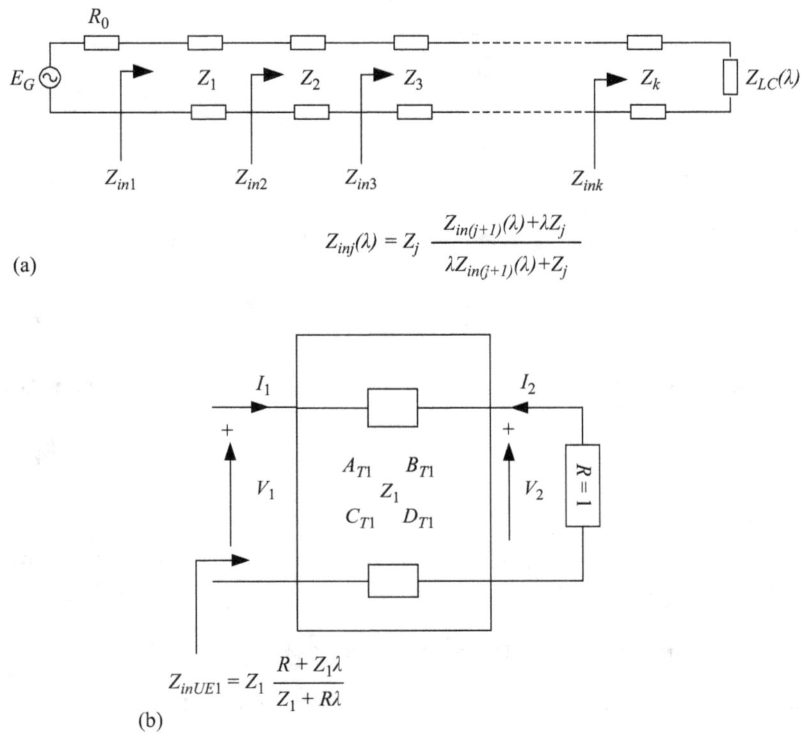

Figure 4.41 (a) Cascade connection of k-UE terminated in $Z_{LC}(\lambda)$, (b) description of a single UE by means of its characteristic impedance Z_1 and chain parameters

In terms of the termination impedance $Z_{in2}(\lambda)$ and the characteristic impedance Z_1, $Z_{in1}(\lambda)$ is given by

$$Z_{in1}(\lambda) = Z_1 \frac{Z_{in2}(\lambda) + \lambda Z_1}{Z_1 + \lambda Z_{in2}(\lambda)} = \frac{a_1(\lambda)}{b_1(\lambda)} \tag{4.114c}$$

At this point, we should emphasize that in (4.114a), the real coefficients $\{a_i, b_i; i = 1, 2, n + 1\}$ of the polynomials $a(\lambda)$ and $b(\lambda)$ should not be confused with the new set of polynomials $a_1(\lambda)$ and $b_1(\lambda)$ of (4.114b) which refers to the driving-point impedance $Z_{in1}(\lambda)$ at the first step of the Richards synthesis.

The termination impedance $Z_{LC}(\lambda)$ of the k-cascade UE is determined after all the UEs are extracted from the initial driving-point impedance $Z_{in1}(\lambda)$.

Furthermore, the even part $R(\lambda)$ of $Z_{in1}(\lambda)$ can be expressed in the following form:

$$R(\lambda) = \frac{1}{2}[Z_{in1}(\lambda) + Z_{in1}(-\lambda)] = (-1)^q \frac{a_0^2 \lambda^{2q} (1 - \lambda^2)^k}{B(\lambda^2)} = \frac{A(\lambda^2)}{B(\lambda^2)} \tag{4.115a}$$

where

$$A(\lambda^2) = A_1\lambda^{2n} + A_2\lambda^{2(n-1)} + \cdots + A_n\lambda + A_{n+1} \tag{4.115b}$$

and

$$B(\lambda^2) = B_1\lambda^{2n} + B_2\lambda^{2(n-1)} + \cdots + B_n\lambda + 1 \tag{4.115c}$$

At $\lambda = 1$, (17a) yields the characteristic impedance Z_1 such that

$$Z_1 = Z_{in1}(1) = Z_1 \frac{Z_{in2}(1) + Z_1}{Z_1 + Z_{in2}(1)} = \frac{a(1)}{b(1)} = \frac{\sum_{i=1}^{n+1} a_k}{1 + \sum_{i=1}^{n} b_k} \tag{4.116a}$$

It should be noted that, as introduced by Levy in [35], $Z_{in1}(\lambda)$ can be expressed as

$$Z_{in1}(\lambda) = \frac{m_1(\lambda) + n_1(\lambda)}{m_2(\lambda) + n_2(\lambda)} \tag{4.116b}$$

where $\{m_1(\lambda), n_1(\lambda)\}$ and $\{m_2(\lambda), n_2(\lambda)\}$ are the even and the odd parts of the numerator and denominator polynomials, respectively. By (4.116a), numerator polynomial of the even part of Z_{in} is

$$A(\lambda^2) = m_1(\lambda)m_2(\lambda) - n_1(\lambda)n_2(\lambda) = (-1)^q a_0^2 \lambda^q (1 - \lambda^2)^k$$

In this case, $A(1) = m_1(1)\, m_2(1) - n_1(1)\, n_2(1) = 0$. It means that

$$\frac{m_1(1)}{n_2(1)} = \frac{n_1(1)}{m_2(1)} \tag{4.116c}$$

or

$$\frac{n_1(1)}{m_1(1)} = \frac{m_2(1)}{n_2(1)} \tag{4.116d}$$

Employing (4.116c) and (4.116d), characteristic impedance $Z_{in1}(1)$ can also be given as

$$z_{in}(1) = \frac{m_1(1) + n_1(1)}{m_2(1) + n_2(1)}$$

or

$$z_{in}(1) = \frac{m_1(1)}{n_2(1)} \left[\frac{1 + \dfrac{n_1(1)}{m_1(1)}}{\dfrac{m_2(1)}{n_2(1)} + 1} \right] = \frac{m_1(1)}{n_2(1)} \tag{4.116e}$$

Similarly, it is shown that

$$Z_{in}(1) = \frac{n_1(1)}{m_2(1)} = \frac{m_1(1)}{n_2(1)} \tag{4.116f}$$

As far as numerical implementation is concerned, Levy generates the characteristic impedance Z_1 as the arithmetic mean of $n_1(1)/m_2(1)$ and $m_1(1)/n_2(1)$; however, we have experienced that the characteristic impedance $Z_1 = m_1(1)/n_2(1)$ yields better numerical error over that of $Z_1 = n_1(1)/m_2(1)$ and $Z_1 = n_1(1)/m_2(1) + m_1(1)/n_2(1)$ as mentioned in [24]. Therefore, in this section we prefer to use

$$Z_1 = Z_{in}(1) = \frac{m_1(1)}{n_2(1)} \tag{4.116g}$$

At this point, step-by-step numerical extraction of the unit elements from the complete driving-point input impedance become crucial. Obviously, the impedance-based Richards extractions can be employed [22]. However, we found that the method proposed by Cameron, Kudia and Mansour results in reasonable numerical stability as detailed in [19,24]. Modified version of this method is outlined in the following section.

4.5.6 UE extractions employing the chain parameters

Referring to [19], the complete network of Figure 4.40 is described in terms of its chain parameters-based transmission matrix $T(\lambda)$ which is specified by

$$[T(\lambda)] = \begin{pmatrix} A_T(\lambda) & B_T(\lambda) \\ C_T(\lambda) & D_T(\lambda) \end{pmatrix} = \begin{pmatrix} m_1(\lambda) & n_1(\lambda) \\ n_2(\lambda) & m_2(\lambda) \end{pmatrix} \tag{4.117a}$$

where

$$A_T(\lambda) = m_1(\lambda), B_T(\lambda) = n_1(\lambda) \tag{4.117b}$$

$$C_T(\lambda) = n_1(\lambda), D_T(\lambda) = m_2(\lambda) \tag{4.117c}$$

and

$$a(\lambda) = m_1(\lambda) + n_1(\lambda) \tag{4.117d}$$

$$b(\lambda) = n_2(\lambda) + m_2(\lambda) \tag{4.117e}$$

In this case, characteristic impedance of the first line is given by

$$Z_1 = \frac{A_T(1)}{D_T(1)} \tag{4.117f}$$

Referring to Figure 4.41(b), let us describe the first UE by means of its driving-point impedance $Z_{inUE1}(\lambda)$ when it is terminated in unit resistance ($R = 1$).

$$Z_{inUE1}(\lambda) = \frac{a_1(\lambda)}{b_1(\lambda)} = Z_1 \frac{1 + Z_1\lambda}{Z_1 + \lambda} = \frac{1 + Z_1\lambda}{1 + (1/Z_1)\lambda} \tag{4.117g}$$

Then, using (20b) and (20c), the transmission matrix $[T_1] = \begin{pmatrix} A_{T1} & B_{T1} \\ C_{T1} & D_{T1} \end{pmatrix}$ of the first extracted line is given by

$$A_{T1}(\lambda) = 1; \quad B_{T1}(\lambda) = Z_1\lambda; C_{T1}(\lambda) = \frac{1}{Z_1}\lambda; \quad D_{T1}(\lambda) = 1 \tag{4.117h}$$

Furthermore:

$$A_{T1}(\lambda)D_{T1}(\lambda) - B_{T1}(\lambda)C_{T1}(\lambda) = \left(1 - \lambda^2\right) \tag{4.117i}$$

When the first line is extracted from the complete structure, the transmission matrix $[T_2(\lambda)]$ of the remaining two-port is given by

$$[T_2(\lambda)] = \begin{pmatrix} A_{2T} & B_{2T} \\ C_{2T} & D_{2T} \end{pmatrix} = [T_1(\lambda)]^{-1}[T] = \frac{1}{\left(1 - \lambda^2\right)} \begin{pmatrix} A_{T1} & -B_{T1} \\ -C_{T1} & D_{T1} \end{pmatrix} \begin{pmatrix} A_T & B_T \\ C_T & D_T \end{pmatrix}$$

or

$$\begin{pmatrix} A_{2T} & B_{2T} \\ C_{2T} & D_{2T} \end{pmatrix} = \frac{1}{\left(1 - \lambda^2\right)} \begin{pmatrix} D_{T1}A_T - B_{T1}C_T & D_{T1}B_T - B_{T1}D_T \\ A_{T1}C_T - C_{T1}A_T & A_{T1}D_T - C_{T1}B_T \end{pmatrix} \tag{4.117j}$$

In this case, the driving-point impedance of $[T_2]$ is given by

$$Z_{in2}(\lambda) = \frac{a_2(\lambda)}{b_2(\lambda)} = \frac{\sum_{i=1}^{k} a_{2i}\lambda^{k-i}}{\sum_{i=1}^{k} b_{2i}\lambda^{k-i}} \tag{4.118a}$$

where

$$a_2(\lambda) = A_{2T} + B_{2T} = \frac{1}{\left(1 - \lambda^2\right)}[D_{T1}(\lambda)a(\lambda) - B_{T1}(\lambda)b(\lambda)] = \frac{a_{in2}(\lambda)}{\left(1 - \lambda^2\right)} \tag{4.118b}$$

and

$$a_{in2}(\lambda) = D_{T1}(\lambda)a(\lambda) - B_{T1}(\lambda)b(\lambda) \tag{4.118c}$$

and

$$b_2(\lambda) = C_{2T} + D_{2T} = \frac{1}{\left(1 - \lambda^2\right)}[A_{T1}(\lambda)b(\lambda) - C_{T1}(\lambda)a(\lambda)] = \frac{b_{in2}(\lambda)}{\left(1 - \lambda^2\right)} \tag{4.118d}$$

and

$$b_{in2}(\lambda) = A_{T1}(\lambda)b(\lambda) - C_{T1}(\lambda)a(\lambda) \tag{4.118e}$$

Close examination of (21) reveals that degree of the polynomials $a_2(\lambda)$ and $b_2(\lambda)$ is one less than that of $a(\lambda)$ and $b(\lambda)$, respectively. Thus, the Richards extraction of a single unit element is accomplished. It should be noted that in (21), division of $a_{in2}(\lambda)$ and $b_{in2}(\lambda)$ by $\left(1 - \lambda^2\right)$ must yield zero remainders $r_a(\lambda)$ and $r_b(\lambda)$, respectively. These remainders are given as follow:

$$r_a(\lambda) = a_{in2}(\lambda) - a_2(\lambda)\left(1 - \lambda^2\right) \tag{4.118f}$$

and

$$r_b(\lambda) = b_{in2}(\lambda) - b_2(\lambda)\left(1 - \lambda^2\right) \tag{4.118g}$$

In *MATLAB* environment, polynomial divisions may be carried out using the function *deconv*. In this case, the pairs $\{a_2(\lambda), r_a(\lambda)\}$ and $\{b_2(\lambda), r_b(\lambda)\}$ are generated as *MATLAB* vectors using the following commands:

$$[a2, ra] = deconv(ain2, Lmdsq) \tag{4.118h}$$

$$[b2, rb] = deconv(bin2, Lmdsq) \tag{4.118i}$$

where $Lmdsq = [-1 \ \ 0 \ \ 1]$ describes the divisor polynomial $(1 - \lambda^2)$ as a *MATLAB* vector and the vectors *a*2, *b*2, *ra*, *rb*, *ain*2, *bin*2 include all the coefficients of the polynomials $a_2(\lambda), b_2(\lambda), r_a(\lambda), r_b(\lambda), a_{in2}(\lambda)$, and $b_{in2}(\lambda)$ in descending order, respectively.

The above process is repeated $k - times$ to extract all the unit elements from the Richards immittance. Eventually, the left-over Richards immittance function $Z_{LC}(\lambda)$ must include the total number of $q - DC$ transmission zeros and $n_L = n - k - q$ number of transmission zeros at infinity. These transmission zeros can be extracted using our previously developed high precision synthesis methods to complete the synthesis [8,14,17].

From the algorithmic implementation point of view, at each step, we initialize $Z_{in2}(\lambda)$ as

$$Z_{in2}(\lambda) = \frac{a(\lambda)}{b(\lambda)} \tag{4.119}$$

where $a(\lambda) = a_2(\lambda)$ and $b(\lambda) = b_2(\lambda)$ are the newly reset polynomials for the extraction of the next UE from $Z_{in2}(\lambda)$.

It has been experienced that the above straightforward chain matrix-based Richards extraction method yields in satisfactory results up to 30-unit element extractions in *MATLAB* environment using 64-bit machine precision. However, depending on the coefficient values of the numerator and the denominator polynomials of $Z_{in}(\lambda) = a(\lambda)/b(\lambda)$, after several steps, accumulated numerical error may become severe and destroys the rational form of the even part of the immittance function as specified by (4.115a)–(4.115c). Therefore, one would end up with either undesirable or no synthesis solution. In this case, it may be useful to employ the immittance correction method to complete the synthesis as described in the following section.

4.5.7 Correction of the Richard impedance after each extraction

In the course of cascade synthesis process, at each step, numerical precision is lost as we continue with multiplication and division operations. This problem is discussed in various publications using transformed techniques to design lumped element filters such as [36–38]. In this section, however, we employ "immittance correction" method using the parametric approach as introduced in [8,14]. In the Richards domain $\lambda = \Sigma + j\Omega$, correction operation differs from that of the correction operation in the classical $p = \sigma + j\omega$ domain. If the total number of elements n is greater than that of the total number of the cascaded unit elements k (i.e., $n > k$), then at each unit element extraction, remaining impedance function changes its character from minimum reactance to minimum susceptance or vice versa. For example, if $n > k$ and if we start the Richards synthesis with a minimum reactance function $F_{in}(\lambda) = a(\lambda)/b(\lambda) = Z_{in1}(\lambda)$, after the first unit element extraction, the remaining driving-point impedance $Z_{in2}(\lambda)$ becomes a minimum susceptance. In this case, we can flip over the function to correct it using our parametric method. Therefore, at the odd steps of the unit element extraction, immittance correction is applied on the admittance function. On the other hand, at the even steps, correction is directly applied on the minimum reactance impedance function. In any case, generic form of the numerator polynomial $A_i(\lambda^2)$ of the even part $R(\lambda^2) = A_i(\lambda^2)/B_i(\lambda^2)$ is forced to be

$$\begin{aligned} A_i(\lambda^2) &= (-1)^q \times (a_{0i}^2) \times (1 - \lambda^2)^{k-1} \\ B_i(\lambda^2) &= b(\lambda)b(-\lambda) \end{aligned} \tag{4.120}$$

Ignoring the remainder terms *ra* and *rb*, the above forms can be generated from the reset polynomials $a(\lambda)$ and $b(\lambda)$ at the end of each step. In the correction process, it may be appropriate to normalize both $A_i(\lambda^2)$ and $B_i(\lambda^2)$ by (a_{0i}^2) so that $A_i(\lambda^2) = (-1)^q \times (1 - \lambda^2)^{k-i}$ is precisely generated at each correction step. In the course of corrections, first, we determine the integers k, q and (a_{0i}^2) from the immittance function $F_{in}(\lambda) = a(\lambda)/b(\lambda)$. This step is completed under the *MATLAB* function

$$[k, q, a0, nA, A1, B1] = Richard\_Numerator(a, b)$$

where the input arguments a and b are the numerator and denominator polynomials of the minimum immittance function $F_{in}(\lambda) = a(\lambda)/b(\lambda)$. Inside the above *MATLAB* function, first, the vectors a and b are normalized with respect to norm of the original a which is specified by $norm(a)$. This process introduces numerical robustness within function *Richard_Numerator*.

The output arguments k and q are the total number of unit elements and DC zeros of $R(\lambda^2) = A_i(\lambda^2)/B_i(\lambda^2)$ which is directly generated from the given a and b. $a0$ is the square root of the leading nonzero coefficient of $A_i(\lambda^2)$. $n = nA - 1$ is the total number of elements of the lossless two-port consisting of commensurate transmission lines which is obtained as the result of synthesis. $A1$ and $B1$ are the corrected normalized numerator and denominator polynomials of the even part function $R(\lambda^2) = A_1(\lambda^2)/B_1(\lambda^2)$. It should be noted that, if the integers k and q are known in advance when specifying the driving-point impedance $Z_{in}(\lambda) = a(\lambda)/b(\lambda)$, then there is no need to use the *MATLAB* function *Richard_Numerator*. In this case, we can just keep track of the unit element extraction to upgrade integers k and q at each correction steps.

4.5.8 Numerical error assessment of the new synthesis software package

One can device several ways to assess the accumulated numerical error as we go along with the synthesis step-by-step.

For example, a practical way to assess the accumulated error may be to compute the difference between the given driving-point input impedance $Z_{in}(j\omega) = a(j\omega)/b(j\omega) = R_{in}(\omega) + jX_{in}(\omega)$ and the actual input impedance $Z_a(j\omega) = R_a(\omega) + jX_a(\omega)$, which is generated from the synthesized network, over a prescribed frequency band. In this case, real and imaginary part-based errors can be defined as

$$\varepsilon_r(\omega) = R_{in}(\omega) - R_a(\omega) \tag{4.121a}$$

$$\varepsilon_x(\omega) = X_{in}(\omega) - X_a(\omega) \tag{4.121b}$$

In addition to the above, for the cascade connection of n-UEs, the size of the accumulated numerical error at each step of the synthesis may be traced by evaluating the norms of the remainder vectors using (4.118h) and (4.118i) as follows:

$$\varepsilon_a = norm(ra) = \sqrt{r_{a1}^2 + r_{a2}^2 + \cdots + r_{a(n+1)}^2} \tag{4.121c}$$

and

$$\varepsilon_b = norm(rb) = \sqrt{r_{b1}^2 + r_{b2}^2 + \cdots + r_{b(n+1)}^2} \tag{4.121d}$$

On the other hand, just for the sake of testing the proposed synthesis method, we can generate the driving-point input impedance from the known characteristic impedances Z_i of a cascade connection of n-unit elements as $Z_{in}(\lambda) = a_{in}(\lambda)/b_{in}(\lambda)$. Then, we can synthesize $Z_{in}(\lambda)$ employing the proposed Richards high-precision synthesis algorithm to end up with the computed characteristic impedances Z_{UEi}. In this case, accumulated error at each extraction step can be computed as

$$\varepsilon_i = Z_i - Z_{UEi} \tag{4.121e}$$

Let us now summarize all the above computational steps under an algorithm called "Richards high precision synthesis."

4.5.9 Algorithm: Richards high-precision synthesis

Let the driving-point Richards impedance be $Z_{in}(\lambda) = a(\lambda)/b(\lambda)$ such that its even part possesses k transmission zeros at $\lambda = \mp 1$, $2q$ transmission zeros at $\lambda = 0$ and $2n_L$ transmission zeros at $\lambda = \infty$ with $n \geq k + q + n_L$.

Inputs:

$\{a_1, a_2, \cdots a_n, a_{n+1}\}$ coefficients of the numerator polynomial $a(\lambda)$.

$\{b_1, b_2, \cdots b_n, b_{n+1}\}$ coefficients of the denominator polynomial $b(\lambda)$.

Computational steps:

Step 1: Form the even part $R(\lambda^2)$ as in (4.115a)–(4.115c) and determine the integers k and q. At this point, we should note that, if $Z_{in}(\lambda)$ is generated using the RFT, then k and q can be considered as the inputs to the algorithm.

Step 2: If $n \leq 8$ extract all the characteristic impedances of UEs employing the straightforward chain-matrix-based Richards extraction method introduced by (4.117g)–(4.119).

Step 3: If $n > 8$, set the integer $k_p = fix(n/2) - 3$ and extract the first k_p unit element's characteristic impedances using (4.117g)–(4.119) without corrections and then apply immittance correction method on the remaining positive real immittance at each step of the unit element extraction using (4.115a)–(4.118i). Then, terminate the computations if $k = n$.

Step 4: If $q > 0$ and if $n_\infty > 0$, extract the transmission zeros at $\lambda = 0$ and $\lambda = \infty$ employing the high-precision synthesis method as introduced in [8,14].

Step 5: Finally, assess the accumulated numerical error occurred in the course of "Richards High Precision Synthesis" process using (4.121).

Let us run a test example to assess the accumulated error of the Richards high-precision synthesis algorithm.

Example 4.16

In [19,35], characteristic impedance tables for the Chebyshev step-line filters are given. Characteristic impedances of the 21 element Chebyshev filter may be expressed by means of an impedance vector $[Z]$ such that $[Z] = [Z_A Z_{11} Z_B]$; where vectors $[Z_A]$ and $[Z_B]$ includes the ten characteristic impedances of the step lines in the symmetrical sections $[A]$ and $[B]$ as shown in Figure 4.42. For the Chebyshev filter under consideration, bandwidth $BW = 1$, voltage standing wave ratio $VSWR = 1.2$ and the commensurate line length is specified as $\theta_0 = BW\pi/4$.

Z_{11} is the characteristic impedance of the mid-unit element which connects symmetrical sections $[A]$ and $[B]$. Vectors $[Z_A]$, $[Z_B]$ and Z_{11} are presented in Table 4.23. Employing Table 4.23, we can generate the driving-point input impedance $Z_{in}(\lambda) = a_{in}(\lambda)/b_{in}(\lambda)$ of the step-line filter as presented in Table 4.24. Thus, in this example, we will synthesize the driving-point input impedance $Z_{in}(\lambda) = a_{in}(\lambda)/b_{in}(\lambda)$ using our proposed Richards high-precision synthesis algorithm for $k = n = 21$ and $q = 0$.

At the end of the computations, extracted characteristic impedances are stored in the impedance vector $[Z_{UE}]$, and then accumulated error is computed as in (4.212e). Close examination of Table 2 reveals that coefficients of the numerator polynomial $a_{in}(\lambda)$ vary between $a_{\min} = 8.131$ and $a_{\max} = 1.157 \times 10^9$. Similarly, coefficients

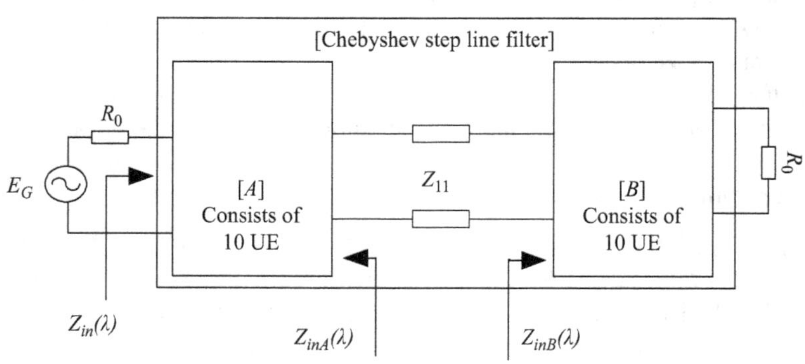

Figure 4.42 n = 21 element step-line Chebyshev filter

Table 4.23 Characteristic impedances of the Chebyshev step line filter for n = 21

Index	$[Z_A]$	$[Z_B]$
1	1.5370	0.4629
2	0.6044	2.5890
3	2.2910	0.4647
4	0.4913	2.5710
5	2.5140	0.4705
6	0.4705	2.5140
7	2.5710	0.4913
8	0.4647	2.2910
9	2.5890	0.6044
10	0.4629	1.5370
Z_{11}	2.5930	

Table 4.24 Driving-point input impedance $Z_{in}(\lambda) = a_{in}(\lambda)/b_{in}(\lambda)$ of a Chebyshev Monotone roll-off step line filter with commensurate transmission lines for n = 21, BW = 1, VSWR = 1.20

1	8.1058e+007	8.1576e−007
2	8.0403e+007	8.0403e+007
3	4.3238e+008	7.9752e+007
4	3.7519e+008	3.7519e+008
5	9.6120e+008	3.1891e+008
6	7.2103e+008	7.2103e+008
7	1.1574e+009	5.2024e+008
8	7.3975e+008	7.3975e+008
9	8.2221e+008	4.4699e+008
10	4.3963e+008	4.3963e+008
11	3.5260e+008	2.1825e+008
12	1.5389e+008	1.5389e+008
13	8.9915e+007	6.1081e+007
14	3.0951e+007	3.0951e+007
15	1.2961e+007	9.4314e+006
16	3.3419e+006	3.3419e+006
17	9.5619e+005	7.3288e+005
18	1.6942e+005	1.6942e+005
19	2.9789e+004	2.3760e+004
20	3.0434e+003	3.0434e+003
21	2.4870e+002	2.0460e+002
22	8.1317e+000	8.1317e+000

of the denominator polynomial $b_{in}(\lambda)$ varies between $b_{\min} = 8.1576 \times 10^{-7}$ and $b_{\max} = 7.3975 \times 10^{8}$. From the numerical computations point of view, coefficient variations are huge and cause over flows and under flows after a few unit element extractions. This drawback, however, is omitted by properly mixing the straightforward chain matrix-based Richards extraction algorithm together with the impedance correction via parametric approach. Thus, we end up with the following synthesis results as presented in Table 4.23.

Remarks

We tried to synthesized the same Richards impedance of Table 4.23 using the straightforward reflection-based extraction techniques of Carlin as detailed by [20,23]. Unfortunately, these classical algorithms were stopped after two extractions due to numerical over flows. As reported by Levy [35], for the same synthesis problem under consideration, the accumulated error was reported about 10^{-1}, whereas for our case, if $Z_{UE1i} = A_T(1)/C_T(1)$ is selected (Table 4.25, column 3) then, it is about 7.863×10^{-7} and if $Z_{UE2i} = B_T(1)/D_T(1)$ is selected (Table 4.25, column 4), then the accumulated error is found as 2.75×10^{-2} at the end of the extraction of the 21st element.

Obviously, the characteristic impedance choice $Z_{UE1i} = A_T(1)/C_T(1)$ yields better accumulated numerical error over $Z_{UE2i} = B_T(1)/D_T(1)$ and Levy's averaging approach. Therefore, we can confidently state that the proposed "High Precision Richards Extraction" method can safely be used to construct lossless two-ports for various microwave design problems.

Table 4.25 Result of synthesis for Example 4.16

Index	$\boldsymbol{Z_i}$	$\boldsymbol{Z_{UE1i} = A_T(1)/C_T(1)}$	$\boldsymbol{Z_{UE2i} = B_T(1)/D_T(1)}$
1	1.5370	1.537000000000000	1.537000000000000
2	0.6044	0.604400000000007	0.604400000000006
3	2.2910	2.291000000000548	2.291000000000619
4	0.4913	0.491300000001708	0.491300000001490
5	2.5140	2.514000000090463	2.514000000107290
6	0.4705	0.470500000127815	0.470500000100916
7	2.5710	2.571000003898722	2.571000005636488
8	0.4647	0.464700002928276	0.464700001108057
9	2.5890	2.589000050665227	2.589000151710818
10	0.4629	0.462900021110268	0.462899946675363
11	2.5930	2.593000209600809	2.593003807422498
12	0.4629	0.462900052305272	0.462898186448551
13	2.5890	2.589000344014129	2.589079489343453
14	0.4647	0.464700065712497	0.464672008685470
15	2.5710	2.571000370230168	2.572046282884658
16	0.4705	0.470500068022497	0.470259485672652
17	2.5140	2.514000363694764	2.521678537045851
18	0.4913	0.491300071082149	0.490256894159464
19	2.2910	2.291000331464653	2.314955678007914
20	0.6044	0.604400087446320	0.602793654881798
21	1.5370	1.537000222378049	1.547934870565496
	ERROR NORM	$\varepsilon_1 = norm(Z - Z_{UE1})$ $\mathbf{7.863 \times 10^{-7}}$	$\varepsilon_2 = norm(Z - Z_{UE2})$ $\mathbf{2.75 \times 10^{-2}}$

All the above computations are gathered under the main *MATLAB* program called *Richards_Example1.m* and its related functions are provided in the appendix of this chapter. Interested readers are encouraged to run this program to reproduce the above results.

In the previous publications [8,14] we have shown that our "high precision LC-ladder synthesis" algorithm is good up to 20 elements with relative error less than 10^{-12}.

In the following example, we test the robustness of the complete synthesis algorithm with the full generic form of the Richards impedance where we use both chain matrix-based unit element extraction with impedance correction and high precision LC ladder synthesis algorithm of [8,14].

Example 4.17

Let =15; $c = 5 \times rand(1,n)$; $k = 2$; $q = 8$; $nL = n-k-q = 5$; $c0 = 1$.

1. Generate the minimum Richards impedance $Z_{in}(\lambda) = a(\lambda)/b(\lambda)$.
2. Synthesize $Z_{in}(\lambda) = a(\lambda)/b(\lambda)$.
3. Comment on the results.

Solution

1. For this example, we developed a *MATLAB* program called *Richards_Example2.m.* In this program, auxiliary polynomial $c(\Omega)$ is generated employing the random number generator as specified by $c = 5 \times rand(1,n)$.

Then, using function[$a1$,$b1$] = Richard_NewMinimumFunction(k,q,c,$c0$), we first generate the Richards impedance. Resulting polynomial coefficients of $c(\Omega)$ and $a(\lambda)$ and $b(\lambda)$ are listed in Table 4.26 (see columns 1–3).

2. Synthesis of $Z_{in}(\lambda) = a(\lambda)/b(\lambda)$ is completed using the *MATLAB* function [*Z_UE, a_LC,b_LC,CT,CV*] = *Richard_CompleteImpedanceSynthesis*(*a,b,k,q,R0,f0*) with $R0 = 1$ and $f0 = 1/2\ \pi$. This function primarily extracts the cascaded unit elements with characteristic impedances $Z\_UE = [6.3422 \times 10^{-4}\ 5.3804 \times 10^{-4}]$. Then, the remaining Richards impedance $Z_{LC}(\lambda) = a_{LC}(\lambda)/b_{LC}(\lambda)$ is synthesized. $Z_{LC}(\lambda) = a_{LC}(\lambda)/b_{LC}(\lambda)$ is listed in Table 4.27.

The above minimum reactance Richards impedance must include $q = 8$ HE and $n_L = 5$ LE out of $n = 15$ total elements. The resulting synthesis of $Z_{LC}(\lambda) = a_{LC}(\lambda)/b_{LC}(\lambda)$ is shown in Figure 4.43 with expected generic layout where the normalized element values are.

$L1 = 36 \times 10^{-6}$	$L5 = 9.9 \times 10^{-6}$	$C9 = 200 \times 10^{3}$	$C13 = 21 \times 10^{3}$
$C2 = 29 \times 10^{3}$	$C6 = 475 \times 10^{3}$	$L10 = 2 \times 10^{-6}$	$R14 = 3. \times 10^{-6}$
$L3 = 13 \times 10^{-6}$	$L7 = 17. \times 10^{-6}$	$C11 = 108 \times 10^{3}$	
$C4 = 132 \times 10^{3}$	$C8 = 3. \times 10^{6}$	$L12 = 982 \times 10^{-9}$	

3. For the example under consideration, the accumulated numerical error of (4.121a) and (4.121b) is computed on the impedances $Z_{in}(j\omega)$ and $Z_a(j\omega)$ over $0 \le \omega \le 1.1$ and it is found that ε_r and ε_x are both less than 10^{-12}.

We should also note that the new Richards synthesis algorithm is employed using a randomly generated Richards impedance function of degree 25 (more

Table 4.26 Richards impedance of Example 4.17. n = 15; c = 5 × rand(1,n); k = 2; q = 8; nL = n − k − q = 5; c0 = 1

c(Ω)	a(λ)	b(λ)
1.8056e+000	0	3.4054e−001
3.1673e+000	4.9367e−004	4.0976e+000
4.9305e+000	5.9402e−003	2.3198e+001
1.0358e+000	3.2923e−002	8.3095e+001
3.7854e+000	1.1195e−001	2.1200e+002
4.4316e+000	2.6042e−001	4.0957e+002
2.3611e+000	4.3663e−001	6.2018e+002
7.9457e−001	5.4307e−001	7.5017e+002
4.0546e+000	5.0916e−001	7.3094e+002
2.3825e+000	3.6148e−001	5.7339e+002
5.8144e−001	1.9251e−001	3.5869e+002
4.3786e+000	7.4719e−002	1.7525e+002
3.1759e+000	1.9899e−002	6.4412e+001
4.8641e−001	3.2081e−003	1.6652e+001
4.5422e+000	2.2788e−004	2.6551e+000
c0 = 1	0	1.8860e−001

Table 4.27 Coefficients of remaining Richards impedance after extraction of cascaded unit elements

$a_{LC}(\lambda)$	$b_{LC}(\lambda)$
0	2.3898e+001
1.1674e−003	2.8755e+002
1.4047e−002	1.5951e+003
7.5512e−002	5.4354e+003
2.3653e−001	1.2691e+004
4.6714e−001	2.1377e+004
5.9250e−001	2.6670e+004
5.0882e−001	2.5046e+004
3.0451e−001	1.7797e+004
1.2648e−001	9.4818e+003
3.4867e−002	3.6811e+003
5.6906e−003	9.8044e+002
4.0423e−004	1.5807e+002
0.00000000	1.1228e+001

specifically, $n = 25$, $k = 10$ and $q = 5$) yielding the accumulated numerical errors less than 10^{-3}. Therefore, we can confidently state that the new synthesis algorithm can safely be employed to synthesize Richards immittance functions up to 25 elements.

It is mentioned that the input data for all the above examples are randomly generated. Therefore, they have no practical meaning. Nevertheless, they are good

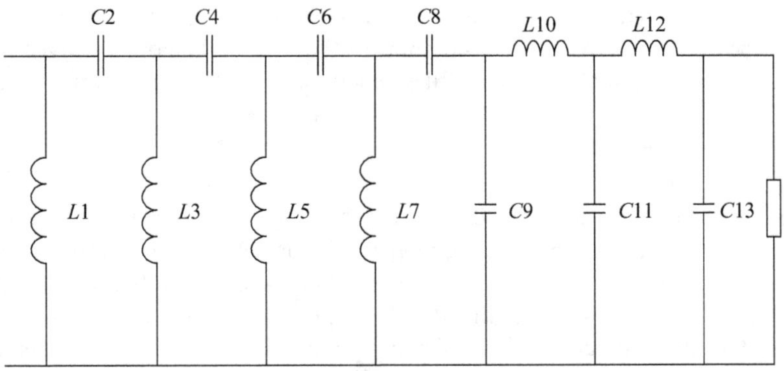

Figure 4.43 Synthesis of the Richards impedance $Z_{LC}(\lambda) = a_{LC}(\lambda)/b_{LC}(\lambda)$ for Example 4.17

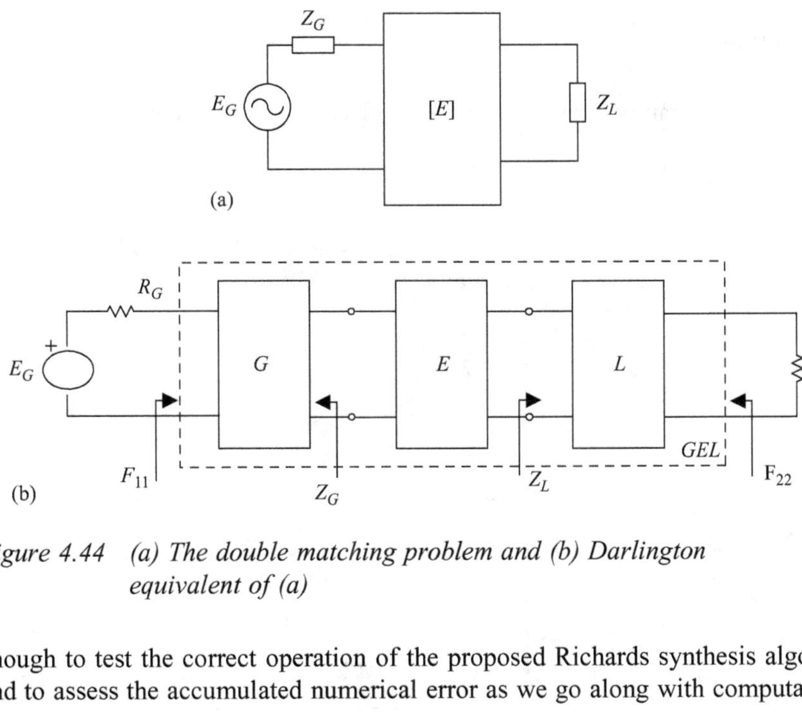

Figure 4.44 (a) The double matching problem and (b) Darlington equivalent of (a)

enough to test the correct operation of the proposed Richards synthesis algorithm and to assess the accumulated numerical error as we go along with computations.

4.5.10 Integration of new Richards synthesis tool with real frequency matching algorithm

The classical double matching problem is described by Figure 4.44(a). The problem is to design a lossless two-port or an equalizer $[E]$ for maximum power transfer from a complex generator with an internal impedance $[Z_G(j\omega) = R_G(\omega) + jX_G(\omega)]$ to a complex load with an impedance $[Z_L(j\omega) = R_L(\omega) + jX_L(\omega)]$ over the frequency band of operation. The power transfer is measured by means of the transducer power gain [*TPG*] which is defined as the ratio of the power delivered to the load (P_L) to the available power of the generator (P_A). In other words:

$$TPG = T(\omega) = \frac{P_L}{P_A} \tag{4.122}$$

The generator and the load impedances are assumed non-Foster positive real functions. Therefore, they can be equivalently represented using Darlington's lossless two ports [*G*] and [*L*] in resistive termination as shown in Figure 4.44(b).

Referring to Figure 4.44, in the RF-DCT, the TPG of the double-matched system is described in terms of the driving-point immittance of the generator $F_G = [Z_G \text{ or } Y_G]$, equalizer $F_B = [Z_B(\lambda) \text{ or } Y_B(\lambda)]$ and the load $F_L = [Z_L \text{ or } Y_L]$.

In the above notation, letter $Z_{(.)}$ designates the impedances and letter $Y_{(.)} = 1/Z_{(.)}$ refers to corresponding admittances.

In this case, *TPG* of Figure 4.44(a) is given by [3]:

$$T(\omega) = \frac{1 - |G_{22}|^2}{|1 - G_{22}S_{in}|^2} T_{[EL]} \tag{4.123}$$

where G_{22} is the unit normalized generator reflectance, and it is given by

$$G_{22} = \frac{Z_G - 1}{Z_G + 1} = \frac{1 - Y_G}{1 + Y_G} \tag{4.124}$$

As proven by the main theorem of Yarman and Carlin [2,13], the unit normalized input reflectance S_{in} of [*EL*] is given by (Figure 4.45):

$$S_{in}(j\omega) = \eta_B(\lambda) \left[\frac{Z_L(j\omega) - Z_B(-\lambda)}{Z_L(j\omega) + Z_B(\lambda)} \right]_{\lambda = jtan(\omega\tau)} \tag{4.125a}$$

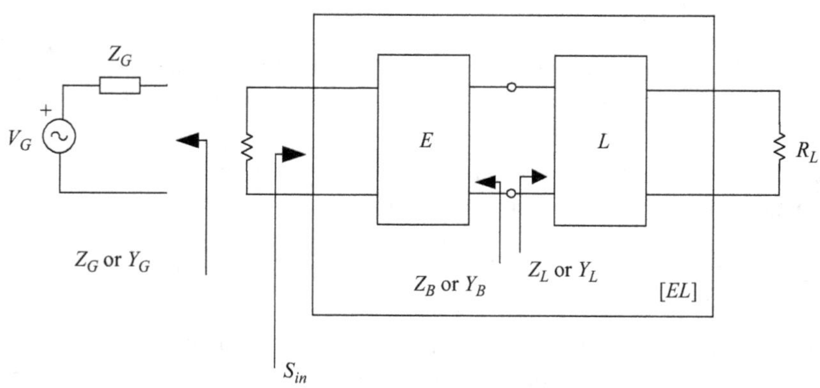

Figure 4.45 Cascade connection of two lossless two-ports [G] and [EL]

or

$$S_{in}(j\omega) = \eta_B(\lambda)\left[\frac{Y_B(\lambda) - Y_L(-j\omega)}{Y_B(\lambda) + Y_L(j\omega)}\right]_{\lambda=jtan(\omega\tau)} \tag{4.125b}$$

In (4.125a) and (4.125b), the all pass function $\eta_B(\lambda)$ is described by

$$\eta_B(\lambda) = \left[\frac{W_B(\lambda)}{W_B(-\lambda)}\right] \tag{4.126}$$

The rational analytic function $W_B(\lambda)$ of (4.126) is constructed on the explicit factorization of the even part $R_B(\lambda^2)$ such that

$$F_B(\lambda) = \frac{a(\lambda)}{b(\lambda)} \tag{4.127}$$

and

$$R_B(\lambda^2) = Even\{F_B(\lambda)\} = \frac{A(\lambda^2)}{B(\lambda^2)} \tag{4.128a}$$

$$= \frac{(-1)^q \lambda^{2q}(1-\lambda^2)^k}{B(\lambda^2)} \tag{4.128b}$$

$$= \frac{n_B(\lambda)n_B(-\lambda)}{b(\lambda)b(-\lambda)}$$

$$= W_B(\lambda)W_B(-\lambda) \tag{4.128c}$$

where the fictitious function $W_B(\lambda)$ is described by

$$W_B(\lambda) = \frac{n_B(\lambda)}{b(\lambda)}$$

The numerator polynomial $n_B(\lambda)$ must include all the proper RHP and $j\Omega - axis$ zeros of $R_B(\lambda^2)$ as described by (4.115a)–(4.115c). Thus, for a cascade connected k UEs with q DC transmission zeros, $W_B(\lambda)$ and $W_B(-\lambda)$ are given as

$$W_B(\lambda) = (-\lambda)^q \frac{(1-\lambda^2)^{k/2}}{b(\lambda)} \tag{4.129a}$$

$$W_B(-\lambda) = (\lambda)^q \frac{(1-\lambda^2)^{k/2}}{b(-\lambda)} \tag{4.129b}$$

Hence, the all pass function $\eta_B(\lambda)$ is given by

$$\eta_B(\lambda) = (-1)^q\left[\frac{b(-\lambda)}{b(\lambda)}\right] \tag{4.130}$$

For the direct method of real-frequency broadband matching, the unknown of the problem is the rational form of the even part $R_B(\lambda^2)$ as specified by (4.115a)–(4.118i).

It is interesting to observe that by (4.123), the TPG T_{EL} of the lossless two-port $[EL]$ is given by

$$T_{EL} = 1 - |S_{in}(j\omega)|^2 = \frac{4R_BR_L}{[R_B + R_L]^2 + [X_B + X_L]^2} \tag{4.131}$$

which is the immittance-based conventional single matching gain. Assuming $F_B(\lambda)$ as a minimum function, TPG is expressed as a function of the real part $R_B(\Omega)$ as

$$T(\omega) = \left[\frac{1 - |G_{22}|^2}{|1 - G_{22}S_{in}|^2}\right]\left\{\frac{4R_BR_L}{[R_B + R_L]^2 + [X_B + X_L]^2}\right\} \tag{4.132}$$

where

$$R_B(\Omega) = \text{Real Part}\{F_B(j\Omega)\}$$

and

$$X_B(\omega) = \text{Imaginary Part}\{F_B(j\Omega)\}.$$

Once the integers k and q are selected by the designer, $R_B(\Omega)$ is described by means of an auxiliary polynomial $c(\Omega) = c_1\Omega^n + c_2\Omega^{n-1} + \cdots + c_n\Omega + c_0$ of (4.109). In this case, unknowns of the matching problem are the arbitrary real coefficients of $c(\Omega)$. When the coefficients $\{c_i; i = 0, 12, \cdots, n\}$ are initialized, $F_B(\lambda)$ is generated using the parametric method of Section II and $T(\omega)$ is computed as in (4.111).

In RF-DCT, *TPG* is optimized over the band of operation employing a nonlinear optimization algorithm. For example, in *MATLAB*, one may wish to employ the nonlinear optimization functions such as *lsqnonlin*, *fminimax*, *fminsearch* from the optimization tool-box. Optimization of *TPG* yields the driving-point immittance $F_B(\lambda) = a(\lambda)/b(\lambda)$. Eventually, $F_B(\lambda)$ is synthesized using the newly proposed Richards synthesis algorithm.

Example 4.18

Let us apply the above-described process to design a wideband matching network.

Referring to Figure 4.46, an impedance-transforming filter is constructed between $R_G = 12\ \Omega$ generator (output of a power amplifier) and a standard load of $R_L = 50\ \Omega$ employing the RF-DCT algorithm. The design is completed over 850–2,100 MHz. At the generator end, the resonance circuit *LG//CG*1 introduces a zero of transmission at 4,200 MHz, which is the second harmonic of the high end of the passband. Furthermore, *CG*2 introduces a zero of transmission at DC at the generator end. Similarly, at load end, the tank circuit *LL/CL* introduces a transmission zero at the third harmonic (6,300 MHz). Thus, Figure 4.46 describes a double-matching problem. Therefore, a lossless matching network is constructed between a complex generator Z_G and a complex load Z_L.

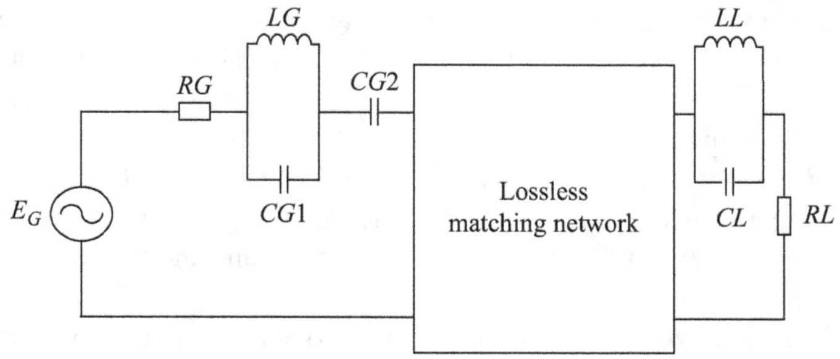

Figure 4.46 Double matching problem for Example 4.19: the element values of the circuit in Figure 4.7 are determined as RG = 12 Ω, LG = 0.947 nH, CG1 = 1.515 pF, CG2 = 7.622 pF, CL = 1.515 pF, LL = 0.410 nH, RL = 50 Ω

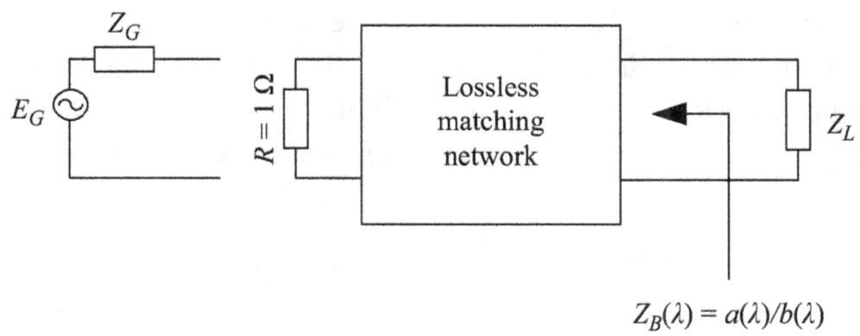

Figure 4.47 Description of the matching network by means of its driving-point input impedance $Z_B(\lambda)$

Referring to Figure 4.47, for the problem under consideration, RF-DCT algorithm is implemented in *MATLAB* under the main program $RichardMain_{ImpTransFilter}.m$.

The matching network is described by means of its driving-point impedance $Z_B(\lambda)$ employing six commensurate transmission lines ($n = 6$). In the course of design, $k = 4$ (total number of cascaded UEs) and $q = 0$ (no DC transmission zero) is selected which in turn yields $n_L = n - k - q = 2$ transmission zeros at infinity. Coefficients $\{c_i; i = 1, 2, .., 6\}$ of the auxiliary polynomial $c(\Omega)$ is initialized in an ad hoc manner. Furthermore, c_0 is fixed as $c_0 = 1$ so that at the far end, the normalized termination resistance R is set to unity.

Using the least square nonlinear optimization tool *lsqnonlin* of *MATLAB*, the driving-point impedance $Z_B(\lambda) = a(\lambda)/b(\lambda)$ is obtained as

$$a(\lambda) = 0\lambda^6 + 0.62\lambda^5 + 0.7259\lambda^4 + 0.2427\lambda^3 + 0.1717\lambda^2 + 0.0169\lambda + 0.0036$$

and

$$b(\lambda) = 2.1530\lambda^6 + 2.5204\lambda^5 + 1.1745\lambda^4 + 0.9848\lambda^3 + 0.1634\lambda^2 + 0.0754\lambda + 0.0036$$

For the interested readers, the above results may be reproduced using the accurate coefficients of the polynomials $c(\Omega)$, $a(\lambda)$ and $b(\lambda)$ as shown in Table 4.28. Coefficients of $c(\Omega)$ is obtained as the result of optimization. Coefficients of the numerator polynomial $a(\lambda)$ and the denominator polynomial $b(\lambda)$ is obtained using our *MATLAB* function [*a*,*b*] = *Richard_NewMinimumFunction*(*k*,*q*,*c*,*czero*) where the input variables are selected as shown at the top and the first column of Table 6. Coefficients of $a(\lambda)$ and $b(\lambda)$ are listed in the second and the third columns of the same table.

$Z_B(\lambda)$ is synthesized using our newly developed Richards synthesis package called [*Z_UE*,*a_LC*,*b_LC*,*CT*,*CV*] = *Richard_CompleteImpedanceSynthesis*(*a*,*b*,*k*,*q*, *R*0,*f*0) with the normalization numbers $R0 = 1, f0 = 1/(2\pi)$. The output, vector Z_UE includes the normalized characteristic impedances of the cascaded transmission lines. Vectors *a_new* and *b_new* include the coefficients of the numerator and the denominator polynomials of the remaining impedance function $Z_{LC}(\lambda) = a_{LC}(\lambda)/b_{LC}(\lambda)$ after cascaded line extractions. Vectors CT and CV includes the synthesis result of $Z_{LC}(\lambda)$. In this representation, Vector CT includes the codes of the Richards components such that *CT*(*i*) = 8 refers to a shunt Richards capacitor, *CT*(*i*) = 1 refers to a series Richards inductor, *CT*(*i*) = 9 is the termination resistor.

List of output vectors are shown in Table 4.29 and final synthesis of $Z_B(\lambda)$ is depicted in Figure 4.48. In summary, characteristic impedances of Figure 4.48 are given by

$$Z_1 = 0.2517, Z_2 = 1.8721, Z_3 = 0.1246, Z_4 = 1.5364$$

Table 4.28 Result of the optimization for Example 4.18. n = 6, k = 4, q = 0, c0 = 1

c(Ω)	**a(λ)**	**b(λ)**
1.0e+02 *[.]		
5.925735	0	2.152976
6.085158	0.620037	2.520421
−2.296435	0.725858	1.174455
−2.380582	0.242668	0.984801
0.157989	0.171746	0.163416
0.175761	0.016854	0.075382
0.010000	0.003633	0.003633

Table 4.29 Result of Richards synthesis

Index	*Z_UE*	*a_LC*	*b_LC*
1	0.251700	0	1.00000
2	1.872104	0.13221	1.17066
3	0.124594	0.15477	0.15477
4	1.536354		

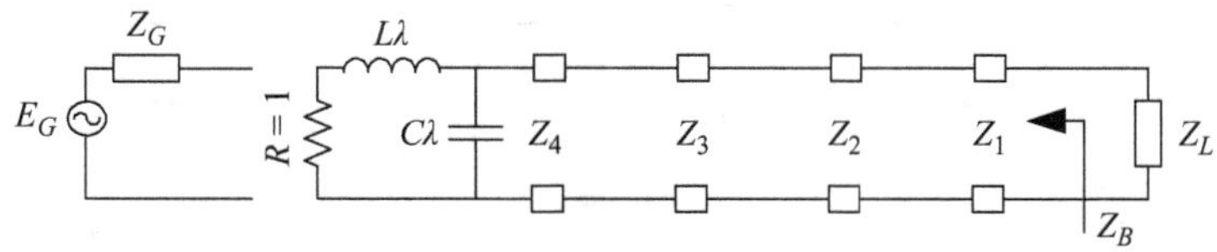

Figure 4.48 Synthesis of $Z_B(\lambda)$

and the Richards components are specified as

$$C = 7.5637, L = 0.8542.$$

By selecting the resistive normalization number $R_0 = 50\,\Omega$, actual element values are given by

$$Z_{UE} = [12.5850 \;\; 93.6052 \;\; 6.2297 \;\; 76.8177\;]\Omega$$

The Richards capacitor C is realized as a shunt open stub with normalized characteristic impedance $Z_{cap} = 1/C$ or with actual characteristic impedance $Z_{cap-act} = R_0/C$.

Similarly, the Richards inductor L is realized as a series short-stub with normalized characteristic impedance $Z_{Ind} = L$ or with actual characteristic impedance $Z_{ind-act} = R_0 L$. Thus, it is found that $Z_{cap-act} = R_0/C = 50/7.5 = 6.6105\ \Omega$ and $Z_{ind-act} = R_0 L = 42.7106\ \Omega$.

As far as practical implementation is concerned, we may prefer to utilize microstrip technology to realize the ideal commensurate transmission lines. Using microstrip technology, the shunt Richards capacitors (i.e., open stubs in shunt configuration) can be easily realized but realization of series Richards inductors (i.e., short stubs in series configuration) may be difficult to produce. Nevertheless, physical implementation problems can be bypassed using the Kuroda identifies [20,39].

In this regard, successive application of the Kuroda identities removes the series short stubs with those of shunt open stubs. Hence, we end up with Figure 4.49 as the final synthesis of $Z_B(\lambda)$.

The Richards capacitor C_{B3} is realized as an open stub with actual characteristic impedance $Z_{cap3-act} = R_0/C_{B3} = 6.7787\ \Omega$. Similarly, C_{B2} is also realized as an open stub with actual characteristic impedance $Z_{cap2-act} = R_0/C_{B2} = 6.9542\ \Omega$.

Thus, final version of the matching network is shown in Figure 4.50, and its characteristic impedances are summarized as

$$ZB2 = 48.7974, ZCap2 = 6.9542,$$

$$ZB3 = 76.9633, ZCap3 = 6.7787,$$

$$Z2 = 93.6050, Z1 = 12.5850.$$

The physical length l of the commensurate transmission lines is fixed at the upper edge of the frequency band $f_{c2} = 2.1$ GHz as $\omega_{c2}\tau_{c2} = 0.5$. Then, we can compute the actual delay length τ_{c2} as follows.

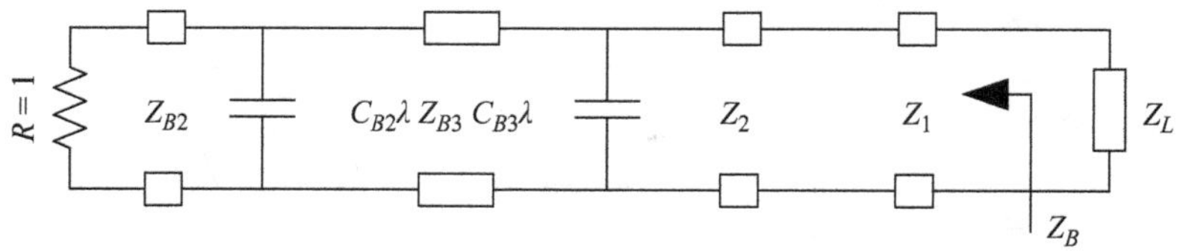

Figure 4.49 Synthesis of $Z_B(\lambda)$ after successive application of Kuroda identities

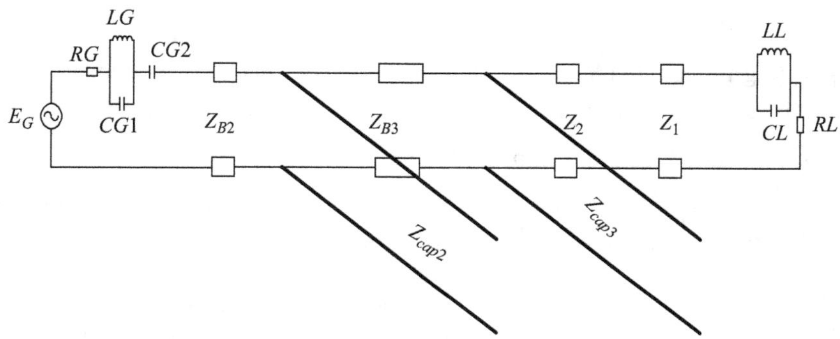

Figure 4.50 Synthesis of $Z_B(\lambda)$ after successive applications of the Kuroda identities

At $f_{c2} = 2.1$ GHz let the normalized angular frequency ω_N be. Then,

$$\omega_{c2}\tau_{c2} = \omega_N\tau_N = 0.5 = 2\pi f_{c2}\tau_{c2}$$

or

$$\tau_{c2} = \frac{0.5}{2 \times \pi \times 2.1 \times 10^9} = 3.7894 \times 10^{-11} \text{ s.}$$

The physical length is computed upon the selection of the substrate using the effective propagation velocity v_{eff} as

$$l = v_{eff}\tau_a$$

Optimized TPG is depicted in Figure 4.51.

4.5.11 Alternative design

We can generate an alternative design with $n = k = 6$ cascaded unit elements of the fixed delay length $\tau = 0.5$. In this case, the coefficients of the auxiliary polynomial $c(\Omega)$ are initialized in an ad hoc manner with fixed $c_0 = 1$. Then, the

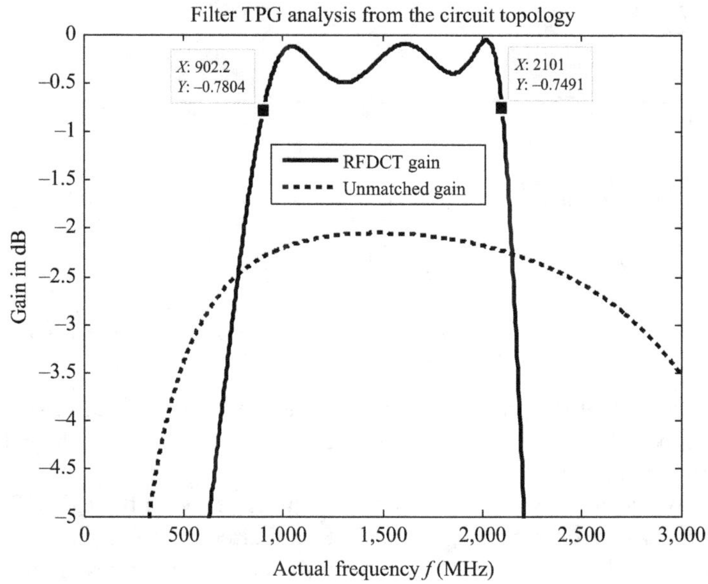

Figure 4.51 Performance of the matched system for Example 4.3

optimization of the TPG results in the following polynomial coefficients for $c(\Omega), a(\lambda)$ and $b(\lambda)$.

$c = [656.8151 \;\; 709.8196 \;\; -248.9498 \;\; -273.0388 \;\; 17.0399 \;\; 19.3983]$ with $c_0 = 1$;

$a = [0.0000 \;\; 0.6071 \;\; 0.7409 \;\; 0.2328 \;\; 0.1679 \;\; 0.0151 \;\; 0.0032]$;

$b = [2.0933 \;\; 2.5548 \;\; 1.1301 \;\; 0.9780 \;\; 0.1528 \;\; 0.0717 \;\; 0.0032]$.

Finally, $Z_B(\lambda) = a(\lambda)/b(\lambda)$ is synthesized using our new Richards synthesis tool yielding the cascaded UE characteristic impedances in the *MATLAB* vector Z_{UE} as

$$Z\_UE = [0.9375 \; 0.1237 \;\; 1.4477 \; 0.1225 \;\; 1.8543 \; 0.2530]$$

and the actual *Z_UE* = [46.8750 6.1850 72.3850 6.1250 92.7150 12.6500].

Employing our newly developed software tool, the resulting Richards synthesis of $Z_B(\lambda) = a(\lambda)/b(\lambda)$ is shown in Figure 4.52 and the performance of the matched system is almost the same as the previous design as depicted in Figure 4.51.

MATLAB Programs for Example 4.19 is provided as open source codes in the appendix. Interested readers are encouraged to reproduce the results presented in this section.

4.5.12 Conclusion

As the continuation of our previous work, in this section, a high precision Richards Synthesis algorithm is presented. The new algorithm utilizes the parametric approach at each Richards's extraction to correct the remaining impedance.

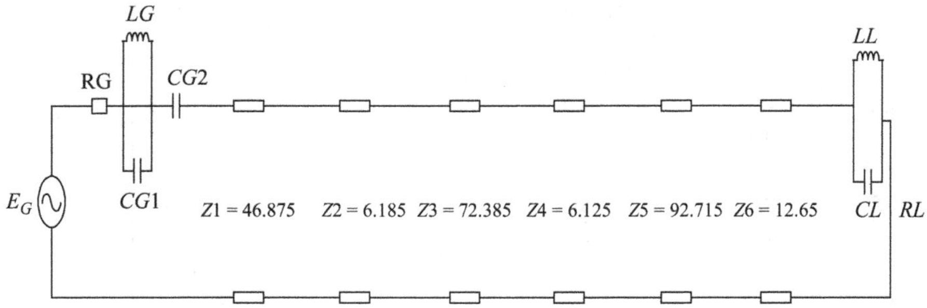

Figure 4.52 Alternative design with $n = 6, k = n, q = 0$ for Example 4.3

It is verified that the proposed algorithm can safely be utilized to synthesize a Richards impedance with 25 commensurate transmission line yielding the accumulated numerical error about 10^{-3}.

The new Richards synthesis algorithm is integrated with the RF-DCT to construct matching networks with optimum TPG and circuit topology.

The new synthesis algorithm is provided as open source codes to all the users.

There are some practical circumstances, which may require mixed element design. In this case, first, one can drive a lumped element prototype and then replaces it with the "almost equivalent mixed element network." In the following section, we cover the construction of lossless mixed networks from a lowpass prototype.

4.6 Practical design of matching networks with mixed lumped and distributed elements

4.6.1 Almost equivalent transmission line model of a CLC-PI section

A symmetrical lowpass CLC-PI section can be approximated by a capacitive loaded transmission line TRL as shown in Figure 4.53 [40,41].

In terms of the specified inductor L characteristic impedance Z_0 of the line is given by

$$Z_0 = \frac{\omega_0 L}{\sin(\omega_0 \tau)} \tag{4.133}$$

If one wishes to fix the characteristic impedance Z_0 in advance, then, the delay length τ of the line can be adjusted at a specified angular frequency $\omega_0 = 2\pi f_0$ yielding

$$\sin(\omega_0 \tau) = \frac{\omega_0 L}{Z_0}$$

or the corresponding delay length τ is determined as

$$\tau = \left(\frac{1}{\omega_0}\right)\sin^{-1}\left(\frac{\omega_0 L}{Z_0}\right) \tag{4.134}$$

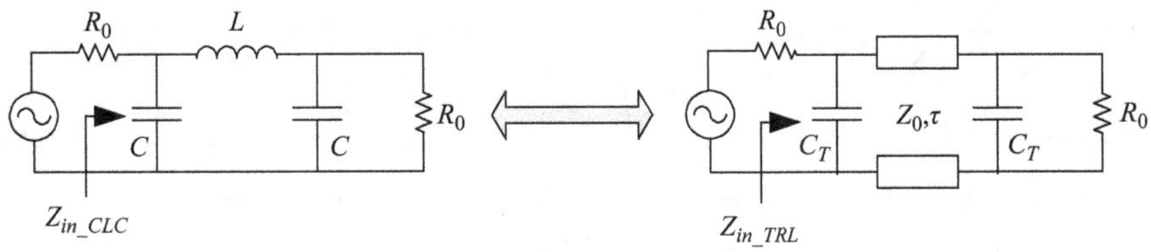

Figure 4.53 CLC PI equivalent is replaced by CT-TRL_CT

On the other hand, in terms of the element values of the PI section, loading capacitor C_T is given by

$$C_T = \frac{\cos(\omega_0\tau) + \omega_0^2 LC - 1}{\omega_0^2 L} \tag{4.135}$$

It is important to emphasize that, for the specified values of L, C, ω_0 and τ, the value of C_T must be nonnegative. Otherwise, *almost equivalent* $CT - TRL - CT$ counterpart of the lumped $CLC - PI$ section does not exist.

In this representation, $\omega_0 = 2\pi f_0$ is chosen as the high end of the useful passband and τ is selected in such a way that at the stopband $\omega_{s2}\tau = 2\pi f_{s2} = \pi/2$.

In many practical situations, stopband frequency f_{s2} is expressed as the multiple of f_0 such that $f_{s2} = mf_0$.

Thus, in terms of the cutoff frequency f_0 of the passband, fixed delay τ is specified as

$$\tau = \frac{\pi}{2\omega_{s2}} = \frac{\pi}{2 \times 2\pi f_{s2}} = \frac{1}{4f_{s2}} = \frac{1}{4mf_0} \tag{4.136}$$

It may be useful to remember that at the stop band frequency $\omega_{s2} = 2\pi f_{s2}$, TPG of Figure 4.53 is practically zero (perhaps, less than -20 dB).

For many applications, it is proper to select ω_{s2} in such a way that it satisfies the below inequality:

$$2\omega_0 \leq \omega_{s2} \leq 4\omega_0 \tag{4.137a}$$

or

$$2 \leq m \leq 4 \tag{4.137b}$$

In other words, the multiplying integer m varies between 2 and 4.

The physical length l of the transmission line can be determined by considering the actual implementation. For example, if the line is printed on substrate with permittivity ε_r and permeability μ_r as a microstrip or coplanar line then, the velocity of propagation is given by

$$v_{sub} = \frac{v_0}{\sqrt{\mu_r \varepsilon_r}} = \frac{l}{\tau} \tag{4.138}$$

where v_0 is the speed of light in free space and it is specified as $v_0 = 3 \times 10^8$ m/s.
Hence, the physical length l is given by

$$l = v_{sub}\tau \tag{4.139}$$

Furthermore, employing microstrip technology, characteristic impedance of the transmission line can be approximated using the Wheeler formulas [27–29]:

$$Z_0 \cong \left(\frac{120\pi}{\varepsilon_{r-eff}}\right)\frac{1}{\left[x + 1.98(x)^{0.172}\right]} \tag{4.140}$$

where $x = W/h$ the ration of width (W) to thickness (h), the effective-relative permittivity ε_{r-eff} is

$$\varepsilon_{r-eff} = 1 + \frac{\varepsilon_{r-1}}{2}\left[1 + \frac{1}{\sqrt{1 + (10/x)}}\right] \tag{4.141}$$

It must be noted that the above formulas are valid for the values of x which satisfies > 0.06.

Obviously, for a fixed characteristic impedance Z_0 and relative permittivity ε_r physical width W of the microstrip line can be found by means of nonlinear equation solving or equivalently or optimization methods.

Example 4.19

Construction of a matching network with mixed lumped and distributed elements from its lumped element prototype.

In Figure 4.54, a typical input matching network of a microwave power amplifier is shown. This network includes a three-element lowpass PI section. It is terminated by the input of an laterally diffused metal oxide semiconductor (LDMOS) power-transistor.

The terminating impedance ZL is measured by means of Agilent load-pull equipment over 330–530 MHz. Real and imaginary parts of the measured load is

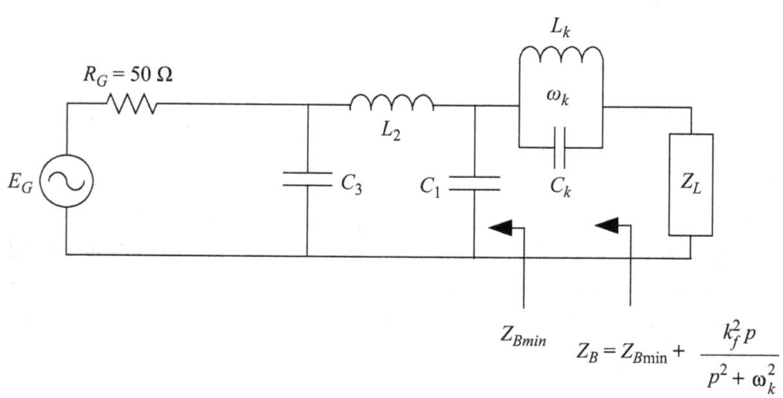

Figure 4.54 Input matching network of a power amplifier

depicted in Figures 4.55 and 4.56, respectively, and the measured impedance data is listed in Table 4.30.

Passband of the amplifier starts at $f_{c1} = 330$ MHz and ends at $f_{c2} = 530$ MHz. The frequency band is normalized with respect to $f_0 = 530$ MHz. The corresponding gain plot is given in Figure 4.57.

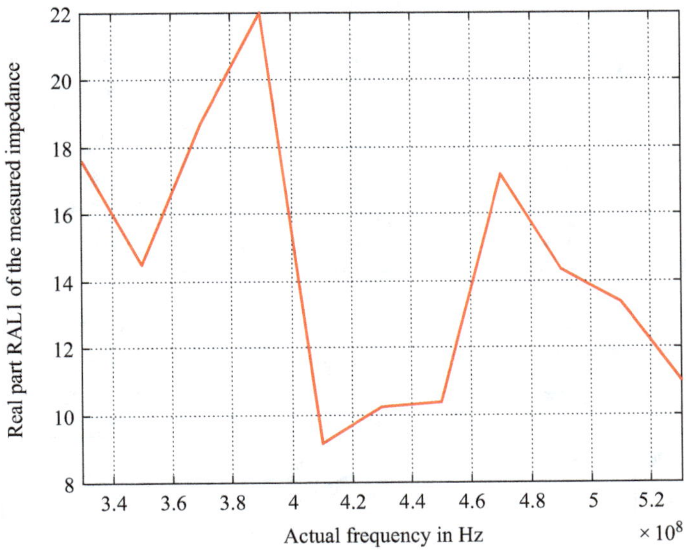

Figure 4.55 Real part of ZL

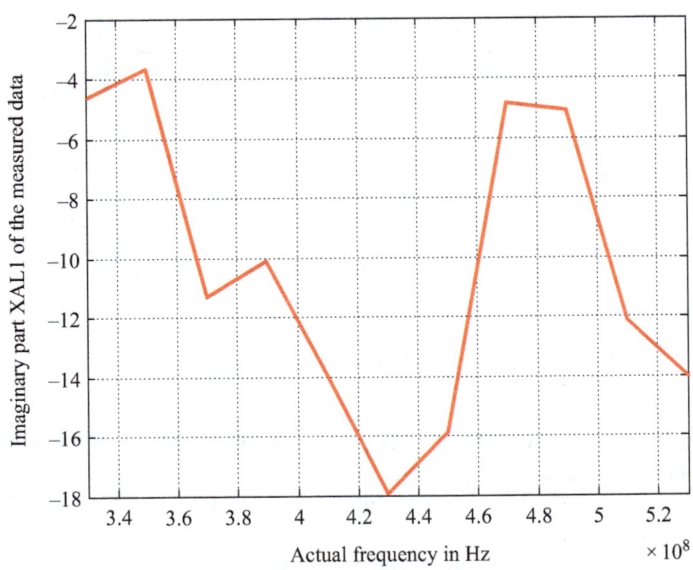

Figure 4.56 Imaginary part of ZL

Table 4.30 Measured input impedance data $Z_{\text{in}}(j\omega) = R_{\text{in}}(\omega) + jX_{\text{in}}(\omega)$

	$R_{in}(\omega)$	$X_{in}(\omega)$
330	17.61	−04.63
350	14.50	−03.67
370	18.70	−11.30
390	22.00	−10.11
410	09.16	−13.96
430	10.23	−17.94
450	10.40	−15.89
470	17.18	−04.85
490	14.33	−05.11
510	13.36	−12.11
530	11.04	−14.03

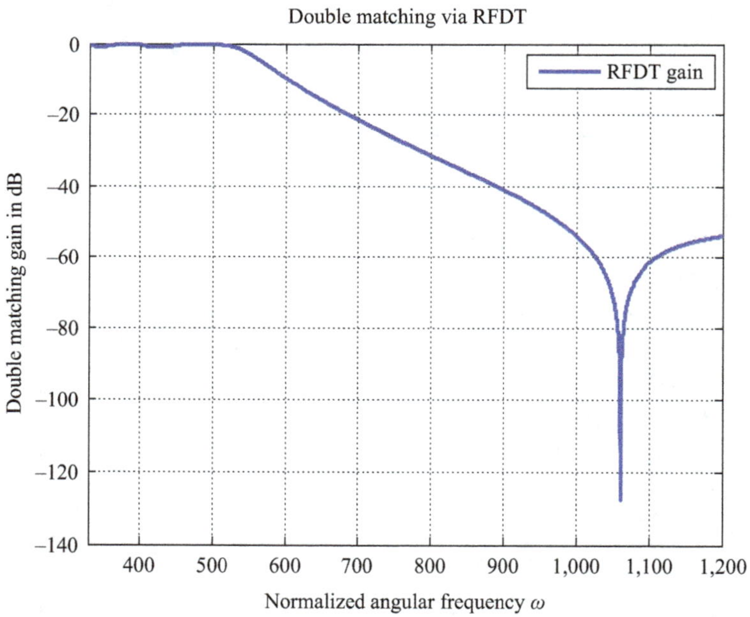

Figure 4.57 Gain performance of the matched system

Lumped element design yields $T_{\min} = -0.9170$ dB which is the minimum of pass band gain occurred at 410 MHz.

The matching network is designed using the RF-DCT with a single foster section (FS) extraction.

In the course of design, the driving-point immittance of the equalizer was chosen as a minimum reactance function designated by $Z_{B\min}$ ($Kflag = 1$).

Therefore, extracted FS is a parallel resonance circuit in series with $Z_{B\min}$ and it is given by

$$Z_F(p) = \frac{k_f^2 p}{p^2 + \omega_k^2} = \frac{[1/C_k]}{p^2 + (1/L_k C_k)} \tag{4.142}$$

Explicit element values of the parallel resonance circuit are driven in terms of the residue $k_f^2 > 0$ and the resonance frequency ω_k:

$$C_k = \frac{1}{k_f^2} \tag{4.143a}$$

$$L_k = \frac{k_f^2}{\omega_k^2} \tag{4.143b}$$

The resonance frequency is selected at $f_k = 2f_{c2} = 2 \times 530 = 1,060$ MHz to suppress the second harmonic of the upper edge frequency of the passband.

Results are summarized below.

$$Z_B(p) = Z_{B\min}(p) + \frac{k_f^2 p}{p^2 + \omega_k^2} = \frac{a(p)}{b(p)} + \frac{k_f^2 p}{p^2 + \omega_k^2}$$

where

$$Z_{B\min}(j\omega) = R_B(\omega) + jH\{R_B(\omega)\}$$

$$R_B(\omega) = \frac{1}{\frac{1}{2}\left[c^2(\omega) + c^2(-\omega)\right]}$$

$$c(\omega) = c_1\omega^n + c_2\omega^{n-1} + \cdots + c_n\omega + 1 \ \ with \ \ n = 3$$

$$k_f = 1.5781; \quad c = [8.2573\ 0.6615\ -5.77501];$$

$$a = [0\ 0.2511\ 0.1024\ 0.1211]; \quad b = [1.0000\ 0.4077\ 0.7793\ 0.1211]$$

Normalized and actual element values of the parallel resonance circuit are computed using (4.143a) and (4.143b) as follows:

$$C_k = \frac{1}{k_f^2} = 0.4015$$

$$C_{Ak} = C_k/2/530/1e6/pi/50;$$

$$C_{Ak} = 2.4115\text{e} - 012 = 2.41 \ \text{pF}$$

$$L_k = 1/Ck/4$$

$$L_k = 0.6226$$

$$L_{Ak} = Lk \times 50/(2pi530e6) = 9.3483\text{e} - 6$$

$$L_{Ak} = 9.3483\text{e} - 009 = 9.35\text{nH}$$

Actual element values of the CLC ladder is given by

$$C_{A1} = 2.3916\text{e} - 011 \cong 23.9 \text{ pF}$$

$$C_{A3} = 1.4731\text{e} - 011 \cong 14.7 \text{ pF}$$

$$L_{A2} = 1.2694\text{e} - 008 \cong 12.7 \text{ nH}$$

Here, it is desired to find the loaded transmission line equivalent of the PI section step-by-step.

It is noted that the new circuit will be fabricated on Rogers 4350B three-layer PCB (printed circuit board) with permittivity $\varepsilon_r = 3.66$ and $h = 0.72$ mm.

Solution

Step 1: Separation of symmetrical CLC-PI section from a given arbitrary $C1$-$L2$-$C3$ section

First, we should distinguish a symmetrical CLC-PI sections from the given matching network so that they are replaced by their almost equivalent CT-TRL-CT sections.

For the example under consideration, we can only separate one PI section from the given matching network. The symmetrical capacitors C may be set to minimum of $\{C1, C3\}$ as shown in Figure 4.58.

In this case, $C = C_3 = 14.7$ pF. Then, we have to include a residual capacitor $C_a = C_1 - C$ to the left of the symmetrical CLC-PI section to preserve the original PI.

Step 2: Construction of almost equivalent CT-TRL-CT section

Once the symmetrical PI is distinguished, it is replaced by its almost equivalent CT-TRL-CT counterpart as shown in Figure 4.59.

In Figure 4.59, resulting load capacitors C_A and C_B are given by

$$C_A = C_a + C_T$$

$$C_B = C_T \tag{4.144}$$

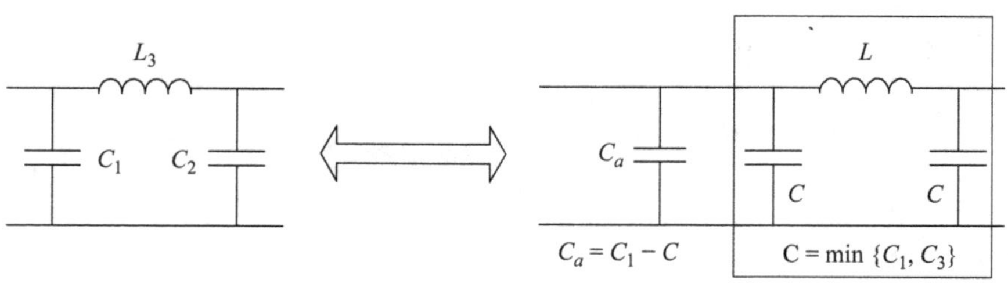

Figure 4.58 Extraction of symmetrical CLC-PI section

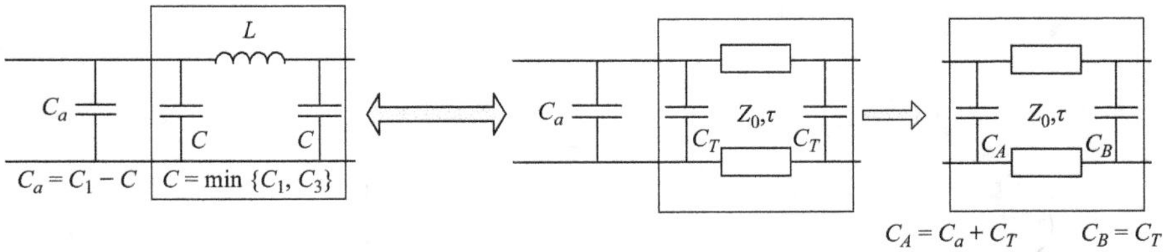

Figure 4.59 Replace CLC by its almost equivalent counterpart CT-TRL-CT

Step 3: Derivation of parameters of CT-TRL-CT section
Now let us calculate the element values of the equivalent circuit.
Let the normalized stop frequency be $\omega_{s2} = m\omega_0 = 2$; $\omega_{s2}\tau = \pi/2$ (i.e., $m = 2$)
Then,

$$\tau = \frac{\pi}{4\omega_0} = \frac{1}{2mf_0} = 2.3585\text{e} - 010 \text{ s}$$

In this case,

$$z_0 = \frac{\omega_0 L}{\sin(\omega_0 \tau)} = \frac{2\pi \times 530 \times 10^6 \times 1.269 \times 10^{-8}}{\sin(\pi/4)}$$

$$z_0 = 59.763\ \Omega$$

$$C = \min(C_1, C_3) = \min(C1 = +2.392\text{e} - 011; C3 = +1.473\text{e} - 011) = 1.473\text{e} - 11$$

$$C_T = \frac{\cos(\omega_0 \tau) + \omega_0^2 LC - 1}{\omega_0^2 L}$$

$$CT = 1.2649\text{e} - 011\ \textit{Farad} = 12.65 \text{ pF}$$

$$C_A = C_1 - C + C_T = 2.1839\text{e} - 011$$

so $C_A = 21.84$ pF

$$C_B = C_3 - C + C_T = 1.2649\text{e} - 011$$

so $C_B = 12.65$ pF

The complete matching network constructed with mixed elements is depicted in Figure 4.60.

In Figure 4.61, gain performance of the lumped element prototype and the mixed element matching networks is depicted. It is observed that both performances are close preserving the minimum gain in the passband.

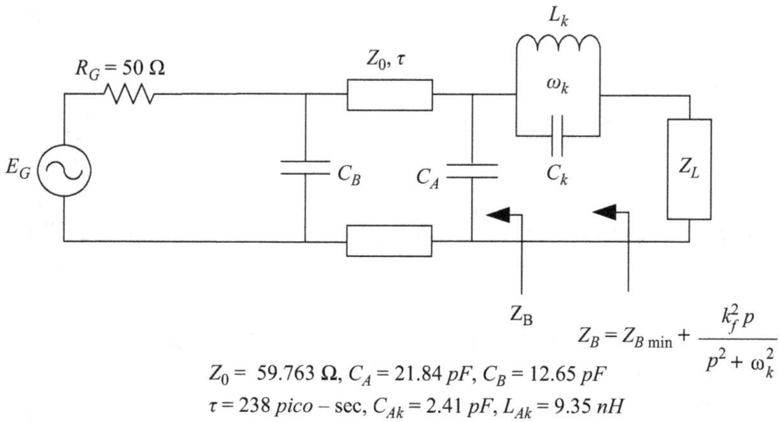

Figure 4.60 Almost equivalent matching network constructed on its lumped element prototype

Step 4: Physical implementation

Once the mixed element structure is obtained using the ideal circuit components, distributed elements may be realized as microstrip lines. In this case, one has to pick a commercially available substrate with a proper dielectric constant to realize the characteristic impedance Z_0 and the delay length τ. For the problem under consideration $\varepsilon_r = 3.66$ and $h = 0.72$ mm.

Hence, using the main program called Microstrip_design.m, the physical width of the line is computed as

$$\text{width} = 0.0012\ \text{m} = 1.2\ \text{mm}$$

This result can easily be checked such that

$$x = \frac{W}{h} = 1.6054$$

$$\varepsilon_{r-eff} = 1 + \frac{\varepsilon_{r-1}}{2}\left[1 + \frac{1}{\sqrt{1 + (10/x)}}\right] = 2.8247$$

$$Z_0 \cong \left(\frac{120\pi}{\varepsilon_{r-eff}}\right)\frac{1}{\left[x + 1.98(x)^{0.172}\right]} = 59.78\ \Omega$$

Eventually, physical length on the substrate is determined as

$$l = v_{sub}\tau = \frac{3 \times 10^8}{\sqrt{3.66}} \times 2.3585\text{e} - 010 = 0.0370 = 37\ \text{mm}$$

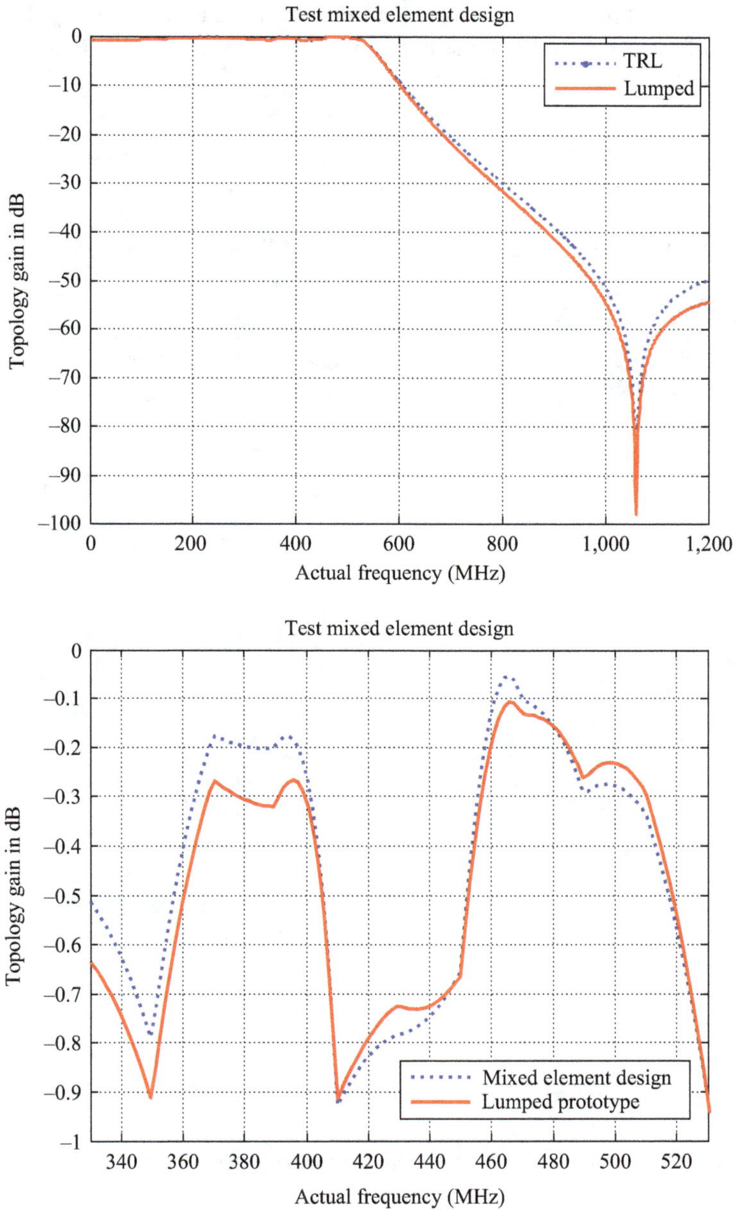

Figure 4.61 Gain performance of matched load using lumped prototype and mixed element two-ports

It should be noted that if the transmission line is realized as a microstrip rather than a parallel plate line then one must use equivalent-permittivity ε_{r-eff} instead of ε_r. In this case, the length l is given by

$$l = v_{sub}\tau \cong \frac{3\times 10^8}{\sqrt{\varepsilon_{r-eff}}} = \frac{3\times 10^8}{\sqrt{2.8247}} = 42 \text{ mm}$$

The MATLAB program is given in the following list:

Design of a Microstrip Line for a specified Characteristic Impedance $\mathbf{Z_0}$, substrate thickness $\mathbf{h}$ and relative di-electric constant (or permittivity) $\boldsymbol{\varepsilon_r}$

```
% Main program Microstrip_design.m
% Inputs:
%         Z0: Fixed Characteristic Impedance of the Microstrip Line
%         epsr: Relative permittivity of the substrate
%         h:Thickness of the substrate
% For a Given Microstrip characteristic impedance Z0, find w/h
close all
clc
clear
%
Z0=input('Characteristic Impedance of the Microstrip Line in ohm  Z0=');
epsr=input('Relative permittivity of the substrate epsr=');
hmm=input('Enter substrate thickness in millimeter h=');
h=hmm/1000;
% Initialize x=W/h
x0=1.0;
% Optimization via Minimax Algorithm
 f=@(x)Microstrip(x,epsr,Z0)
[x,fval] = fminimax(f,x0);
% Computation of a microstrip Characteristic impedance for given width and epsr
% See the book Fields and Waves in Communication Electronics by Simon Ramo,
% John R. Whinnery, Theodore Van Duzer, Third Edition, John Wiley,1994,
% pp.412
width=x*h
woverh=x;
if woverh>0.06
Z00=120*pi/(x+1.98*(x)^0.172);
epseff=1+((epsr-1)/2)*(1+1/sqrt(1+10/x))
Z_Microstrip=Z00/sqrt(epseff)
end
if woverh<0.06
    'Microstrip impedance can not to be realized'
end
```

4.6.2 Physical model of an inductor using ideal parallel plate transmission line

A single inductor can be realized using an ideal parallel plate TEM line with characteristic impedance Z_0 and physical length l printed on substrate of relative permittivity ε_r.

In this case, ideal TEM line inductance is given by [3,42] as

$$L_{TRL} = \frac{\emptyset(t,z)}{I(t,z)} = \mu\frac{h}{W}l = 4\pi\times 10^{-7}\frac{h}{W}l \tag{4.145}$$

However, as a parallel plate TEM line, this inductor is associated with a total capacitance C_{TRL} which is specified by

$$C_{TRL} = \frac{Q(t,z)}{V(t,z)} = \varepsilon \frac{W}{h} l \tag{4.146}$$

As far as the physical model is concerned, this capacitor may be evenly distributed on left and right end of the line as shown in Figure 4.59.

Ideally, the characteristic impedance Z_0 is

$$Z_0 = \sqrt{\frac{L_{TRL}}{C_{TRL}}} = \sqrt{\frac{\mu}{\varepsilon}\frac{h}{W}} = 120\pi\sqrt{\frac{\mu_r}{\varepsilon_r}\frac{h}{W}}$$

For many practical cases, relative permeability μ_r is unity. Therefore, we can confidently state that

$$Z_0 \cong \frac{120\pi}{\sqrt{\varepsilon_r}}\frac{h}{W} \tag{4.147}$$

In the above representations, h and W refer the thickness and the width of the parallel plate line as printed on a substrate.

It is important to note that for a specified actual inductance value L_A, once the substrate is selected among the manufacturers specified thickness h and permeability μ, width W is determined by means of (4.145) such that

$$W = \mu \frac{h}{L_A} l = 4\pi \times 10^{-7} \frac{h}{L_A} l \tag{4.148}$$

where the physical length of the line is specified by

$$l = v_{sub}\tau = \frac{3 \times 10^8}{\sqrt{\varepsilon_r}} \frac{1}{4mf_0} \text{(in meter)} \tag{4.149}$$

Hence, the actual line width is given by

$$W = \mu \frac{h}{L_A} l = \frac{30\pi h}{\sqrt{\varepsilon_r} m f_0 L_A} \text{(in meter)} \tag{4.150}$$

In terms of the actual inductor L_A, the characteristic impedance Z_0 is given by

$$Z_0 = 4mf_0L_A \tag{4.151}$$

It must be kept in mind that, an inductor as a parallel plate ideal transmission line must be loaded with symmetrical capacitors C_p as shown in Figure 4.60. In this case, C_p is determined by means of (4.152a) and (4.152b) such that

$$C_p = \frac{C_{TRL}}{2} = \varepsilon \frac{W}{2h} l = \mu\varepsilon \frac{l^2}{2L_A} = \frac{1}{v_{sub}^2} \frac{v_{sub}^2 \tau^2}{2L_A} = \frac{(\omega_{s2}\tau)^2}{2\omega_{s2}^2 L_A} = \frac{(\pi/2)^2}{2(2\pi m f_0)^2 L_A} \tag{4.152a}$$

$$C_P = \frac{1}{32(mf_0)^2 L_A} \tag{4.152b}$$

Hence, we can confidently state that an actual inductor L_A, which is associated with symmetrically loaded capacitors C_p specified by (4.152a) and (4.152b), can be replaced by an ideal transmission line as shown in Figure 4.62.

In this case, idealized characteristic impedance is computed using (4.151)

$$Z_0 = 4mf_0L_A$$

and the idealized delay length τ is generated employing (4.136) such that

$$\tau = \frac{1}{4mf_0}$$

Hence, as in the above section, any $CLC - PI$ section can be approximated by means of capacitively loaded transmission line $CA - TRL - CB$ using the formulas given in Table 4.31.

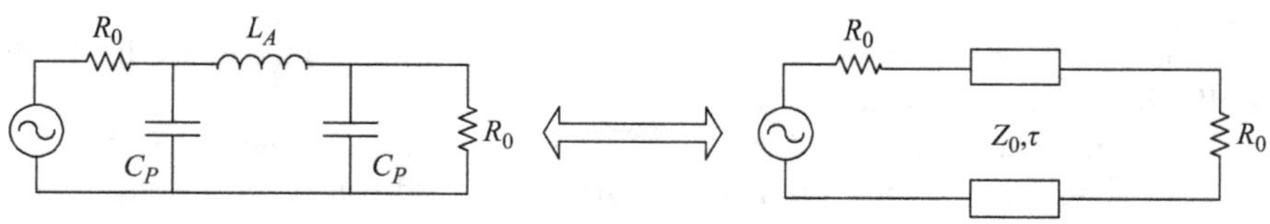

Figure 4.62 Approximation of an ideal inductor with a transmission line

Table 4.31 Summary of the design equations to replace CLC – PI section with an ideal physical model of CA – TRL – CB section

Given parameters for lumped *PI* section	**Computed parameters for *CA* – *TRL* – *CB* section**
C1 – L2 –C3 PI section 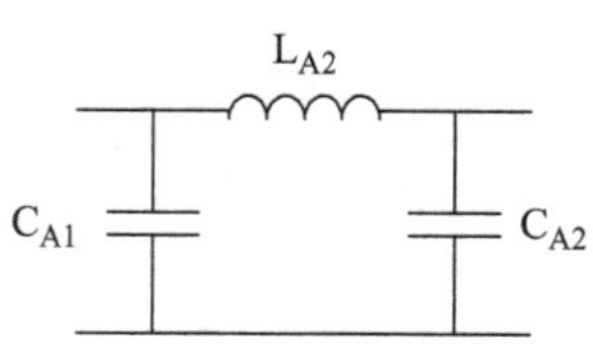	*CA – TRL – CB* SECTION GENERATED BY MEANS OF IDEAL VALUES OF LUMPED COMPONENTS 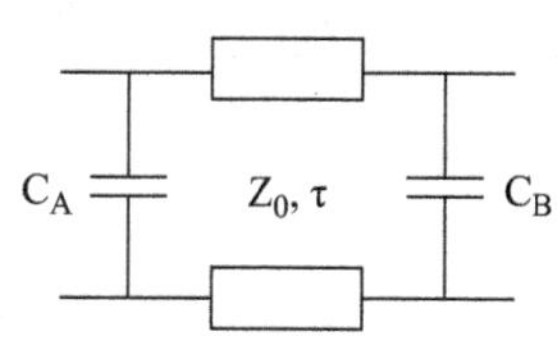
CA1: GIVEN ACTUAL CAPACITANCE LA2: GIVEN ACTUAL INDUCTOR CA3: GIVEN ACTUAL CAPACITOR	$C_p = 1/32m^2f_0^2L_{A2}$ $C_A = C_{A1} - C_p > 0$ $C_B = C_{A2} - C_p > 0$ $Z_0 = 4mf_0L_{A2}$ $\tau = 1/4mf_0$ $l = v_{sub}\tau = \left((3 \times 10^8)/\sqrt{\varepsilon_r}\right)(1/4mf_0)$

Example 4.20

It is desired to realize an actual inductor $L_A = 1.2694\text{e} - 008 = 12.69$ nH as a microstrip transmission line with thickness $h = 0.72$ mm, permittivity $\varepsilon_r = 3.66$, stop band multiplier $m = 4$; cutoff frequency $f_0 = 530$ MHz.

Compute the following quantities:

1. The physical length l of the microstrip line
2. The characteristic impedance of the microstrip line Z_0
3. The width W
4. Accompanied capacitor C_p
5. Line loading capacitor C_A
6. Line loading capacitor C_B.

Referring to Figure 4.63, develop a MATLAB program to compute the driving-point input impedances of lumped prototype $C_p - L_{A2} - C_p$ section and the physical transmission line model when they are terminated in 50 Ω.

Solution

The physical length of the transmission line is given by (4.149). Hence,

$$l = v_{sub}\tau = \frac{3 \times 10^8}{\sqrt{\varepsilon_r}} \frac{1}{4mf_0} = 0.0185 \text{ m} = 1.85 \text{ cm}$$

The characteristic impedance Z_0 is given by (4.151) as

$$Z_0 = 4mf_0 L_A$$

Then, it is found as

$$Z_0 = 107.6451 \ \Omega$$

The line width is specified by (4.148) or (4.150). Then,

$$W = \mu \frac{h}{L_A} l = 4\pi \times 10^{-7} \frac{h}{L_A} l = \frac{30\pi h}{\sqrt{\varepsilon_r} m f_0 L_A} = 0.0037 = 3.7 \text{ mm}$$

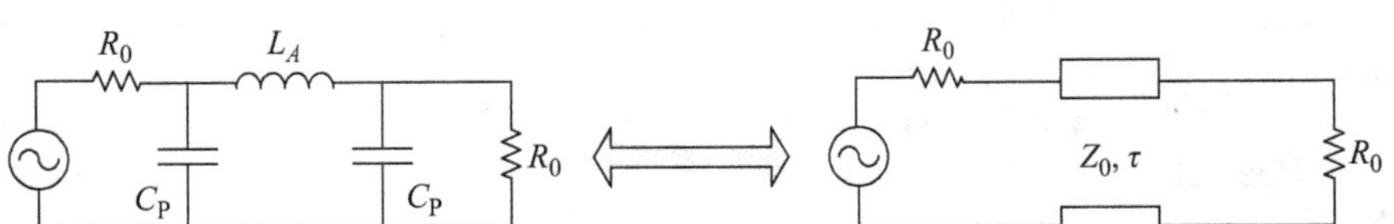

Figure 4.63 Comparison of input impedances

The capacitor is given by (4.152a) and (4.152b). Hence,

$$C_P = \frac{1}{32(mf_0)^2 L_A} = 5.4775\text{e} - 013 = 0.548 \text{ pF}$$

$$C_A = C_{A1} - C_p = 2.3368\text{e} - 011 = 23.368 \text{ pF}$$

$$C_B = C_{A3} - C_p = 1.4183\text{e} - 011 = 14.183 \text{ pF}$$

Comparison of input impedances

Input impedance of the lumped element PI section $C_P - L_{A2} - C_P$ terminated in 50 Ω is given by

$$Z_{inLump}(p) = \frac{1}{Y_{inLump}(p)}$$

On the *jω axis*

$$Z_{inLump}(j\omega) = R_{inLump}(\omega) + jX_{inLump}(\omega)$$

where

$$Y_{inLump} = pC_p + \frac{1}{pL_A + (1/(pC_p + (1/50)))}$$

Input impedance of the transmission line terminated in 50 Ω is given by

$$Z_{inTRL}(j\omega) = R_{inTRL}(\omega) + jX_{inTRL}(\omega) = Z_0 \frac{50 + jZ_0 \tan(\omega\tau)}{Z_0 + j50\tan(\omega\tau)}$$

Above equations are programmed on MATLAB. Comparative real and imaginary parts are depicted in Figures 4.64 and 4.65, respectively.

Close examination of the above figures reveals maximum relative error between the curves.

Hence,

$$Error_{Rin} = \frac{\max(R_{inLump} - R_{inTRL})}{\max(R_{inLump})} = 0.028 < 3\%$$

$$Error_{Xin} = \frac{\max(X_{inLump} - X_{inTRL})}{\max(X_{inLump})} = 0.0362 < 4\%$$

As it is observed, the fit between real and imaginary parts can be acceptable for many practical implementations.

On the other hand, if we plot the above figures over a wide frequency band, we should be able to see the periodic behavior of the transmission line impedance as shown in Figures 4.66 and 4.67.

All the above computations are performed using the MATLAB program given as an attachment to this book.

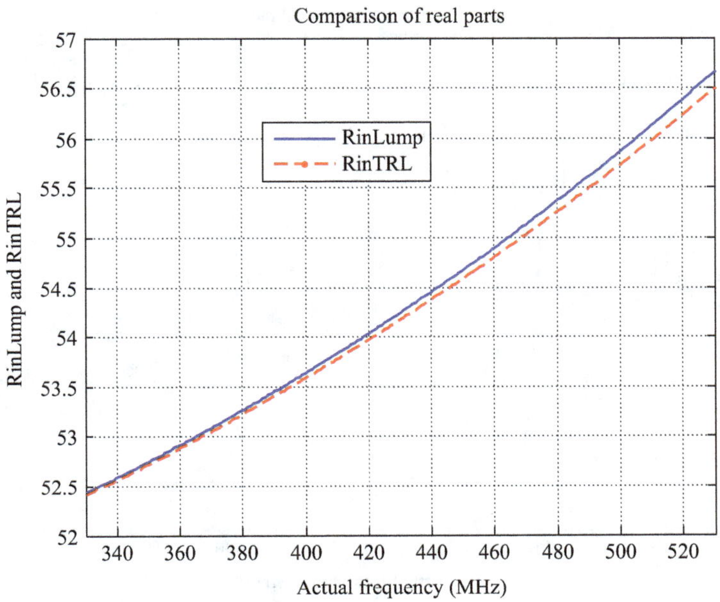

Figure 4.64 Comparison of real parts of the input impedances

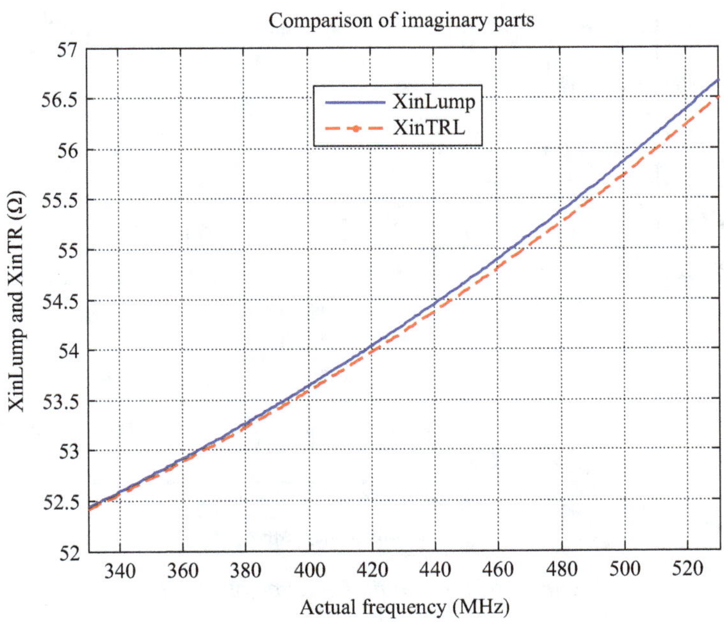

Figure 4.65 Comparison of imaginary parts of the input impedances

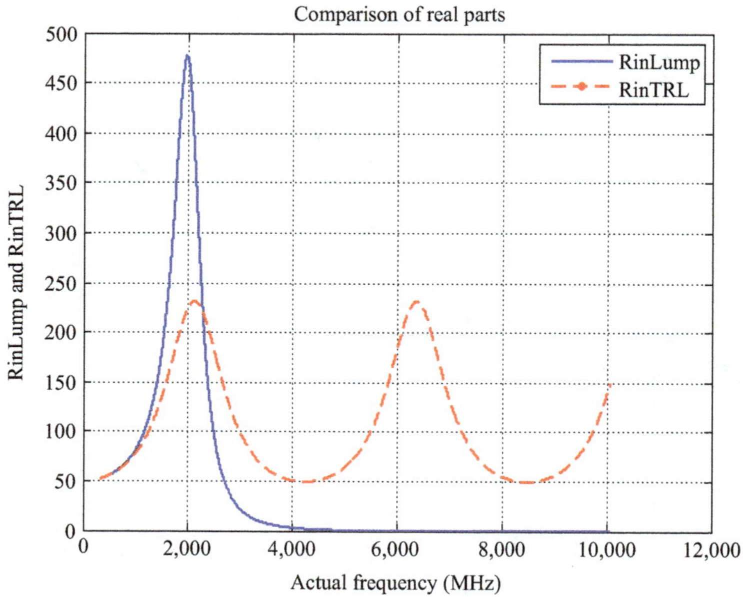

Figure 4.66 Periodic behavior of RinTRL

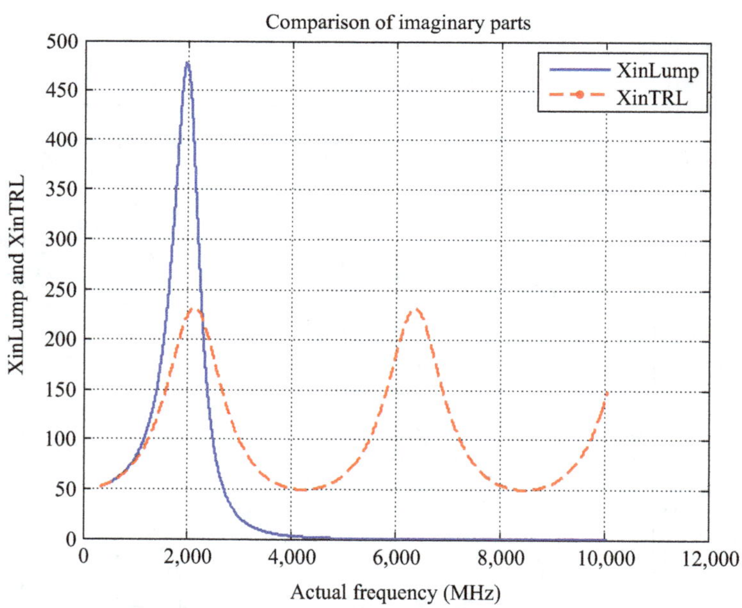

Figure 4.67 Periodic behavior of XinTRL

Example 4.21

Using the single matching lumped element prototype circuit of Example 4.1, generate the mixed element design using physical model method introduced in this section.

The lumped element CLC PI section consists of $C_{A1} = 2.3916\mathrm{e}-011 \cong 23.9$ pF, $C_{A3} = 1.4731\mathrm{e}-011 \cong 14.7$ pF, $L_{A2} = 1.2694\mathrm{e}-008 \cong 12.7$ nH. The parallel resonance circuit in series configuration has $C_{Ak} = 2.41$ pF and $L_{Ak} = 9.35$ nH.

Employing $m = 2, f0 = 530$ MHz, $h = 0.72$ mm, $epsr = 3.66$.

Solution

Step 1: Computation of line parameters

First of all, inductor L_{A2} must be associated with its counterpart capacitor C_p which is due to physical width of the microstrip line. Then, the characteristic impedance, length and width of the microstrip line is determined. For this purpose, we developed a MATLAB program called TRL_Model which yields

$$C_p = 2.1910\mathrm{e}-012; \quad \mathrm{width} = 0.0073; \quad \mathrm{length} = 0.0370;$$

$$Z0 = 53.8226; \quad \mathrm{tau} = 2.3585\mathrm{e}-010$$

As the result of approximation, the relative errors between real and imaginary parts are given as

$$Rin\_error = 0.0724; \quad Xin\_error = 0.5578;$$

Step 2: Computation of loading capacitors C_A and C_B

In the second step, one must examine the situation if the inductor-related capacitor C_p is less than C_1 and C_2. If the answer is no, the solution does not exist. If the answer is yes,

$$C_A = C_1 - C_p$$

and

$$C_B = C_3 - C_p$$

For our case, indeed, $C_p < (C_1 \text{ and } C_3)$.

Hence,

$$CA = 2.1725\mathrm{e}-011; \quad CB = 1.2540\mathrm{e}-011$$

Step 3: Computation of the gain from the physical model

At this step, gain of the single matching problem is computed using the physical model circuit elements. For comparison purposes performance of the lumped prototype and the mixed element model is depicted in Figure 4.68.

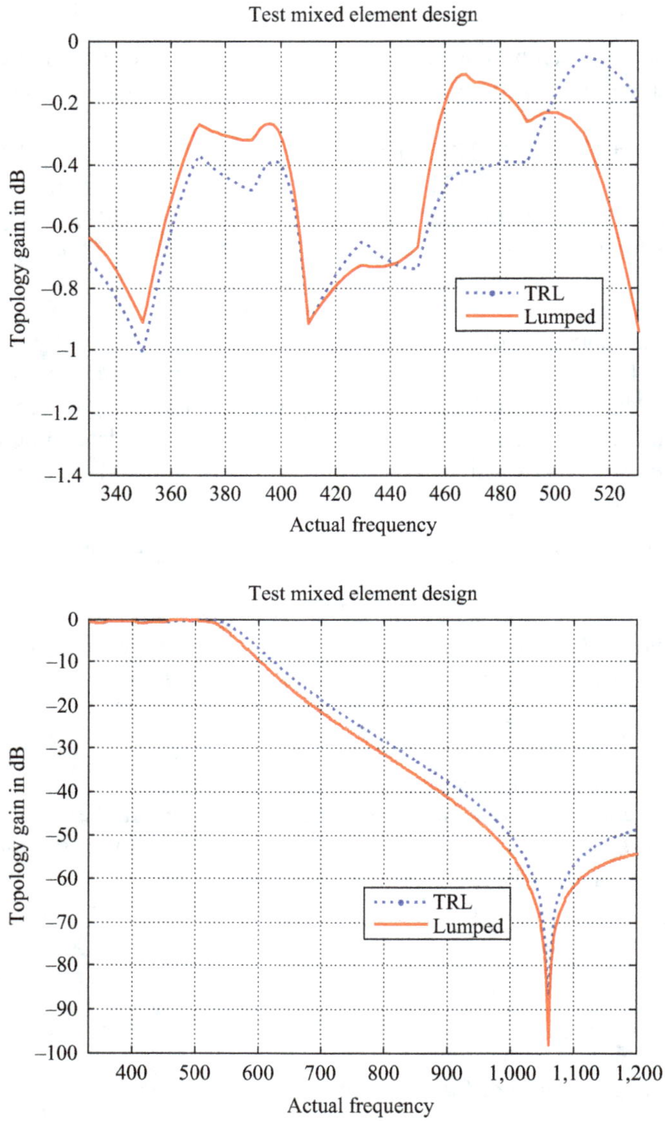

Figure 4.68 Comparison of the gain performances obtained from the circuit topologies using lumped element prototype and physical model

It is clearly seen that physical model follows the gain performance of the lumped prototype at the lower end of the passband. However, it deviated from the original as the frequency increases.

MATLAB programs written for the gain generation is given as an attachment to this book.

Appendix Computation of the element values of CT-TRL-CT from the given lumped element C-L-C PI section

Chain parameters of a unit element (UE)

Referring to Figure 4.69, output voltage and current pair of a unit element (UE) can be expressed as

$$\begin{aligned} V_2 &= A + B \\ Z_0 I_2 &= A - B \end{aligned} \tag{4.153}$$

where A and B are the "voltage-based" incident and the reflected waves of the output port of the unit element (see [3]; Ch. 3, Eq. (3.17) pp.127).

Similarly, the input voltage and current pair are given as follows:

$$\begin{aligned} V_1 &= Ae^{+j(\omega\tau)} + Be^{-j(\omega\tau)} \\ Z_0 I_1 &= Ae^{+j(\omega\tau)} - Be^{-j(\omega\tau)} \end{aligned} \tag{4.154}$$

Delay length τ is expressed in terms of the physical length l and the propagation velocity v_p as

$$\tau = \frac{l}{v_p} \tag{4.155}$$

At a specified frequency f_s physical length l is given by

$$l = \frac{v_p}{f_s} \tag{4.156}$$

If f_0 is the cutoff frequency of the passband, usually f_s is selected as

$$f_s = mf_0; \quad m > 1 \tag{4.157}$$

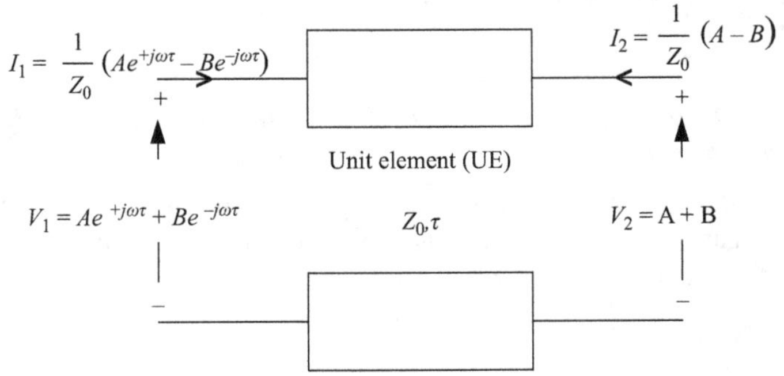

Figure 4.69 An ideal transmission line with characteristic impedance Z_0 and delay length τ as a unit element

It should be noted that when the output port is open, $I_2 = 0$. In this case, $B = A$.

On the other hand, when output port is shorted, $V_2 = 0$. In this case, $B = -A$.

By means of actual port voltages and currents, the chain parameters $\boldsymbol{A}$, $\boldsymbol{B}$, $\boldsymbol{C}$ and $\boldsymbol{D}$ are defined as follows:

$$\begin{bmatrix} V_1 \\ I_1 \end{bmatrix} = \begin{bmatrix} \boldsymbol{A} & \boldsymbol{B} \\ \boldsymbol{C} & \boldsymbol{D} \end{bmatrix} \begin{bmatrix} V_2 \\ -I_2 \end{bmatrix} = T_{UE} \begin{bmatrix} V_2 \\ -I_2 \end{bmatrix} \tag{4.158}$$

where $T_{UE} = \begin{bmatrix} \boldsymbol{A} & \boldsymbol{B} \\ \boldsymbol{C} & \boldsymbol{D} \end{bmatrix}$ is called chain or transmission matrix of the unit element UE and its entrees $\boldsymbol{A}$, $\boldsymbol{B}$, $\boldsymbol{C}$ and $\boldsymbol{D}$ are given by

$$\begin{aligned}
\boldsymbol{A} &= \frac{V_1}{V_2}\Big|_{I_2=0} = \frac{A\left(e^{+j(\omega\tau)} + e^{-j(\omega\tau)}\right)}{2A} = \cosh(j\omega\tau) = \cos(\omega\tau) \\
\boldsymbol{B} &= \frac{V_1}{-I_2}\Big|_{V_2=0} = Z_0\frac{A\left(e^{+j(\omega\tau)} - e^{-j(\omega\tau)}\right)}{2A} = Z_0\sinh(j\omega\tau) = jZ_0\sin(\omega\tau) \\
\boldsymbol{C} &= \frac{I_1}{V_2}\Big|_{I_2=0} = \frac{A\left(e^{+j(\omega\tau)} - e^{-j(\omega\tau)}\right)}{2Z_0A} = Y_0\sinh(j\omega\tau) = jY_0\sin(\omega\tau) \\
\boldsymbol{D} &= \frac{I_1}{-I_2}\Big|_{V_2=0} = \frac{Z_0A\left(e^{+j(\omega\tau)} + e^{-j(\omega\tau)}\right)}{2Z_0A} = \cosh(j\omega\tau) = \cos(\omega\tau)
\end{aligned} \tag{4.159}$$

It should be noted that

$$\begin{aligned}
\cosh(j\omega\tau) &= \frac{e^{+j(\omega\tau)} + e^{-j(\omega\tau)}}{2} = \cos(\omega\tau) \\
\sinh(j\omega\tau) &= \frac{e^{+j(\omega\tau)} + e^{-j(\omega\tau)}}{2} = j\sin(\omega\tau)
\end{aligned}$$

At this point, we note that "voltage-based" incident and reflected wave notations A and B should not be confused with the classical notation of the chain parameters $\boldsymbol{A}$, $\boldsymbol{B}$, $\boldsymbol{C}$ and $\boldsymbol{D}$.

Derivation of chain parameters for CLC lumped and CT-UE-CT distributed circuit

Referring to Figure 4.70, let T_A designate the chain matrix of CLC-π circuit. Let T_C and T_L be the chain matrices of individual circuit elements, namely, shunt capacitor C and series inductor L, respectively. Then,

$$T_A = [T_C][T_L][T_C] \tag{4.160}$$

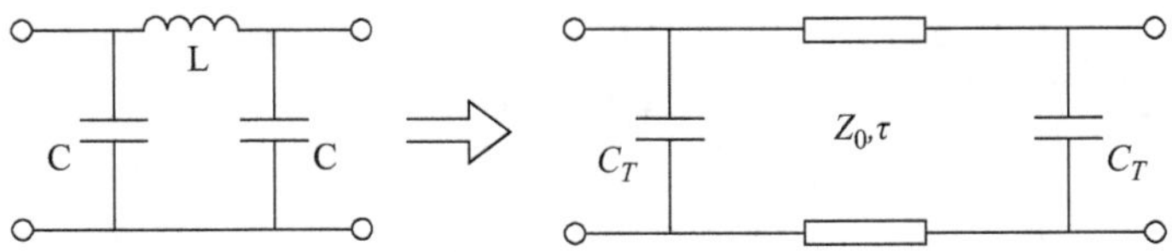

Figure 4.70 Almost equivalent circuit of a symmetrical lowpass C-L-C π-section by means of a CT-UE-CT section

where

$$T_C = \begin{bmatrix} 1 & 0 \\ pC & 1 \end{bmatrix}$$

$$T_L = \begin{bmatrix} 1 & pL \\ 0 & 1 \end{bmatrix} \tag{4.161}$$

Therefore,

$$T_A = [T_C][T_L][T_C] = \begin{bmatrix} 1 + LCp^2 & pL \\ LC^2p^3 + 2Cp & 1 + LCp^2 \end{bmatrix} \tag{4.162}$$

on $j\omega$ *axis*

$$T_A = \begin{bmatrix} 1 - LC\omega^2 & j\omega L \\ j\omega C(2 - LC\omega^2) & 1 - LC\omega^2 \end{bmatrix} \tag{4.163}$$

Similarly, let T_B be the chain matrix of CT-UE-CT section,

$$T_B = [T_{CT}][T_{UE}][T_{CT}] \tag{4.164}$$

where

$$T_{CT} = \begin{bmatrix} 1 & 0 \\ j\omega C_T & 1 \end{bmatrix} \tag{4.165a}$$

$$T_{UE} = \begin{bmatrix} \cos(\omega\tau) & jZ_0\sin(\omega\tau) \\ jY_0\sin(\omega\tau) & \cos(\omega\tau) \end{bmatrix} \tag{4.165b}$$

Then,

$$T_B = [T_{CT}][T_{UE}][T_{CT}] \tag{4.166a}$$

$$= \begin{bmatrix} \cos(\omega\tau) - \omega Z_0 C_T \sin(\omega\tau) & jZ_0\sin(\omega\tau) \\ \frac{j}{Z_0}(2Z_0C_T\cos(\omega\tau)) + \left(1 - \omega^2 Z_0^2 C_T^2\right)\sin(\omega\tau) & \cos(\omega\tau) - \omega Z_0 C_T \sin(\omega\tau) \end{bmatrix} \tag{4.166b}$$

Obviously, at $\omega = 0$, chain matrices $T_A(0)$ and $T_B(0)$ are equal to each other and they are given by

$$T_A(0) = T_B(0) = \begin{bmatrix} 1 & 0 \\ 0 & 1 \end{bmatrix} \tag{4.167}$$

On the other hand, we can equate chain matrices T_A and T_B at a given frequency ω_0 such that

$$T_{A11}(j\omega_0) = T_{B11}(j\omega_0) yields 1 - \omega_0^2 LC = \cos\omega\tau - \omega Z_0 C_T \sin(\omega\tau) \tag{4.168a}$$

and

$$T_{A12}(j\omega_0) = T_{B12}(j\omega_0) yields j\omega L = jZ_0 \sin(\omega\tau) \tag{4.168b}$$

Solving above equations for Z_0 and C_T, we have

$$Z_0 = \frac{\omega_0 L}{\sin(\omega_0\tau)} \tag{4.169a}$$

$$C_T = \frac{\cos(\omega\tau) + \omega_0^2 LC - 1}{\omega_0^2 L} \tag{4.169b}$$

Using (4.168a) and (4.168b), it is seen that at $\omega_0 = 1, T_{A21}(j\omega_0) = T_{B21}(j\omega_0)$ and $T_{A22}(j\omega_0) = T_{B22}(j\omega_0)$.

Moreover, it is expected that between $\omega = 0$ and $\omega = 1$, $C - L - C$ *and* $CT - UE - CT$ sections exhibit similar electrical performances due to their smooth lowpass behavior over the entire frequency axis.

Do CLC lumped and CT-UE-CT sections have equal chain matrices at ω_0?

Let us verify if the elements T_A and T_B are equal to each other.

We have already derived Z_0 and C_T to make $T_B(2,2) = T_A(2,2)$ as analyzed below,

$$\begin{aligned} T_B(2,2) &= \cos\ \omega\tau - \omega C_T Z_0 \sin\ \omega\tau \Big|_{\substack{C_T = (\cos\ \omega\tau + \omega^2 LC - 1)/\omega^2 L \\ Z_0 = \omega L/\sin\ \omega\tau}} \\ &= \cos\ \omega\tau - \omega \frac{\cos\ \omega\tau + \omega^2 LC - 1}{\omega^2 L} \frac{\omega L}{\sin\ \omega\tau} \sin\ \omega\tau \\ &= 1 - \omega^2 LC \equiv T_A(2,2) \end{aligned}$$

and substituting Z_0 and C_T in $T_B(2,1)$ we show that $T_B(2,1) = T_A(2,1)$ as seen in the following derivations.

$$T_B(2,1) = \frac{j}{Z_0}\left\{2\omega C_T Z_0 \cos\omega\tau + \sin\omega\tau\left(1 - {Z_0}^2\omega^2{C_T}^2\right)\right\}$$

$$= \left\{ j2\omega C_T \cos\omega\tau + \frac{j}{Z_0}\sin\omega\tau - jZ_0\omega^2{C_T}^2\sin\omega\tau \right\}_{\substack{C_T=\left(\cos\omega\tau+\omega^2LC-1\right)/\omega^2L \\ Z_0=\omega L/\sin\omega\tau}}$$

$$= j\left\{ 2\omega\frac{\cos\omega\tau+\omega^2LC-1}{\omega^2L}\cos\omega\tau + \frac{\sin^2\omega\tau}{\omega L} - \frac{\omega^3L}{\sin\omega\tau}\left(\frac{\cos\omega\tau+\omega^2LC-1}{\omega^2L}\right)^2\sin\omega\tau \right\}$$

$$= j\left\{ 2\frac{\cos\omega\tau+\omega^2LC-1}{\omega L}\cos\omega\tau + \frac{\sin^2\omega\tau}{\omega L} - \frac{\omega^3L}{\omega^4L^2}\left(\cos\omega\tau+\omega^2LC-1\right)^2 \right\}$$

$$= j\left\{ 2\frac{\cos\omega\tau+\omega^2LC-1}{\omega L}\cos\omega\tau + \frac{\sin^2\omega\tau}{\omega L} - \frac{1}{\omega L}\left(\cos\omega\tau+\omega^2LC-1\right)^2 \right\}$$

$$= j\left\{ \frac{\left(\cos\omega\tau+\omega^2LC-1\right)}{\omega L}\left[2\cos\omega\tau - \left(\cos\omega\tau+\omega^2LC-1\right)\right] + \frac{\sin^2\omega\tau}{\omega L} \right\}$$

$$= j\left\{ \frac{\left(\cos\omega\tau+\omega^2LC-1\right)\left(\cos\omega\tau-\left(\omega^2LC-1\right)\right)}{\omega L} + \frac{\sin^2\omega\tau}{\omega L} \right\}$$

$$= j\left\{ \frac{\cos^2\omega\tau - \left(\omega^2LC-1\right)^2 + \sin^2\omega\tau}{\omega L} \right\}$$

$$= j\left\{ \frac{1-\left(\omega^2LC-1\right)^2}{\omega L} \right\}$$

$$= j\left\{ \frac{\left(1-\left(\omega^2LC-1\right)\right)\left(1+\left(\omega^2LC-1\right)\right)}{\omega L} \right\}$$

$$= j\left\{ \frac{\left(2-\omega^2LC\right)\omega^2LC}{\omega L} \right\}$$

$$= j\left\{ \left(2-\omega^2LC\right)\omega C \right\}$$

$$= j\omega C\left(2-\omega^2LC\right) \equiv T_A(2,1)$$

References

[1] H. Carlin, "A New Approach to Gain–Bandwidth Problems," *IEEE Trans CAS*, vol. 23, no. 4, pp. 170–175, 1977.

[2] B. Yarman, Broadband matching from a complex source to a complex load, Ph.D. Thesis, Ithaca, NY: Cornell University, 1982.

[3] B. Yarman, Design of Ultra Wide-Band Matching Networks via Real Frequency Techniques, London: Wiley, 2010.

[4] B. Yarman, Design of Ultra Wideband Matching Networks via Simplified Real Frequency Technique, Netherlands: Springer, 2008.

[5] S. Darlington, "Synthesis of Reactance 4-Poles Which Produce Prescribed Insertion Loss Characteristics," *J. Math Phys*, vol. 18, no. 9, pp. 257–353, 1939.

[6] H. Carlin, "Darlington Synthesis Revisited," *IEEE Trans Circuits Syst I, Fundam Theory Appl*, vol. 46, no. 1, pp. 14–21, 1999.

[7] B. Gordon, G. Scarth and G. Martens, "Synthesis of Complex Lossless Two-Ports," *Int J Circuit Theory Appl*, vol. 22, pp. 121–143, 1994.

[8] A. Kilinc and B. Yarman, "High Precision LC Ladder Synthesis Par I: Lowpass Ladder Synthesis via Parametric Approach," *IEEE Trans CAS I: Regul Pap*, vol. 60, no. 8, pp. 2074–2083, 2013.

[9] A. Fettweis, "Parametric Representation of Brune Functions," *Int J Circuit Theory Appl*, vol. 7, pp. 113–119, 1979.

[10] B. S. Yarman and A. Fettweis, "Computer Aided Double Matching via Parametric Representation of Brune Functions," *IEEE Trans CAS*, vol. 37, no. 2, pp. 212–222, 1990.

[11] J. Pandel and A. Fettweis, "Broadband Matching Using Parametric Representation," in *IEEE International Symposium on Circuit and Systems*, 1985.

[12] MatLab, MatLab S/W Package, Natick, MA, USA: The MathWorks Inc., 2013b.

[13] H. Carlin and B. Yarman, "The Double Matching Problem: Analytic and Real Frequency Solutions," *IEEE Trans CAS*, vol. 30, no. 1, pp. 15–28, 1983.

[14] B. S. Yarman and A. Kilinc, "High Precision LC Ladder Synthesis Pat II: Immittance Synthesis with Transmission Zeros at DC and Infinity," *IEEE Trans CAS I*, vol. 60, no. 10, pp. 2719–2729, 2013.

[15] B. S. Yarman, A. Aksen, R. Kopru, *et al.*, "Computer Aided Darlington Synthesis of an All Purpose Immittance Function," *Istanbul Univ J Electr Electron Eng (IU-JEE)*, vol. 16, no. 1, pp. 2027–2037, 2016.

[16] N. Ballabanian, Network Synthesis, Englewood Cliffs, NJ, USA: Prentice Hall Inc., 1958.

[17] B. S. Yarman, A. Aksen, R. Kopru, C. Aydın, and C. Atilla, "Computer Aided High Precision Darlington Synthesis for Real Frequency Matching," in *IEEE Benjamin Franklin Symposium on Microwave and Antenna Sub-Systems*, Philadelphia, 2014.

[18] D. C. Youla, "A New Theory of Cascade Synthesis," *IRE Trans Circuit Theory*, vol. 8, no. 3, pp. 244–260, 1961.

[19] R. Cameron, C. Kudia and R. R. Mansour, Microwave Filters for Communication Systems: Fundamentals, Design, and Applications, Hoboken, NJ, USA: Wiley Interscience, 2007.

[20] H. J. Carlin and P. Civaleri, Design of Broadband Networks, Milton Park: Taylor and Francis, CRC Press Inc., 1998.

[21] V. Belevitch, Classical Network Theory, San Francisco, CA: Halden Day, 1968.

[22] P. I. Richards, "Resistor-Transmission Line Circuits," *Proc IRE*, vol. 36, no. 2, pp. 217–220, 1948.

[23] H. J. Carlin, "Distributed Circuit Design with Transmission Lines," *Proc IEEE*, vol. 59, no. 7, pp. 1059–1081, 1971.

[24] B. S. Yarman, R. Kopru, N. Kumar and C. Prakash, "High Precision Synthesis of a Richards Immittance via Parametric Approach," *IEEE Trans CAS I: Regul Pap*, vol. 61, no. 4, pp. 1055–1067, 2014.

[25] I. Bahl, Lumped Elements for RF Microwave Circuits, London: Artech House, 2003.

[26] I. Bahl and P. Bhartia, Microwave Solid State Circuit Design, Hoboken, NJ, USA: Wiley Interscience, 2003.

[27] H. A. Wheeler, "Transmission Line Properties of Parallel Wide-Strips by Conformal-Mapping Approximation," *IEEE MTT*, vol. 12, no. 5, pp. 280–289, 1964.

[28] H. A. Wheeler, "Transmission Line Properties of Parallel Strips Separated by a Dielectric Sheet," *IEEE MTT*, vol. 13, no. 3, pp. 172–185, 1965.

[29] H. A. Wheeler, "Transmission Line Properties of a Strip on a Dielectric Sheet on a Plane," *IEEE MTT*, vol. 25, no. 8, pp. 631–647, 1977.

[30] H. J. Carlin and P. Amstutz, "On Optimum Broadband Matching," *IEEE Trans CAS*, vol. 28, no. 5, pp. 401–405, 1981.

[31] B. S. Yarman, "Modern Approaches to Broadband Matching Problems," *Proc IEE Antenna Propag*, vol. 132, no. 4, pp. 87–92, 1985.

[32] R. M. Foster, "A Reactance Theorem," *Bell Syst Tech J*, vol. 3, no. 2, pp. 259–267, 1924.

[33] B. S. Yarman and H. J. Carlin, "A Simplified Real Frequency Technique is Applied to Broadband Multi-Stage Microwave Amplifiers," *IEEE MTT*, vol. 30, no. 12, pp. 2216–2222, 1982.

[34] B. S. Yarman, "A Simplified Real Frequency Technique for Broadband Matching Complex Generator to Complex Load," *RCA Rev*, vol. 43, no. 9, pp. 529–541, 1982.

[35] R. Levy, "Tables of Element Values for the Distributed Lowpass Prototype," *IEEE MTT*, vol. 13, no. 9, pp. 514–536, 1965.

[36] R. Saal and E. Ulbrich, "On the Design of Filters by Synthesis," IRE Trans Circuit Theory, *vols. CT*-5, pp. 284–327, 1958.

[37] J. A. C. Bingham, "A New Method of Solving the Accuracy Problem in Filter Design," IEEE Trans Circuit Theory, *vols. CT*-11, pp. 327–341, 1964.

[38] H. J. Orchard and G. C. Temes, "Filter Design Using Transformed Variables," IEEE Trans Circuit Theory, *vols. CT*-15, pp. 385–407, 1968.

[39] K. Kuroda, “A Method to Drive Distributed Constant Filters from Lumped Constant Filters,” in *Joint Convention of Electrical Engineering Institute of Japan (IECE)*, Kansai, October 1952.

[40] B. S. Yarman and A. Aksen, “An Integrated Design Tool to Construct Lossless Matching Networks with Mixed Lumped and Distributed Elements,” *IEEE Trans CAS*, vol. 39, no. 9, pp. 713–723, 1992.

[41] B. S. Yarman, A. Aksen and A. Fettweis, “An Integrated Design Tool to Construct Lossless Two-Ports with Mixed Lumped and Distributed Elements for Matching Problems,” in *Symposium on Recent Advances in Microwave Tech.*, August 1991.

[42] S. Ramo, J. R. Whinnery and T. Van Duzer, Fields and Waves in Communication *Electronics*, New York: John Wiley and Sons Inc., 1994.

MATLAB® program lists

MATLAB PROGRAMS OF CHAPTER 4

MatLab Program Developed for Example 4.20 :

Design of an inductor employing a transmission line

```
% Main Program: INDviaTRL
% Inputs:
%       LA: actual impedance to be realized by a single
transmission line
%       f0: actual operating frequency
%       h: substrate thickness
%       epsr: Relative permittivity
%       m: pass band multiplier
% Outputs:
%       l: find physical length (meter)
%       Z0: Characteristic impedance (ohm)
%       W: Actual line width (meter)
%       Cp: Parasitic capacitance of the given actual inductor LA
%

close all
clc
clear
%
% Inputs:
LA= 1.2694e-008;f0=530e6;epsr=3.66;h=2e-3;m=4
%
% Outputs
tau=1/4/m/f0;
c=3*1e8;
mu0=4*pi*1e-7;
eps0=1e-9/36/pi;
v0=1/sqrt(mu0*eps0)
```

```
vsub=v0/sqrt(epsr);
length=vsub/4/m/f0;
width=mu0*(h/LA)*length;
W=30*pi*h/sqrt(epsr)/m/f0/LA;
Z0=4*m*f0*LA;
Chr_imp=120*pi*h/W/sqrt(epsr)
Cp=epsr*eps0*W*length/2/h
Cap_Parasitic=LA/2/Z0/Z0
Cpt=1/32/m/f0/m/f0/LA
% Generate input impedance of the transmission line as it is
terminated in
% 50 ohms:
fs1=330e6;fs2=10060e6;
N=10000;
df=(fs2-fs1)/N;DW=2*pi*df;
f=fs1;w=2*pi*f;
RN=50;
j=sqrt(-1);
for i=1:N
    FA(i)=f/1e6;
    p=j*w;
    YN=1/RN;Y1=p*Cp+YN;
    Z2=p*LA+1/Y1;Y3=p*Cp+1/Z2;
    ZinLump=1/Y3;
    RinLump(i)=real(ZinLump);
    XinLump(i)=imag(ZinLump);
    ZinTRL=Z0*(RN+Z0*j*tan(w*tau))/(Z0+RN*j*tan(w*tau));
    RinTRL(i)=real(ZinTRL);
    XinTRL(i)=imag(ZinTRL);
    w=w+DW;
    f=f+df;
end
```

```
figure (1)
plot(FA,RinLump,FA,RinTRL)
title('Comparison of real parts');xlabel('Actual Frequency (MHz)')
ylabel('RinLump & RinTRL');legend('RinLump','RinTRL')
%
figure (2)
plot(FA,RinLump,FA,RinTRL)
title('Comparison of imaginary parts');xlabel('Actual Frequency
(MHz)')
ylabel('XinLump & XinTRL');legend('XinLump','XinTRL'
```

Matlab Program developed for Example 4.21

```
% Main Program Mixed_Design.m
% Inputs:
% Element Values:
close all
clc
clear
%
 RGEN =[
   0       50      0
   330     50      0
   350     50      0
   370     50      0
   390     50      0
   410     50      0
   430     50      0
   450     50      0
   470     50      0
   490     50      0
   510     50      0
```

```
   530      50      0
   2000     50      0];
%
Load =[
000      19.00   -0.000
330      17.61   -04.63
350      14.50   -03.67
370      18.70   -11.30
390      22.00   -10.11
410      09.16   -13.96
430      10.23   -17.94
450      10.40   -15.89
470      17.18   -04.85
490      14.33   -05.11
510      13.36   -12.11
530      11.04   -14.03
2000     7.00    -17.00];
%
A1=RGEN; % Data Matrix For Gen_Data1
A2=Load; % Data Matrix Load_Dat1
%
NA1=length(A1);
NA2=length(A2);
%


% Resistive Generator Data for Single Matching Problem
for i=1:NA1
    FAG(i)=A1(i,1)*1e6;WAG(i)=2*pi*FAG(i);
    RAG1(i)=A1(i,2);
    XAG1(i)=A1(i,3);
end
```

```
% Actual Measurement for the Input Load Data 3; Single Matching
for i=1:NA2
    FAL(i)=A2(i,1)*1e6;WAL(i)=2*pi*FAL(i);
    RAL1(i)=A2(i,2);
    XAL1(i)=A2(i,3);
end
%
figure
plot(FAL,RAL1)
figure
plot(FAL,XAL1)
%
m=2;
f0=530e6;fs1=06;fs2=3*f0;epsr=3.66
ws1=2*pi*fs1;ws2=2*pi*fs2;KFlag=1;
%-----------------------------------------------------------
CAk=2.4115e-012;LAk=9.3483e-009;%Lumped Element Prototype Resonance
Circuit
CA1=2.3916e-011;CA3= 1.4731e-011;LA2=1.2694e-008;% Lumped element CLC
CA=2.1725e-011;CB=1.2540e-011;Z0=53.8226;tau= 2.3585e-010%Physical TRL
Model


%CA=2.1839e-011;CB=1.2649e-011;Z0=59.763;tau= 2.3585e-010;%Parameters
of
%almost equivalent circuit
%-----------------------------------------------------------
%
%Z_fixed=0.0;
%[Z0,tau,length,C,Ca,CT,CA,CB]=CLCPItoTRL(f0,m,CA1,LA2,CA3,epsr,Z_fixe
d);
Nprint=1001;
cmplx=sqrt(-1);
DW=(ws2-ws1)/(Nprint-1);
```

```
w=ws1;
for j=1:Nprint
            WA(j)=w;
            p=cmplx*w;
            [RG,XG]=Line_Impedance(w,WAG, RAG1, XAG1,KFlag);
            [RL,XL]=Line_Impedance(w,WAL, RAL1, XAL1,KFlag);
            ZL=complex(RL,XL);
            ZG=complex(RG,XG);
% Impedance of the resonance circuit:
            YRes=p*CAk+1/p/LAk;
            ZRes=1/YRes;
% Transmission Line input impedance
        %  Load impedance of the Line: CB//ZL
            ZLoad=ZRes+ZL;
            YLA=p*CA+1/ZLoad;% Line load: CB//ZL
            ZLA=1/YLA;% TRL Load
            Num=ZLA+cmplx*Z0*tan(w*tau);
            Den=Z0+cmplx*ZLA*tan(w*tau);
            ZTRL=Z0*Num/Den;% Input impedance of the loaded line in ZLB
            Yin=1/ZTRL+p*CB;
            Zin=1/Yin;
            SinTrl=(Zin-conj(ZG))/(Zin+ZG);
            GainTrl=1-abs(SinTrl)*abs(SinTrl);
             T_Double=GainTrl;
             TATrl(j)=10*log10(T_Double);
             % Computation of gain with lumped elements
             YC1=p*CA1+1/(ZLoad);
             ZL2=p*LA2+1/YC1;
             YC3=p*CA3+1/ZL2;
             ZinLmp=1/YC3;
             SinLmp=(ZinLmp-conj(ZG))/(ZinLmp+ZG);
```

```
            GainLmp=1-abs(SinLmp)*abs(SinLmp);
            TALmp(j)=10*log10(GainLmp);
    w=w+DW;
end
 %
  FA=WA/2/pi/1e6;
 %
        figure
        plot(FA,TATrl,FA,TALmp)
        title('Test Mixed Element Design')
        xlabel('Actual Frequency')
        ylabel('Topology Gain in dB')
        legend('TRL','Lumped')
```

MatLab Function to generate CT-TRL-CT Model from the Lumped Element C-L-C Pi Section

```
function [ Z0,tau,length,C,Ca,CT,CA,CB] =
CLCPItoTRL(f0,m,CA1,LA2,CA3,epsilonr,Z_fixed)
%Generate almost equivalent CT-TRL-CT from the given C1-L2-C3 PI
section
% ----------------------------------------------------------------------
-----
%   Inputs: Element values of the lumped PI Section
%           f0: Actual normalization frequency
%           m: Integer which specifies the stop-band frequency as
fs2=m*f0
%              Note m cannot be less 2. Perhaps it is chosen as 2, 3
or 4
%           RN: Normalization Resistance. It may be 50 ohm.
%           CA1: Input Capacitor
%           LA2: Mid Inductor
%           CA3: Output Capacitor
```

```
%           epsilon_r: Di-electric constant of the substrate
%           Z_fixed: fixed characteristic impedance of the line.
% Note: if Z_fixed==0, characteristic impedance Z0 is computed as a
% function of tau.
% If Z_fixed>0 then, tau is adjusted to yield the desired
characteristic
% impedance Z_fixed.
%   Output:
%           Z0: Characteristic Impedance of the transmission Line
%           tau: Delay length of the transmission line
%           length: Physical length of the transmission line
%           C=min(C1,C3)
%           Ca: Residual Capacitor for the PI section to make it
%           symmetrical
%           CT: Symmetrical loading capacitor of the transmission line
%           (CT_TRL_CT)
%           CA: Equivalent left loading capacitance of the line
%           CB: Equivalent left loading capacitance
% -----------------------------------------------------------------
-----
% Normalization:
fs2=m*f0;
if Z_fixed==0
    C=min(CA1,CA3);% find symmetrical Capacitors C
    Ca=max(CA1,CA3)-C;% Generate residual capacitor Ca
    tau=1/4/fs2;% Compute delay length tau
    Z0=2*pi*f0*LA2/sin(2*pi*f0*tau);
    CT=(cos(2*pi*f0*tau)+(2*pi*f0)*(2*pi*f0)*LA2*C-
1)/(2*pi*f0)/(2*pi*f0)/LA2;
    if CA1>CA3;CA=Ca+CT;CB=CT;end
    if CA3>CA1; CA=CT;CB=Ca+CT;end
end
```

```
if Z_fixed>0
    Z0=Z_fixed;
    C=min(CA1,CA3);% find symmetrical Capacitors C
    Ca=max(CA1,CA3)-C;% Generate residual capacitor Ca
    Q=2*pi*f0*LA2/Z0;
    teta=asin(2*pi*f0*LA2/Z0);
    tau=(1/2/pi/f0)*teta;
     CT=(cos(2*pi*f0*tau)+(2*pi*f0)*(2*pi*f0)*LA2*C-
1)/(2*pi*f0)/(2*pi*f0)/LA2;
    if CA1>CA3;CA=Ca+CT;CB=CT;end
    if CA3>CA1; CA=CT;CB=Ca+CT;end
end
v0=3*1e8; vr=v0/sqrt(epsilonr);
length=vr*tau;


end% end for the function
```

PROGRAM LIST 4.1. EXAMPLE 4.1 OF CHAPTER 4

```
% Main Program: Main_Example4_1.m
% This Program is developed by B.S. Yarman on September 20, 2018.
% Vanikoy, Istanbul, Turkey
% IET Book 2018
clear;clc;close all
% Given Z(p)=a(p)/b(p)
%Z(p)=(0p^3+0.2755p^2+0.1051p+0.1182)/(1p^3+0.3816p^2+0.7390p+0.1182)
% Given a(p) and b(p)
a=[0 0.2755 0.1051 0.1182]
b=[1 0.381 0.7390 0.1182]
% Find even part R(p)=A(p)/B(p)of Z(p)as an even-rational function:
format long e
[A,B]=even_part(a,b)
```

PROGRAM LIST 4.2. EXAMPLE 4.2 OF CHAPTER 4

```
% Main Program: Main_Example4_2.m
% This Program is developed by B.S. Yarman on September 20, 2018.
% Vanikoy, Istanbul, Turkey
% IET Book Example 4.2 of Chapter 4
clear;clc;close all
% Given R(p)=A(p)/B(p);
A1=[0 0 0 0.0140];% Exact Lowpass Ladder form of the numerator of
R(p)
B1=[-1 -1.3328 -0.4561 0.0140]; % Denominator of R(p)
a=[0 0.2755 0.1051 0.1182]; % Given for Example 1
b=[1 0.381 0.7390 0.1182];% Given for Example 1
% Find even part R(p)=A(p)/B(p)of Z(p)as an even-rational function:
format long
[p,k]=residues_evenpart(A1,B1)
[a1,b1]=evenpartTo_MinFunc(A1,B1)
eps_a=norm(a-a1)
eps_b=norm(b-b1)
```

PROGRAM LIST 4.3. EXAMPLE 4.3 OF CHAPTER 4

```
% Main Program: Main_Example4_3.m
% This Program is developed by B.S. Yarman on September 20, 2018.
% Vanikoy, Istanbul, Turkey
% IET Book 2018
clear;clc;close all
% Example 3:
% Given Exact Lowpass LC Ladder form for Z1(p)=a1(p)/b1(p)
%                                        as obtained in Example 2;
a1=[0 0.275780548962448 0.105214850053913 0.118321595661992]
b1=[1 0.381516573412289 0.739177447894127 0.118321595661992]
% Generated Even Part R(p)=A1(p)/B1(p)from Z1(p) and observe the
numerical
% precision in R(p):
%Data given for Example2
A=[0 0 0 0.0140];% Exact Lowpass LC Ladder form of the numerator of
R(p)
B=[-1 -1.3328 -0.4561 0.0140]; % Denominator of R(p)
% Generate even part of Z1(p);
format long e
[A1,B1]=even_part(a1,b1)
```

PROGRAM LIST 4.4. EXAMPLE 4.4 OF CHAPTER 4

```
% Main Program: Main_Example4_4.m
% This Program is developed by B.S. Yarman on September 20, 2018.
% Vanikoy, Istanbul, Turkey
% IET Book 2018
clear;clc;close all
```

```
% Given F(p)=a(p)/b(p) generate Exact Lowpass LC
% Ladder function
% F(p)=a(p)/b(p) via parametric approach.
%
% Inputs:
a=[0 0.275685028606870 0.105258913364013 0.118321595661992];
b=[1 0.381808594742783 0.739088901509729 0.118321595661992];
eps_zero=1e-8;% Threshold for Algorithmic zero
% Step 1: Generate Exact Lowpass LC Ladder Function F(p)=a(p)/b(p)
[a,b,ndc]=Check_immittance(a,b);
aa=a;bb=b;% Save the original form of a(p) and b(p)
% Check even part if it belongs to a ladder network.
% Part (a)
[A,B]=even_part(aa,bb)
% End of Part (a)
% Step 2:Remove poles at infinity which in turn yields the residues
q(i)
% Part (b)
nplus1=length(b);n=nplus1-1;
%
for i=1:n-1 [Q,a1,b1,ndc]=ExtractTrZero_infinity(eps_zero,a,b);
    q(i)=Q;
    clear a;clear b
    a=a1;b=b1;
end
% Now, we have least degree F1(p)=b1(p)/a1(p)
% Compute the last residue q(n)
q(n)=b1(1)/a1(2)
% Compute terminating constant q(n+1)
q(n+1)=b1(2)/a1(2)
% End of Part (b)
% Check the result
% Part (c): Generate the driving point input impedance from the
topology.
[a2,b2]=ExactLowpassLadder(q)
eps_a=norm(aa-a2)
eps_b=norm(bb-b2)
```

PROGRAM LIST 4.5. EXAMPLE 4.5 OF CHAPTER 4

```
% Main Program: Main_Example4_5.m
% This Program is developed by B.S. Yarman on September 20, 2018.
% Vanikoy, Istanbul, Turkey
% IET Book 2018
clear;clc;close all
% Given F(p)=a(p)/b(p) generate Exact Lowpass LC Ladder function
% F(p)=a(p)/b(p) via parametric approach.
% Inputs:
%a=[0 0.2755 0.1051 0.1182]; % Given by Example 1
%b=[1 0.381 0.7390 0.1182];% Given by Example 1
a1=1.0e+003*[0.00000 0.000243570982330 0.005942391196503
```

```
0.048441163672809 0.223771800321183 0.703166932444413
1.647563920173623 3.038406878823534];
a2=1.0e+003*[4.561480281228956 5.691843698397583 5.973794366516633
5.301481077928296 3.977617930832275 2.509345559081507
1.315779401254989];
a3=1.0e+003*[0.562150827666278 0.189558067896765 0.048032011167119
0.008431529662418 0.000827704427793];
a=[a1,a2,a3];
b1= 1.0e+004 *[0.0001 0.002439695870029    0.019914726153765
0.092525662299536 0.294017502515733 0.700877094923677
1.323370333662979];
b2=1e4*[2.047310741949944 2.650311854053223 2.907173323250695
2.719656869566401 2.173275287536554 1.479042415464306
0.850450799464403];
b3=1e4*[0.407297891494083 0.158774880981138 0.048562164264357
0.010905748934925 0.001542112526945 0.000082770442779];
b=[b1,b2,b3];
eps_zero=1e-8;% Threshold for Algorithmic zero
% Step 1: Generate Exact Lowpass LC Ladder Function F(p)=a(p)/b(p)
[a,b,ndc]=Check_immittance(a,b);
aa=a;bb=b;% Save the original form of a(p) and b(p)
% Check even part if it belongs to a ladder network.
% Part (a)
[A,B]=even_part(aa,bb)
% End of Part (a)
% Step 2:Remove poles at infinity which in turn yields the residues
q(i)
% Part (b)
nplus1=length(b);n=nplus1-1;
%
for i=1:n-1
    [Q,a1,b1,ndc]=ExtractTrZero_infinity(eps_zero,a,b);
    q(i)=Q;
    clear a;clear b
    a=a1;b=b1;
end
% Now, we have least degree F1(p)=b1(p)/a1(p)
% Compute the last residue q(n)
q(n)=b1(1)/a1(2);
% Compute terminating constant q(n+1)
q(n+1)=b1(2)/a1(2);
% End of Part (b)
% Check the result
% Part (c): Generate the driving point input impedance from the
topology.
[a2,b2]=ExactLowpassLadder(q);
eps_a=norm(aa-a2)
eps_b=norm(bb-b2)
% End of Part (c)
```

PROGRAM LIST 4.6. EXAMPLE 4.6 OF CHAPTER 4

```
% Main Program: Main_Example4_6.m
% This Program is developed by B.S. Yarman on September 20, 2018.
% Vanikoy, Istanbul, Turkey
% IET Book 2018
% Double Matching via RFDCT Parametric Approach
% February 11, 2012, Vanikoy, Istanbul.
% This program determines the driving point input immittance
ZB(p)=a(p)/b(p)from the back-end:
% Generator immittance is computed using function
[RGen,XGen]=Gen(w,RG,KFlag)
% Load immittance is computed using function
[RL,XL]=Load(w,RLoad,KFlag)
%-------------------------------------------------------------------
-----
%
% Design of impedance transforming filter from fc1=850 MHz to fc2=2100
MHz
% For the case under consideration:
%                   RG=12ohm+LG//CG+Cx
% where Normalized Value of CG=1 or CG=1.515761362779956e-012 (Actual)
% LG=1/wres/wres/CG LG=9.473508517374721e-010 (Actual)
%                   Cx=7.6228e-012 (Actual) or Cx=5.029 (Normalized)
% Resonance frequency frG=4200 MHz (Normalized wrG=2)
%                   RL=50ohm+LL//CL
% where Normalized Value of CL=1 or CL=1.515761362779956e-012 (Actual)
% LL=1/wres/wres/CL LL=4.210448229944320e-010 (Actual)
% Resonance frequency frG=6300 MHz (Normalized wrL=3)
%
% Inputs:
%           x0: Initialized unknowns
%           T0: Flat gain level
%           wc1: Beginning of optimization
%           wc2: End of optimization
%           KFlag: =1>Work with impedance functions
%           KFlag: =0>Work with admittance functions
%           ntr:=1> work with transformer
%           ntr:=0> work without transformer. For sure you have LC Low
pass
%           structure for the equalizer.
%           ndc: Zero of transmissions at DC
%           W: Zero of transmissions at finite frequencies. It is
usually
%           zero.
%           a0: Initial for R(w)=a0*a0*W/B(w^2)
%
clc
clear
close all
Program='Main_RFDCT.m';
```

```
% Inputs: Lower and the upper cut-off frequencies subject to
optimization
fc1=850e6;fc2=2100e6;
% Inputs: lower and upper frequencies to print performance.
fs1=0;fs2=7000e6;
% Normalized frequencies
ws1=fs1/fc2;ws2=fs2/fc2;wc1=fc1/fc2;wc2=fc2/fc2;
%Inputs:
% Generator and Load networks:
RGEN=12/50; RLoad=1;
% Input: Flat Gain level subject to optimization
T0=1;
%Inputs:
KFlag=1;ndc=0;W=0;ntr=0;a0=1;W=0;
%
format short
c0=[-14.36823865790642  -18.04165545696005      19.01759497233722
26.41763565625652 -6.132379848456734  -8.20957326682746]
c1=c0;
Koptimization=1;
input=('Koptimization=1>minimax, Koptimization=0>least square: Enter
Koptimization=');
%-----------------------------------------------------------------
for i=1:15
[ c ] = nonlinear_optimization(
c0,wc1,wc2,T0,KFlag,W,ndc,a0,ntr,Koptimization,RGEN,RLoad );
c0=c;
end
% Generate analytic form of Fmin(p)=a(p)/b(p)
[a,b]=Minimum_Function(ndc,W,a0,c)
q=LowpassLadder_Yarman(a,b);% Normalized element values
% Plot LC Ladder with normalized element values
f0=1/2/pi;% w0=1> f0=1/2/pi> Normalization frequency at w0=1
R0=1;% Unit normalization resistance
 [CT,CV] = CircuitPlot_Yarman(KFlag,R0,f0,a,b,ndc);% Generate Circuit
Codes
 Plot_Circuitv1(CT,CV); % Plot LC Ladder from the circuit codes

% Compute Double Matching Gain and plot the results:
% Step 6: Print and Plot results
Nprint=1001;
DW=(ws2-ws1)/(Nprint-1);
w=ws1;
Tmax=-1000;
Tmin=0;
RGEN=12/50; RLoad=1;
for j=1:Nprint
          WA(j)=w;
          [RG,XG]=Gen(w,RGEN,KFlag);
          [RL,XL]=Load(w,RLoad,KFlag);
```

```
 [ T_Double ] = Gain_DoubleMatching( w,a,b,ndc,KFlag,RG,XG,RL,XL );
 TA(j)=10*log10(T_Double);
%
% Compute the performance parameters:Tmax,Tmin,Tave and detT
    if max(TA(j))>Tmax
        wmax=WA(j);Tmax=TA(j);
    end
if w>=wc1
    if w<=wc2
    if min(TA(j))<Tmin
        wmin=WA(j);Tmin=TA(j);
    end
    end
end
    w=w+DW;
end
%
        figure
        plot(WA*fc2/1e6,TA)
        title('Double Matching Via RFDT')
        xlabel('Actual Frequency (MHz)')
        ylabel('Double Matching Gain in dB')
        legend('RFDCT Gain')
  %
  f0=2100e6;R0=50;% De-normalization numbers
 [CT,CV] = CircuitPlot_Yarman(KFlag,R0,f0,a,b,ndc);% Compute the
actual element values for the lowpass LC Ladder
 Plot_Circuitv1(CT,CV)
% Print Performance parameters:
wmax, Tmax,
wmin,Tmin,
T_Average=(Tmax+Tmin)/2
delT=T_Average-Tmin
DrawLCLadder_withGenLoad(KFlag,ndc,a,b,R0,f0);%Draw the circuit
schematic for Lowpass LC Ladder
% Compute the errors accumulated during the synthesis process
[a2,b2]=ExactLowpassLadder(q);% Zin(p)=a2(p)/b2(p) from the normalized
element values
eps_a=norm(a-a2);% norm error in a(p)
eps_b=norm(b-b2);% norm error in b(p)
```

PROGRAM LIST 4.7. FUNCTION EVEN_PART

```
function [A,B]=even_part(a,b)
% Generate even part R(p)=A(p)/B(p) of a given immittance function
F(p)=a(p)/b(p)
% Computation of Numerator of Even Part
na=length(a);
nb=length(b);
    sign=-1;
```

```
    for i=1:na
        sign=-sign;
        a_(na-i+1)=sign*a(na-i+1);
    end
    sign=-1;
    for i=1:nb
        sign=-sign;
        b_(na-i+1)=sign*b(nb-i+1);
    end
    Num_Even=(conv(a,b_)+conv(a_,b))/2;
      A=clear_oddpower(Num_Even);

    n_Even=length(Num_Even);
       BB=conv(b,b_);
    B=clear_oddpower(BB);
end
```

PROGRAM LIST 4.8. FUNCTION RESIDUES_EVENPART

```
function [p,k]=residues_evenpart(A,B)
% This function computes the residues k(j) of a given even part
R(p^2)
% R(p^2)=A(p^2)/B(p^2)
% Inputs:
% ------- A(p^2); full coefficient numerator polynomial
% ------- B(p^2); full coefficient denominator polynomial
% Step 1: Find the LHP poles of R(p^2)
x=roots(B);
p=-sqrt(x);
% Step 2:Compute the product terms
prd=product(p);
n=length(p);
% Step 3: Generate residues
for j=1:n
    y=p(j)*p(j);
    Aval=polyval(A,y);
    k(j)=(-1)^n*Aval/p(j)/B(1)/prd(j);
end
end
```

PROGRAM LIST 4.9. FUNCTION EVENPARTTO_MIN_FUNC

```
function [a,b]=evenpartTo_MinFunc(A,B)
% function [a,b]=evenpartTo_MinFunc(A,B)
% This MatLab function generates a minimum function Z(p)=a(p)/b(p)
%      from its even part specified as R(p^2)=A(p^2)/B(p^2)
%      via Bode (or Parametric)approach
% Inputs: In p-domain, enter A(p) and B(p)
% A=[A(1) A(2) A(3)...A(n+1)]; for many practical cases we set
A(1)=0.
% B=[B(1) B(2) B(3)...B(n+1)]
% Output:
%        Z(p)=a(p)/b(p) such that
%        a(p)=a(1)p^n+a(2)p^(n-1)+...+a(n)p+a(n+1)
%        b(p)=b(1)p^n+b(2)p^(n-1)+...+b(n)p+a(n+1)
% Generation of an immittance Function Z by means of Parametric
Approach
% In parametric approach Z(p)=Z0+k(1)/[p-p(1)]+...+k(n)/[p-p(n)]
% R(p^2)=Even{Z(p)}=A(-p^2)/B(-p^2) where Z0=A(n+1)/B(n+1).
%
% Given A(-p^2)>0
% Given B(-p^2)>0
%
% Computational Steps
% Given A and B vectors. A(p) and B(p) vectors are in p-domain
% Compute poles p(1),p(2),...,p(n)and the residues k(i) at poles
p(1),p(2),...,p(n)
[p,k]=residues_evenpart(A,B);
%
% Compute numerator and denominator polynomials
Z0=abs(A(1)/B(1));
[num,errorn]=num_Z0(p,Z0,k);
num=abs(num);
[denom,errord]=denominator(p);
%
a=num;
b=denom;
end
```

PROGRAM LIST 4.10. FUNCTİON NUM_Z0

```
function [num,errorn]=num_Z0(p,Z0,k)
% This function computes the numerator polynomial of an
% immittance function: Z(p)=Z0+sum{k(1)/[p-p(i)]
% where we assume that Z0=A(1)/B(1)which is provided as input.
%
% Input:
%-------- poles p(i) of the immittance function Z(p)
%         as a MatLab row vector p
%-------- Residues k(i) of poles p(i)
% Output:
%-------- num; MatLab Row-Vector
% which includes coefficients of numerator polynomial of an
immittance function.
% num=Sum{k(j)*product[p-p(i)]} which skips the term when j=i
%
%----- Step 1: Determine total number of poles n
%
n=length(p);
nn=n-1;
%
%----- Step 2: Generation of numerator polynomials:
% numerator polynomial=sum of
% sum of
% {Z0*[p-p(1)].[p-p(2)]...(p-p(n)]; n the degree-full product
% +k(1)*[p=p(2)].[p-p(3)]..[p-p(n)];degree of (n-1); the term with
p(1)is skipped.
% +k(2)*[p-p(1)].[p-p(3)]..[p-p(j-1)].[p-p(j+1)]..[p-p(n)];degree
of(n-1)-the term with p(2)is skipped
% +.........................................
% +k(j)*[p-p(1)].[p-p(2)]..[p-p(j-1)].[p-p(j+1)]..[p-p(n)];degree
of (n-1)-the term with p(j)is skipped.
% +.........................................
% +k(n)[p-p(1)].[p-p(2)]...[p-p(n-1)];degree of (n-1)-the term with
p(n)is
% skipped.
%
% Note that we generate the numerator polynomial within 4 steps.
% In Step 2a, product polynomial pra of k(1)is evaluated.
% In Step 2b, product polynomial prb of k(j)is evaluated by
skipping the term when i=j.
% In Step 2c, product polynomial prc of k(n)is evaluated.
% In Step 2d, denominator of Z0 is generated.
%----------------------------------------------------------------------
%
% Step 2a: Generate the polynomial for the residue k(1)
pra=[1];
for i=2:n

  simpA=[1 -p(i)];
```

```
% pra is a polynomial vector of degree n-1; total number of entrees
are n.
  pra=conv(pra,simpA);% This is an (n-1)th degree polynomial.
end
na=length(pra);
% store first polynomial onto first row of A i.e. A(1,:)
for r=1:na
  A(1,r)=pra(r);
end
% Step 2a: Compute the product for 2<j<(n-1)

for j=2:nn
    prb1=[1];
        for i=1:j-1
        simpB=[1 -p(i)];
        prb1=conv(prb1,simpB);
        end
    % Skip j th term
    prb2=[1];
        for i=(j+1):n
            simpB1=[1 -p(i)];
            prb2=conv(prb2,simpB1);
        end
        prb=conv(prb1,prb2);
        nb=length(prb);
%
% Store j polynomials on to j th row of A; i.e. A(j,:)
for r=1:nb
A(j,r)=prb(r);
end
%
    end
% Step 2c: Compute the product term for j=n
prc=[1];
for i=1:nn
    simpC=[1 -p(i)];
    prc=conv(prc,simpC);
end
nc=length(prc);
% store n the polynomial onto n the row of A(n,:)
for r=1:nc
A(n,r)=prc(r);
end
```

PROGRAM LIST 4.11. FUNCTION DENOMINATOR

```
function [denom,errord]=denominator(p)
% This function computes the denominator polynomial of an
% immittance function from the given poles.
% It should be noted that this form of the denominator is normalized
with
% the leading coefficient b(1)=1: denom=(p-p(1))(p-p(2))....(p-p(n))
% Input:
%-------- poles p(i) as a MatLab row vector p
%-------- Residues k(i) of poles p(i)
% Output:
%-------- denom; MatLab Row-Vector
% ---which includes coefficients of the denominator polynomial of an
immittance function.
% denom=product[p-p(i)]
%
% --------- Step 1: Determine n
n=length(p);
% -------- Step 2: Form the product term.
 pr=[1]; %Define a simple polynomial pr=1.
for j=1:n
     simple=[1 -p(j)];%this is the simple polynomial [p-p(j)]
% Generate multiplication of polynomials starting with 1.
      pr=conv(pr,simple);
end
denom=real(pr);
errord=imag(pr);
end
```

PROGRAM LIST 4.12. FUNCTION CHECK_IMMITTANCE

```
function [a1,b1,ndc]=Check_immittance (a,b)
% Given F(p)=a(p)/b(p)as a minimum function
% F(p) is free of finite jW zeros but it may have zeros at dc
% Find zeros of transmissions at dc
% Re-compute a(p) and b(p) to make F(p) minimum reactance
na=length(a);
nb=length(b);
% Check the leading and the last coefficient of a(p)if they are zero
eps_zero=1e-5;
if abs(a(1))<eps_zero
    if b(1)>0
    Comment_1='Degree of numerator is one less than denominator and
F(p) is minimum function';
    end
    if a(na)==0
        Comment_2='There is at least one DC transmission zeros';
    end
```

```
end
%
    if b(1)<0
    Comment_3='Given F(p) is not positive real';
    end
    % Computation of Numerator of Even Part
    sign=-1;
    for i=1:na
        sign=-sign;
        a_(na-i+1)=sign*a(na-i+1);
    end
    sign=-1;
    for i=1:nb
        sign=-sign;
        b_(na-i+1)=sign*b(nb-i+1);
    end
    A1=conv(a,b_);A2=conv(a_,b);
    Num_Even=(A1+A2)/2;
    normA=norm(Num_Even);
    n_Even=length(Num_Even);
    %
    if Num_Even(n_Even)==0
    for i=1:n_Even
        if abs(Num_Even(i))/normA<eps_zero
            Num_Even(i)=0;
        end
    end
    end
    %
    for i=1:n_Even
        if abs(Num_Even(n_Even+1-i))/normA>eps_zero
            ndc=(i-1)/2;
            a0_sq=Num_Even(n_Even+1-i);
        end
    end
    if Num_Even(n_Even)/normA>eps_zero
        ndc=0;
        a0_sq=Num_Even(n_Even);
        Comment_3='No dc transmission zero; ndc=0';
    end
    BB=conv(b,b_);
    B1=clear_oddpower(BB);
    nB=length(B1);
    %
    A1=a0_sq*R_Num(ndc,0);% A is specified in p-domain
        nA=length(A1);
        if (abs(nB-nA)>0)
        A1=fullvector(nB,A1);% work with equal length vectors
        end
        %
        if na>1;[a1,b1]=RtoZ(A1,B1);end
```

```
        if na==1;a1=a;b1=b;end
    a1=abs(a1);b1=abs(b1);
end
```

PROGRAM LIST 4.13. FUNCTION FULLVECTOR

```
function B=fullvector(n,A)
na=length(A);
for i=1:na
    B(n-i+1)=A(na-i+1);
end
for i=1:(n-na-1)
        B(i)=0;
end
end
```

PROGRAM LIST 4.14. CASE STUDY 1

```
% High Precision LC Ladder Synthesis Part II:
% immittance Synthesis with transmission zeros at DC and infinity
% Main Program: CaseStudy1.m
clc;clear all;close all;
format long e
% Enter rational form of F(p)=a(p)/b(p)as in Table VIII
a=[0
0.513976968750998
1.157898233543669
1.213142248104650
0.694401166177243
0.188516938299314
0]';
b=[1.000000000000000
4.198433713455139
7.913422823138356
9.268979838320304
7.284067053656422
3.683494822490302
1.000000000000000]';
pa=roots(a);% roots of a(p)as in Table IX
pb=roots(b);% roots of b(p)as in Table IX
%
[A,B]=even_part(a,b);% Even part R(p)=A(p)/B(p)
% Correct ladder structure:
[a1,b1,ndc]=Check_immittance(a,b)
% Extract transmission zeros at DC and then at infinity
[k,q,Highpass_Elements,Lowpass_Elements]=GeneralSynthesis_Yarman(a1,b1
)
% Plot the circuit
```

```
KFlag=1;w0=1;f0=1/2/pi;R0=1;
[CT,CV] = CircuitPlot_Yarman(KFlag,R0,f0,a,b,ndc)
```

PROGRAM LIST 4.15. FUNCTİON GENERALSYNTHESİS_YARMAN

```
function
[k,q,Highpass_Elements,Lowpass_Elements]=GeneralSynthesis_Yarman(a,b)
% This is a general sythesis program developed by yarman for given
Minimum
% Function F(p)=a(p)/b(p)
na=length(a);nb=length(b);eps_zero=1e-8;na1=na-1;
if(na-nb)==0;n=nb;end
if abs(na-nb)>0
    comment='nb is different than na. Your Function F(p)=a(p)/b(p)is
not proper for our synthesis package'
end
aa=a;bb=b;
[a,b,ndc]=Check_immittance(aa,bb);
n1=n-1;

%-------------------------------------------------------------------
if ndc==0
    Comment='ndc=0, Lowpass Ladder '
    q=LowpassLadder_Yarman(a,b);
    k=0;Comment=('No transmission zero at DC. Therefore, we set k=0')
end
    for i=1:ndc
            [kr,R,ar]=Highpass_Remainder(a,b);
                        k(i)=abs(kr);
%-----Fr(p)=a(p)/b(p)
            clear a;clear b
            a=abs(R);b=abs(ar);
%-------------------------------------------------------------------
%'Corrections on the Ladder Coefficients'
if i<ndc;[a,b]=Laddercorrection_onF(eps_zero,a,b);end
%--------------------------------------------------------------------
if (ndc-1)>0;R(ndc-1)=0.0;end
    end
 % -----Patch made on April 4, 2011
 if (n-ndc)>0
   if abs(a(1))<1e-8
       % Design='Lowpass Design with a(1)=0:
k(ndc)/p+{a(p)/b(p)}=k(ndc)/p+1/{q(1)p+..'
         q=LowpassLadder_Yarman(a,b);
   end
   if abs(b(1))<1e-8
      %Design='Lowpass Design with b(1)=0:
k(ndc)/p+{a(p)/b(p)}=k(ndc)/p+q(1)p+..'
      q=LowpassLadder_Yarman(b,a);
```

```
   end
 end
 % end of Patch made on April 4, 2011
%
 if (n1-ndc)==0;q=0;%'Synthesis with
n=ndc',Constant_Termination=a(1)/b(1),
 end
 if ndc>0
 Highpass_Elements=1./k;
 end
 if ndc==0
     Highpass_Elements='There is no highpass elment'
 end
 Lowpass_Elements=q;
 if q==0,q=0; %'highpass design; there is no lowpass component'
 end
 if na1==ndc; Highpass_Elements(ndc+1)=a(1)/b(1);end
end
```

PROGRAM LIST 4.16. FUNCTION CIRCUITPLOT_YARMAN

```
function [CT,CV] = CircuitPlot_Yarman(KFlag,R0,f0,a,b,ndc)
% This function generates the circuit code vector for a given
Component matrix
% CVal(i,j) describes the rows and columns of CVal in actual element
values
% -------------------------------------------------------------------
% Series Capacitors(CHP)    Shunt Inductors(LHP)      Shunt
Capacitors(CLP)    Series Inductors(LLP)        Termination
% -------------------------------------------------------------------
% j=1> First Column of CVal gives the series capacitors C_HP; CT=2
% j=2> Second Column of CVal gives the shunt inductors L_HP; CT=7
% j=3> Third Column of CVal(:;3) gives the shunt capacitors C_LP; CT=8
% j=4> Fourth Column of CVal(:;4) gives the series inductors L_LP;
CT=1
% j=5> First Row (i=1) and Fifth Column (j=5) gives the termination
Ter=9
% ----------------------------------------------------------------------
-----
if KFlag==1;[CT,CV] =
SynthesisMinimumReactance_Yarman(R0,f0,a,b,ndc);end
if KFlag==0;[CT,CV] =
SynthesisMinimumSusceptance_Yarman(R0,f0,a,b,ndc);end
end
```

PROGRAM LIST 4.17. FUNCTİON SYNTHESIS MINIMUMREACTANCE

```
function [CT,CV] = SynthesisMinimumReactance_Yarman(R0,f0,a,b,ndc)
% This function generates the circuit codes for CT and the element
values CV for a minimum reactance impedance function
% This function generates the circuit code vector for a given
Component matrix
% CVal(i,j) describes the rows and columns of CVal in actual element
values
% ---------------------------------------------------------------
% Series Capacitors(CHP)    Shunt Inductors(LHP)      Shunt
Capacitors(CLP)    Series Inductors(LLP)        Termination
% ---------------------------------------------------------------
% j=1> First Column of CVal gives the series capacitors C_HP; CT=2
% j=2> Second Column of CVal gives the shunt inductors L_HP; CT=7
% j=3> Third Column of CVal(:;3) gives the shunt capacitors C_LP; CT=8
% j=4> Fourth Column of CVal(:;4) gives the series inductors L_LP;
CT=1
% j=5> First Row (i=1) and Fifth Column (j=5) gives the termination
Ter=9
% -------------------------------------------------------------------
-----
nb=length(b);% Total number elements with termination
nL=nb-1-ndc;% Total number of lowpass elements (q)
% Step 1; Generate CVal which is the matrix consist of actual element
% values:
%---------------------------------------------------------------
% Case A: Minimum Reactance Design
%
 KFlag=1; % Minimum Reactance Design start with shunt inductor j=2
             % Start with 1=1, j=2 then go to C(2,1).
             % Notice that here, inductors are odd rows, capacitors
are even rows.
% ---------------------------------------------------------------
[A,CVal] = LadderSynthesis_BSYarman( KFlag,R0,f0,ndc,a,b );
%
%             Extraction of Actual Highpass Elements
% For inductors use odd values of CT; for capacitors use even values
of CT
%---------------------------------------------------------------
%j=2;% extraction starts with a shunt inductor; second column of CVal
% Extraction of Highpass Elements
if ndc==1; CT(1)=7; CV(1)=CVal(1,2);% First element is a shunt
inductor
    if ndc==nb-1;CT(2)=9;CV(2)=R0*a(1)/b(1);end; % Termination is a
resistor
end
if ndc>1
     k=even_odd(ndc);
   if k==0; % ndc=even case
     for i=1:ndc/2
```

```
        CT(2*i-1)=7;% Start with a highpass shunt inductor L_HP
        CV(2*i-1)=CVal(2*i-1,2);% Value of the shunt inductor
        CT(2*i)=2;% Second element is a series capacitor
        CV(2*i)=CVal(2*i,1);% Value of the series capacitor
    end
     if (ndc==nb-1);CT(ndc+1)=9; CV(ndc+1)=R0*a(1)/b(1);end; %
Termination is a resistor.
  end
  if k==1; % ndc=odd case ·
 %
     for i=1:(ndc-1)/2+1
        CT(2*i-1)=7;% Start with a highpass shunt inductor L_HP
        CV(2*i-1)=CVal(2*i-1,2);% Value of the shunt inductor
        if (2*i)<ndc
        CT(2*i)=2;% Second element is a highpass series capacitor
        CV(2*i)=CVal(2*i,1);% Value of the series capacitor
        end
     end
%
      if ndc==1; CT(1)=7; CV(1)=CVal(1,2);end
 if (ndc==nb-1);CT(ndc+1)=9; CV(ndc+1)=R0*a(1)/b(1);end
  end
end
%
% End of extraction of highpass elements.
% KFlag is still one and ndc is still greater than zero: extraction of
lowpass elements
        %j=3;% extraction starts with a shunt capacitor CT=8
        if nL==1;
            CT(ndc+1)=8; CV(ndc+1)=CVal(1,3);% First Lowpass element
is a shunt capacitor.
            CT(ndc+2)=9; CV(ndc+2)=CVal(1,5);% Termination is always
is given as a resistor.
        end; % Here CV(ndc+2) is a resistor in ohms.
        if nL>1
            k=even_odd(nL);
          if k==0; % nL=even case
            for i=1:nL/2
                CT(ndc+2*i-1)=8;% First Lowpass element is a shunt
capacitor
                CV(ndc+2*i-1)=CVal(2*i-1,3);% Value of the shunt
capacitor
                %
                CT(ndc+2*i)=1; % Code of a series inductor
                CV(ndc+2*i)=CVal(2*i,4);% Value of the series inductor

            end
            % Extract of the termination
            CT(ndc+nL+1)=9;R=CVal(1,5);
            CV(ndc+nL+1)=R;% Termination is always given as a resistor
in ohms
```

```
            end
             if k==1; % nL=odd case
              for i=1:(nL-1)/2+1
                  CT(ndc+2*i-1)=8;% First Lowpass element is a shunt
capacitor
                  CV(ndc+2*i-1)=CVal(2*i-1,3);% Value of the shunt
capacitor
                  if (2*i)<nL
                  CT(ndc+2*i)=1; % Code of a series inductor
                  CV(ndc+2*i)=CVal(2*i,4);% Value of the series inductor
                  end
              end
              % Extract the termination which is always set to be a
resistance R in
              % the synthesis algorithm.
              CT(ndc+nL+1)=9;R=CVal(1,5);
              CV(ndc+nL+1)=R;% Termination is always given as a resistor
in ohms
             end

         end
         Plot_Circuitv1(CT,CV)

end
```

PROGRAM LIST 4.18. FUNCTION SYNTHESISMINIMUMSUSCEPTANCE_YARMAN

```
function [CT,CV] = SynthesisMinimumSusceptance_Yarman(R0,f0,a,b,ndc)
% Minimum Susceptance Case
% This function generates the circuit code vector for a given
Component matrix
% CVal(i,j) describes the rows and columns of CVal in actual element
values
% ------------------------------------------------------------------
% Series Capacitors(CHP)    Shunt Inductors(LHP)       Shunt
Capacitors(CLP)     Series Inductors(LLP)         Termination
% ------------------------------------------------------------------
% j=1> First Column of CVal gives the series capacitors C_HP;
CT=2
% j=2> Second Column of CVal gives the shunt inductors L_HP; CT=7
% j=3> Third Column of CVal(:;3) gives the shunt capacitors C_LP;
CT=8
% j=4> Fourth Column of CVal(:;4) gives the series inductors L_LP;
CT=1
% j=5> First Row (i=1) and Fifth Column (j=5) gives the termination
Ter=9
% ------------------------------------------------------------------
nb=length(b);% Total number elements with termination
nL=nb-1-ndc;% Total number of lowpass elements (q)
% Step 1; Generate CVal which is the matrix consist of actual element
%------------------------------------------------------------------
```

```
% Case B: Minimum Reactance Design
%
KFlag=0; % Minimum Reactance Design start with shunt inductor j=2
          % Start with l=1, j=2 then go to C(2,1).
          % Notice that here, inductors are odd rows, capacitors
are even rows.
% ------------------------------------------------------------------
% values:
[A,CVal] = LadderSynthesis_BSYarman( KFlag,R0,f0,ndc,a,b );

%              Extraction of Actual Highpass Elements
% For inductors use odd values of CT; for capacitors use even values
of CT
%-------------------------------------------------------------------
% j=1; % extraction starts with a series capacitor (CT=2); first
column of CVal
% Extraction of Highpass Elements
if ndc==1; CT(1)=2; CV(1)=CVal(1,1);% First element is a series
capacitor
    if ndc==nb-1;CT(2)=9;G=1/R0; G1=G*a(1)/b(1); CV(2)=1/G1;end;%
Here, however, termination is given as a resistor in ohms
end
if ndc>1
    k=even_odd(ndc);
  if k==0; % ndc=even case
    for i=1:ndc/2
        CT(2*i-1)=2;% Start with a series highpass capacitor C_HP
        CV(2*i-1)=CVal(2*i-1,1);% Value of the series capacitor
        CT(2*i)=7;% Second element is a shunt inductor (CT=7)
        CV(2*i)=CVal(2*i,2);% Value of the shunt inductor
    end

    if (ndc==nb-1);CT(ndc+1)=9;
G=1/R0;G1=G*a(1)/b(1);CV(ndc+1)=1/G1;end; % Termination is conductor
but we use Resistor in ohms
  end
  if k==1; % ndc=odd case
 %
     for i=1:(ndc-1)/2+1
        CT(2*i-1)=2;% Start with a highpass series capacitor C_HP
        CV(2*i-1)=CVal(2*i-1,1);% Value of the shunt inductor
        if (2*i)<ndc
        CT(2*i)=7;% Second element is a shunt inductor
        CV(2*i)=CVal(2*i,2);% Value of the shunt inductor
        end
     end
%
      if ndc==1; CT(1)=2; CV(1)=CVal(1,1);end;% First element is a
series capacitor
 if (ndc==nb-1);CT(ndc+1)=9;
G=1/R0;G1=G*a(1)/b(1);CV(ndc+1)=1/G1;end; % Termination is conductor
```

```
but we use Resistor in ohms
  end
end
%
% End of extraction of highpass elements.
%
% KFlag is zero and ndc is greater than zero: extraction of lowpass
elements
        %j=4;% extraction starts with a series inductor CT=1
        if nL==1;
            CT(ndc+1)=1; CV(ndc+1)=CVal(1,4);
            CT(ndc+2)=9; CV(ndc+2)=CVal(1,5);
        end; % Here CV(ndc+2) is a resistor in ohms.
        if nL>1
            k=even_odd(nL);
          if k==0; % nL=even case
            for i=1:nL/2
                CT(ndc+2*i-1)=1;% First Lowpass element is a series
inductor
                CV(ndc+2*i-1)=CVal(2*i-1,4);% Value of the series
inductor
                %
                CT(ndc+2*i)=8; % Code of a shunt capacitor
                CV(ndc+2*i)=CVal(2*i,3);% Value of the shunt
capacitor

            end
            % Extract of the termination
            CT(ndc+nL+1)=9;R=CVal(1,5);
            CV(ndc+nL+1)=R;% Termination is always given as a
resistor in ohms
          end
           if k==1; % nL=odd case
            for i=1:(nL-1)/2+1
                CT(ndc+2*i-1)=1;% First Lowpass element is a series
inductor
                CV(ndc+2*i-1)=CVal(2*i-1,4);% Value of the series
inductor
                if (2*i)<nL
                CT(ndc+2*i)=8; % Code of a shunt capacitor
                CV(ndc+2*i)=CVal(2*i,3);% Value of the series
inductor
                end
            end
            % Extract the termination
            CT(ndc+nL+1)=9;R=CVal(1,5);
            CV(ndc+nL+1)=R;% Termination is always given as a
resistor in ohm
           end

       end
```

```
    Plot_Circuitv1(CT,CV)

end
```

PROGRAM LIST 4.19. FUNCTION PLOT_CIRCUITV1

```
function [h]=Plot_Circuitv1(varargin)
%
% Plot_Circuit1(CType, elm_val, h);
%
% inputs
% CType: component type vector
% elm_val: Values vector
% h: figure handles
%
% Component types:
% To construct series L alone use the following code:
%         case 1,    'L in Serial branch                    ';
% To construct series C alone use the following code:
%         case 2,    'C in Serial branch                    ';
% To construct series R alone use the following code:
%         case 3,    'R in Serial branch                    ';
% ------------------------------------------------------------------
-----
% To construct parallel L//C//R use the following codes:
%         case 4,    'L of parallel R&L&C in Serial branch';
%         case 5,    'C of parallel R&L&C in Serial branch';
%         case 6,    'R of parallel R&L&C in Serial branch';
% To construct shunt L alone use the following code:
%         case 7,    'L in shunt branch                     ';
% To construct shunt C alone use the following code:
%         case 8,    'C in shunt branch                     ';
%         case 9,    'R in shunt branch                     ';
% ---------------------------------------------------------------
% To construct series L,C, R circuit use the following codes:
%         case 10,   'L of Serial R&L&C in shunt branch   ';
%         case 11,   'C of Serial R&L&C in shunt branch   ';
%         case 12,   'R of Serial R&L&C in shunt branch   ';


    CType=varargin{1};
    CVal=varargin{2};

    %open figure
    if nargin==2
        h=figure ();
    else
        h=varargin{3};
    end

    if length(CType)==0
```

```
        return
    end
    if nargin>3
        tol=varargin{4};
    else
        tol=0.0001;
    end

    % starting location of overall figure,
    xr0=0; yr0=20;
    % input offset for circuit
    xd=5; yd=40;
    % draw
    [xx,cn]=draw_circuit2(h, CType, xr0+xd, yr0+yd, 0);

    % connection line
    X=[xr0 xr0+xd;  xr0 xr0+xd ]';
    Y=[yr0 yr0;     yr0+yd yr0+yd ]';

    ct=CType;
    tx=CVal;

    % set figure labels, borders, ...
    line(X,Y,'Color','k','LineWidth',2);
    title('Circuit Schematic');
    set(gca,'XGrid','off');
    set(gca,'YGrid','off');
    set(gca,'XTickLabel',{''});
    set(gca,'YTickLabel',{''});
    set(gca,'XTick',[]);
    set(gca,'YTick',[]);
    set(gca,'box', 'on');

    % print component values
    pos=get(h,'Position');

    ts=length(tx); % how many label

    % label count for each row
    if ts>12
        cs=4;
    else
        cs=3;
    end
    rs=round(ts/cs+0.4999); % how many rows

    axis([-5 xx -2*(rs-1) yr0+yd+yd/2+2])

    etx=ig_selti(ct,tx,1,tol);
    rw= fix(xx/(cs))+1;       %  column width
    rh=5;             % row height
```

```
    row=-1;    col=0;
    for i=1:ts
        row=row+1;
        if row>=rs,
            col=col+1;
            row=0;
        end
        if tx(i)>0, vcl='k'; else, vcl='r'; end
        text( 0+col*rw, 15-row*rh,
etx(i),'HorizontalAlignment','left','FontSize',8,'color',vcl);
    end

%     % increase picture size if required
%     if (xx>300)||(cs>3),
%         pos(3)=pos(3)+250;
%     elseif xx>200
%         pos(3)=pos(3)+150;
%     end

    set(h,'Position',pos);
  return
%-----------------------------------------------------------------
function [x,ii]=draw_circuit2(hf,ctype,x0,y0,Num_offset)
% Draw circuit schematic
% Dr. Ali KILINÇ, 17-11-2003
%
%
% inputs
% ctype: circuit components, given by the types defined in synthesis
prog.
%   One resonant circuit components must be given in order of L-C-R.
% x0,y0: drawing start position
% Num_offset : offset for component index, used for display

wid=5; % component width
ls=17; % serial comp length
lp=40; % parallel comp length
lw=2; % line width
nc=length(ctype); % number of component
fs=8;%fix(1000/(1+nc*ls)+8); % font size

cl='k'; % line color
cL='r'; % inductance color
cC='b'; % cap. color
cR='k'; % resistor color

x=x0+wid;
y=y0;
y2=lp/3;

X=[x0 x; x0   x]';
```

```
Y=[y  y; y-lp y-lp]';
line(X,Y,'Color',cl,'LineWidth',lw);

ii=1;
%for i=1:nc,
while (ii<=nc),
    T='';
    switch ctype(ii)
        case 1,    %            z='#1-Serial L
            A=draw_L( x,y,0,wid,ls,cL,lw);
            ts=['L' num2str(ii+Num_offset)]; tj='center';
            T='s';
        case 2,    %            z='#2-Serial C
            A=draw_C( x,y,0,wid,ls,cC,lw);
            ts=['C' num2str(ii+Num_offset)]; tj='center';
            T='s';
        case 3,    %            z='#3-Serial R
            A=draw_R( x,y,0,wid,ls,cR,lw);
            ts=['R' num2str(ii+Num_offset)]; tj='center';
            T='s';

        case 4,    %            z='#1-Serial L
            A=draw_L( x,y+y2,0,wid,ls,cL,lw);
            ts=['L' num2str(ii+Num_offset)]; tj='center';
            T='s';
        case 5,    %            z='#2-Serial C
            A=draw_C( x,y-y2,0,wid,ls,cC,lw);
            ts=['C' num2str(ii+Num_offset)]; tj='center';
            T='s';
        case 6,    %            z='#3-Serial R
            A=draw_R( x,y,0,wid,ls,cR,lw);
            ts=['R' num2str(ii+Num_offset)]; tj='center';
            T='s';

        case 7,    %            z='#7-Shunt L
            A=draw_L( x,y,01,wid,lp,cL,lw);
            ts=['L' num2str(ii+Num_offset)]; tj='left';
            T='p';
        case 8,    %            z='#8-Shunt C
            A=draw_C( x,y,01,wid,lp,cC,lw);
            ts=['C' num2str(ii+Num_offset)]; tj='left';
            T='p';
        case 9,    %            z='#9-Shunt R
            A=draw_R( x,y,01,wid,lp,cR,lw);
            ts=['R' num2str(ii+Num_offset)]; tj='left';
            T='p';
        case 10,    %            z='#7-Shunt L
            A=draw_L( x,y,01,wid,lp/3,cL,lw);
            ts=['L' num2str(ii+Num_offset)]; tj='left';
            T='p';
        case 11,    %            z='#8-Shunt C
```

```
            A=draw_C( x,y-y2,01,wid,lp/3,cC,lw);
            ts=['C' num2str(ii+Num_offset)]; tj='left';
            T='p';
        case 12,    %              z='#9-Shunt R
            A=draw_R( x,y-2*y2,01,wid,lp/3,cR,lw);
            ts=['R' num2str(ii+Num_offset)]; tj='left';
            T='p';
        otherwise  %              z='non-defined element        ';
            A=[x,y];
            ts='';tj='center';
            T='p';
    end
    % write component name
    text(A(1),A(2),ts,'HorizontalAlignment',tj,'FontSize',fs);

    % for leaving free space after a component
    % check the type of next component
    if ii<nc,
        nx=ctype(ii+1); % next component exist
    else
        nx=-1;
    end
    if (nx>0)&(nx<7)
        Tnx='s';
    elseif (nx>6)&(nx<12)
        Tnx='p';
    else
        Tnx='';
    end

    %elemanýn ana hat üzerindeki baðlantý çizgileri
    % draw wirings
    if T=='s',
        % seri kolda parlel rezonans devresi varsa
        % sonraki elemaný daha uzaða çiz
%       if (ii<nc)&& ...
%           ((ctype(ii)==4)||(ctype(ii)==5)||(ctype(ii)==6)) && ...
%           ((ctype(ii+1)~=4)&&(ctype(ii+1)~=5)&&(ctype(ii+1)~=6))
%           % uzun artýþ
            x1=x+ls+wid;
%       else
%           x1=x+ls+wid/2;
%       end
        % seri uç uzatma ve nötr hattý çizme
        if ii<nc
            X=[x+ls x1; x    x1]';
            Y=[y    y ; y-lp y-lp]';
        end
        % seri kolda paralel rezonans devresinin baðlantýlarý
        if (ctype(ii)==4),
            X=[X, [x, x+ls; x, x+ls]];
```

```
            Y=[Y, [y, y;y+y2, y+y2]];
        elseif (ctype(ii)==5),
            X=[X, [x, x+ls; x, x+ls]];
            Y=[Y, [y, y;y-y2, y-y2]];
        end

    else % T=='p'
        % sonraki elemaný çizme mesafesi
        if Tnx=='s',
            x1=x+wid;
        else
            x1=x+ls;
        end

        % sonraki eleman için seri uç uzatma ve nötr hattý çizme
        if (ii<nc),
            if (ctype(ii)~=ctype(ii+1))&& ...
                ( ...
                  ( ...
                    ((ctype(ii)==4)||(ctype(ii)==5)) && ...
                    ((ctype(ii+1)==5)||(ctype(ii+1)==6)) ...
                  ) || ( ...
                    ((ctype(ii)==10)||(ctype(ii)==11)) && ...
                    ((ctype(ii+1)==11)||(ctype(ii+1)==12)) ...
                  )...
                )
                %seri rezonans devresi için
                X=[];
                Y=[];
            else
                X=[x x1; x   x1]';
                Y=[y  y; y-lp y-lp]';
            end
        end
        % paralel kolda seri rezonans devresinin baðlantýlarý
        if (ctype(ii)==10)&&(ii<nc)
            %LR devresi ise, Bu eleman L, ve C'nin yerini çiz
            if (ctype(ii+1)~=11),
                X=[X, [x ; x  ]];
                Y=[Y, [y-y2 ;y-2*y2]];
            end
        elseif (ctype(ii)==11)
            % CR devresi
            if (ii==1)
                % sentezin ilk elemaný C ise L'nin yerini çiz
                    X=[X, [x; x  ]];
                    Y=[Y, [y;y-y2]];
            elseif (ii<nc)
                % ilk eleman rezonansýn L'si deðilse
                if (ctype(ii-1)~=10),
                    X=[X, [x; x  ]];
```

```
                    Y=[Y, [y;y-y2]];
                end
            end
            % LC devresi ise, R'nin yerini çiz
            if (ii>1)&&(ii<nc)
                if (ctype(ii-1)==10)&&(ctype(ii+1)~=12),
                    X=[X, [x   ; x     ]];
                    Y=[Y, [y-3*y2; y-2*y2]];
                end
            elseif (ii>1)&&(ii==nc)
                if (ctype(ii-1)==10),
                    X=[X, [x   ; x     ]];
                    Y=[Y, [y-3*y2; y-2*y2]];
                end
            end
        end

    end

    % draw line
        line(X,Y,'Color',cl,'LineWidth',lw);

%       % terminate serial circuit
%       if ((ii==nc)&(ctype(end)=0)),
%           line(X,Y,'Color',cl,'LineWidth',lw);
%           line([x1 x1],[y y-lp],'Color',cl,'LineWidth',lw);
%       end

    % update cursor
    if (ii<nc)
        % paralel rezonans devresi için
        if (ctype(ii)~=ctype(ii+1))&& ...
            ( ...
              ( ...
                ((ctype(ii)==4)||(ctype(ii)==5)) && ...
                ((ctype(ii+1)==5)||(ctype(ii+1)==6)) ...
              ) || ( ...
                ((ctype(ii)==10)||(ctype(ii)==11)) && ...
                ((ctype(ii+1)==11)||(ctype(ii+1)==12)) ...
              )...
            )
        %seri rezonans devresi için

            % artýþ yok
        else
            x2=x;
            x=x1;
        end
    else
            x2=x;
            x=x1;
```

```
    end
    %
    ii=ii+1;
end


return

%---------------------------------------------------------------------
 function A=draw_C(x,y,dir,wid,len,c,lw)
% grafik ekrana kondansator seklinin
% yatay veya düsey eksende cizilmesi

wid=abs(wid); % genislik
len=abs(len); % uzunluk
dist=(wid/4); % yapraklar arasi aralik

% yatay veya düsey eksende cizim icin
if dir==0,
    x0=x; % yatay
    y0=y;
    s=1;
else
    x0=y; % düsey
    y0=x;
    s=-1;
end

a=((len-dist)/2);
b=(wid/2);

x1=x0+s*a;
x2=x0+s*(len-a);
x3=x0+s*len;

X(:,1)=[x0 x1]';
Y(:,1)=[y0 y0]';

X(:,2)=[x2 x3]';
Y(:,2)=[y0 y0]';

X(:,3)=[x1 x1]';
Y(:,3)=[y0-b y0+b]';

X(:,4)=[x2 x2]';
Y(:,4)=[y0-b y0+b]';

% text loc
t1=((min(x0)+max(x3))/2);
t2=y0+wid/2;
to=wid/3;
```

```
% draw
if dir==0,
    line(X,Y,'Color',c,'LineWidth',lw);
    A=[t1 t2+to];
else
    line(Y,X,'Color',c,'LineWidth',lw);
    A=[t2+to t1];
end

%-------------------------------------------------------------------
function A=draw_L(x,y,dir,wid,len,c,lw)
% grafik ekrana bobin seklinin
% yatay veya düsey eksende cizilmesi

wid=abs(wid); % genislik
len=abs(len); % uzunluk
n=10; % daire uzerindeki adim sayýsý
tur=4; % bobin tur sayýsý

% yatay veya düsey eksende cizim icin
if dir==0,
    x0=x; % yatay
    y0=y;
    s=1;
else
    x0=y;  % düsey
    y0=x;
    s=-1;
end

% tur capi
r=(wid/3);

% tur sayýsý verilen uzunluga sýðmýyorsa, azalt
while (tur*r*2>=(len-wid/2)), tur=tur-1; end

% kol uzunlugu
a=((len-tur*2*r)/2);

x1=x0+s*a;
X=[x0 x1]';
Y=[y0 y0]';

for i=n:-1:1,
    xa(i)=s*r*(1-cos(pi/n*i));
    ya(i)=y0+s*r*sin(pi/n*i);
end
for i=0:tur-1
    X=[X' x1+xa+2*s*r*i]';
    Y=[Y' ya]';
end
```

```
X=[X' x0+s*len]';
Y=[Y' y0]';

% text loc
t1=((min(X)+max(X))/2);
t2=y0+wid/2;
to=wid/3;
% draw
if dir==0,
    line(X,Y,'Color',c,'LineWidth',lw);
    A=[t1 t2+to];
else
    line(Y,X,'Color',c,'LineWidth',lw);
    A=[t2+to t1];
end
 %-----------------------------------------------
function A=draw_R(x,y,dir,wid,len,c,lw)
% grafik ekrana kondansator seklinin
% yatay veya düsey eksende cizilmesi

wid=abs(wid); % genislik
len=abs(len); % uzunluk

% yatay veya düsey eksende cizim icin
if dir==0,
    x0=x; % yatay
    y0=y;
    s=1;
else
    x0=y; % düsey
    y0=x;
    s=-1;
end

% kutunun eni
a=((wid)/3);
% kutunun boyu
b=((len-1.5*wid)/2);

x1=x0+s*b;
x2=x0+s*(len-b);
x3=x0+s*len;

X(:,1)=[x0 x1]';
Y(:,1)=[y0 y0]';

X(:,2)=[x2 x3]';
Y(:,2)=[y0 y0]';

X(:,3)=[x1 x1]';
```

```
Y(:,3)=[y0-a y0+a]';

X(:,4)=[x2 x2]';
Y(:,4)=[y0-a y0+a]';

X(:,5)=[x1 x2]';
Y(:,5)=[y0-a y0-a]';

X(:,6)=[x1 x2]';
Y(:,6)=[y0+a y0+a]';

% text loc
t1=((min(x0)+max(x3))/2);
t2=y0+wid/2;
to=wid/3;
% draw
if dir==0,
    line(X,Y,'Color',c,'LineWidth',lw);
    A=[t1 t2+to];
else
    line(Y,X,'Color',c,'LineWidth',lw);
    A=[t2+to t1];
end
%-----------------------------------------------------------------

function [out, out_dsc]=ig_selti(x,varargin)
% adding name and components value to right bottom of circuit
% Sentez Sirasinda kullanilan eleman Tipleri ve konumu
% Ali Kilinç - 07*09*2001
n=length(x); z=[]; e=[]; ot=0;
if nargin >=2,
    e=varargin{1};
end
if nargin>2,
    ot=varargin{2};
end
if nargin>3,
    tol=varargin{3};
else
    tol=0.001;
end
% sayýlarý string/karakter dizisi olarak yazým formatý
ns=fix(log10(1/tol)+2);
if ns<3
    ns=3;
elseif ns>8
    ns=8;
end
frm=['%1.' num2str(ns-1) 'f'];

for i=1:n,
```

```
    %eleman deðerini K(kilo),M(Mega),m(mili),u(micro) ile yaz
    nsi=ns;
    if (abs(e(i))>=1e6)
        kt=1e6;
        un='M';
    elseif (abs(e(i))>=1e3)
        kt=1000;
        un='k';
    elseif (abs(e(i))<1)&&(abs(e(i))>1e-3)
        kt=1e-3;
        un='m';
    elseif (abs(e(i))<1e-3)&&(abs(e(i))>=1e-6)
        kt=1e-6;
        un='\mu';
    elseif (abs(e(i))<1e-6)&&(abs(e(i))>=1e-9)
        kt=1e-9;
        un='n';
    elseif (abs(e(i))<1e-9)&&(abs(e(i))>=1e-12)
        kt=1e-12;
        un='p';
    elseif (abs(e(i))<1e-12)&&(abs(e(i))>=1e-15)
        kt=1e-15;
        un='f';
    else
        kt=1;
        un='';
    end
    %ee=num2str(e(i)/kt,frm);
    ee=num2str(e(i)/kt,ns);

    switch x(i)
        case 1,    z='L'; br='H';        % ed='L in Serial branch
';
        case 2,    z='C'; br='F';        % ed='C in Serial branch
';
        case 3,    z='R'; br='\Omega';   % ed='R in Serial branch
';
        case 4,    z='L'; br='H';        % ed='L of parallel R&L&C in
Serial branch';
        case 5,    z='C'; br='F';        % ed='C of parallel R&L&C in
Serial branch';
        case 6,    z='R'; br='\Omega';   % ed='R of parallel R&L&C in
Serial branch';
        case 7,    z='L'; br='H';        % ed='L in shunt branch
';
        case 8,    z='C'; br='F';        % ed='C in shunt branch
';
        case 9,    z='R'; br='\Omega';   % ed='C in shunt branch
';
        case 10,   z='L'; br='H';        % ed='L of Serial R&L&C in
shunt branch  ';
```

```
        case 11,    z='C'; br='F';        % ed='C of Serial R&L&C in
shunt branch  ';
        case 12,    z='R'; br='\Omega';   % ed='R of Serial R&L&C in
shunt branch  ';
        otherwise   z='X'; br='#';        % ed='Non-defined element
';
    end
    out{i} = [z num2str(i) '= ' ee ' ' un '' br];
end

return
```

PROGRAM LIST 4.20. *CASE STUDY 2*

```
% High Precision LC Ladder Synthesis Part II:
% immittance Synthesis with transmission zeros at DC and infinity
% Main Program: CaseStudy2.m
clc;clear all;close all;
format long e
% Enter rational form of F(p)=a(p)/b(p)as in Table VIII
a=[0
0.513976968750998
1.157898233543669
1.213142248104650
0.694401166177243
0.188516938299314
0]';
b=[1.000000000000000
4.198433713455139
7.913422823138356
9.268979838320304
7.284067053656422
3.683494822490302
1.000000000000000]';
%
% Step 1: First extract a pole at infinity from H(p)=b(p)/a(p)
[kinf,R,br]=removepole_atinfinity(b,a)
% Step 2:Remove poles at DC and compute the element values of the
LC Ladder
% as shown in Table X
clear a; clear b;a=br;b=R;
for i=1:5
[kr,R,ar]=Highpass_Remainder(a,b)
k(i)=kr;
clear a; clear b;
a=R;b=ar;
end
k(6)=R/ar
Highpass_Components=1./k
% Termination conductance G=R/ar
G=R/ar
```

```
Resistance=1/G
% Plot Circuit
CT(1)=8;CV(1)=kinf;
CT(2)=7;CV(2)=1/k(1);CT(3)=2;CV(3)=1/k(2);CT(4)=7;CV(4)=1/k(3);
CT(5)=2;CV(5)=1/k(4);
CT(6)=7;CV(6)=1/k(5);CT(7)=9;CV(7)=Resistance;
Plot_Circuitv1(CT,CV)
```

PROGRAM LIST 4.21. LADDERSYNTHESIS_BSYARMAN

```
function [A,CVal] = LadderSynthesis_BSYarman( KFlag,R0,f0,ndc,a,b )
[C_HP,L_HP,C_LP,L_LP,Termination] =
BandPassLadderSynthesis_Yarman(KFlag,R0,f0,ndc,a,b);
nHP1=length(C_HP);nHP2=length(L_HP);nLP1=length(C_LP);nLP2=length(L_LP
);
A=[nHP1 nHP2 nLP1 nLP2];
nmax=max(A);
CHP=fullvector_topdown(nmax,C_HP);
LHP=fullvector_topdown(nmax,L_HP);
CLP=fullvector_topdown(nmax,C_LP);
LLP=fullvector_topdown(nmax,L_LP);
Ter=fullvector_topdown(nmax,Termination);
A='    Series Capacitors(CHP)     Shunt Inductors(LHP)       Shunt
Capacitors(CLP)     Series Inductors(LLP)         Termination';
format long e
CVal=[CHP', LHP', CLP', LLP', Ter'];


end
```

PROGRAM LIST 4.22. FUNCTION REMOVEPOLE_ATINFINITY

```
function [kinf,R,br]=removepole_atinfinity(a,b)
% a(1)>0,b(1)=0,b(2)>0
eps_zero=1e-8;
if a(1)>eps_zero
    if abs(b(1))<eps_zero
kinf=a(1)/b(2);
V=conv(b,[kinf 0]);
nv=length(V);
Remainder=fullvector(nv,a)-conv(b,[kinf 0]);
nR=length(Remainder);
for i=1:nR-2
    R(i)=Remainder(i+2);
end
    for i=1:nR-2
        br(i)=b(i+1);
    end
```

```
    end
else 'F(p)=a(p)/b(p) does not have a pole at infinity'
end
end
```

PROGRAM LIST 4.23. FUNCTION HIGHPASS_REMAINDER

```
function [kr,R,ar]=Highpass_Remainder(a,b)
% Computation of remainder for highpass element extraction
%  Here it is assumed that a(n)=0, a(1)>0,b(1)>0
% Given a minimum function F(p)=a(p)/b(p) such that a(p)=p*a^(p).
% Then we express H(p)=b(p)/a(p)=[1/F(p)]=kr/p+R(p)/ar(p) where
R(p)=b(p)-kr*a^(p)
nb=length(b);
kr=b(nb)/a(nb-1);
for i=1:nb-1
    ar(i)=a(i);
end
ka=fullvector(nb,kr*ar);
for i=1:nb-1
    R(i)=b(i)-ka(i);
end
end
```

PROGRAM LIST 4.24. EXAMPLE 4.7 OF CHAPTER 4-SECTİON 4.3

```
% Main Program: Main_Example4_6.m
% This Program is developed by B.S. Yarman on September 20, 2018.
% Vanikoy, Istanbul, Turkey
% IET Book 2018
% This program synthesizes an impedance transforming filter between 12
ohm
% and 50 ogn over the frequency band of 850 MHz to 2100 MHz
% High Precision LC Ladder Synthesis Part I: Lowpass LC Ladder
Synthesis
clc
clear
close all
% Description of the minimum reactance part of the driving point
impedance:
% Inputs to describe LC Lowpass Ladder
KFlag=1;ndc=0;W=0;a0=1;
R0=1;f0=1/2/pi;n=40;
c=10*rand(1,n);
% Note that random number generator c=10*rand(1,n) generates different
% numbers at each time when it is run.
% In the following we have selected a c-vector of 40 entrees which
```

```
produces
% a minimum function F(p)=a(p)/b(p) that is realizable numerically when
synthesized using our programs.
%------------- Realizable c=[c(1) c(2) c(3).... c(40)] with 40
% entrees
c1=[ 1.547523486560447 1.998628228574521 4.069548371389068
7.487057182156915 8.255838157861556 7.899630299445308];
c2=[3.185242453989922 5.340641273707264 0.8995067877058105
1.117057441932035 1.362925489382987 6.786523048001882];
c3=[4.951770190896606 1.897104060175802 4.950058249902209
1.476082219766887 0.5497414690618818 8.507126742890074];
c4=[5.605595273548848 9.296088667566632 6.966672005552276
5.827909651758400 8.153972114774211 8.790139045971777];
c5=[9.889116160795892 5.223753569447709e-003 8.654385910130246
6.125664694839987 9.899502057088309 5.276800693384423];
c6=[4.795233852102188 8.013476055219522 2.278429357060419
4.980942911963897 9.008524885320048 5.746612191301876];
c7=[8.451781850540366 7.386402919954018 5.859870358264758
2.467345259859749];
%c=[c1 c2 c3 c4 c5 c6 c7];
% Generate Minimum Function
[a,b]=Minimum_Function(ndc,W,a0,c);
% High Precision Lowpass LC Ladder Synthesis:
q=LowpassLadder_Yarman(a,b);
%---------------------------------------------------------------
-----
% Draw result of synthesis:
[CT,CV] = CircuitPlot_Yarman(KFlag,R0,f0,a,b,ndc);
Plot_Circuitv1(CT,CV)
% Generate the driving point input impedance from the topology.
[a1,b1]=ExactLowpassLadder(q);
% Computer errors
eps_a=norm(a-a1)/norm(a)
eps_b=norm(b-b1)/norm(b)
% end of Part (c)
```

PROGRAM LIST 4.25. FUNCTION ZEROSHIFTING_STEP1

```
function [Even_Part, La ] =Zeroshifting_Step1( a,b,wa,eps_zero )
% Zeroshifting Step 1: Computation of La:
% Inputs:
%       a(p) of Z(p)=a(p)/b(p) Minimum reactance
%       b(p) of Z(p)=a(p)/b(p) Minimum reactance
% Output:
%       La
% Step 1
% Finite Transmission zero for which R_even=Even{Z(pa)}=0
```

```
pa=sqrt(-1)*wa ; % Laplace complex variable at p=pa=jwa
% Check the even part if it is less than that of eps_zero
[A,B]=even_part(a,b);% Compute even part of the given impedance Z(p)
Bnorm=norm(B);
pa_sq=pa*pa;% Compute p*p
Num_Even=polyval(A,pa_sq); % Compute numerator of the even part at
p=pa
%                           if it is less than that of algorithmic
zero
if abs(Num_Even/Bnorm)<eps_zero; % if yes, go ahead with zero
shifting
    Na=polyval(a,pa);
    Da=polyval(b,pa);
    Xa=Na/Da;
    % Define La as a series inductor extracted from
    % Z(p)=a(p)b(p)=pLa+a1(p)/b1(p)
    La=imag(Xa/wa);
end
Even_Part=abs(Num_Even);
if abs(Num_Even/Bnorm)>eps_zero;Attention1='wa is not a transmission
zero';end

%


end
```

PROGRAM LIST 4.26. FUNCTION ZEROSHIFTING_STEP2

```
function [ Lb,Cb,kb,a2,b2, r_norm ] =
Zeroshifting_Step2(wa,La,a,b,eps_zero )
% Step 2: Computation of polynomial a1(p) of Z1(p)=a1(p)/b(p)
% Note: In zero shifting Z(p)=pLa+Z1(p); Z1(p)=a1(p)/b(p)
% therefore, a1(p)=a(p)-pLa*b(p)
pa=sqrt(-1)*wa;
bnorm=norm(b);
p=[1 0];% Define MatLab polynomial p
Poly_La=p*La; % Define p*La as a MatLab polynomial.
% At the end of Step 1, we compute Z1(p)=a1(p)/b1(p)=a1(p)/b(p)
% where a1(p) must be zero for p^2=-wa^2.
% In other words a1(p) must be (p^2+wa^2)a2(p).
aa=-conv(Poly_La,b); % aa(p)=-p*La*b(p)
a1=vectorsum(a,aa);% a1(p)=a(p)-p*La*b(p)
u=[1 0 wa*wa]; % Define polynomial p^2+wa^2
[a2,r1]=deconv(a1,u);% Divide a1 by p^2+wa^2
r_norm=norm(r1)/bnorm;
if r_norm<eps_zero; % Algorithm is OK. Go ahead with computations
% Step 2: At this step we define Z2(p)=a2/b(p) which has a zero at
p=pa and
% p=conjugate(pa). Therefore, Y2=b(p)/a2 has a pole at p=(+/-)jwa.
% Hence, Y2(p)=ka*p/(p^2+wa^2)+b2(p)/a2(p).
```

```
% Compute residue ka at p=jwa. In must be pure real number
kb=polyval(b,pa)/pa/polyval(a2,pa);
imag_kb=imag(kb);
kb=real(kb);
% Computation of Lb. Lb is part of a series resonance circuit in
shunt
% configuration: (Lb+Cb)
Lb=1/kb;
% Computation of Cb as a series connection to Lb in shunt
configuration
Cb=1.0/Lb/wa/wa;
       if abs((imag_kb)/norm(kb))<eps_zero; % Algorithm is OK. Go
ahead with computations
            p=[1 0];
              br=kb*p;
              % Perform bb=-kb*p*a2(p)
              bb1=-conv(br,a2);
              % Perform b(p)-kb*p*a2(p)
              b2r=vectorsum(b,bb1);
              % perform division by (p^2+wa^2)
              [b2,r2]=deconv(b2r,u);
       end
end
if r_norm/bnorm>eps_zero; Attention2='Remainder |r1|>0, zero
shifting is not possible';end
if abs((imag_kb/kb))>eps_zero;Attention3='|imag(kb)|>0 zero
shifting is not possible';end



end
```

PROGRAM LIST 4.27. FUNCTİON ZEROSHIFTING_STEP3

```
function [ Lc, a3,b3, L1, L2,M ] =Zeroshifting_Step3(La, Lb, a2,b2 )
% Step 3: Computation of the third series inductor Lc
p=[1 0];
    Lc=a2(1)/b2(1);
    bb3=-conv(Lc*p,b2);
    a3=vectorsum(a2,bb3);
    b3=b2;
    L1=La+Lb;L2=Lb+Lc;M=Lb;
   % ------------------------------------------------------------------
  % Note: a3 has its two leading term zero. Therefore, we have to
erase the
  % first zero-leading term using our function
vector_degreereduction:
   [a3]=vector_degreereduction(1,a3);
```

```
 end
```

PROGRAM LIST 4.28. FUNCTION VECTOR_DEGREEREDUCTION

```
function [B]=vector_degreereduction(n,A)
% This function reduces the degree of vector of nA down to nA-n
starting
% by shifting vector A to the left cutting the leading terms.
% We should note that the leading terms presumably are equal to
zero.
% This function is developed for zero shifting algorithm to shift
the
% remaining vector a_new one block to the left i.e. n=1
nA=length(A);
for i=1:nA-n
    B(i)=A(i+n);
end
```

PROGRAM LIST 4.29. FUNCTION SYNTHESISBYRHPTRANSZEROS

```
function [ CTFinal, CVFinal,LL1,LL2,MM,aT,bT ] =
SynthesisbyRHPTranszeros(KFlag,W,S,ndc,a,b,eps_zero)
% This function is modified on October 14, 2013 by BS Yarman to
include RHP
% transmission zeros.
% S is a MatLab vector which includes all the RHP real transmission
zeros.
% -----------------------------------------------------------------
% This function synthesizes general form of an input impedance
% Z(p)=a(p)/b(p).
%
% -----------------------------------------------------------------
aa=a;bb=b;
if norm(S)>0
if KFlag==1;
[ CTFinal, CVFinal,LL1,LL2,MM,aT,bT ] =
ZeroShifting_RHPAccurateImpedanceSynthesis(1,W,S,ndc,a,b,eps_zero
);
end
if KFlag==0,
    a=bb;b=aa; %Y(p)=aa/bb; Z(p)=bb(p)/aa(p)
    [ index ] = CheckIfab_samedegreewithnonzeroterms( a,b );
    if index==1;% Flip over the admittance function Y(p)=a(p)/b(p)
to synthesize impedance Z(p)=b(p)/a(p)
        [ CTFinal, CVFinal,LL1,LL2,MM,aT,bT ] =
ZeroShifting_RHPAccurateImpedanceSynthesis(1,
W,S,ndc,bb,aa,eps_zero );
    end
    if index==0; % Start with admittance function to extract poles
```

```
of Y(p)=a(p)/b(p)at infinity and DC
        [ CTFinal, CVFinal,LL1,LL2,MM,aT,bT ] =
ZeroShifting_RHPAccurateImpedanceSynthesis(0,
W,S,ndc,aa,bb,eps_zero );
    end
end
% -------------------------------------------------------------------
end
%
if S==0
    LL1=' There is no finite transmission zero'
    LL2='There is no finite transmission zero'
    MM='There is no finite transmission zero'
    R0=1;f0=1/2/pi;
  [CTFinal,CVFinal] = CircuitPlot_Yarman(KFlag,R0,f0,a,b,ndc);
end
n=length(a)-1;
if n>2;Plot_Circuitv1(CTFinal,CVFinal);end
if S==0;aT='No RHP zeros';bT='Nor RHP zeros';end
end
```

PROGRAM LIST 4.30. ***SYNTHESISBYTRANSMISSIONZEROS***

```
function [ CTFinal, CVFinal,LL1,LL2,MM ] =
SynthesisbyTranszeros(KFlag,W,ndc,a,b,eps_zero)
% This function synthesizes general form of an input impedance
% Z(p)=a(p)/b(p).
%
% -------------------------------------------------------------------
aa=a;bb=b;
if norm(W)>0
if KFlag==1;
[ CTFinal, CVFinal,LL1,LL2,MM ] =
ZeroShifting_AccurateImpedanceSynthesis(1,W,ndc,a,b,eps_zero );
end
if KFlag==0,
    a=bb;b=aa; %Y(p)=aa/bb; Z(p)=bb(p)/aa(p)
    [ index ] = CheckIfab_samedegreewithnonzeroterms( a,b );
    if index==1;% Flip over the admittance function Y(p)=a(p)/b(p)
to synthesize impedance Z(p)=b(p)/a(p)
        [ CTFinal, CVFinal,LL1,LL2,MM ] =
ZeroShifting_AccurateImpedanceSynthesis(1, W,ndc,bb,aa,eps_zero );
    end
    if index==0; % Start with admittance function to extract poles
of Y(p)=a(p)/b(p)at infinity and DC
        [ CTFinal, CVFinal,LL1,LL2,MM ] =
ZeroShifting_AccurateImpedanceSynthesis(0, W,ndc,aa,bb,eps_zero );
    end
end
% -------------------------------------------------------------------
end
```

```
%
if W==0
    LL1=' There is no finite transmission zero'
    LL2='There is no finite transmission zero'
    MM='There is no finite transmission zero'
    R0=1;f0=1/2/pi;
  [CTFinal,CVFinal] = CircuitPlot_Yarman(KFlag,R0,f0,a,b,ndc);
end

end
```

PROGRAM LIST 4.31. FUNCTION ZEROSHIFTINGVIACHAINMATRIX

```
function [ a3,b3,CT,CV,La,Lb,Lc,Cb ] = ZeroshiftingViaChainMatrix(
a,b,wa,n,ndc,R0,eps_zero )
% Darlington Type-C Section Extraction from the impedance function
% Z(p)=a(p)/b(p) via Chain Matrix Approach
%   IMPORTANT NOTE: THIS FUNCTION WORKS ONLY ON THE MINIMUM IMPEDANCE
%   FUNCTIONS

% ------------------------------------------------------------------
% Chain Matrix of a Darlington C Section (Type C Section)
% Given Z(p)=a(p)/b(p)
% IMPORTENT NOTE: THIS FUNCTION WORKs ONLY ON THE MINIMUM IMPEDANCES
% ------------------------------------------------------------------
% Step 1: Generate ABCD Parameters from the given impedance
[ A,B,C,D ] = Chain_Matrix( a,b );
%
% ------------------------------------------------------------------
% Step 2:Generation of La, Lb,Lc,Cb a Type C section:
[CT,CV,a31,b31,La,Lb,Cb,Lc,L1,L2,M]=zeroshifting(wa,n,ndc,a,b,eps_zero
);
% ------------------------------------------------------------------
% Step 3: Generation of Chain Parameters for the Type-C Section
A1=[R0*Cb*(La+Lb) 0 R0];
B1=[0 (La+Lc) 0];
C1=[0 R0*Cb 0];
D1=[Cb*(Lb+Lc) 0 1];
% ------------------------------------------------------------------
% Step 4: f1=Lb*Cb (p^2+wa^2 )
f1=Lb*Cb*[1 0 wa*wa];
f1sq=conv(f1,f1);
% ------------------------------------------------------------------
% Step 5: Generate Z2(p)=a2(p)/b2(p)
% a2(p)=   [1/f1^2]*[a(p)D1-b(p)B1]
% b2(p)=   [1/f1^2]*[b(p)A1-a(p)C1]
a2=conv(a,D1)-conv(b,B1);
%
b2=conv(b,A1)-conv(a,C1);
[a3,ra]=deconv(a2,f1sq);
[b3,rb]=deconv(b2,f1sq);
```

```
end
```

PROGRAM LIST 4.32. MAIN_EXAMPLE4_9 OF SECTION 4.3 OF CHAPTER 4

```
% Main_Example4_9_MonopoleAntenna.m
% This Program is developed by B.S. Yarman on September 20, 2018.
% Vanikoy, Istanbul, Turkey
% IET Book 2018
clear
clc
close all
% User Information
display('Example:Monopole Antenna Matching Problem via RFDT Single
Matching')
display('The measured load Data is directly embedded into function
Load(w,KFlag)')
%
% Normalization numbers for the measured load
% Measured antenna data is normalized with respect to R0=50 ohm,
f0=100 MHz

R0=50;
f0=100e6;
%
% Program inputs: Real Frequency Direct Computational Technique
ws1=0.2,ws2=1,
display('ws1 and ws2 are the normalized stop band frequencies')
display('Normalization frequency f0=100MHz')
% Step 2: Enter zeros of the RB
T0=input('Enter Flat Gain Level T0=')
KFlag=input('KFlag=0>Admittance, KFlag=1>Impedance..Enter KFlag=')
ndc=input('Enter transmission zeros at DC ndc=')
W=input('Enter finite transmission zeros at [W]=[....]=')
a0=input('Enter a0=')
[c0]=input('Enter initials for [c0]=[....]=')%sigma(i): RBAl roots of
b(p)

ntr=input('ntr=1, Yes, model with Transformer, ntr=0 No Transformer
ntr=')
%
wc1=input('Enter wc1=')
wc2=input('Enter wc2=')
%
  Nc=length(c0);
% -------------------------------------------------------------------
Antenna_Data =[

           20          0.6          -6
           30          0.8        -2.2
           40          0.8           0
           45            1         1.4
```

```
            50          2          2.8
            55        3.4          4.6
            60          7          7.6
            65         15          8.8
            70       22.4         -5.4
            75         11          -13
            80          5        -10.8
            90        1.6         -6.8
           100          1         -4.4];

FR=Antenna_Data(:,1);RLA=Antenna_Data(:,2);XLA=Antenna_Data(:,3);
      figure (1)
      plot(FR,RLA)
      xlabel('Actual Frequency')
      ylabel('Actual Real Part RLA')
      %
      figure (2)
      plot(FR,XLA)
      xlabel('Actual Frequency')
      ylabel('Actual Imaginary Part XLA')
      %

  % Optimization

% Step 4: Call nonlinear data fitting function
%%%%%%%%%% Preparation for the optimization %%%%%%%%%%%%%%%%%%%%%%
%
 OPTIONS=optimset('MaxFunEvals',20000,'MaxIter',50000);
%
%%%%%%%%%%%%%%%%%%%%%%%%%%%%%%%%%%%%%%%%%%%%%%%%%%%%%%%%%%%%%%%%%%
%
if ntr==1; x0=[c0 a0];Nx=length(x0);end;%Yes transformer case
if ntr==0; x0=c0;Nx=length(x0);end;%No transformer case
%
%f = @(x)direct_Monopole(x,KFlag,T0,wc1,wc2,ntr,ndc,W,a0);
%[x,fval] = fminimax(f,x0);
% Call optimization function lsqnonlin:

x=lsqnonlin('direct_Monopole',x0,[],[],OPTIONS,KFlag,T0,wc1,wc2,ntr,nd
c,W,a0);
% fun=          direct_Monopole (x0,
KFlag,T0,wc1,wc2,ntr,ndc,W,a0)
%
%
% After optimization separate x into coefficients
%
if ntr==1; %Yes transformer case
    Nx=length(x);
        for i=1:Nx-1;
            c(i)=x(i);
        end;
```

```
        a0=x(Nx);
end
if ntr==0;% No Transformer case
    Nx=length(x);
    for i=1:Nx;
        c(i)=x(i);
    end;
end
%
% Step 5: Generate analytic form of Fmin(p)=a(p)/b(p)
% Generate B(-p^2)
        C=[c 1];
        BB=Poly_Positive(C);% This positive polynomial is in w-domain
        B=polarity(BB);% Now, it is transferred to p-domain
% Generate A(-p^2) of R(-p^2)=A(-p^2)/B(-p^2)
nB=length(B);
        A=(a0*a0)*R_Num(ndc,W);% A is specified in p-domain
nA=length(A);
if (abs(nB-nA)>0)
    A=fullvector(nB,A);
% Note that function RtoZ requires same length vectors A and B
end
        [a,b]=RtoZ(A,B);% Here A and B are specified in p-domain
%
% Initialize the performance measures.
Tmax=0;
Tmin=1;
NPrint=100;
DW=(ws2-ws1)/(NPrint-1);
w=ws1;
       cmplx=sqrt(-1);
%
       for j=1:NPrint
           WA(j)=w;
           p=cmplx*w;
           aval=polyval(a,p);
           bval=polyval(b,p);
           Z=aval/bval;
           RB=real(Z);
           XB=imag(Z);
           [RL,XL]=Load(w, KFlag)
           T=4*RB*RL/((RL+RB)^2+(XB+XL)^2);
           TA(j)=T;
           TdB(j)=10*log10(TA(j));
%
% Compute the performance parameters:Tmax,Tmin,Tave and detT
    if max(TA(j))>Tmax
        wmax=WA(j);Tmax=TA(j);
    end
if w>=wc1
    if w<=wc2
```

```
    if min(TA(j))<Tmin
        wmin=WA(j);Tmin=TA(j);
    end
    end
end
% End of Performance measure loop
%
           w=w+DW;
       end
%

 figure(3)
 plot(WA,TA)
 xlabel('Normalized angular frequency w')
 ylabel('Transducer Power Gain')
 title('Matched Monopole Antenna Gain')
 %
 figure(4)
 plot(WA,TdB)
 xlabel('Normalized angular frequency w')
 ylabel('TPG in dB')
  title('Matched Monopole Antenna Gain in dB')
  %
  f0=100e6;R0=50;n=length(b)-1;wa=W(1);eps_zero=1e-5;
  Wnorm=norm(W);
  if Wnorm>0

[a3,b3,LA,LB,LC,LL1,LL2,MM,CT,CV]=SynthesiswithC_Section(KFlag,W,ndc,a
,b,eps_zero) % [CT,CV] = CircuitPlot_Yarman(KFlag,R0,f0,a,b,ndc);
  end
  if Wnorm==0;[CT,CV] = CircuitPlot_Yarman(KFlag,R0,f0,a,b,ndc);end

  %Synthesis of the driving point immittance FB(p)

    % Print Performance parameters:
wmax, Tmax,
wmin,Tmin,
T_Average=(Tmax+Tmin)/2
delT=T_Average-Tmin
dbmax=10*log10(Tmax)
dbmin=10*log10(Tmin)
dB_Average=(dbmax+dbmin)/2
dbdel=dB_Average-dbmin
```

PROGRAM LIST 4.33. SYNTHESIS_IMPEDANCEBASED

```
function [ CT, CV ] = Synthesis_ImpedanceBased( a,b,ndc,R0,f0 )
% This function completes the impedance based synthesis
%   F(p)=a(p)/b(p) is an impedance.
n1=length(a);zero=1e-8;
norma=norm(a);
```

```
normb=norm(b);
norma=1;normb=1;
%norma=1;normb=1;
%-----------------------------------------------------------------------
% Case 1:a(1)=0,b(1)>0. Then, F(p) is a minimum
% reactance function. Therefore we set KFlag=1.
%[ ndc] = DCZeros_Evenpart( a,b )
if abs(a(1))/norma<=zero
        if abs(b(1))/normb>zero
            if abs(b(n1))/normb>zero
                KFlag=1;

                [CT,CV] = CircuitPlot_Yarman(KFlag,R0,f0,a,b,ndc);
Case='Case 1> a(1)=0,b(1)>0,b(n1)>0: Minimum Reactance Input
Impedance F(p)'
            end
        end
end
%End of Case 1
%-----------------------------------------------------------------------
% Case 2:a(1)>0, or b(1)=0, b(n1)=0. Then, F(p) is a minimum
% suseptance function. Therefore we set KFlag=0 and flip over
% F(p) as H(p)=b(p)/a(p)=1/F(p).
if abs(a(1))/norma>zero
        if abs(b(1))/normb<=zero
            if abs(b(n1))/normb<=zero
                KFlag=0;
                %[ ndc] = DCZeros_Evenpart( b,a )
                [CT,CV] = CircuitPlot_Yarman(KFlag,R0,f0,b,a,ndc);
Case='Case 2> a(1)>0,b(1)=0,b(n1)=0: Minimum Suseptance Input
Impedance F(p)'
            end
        end
end
%End of Case 2
%-----------------------------------------------------------------------
% Case 3:a(1)>0, b(1)=0, b(n1)>0. Then, F(p) is not a minimum
%  reactance. It has a pole only at infinity. Therefore, we extract
the
%  pole.
if abs(a(1))/norma>zero
        if abs(b(1))/normb<=zero
            if abs(b(n1))/normb>zero
                [L0,a1,b1]=removepole_atinfinity(a,b);
                L_Act=L0*R0/2/pi/f0;
                KFlag=1;
                %[ ndc] = DCZeros_Evenpart( a1,b1 )
                [CT,CV] =
CircuitPlot_Yarman(KFlag,R0,f0,a1,b1,ndc);
                CT=[1 CT]; CV=[L_Act CV];
Case='Case 3> a(1)>0,b(1)=0,b(n1)>0: F(p) has a pole at infinity'
```

```
            end
        end
end
%End of Case 3
%-----------------------------------------------------------------
% Case 4:a(1)=0, or b(1)>0, b(n1)=0. Then, F(p) is not a minimum
%  reactance function. It has pole at p=0. Therefore, we extract
the
%  pole at p=0.
if abs(a(1))/norma<=zero
        if abs(b(1))/normb>zero
            if abs(b(n1))/normb<=zero
                [C0,a1,b1]=removepole_atzero(a,b);
                C_Act=C0/R0/2/pi/f0;
                KFlag=1;
                ndc1=ndc-1;
                %[ ndc1] = DCZeros_Evenpart( a1,b1 )
                [CT,CV] =
CircuitPlot_Yarman(KFlag,R0,f0,a1,b1,ndc1);
                CT=[2 CT]; CV=[C_Act CV];
Case='Case 4> a(1)=0,b(1)>0,b(n1)=0: F(p) has a pole at p=0'

            end
        end
end
%End of Case 4
%-----------------------------------------------------------------
% Case 5:F(p)>Highpass Circuit Structure as a minimum susceptance
function
if abs(a(1))>zero
    if abs(b(1))>zero
        if abs(b(n1))/normb<=zero
            KFlag=0;
            %[ ndc] = DCZeros_Evenpart( b,a )
            [CT,CV] = CircuitPlot_Yarman(KFlag,R0,f0,b,a,ndc);
Case='Case 5> a(1)>0,b(1)>0,b(n1)=0: Highpass F(p) as Minimum
Susceptance'
        end
    end
end
% End of Case 5
%-----------------------------------------------------------------
% Case 6:F(p)> Highpass Circuit Structure as a minimum reactance
function
if abs(a(1))>zero
    if abs(b(1))>zero
        if abs(a(n1))/norma<=zero
            KFlag=1;
            %[ ndc] = DCZeros_Evenpart( a,b )
            [CT,CV] = CircuitPlot_Yarman(KFlag,R0,f0,a,b,ndc);
Case='Case 6> a(1)=0,b(1)>0,a(n1)=0: Highpass F(p) as Minimum
```

```
Reactance'
        end
    end
end
% End of Case 6
%-------------------------------------------------------------------
% Plot_Circuit4(CT,CV);
end
%End of function
```

PROGRAM LIST 4.34. FUNCTION PLOT_CIRCUIT4

```
function [h]=Plot_Circuit4(varargin)
%
% COMPONENT TYPES
% ------------------------------------------------------------------
% CT=20> Unit Element
% CT=21> Series-Short Stub
% CT=22> Series-Open Stub
% CT=23> Shunt-Short Stub
% CT=24> Shunt-Open Stub
% ------------------------------------------------------------------
% CT=1> Series Inductor
% CT=7> Shunt Inductor
% CT=8> Shunt Capacitor
% CT=2> Series Capacitor
% CT=9> Terminating Resistor
% Series R//L//C: CT=4> Inductor, CT=5> Capacitor,CT=6> Resistor,
% Shunt R+L+C: CT=10>Inductor,CT=11> Capacitor,CT=12>Resistor,
% ------------------------------------------------------------------
%
% Plot_Circuit(CType, CValue, h, tol);
%
% inputs
% CType: components types list vector (given below)
% CValue: Components Values list vector
% h: figure handles
% tol: tolerance (component value meaningful digit)
%
% Component types:
% ------------------------------------------------------------------
%   Lumped Components
%         case 1,    'L in Serial branch                  ';
%         case 2,    'C in Serial branch                  ';
%         case 3,    'R in Serial branch                  ';
%         case 4,    'L of parallel L&C&R in Serial branch';
%         case 5,    'C of parallel L&C&R in Serial branch';
%         case 6,    'R of parallel L&C&R in Serial branch';
%         case 7,    'L in shunt branch                   ';
%         case 8,    'C in shunt branch                   ';
%         case 9,    'R in shunt branch                   ';
```

```
%         case 10,   'L of Serial L&C&R in shunt branch  ';
%         case 11,   'C of Serial L&C&R in shunt branch  ';
%         case 12,   'R of Serial L&C&R in shunt branch  ';
% ----------------------------------------------------------------------------
%   Transmission Line
%         case 20,   'U Unit Element in Cascade          ';
%         case 21,   'U Unit Element in Serial, short circuit
termination ';
%         case 22,   'U Unit Element in Serial, open circuit
termination ';
%         case 23,   'U Unit Element in Parallel, short circuit
termination ';
%         case 24,   'U Unit Element in Parallel, open circuit
termination ';
%
nargin=length(varargin);
    CType=varargin{1};
    CVal=varargin{2};

    %open figure
    if nargin==2
        h=figure ();
    else
        fh=varargin{3};

        if (isempty(fh))||(fh==0)
            h=figure ();
        else
            h=fh;
        end
    end

    if length(CType)==0
        return
    end

    if nargin>3
        tol=varargin{4};
    else
        tol=0.0001;
    end

    % circuit drawing colors
    dc=struct(  'cl','k',... % line color
                'cL','r',... % inductance color
                'cC','b',... % cap. color
                'cR','m',... % resistor color
                'cU','g');   % tline unit element


    % starting location of overall figure,
```

```
    %   input port connection point y axis = (gnd level) +
(ComponentSize)

    dwx0=struct('xn',0,      ...      %   output port cascade
connection point x axis
                'yn',70,     ...      %   output port cascade
connection point y axis
                             ...      %   (y axis)= (Component Size) +
( y offset)
                'n',0,       ...      %   component number index
                'ymax',0,    ...      %   drawing size y axis highest
level
                'ComponentSize',60,...  %
                'FontSize',8,...      %   Label Font Size
                'CT',CType,...        %   Component Types List
                'dc',dc,...           %   drawing colors
                'FuncName','Z_{in}');

    draw_ladder_circuit(h, dwx0,CVal,tol);

end
```

PROGRAM LIST 4.35. MAIN_SPARBASEDSYNTHESIS

```
% Program Main_SParBasedSynthesis.m
% This program tests the scattering parameters based synthesis
clc
clear
close all
f0=1/2/pi;  R0=1;
format short
n=input('Enter number of h(p) coefficients n=')
%h=input(' Enter coefficients of h(p) in MatLab Format: h=[...]=')
h=rand(1,n)
ndc=input(' Enter transmission zeros of the lossless LC ladder at DC
ndc=')
%---------------------------------------------------------------------
% Swap h(p)as a text book polynomial vector in
hF(p)=hF0+hF1p+hF2p^2+...+hFnp^n
nF=length(h);
for i=1:nF
hF(i)=h(nF-i+1);
end
% Generate gF as a text book
polynomial:gF(p)=gF0+gF1p+gF2p^2+...+gFnp^n
    gF=poly_g(ndc,hF);
% Generate standard MatLab polynomial g(p):
for i=1:nF
g(i)=gF(nF-i+1);
end
% Computation of the actual element values of the lossless LC Ladder
```

```
            a=g+h;b=g-h;
%Swap a(p) and b(p)to end up with standard MatLab Polynomials
%
[ ndcA] = DCZeros_Evenpart( a,b )
%-------------------------------------------------------
[ CT CV ] = Synthesis_ImpedanceBased( a,b,ndc,R0,f0 );
%--------------------------------------------------------------------------
format short
```

PROGRAM LIST 4.36.

```
function [ k,q,a0,nA,A,B ] = Richard_Numerator( a,b )
% This function generates k, q and a0 of the even part F=a/b
specified in
% lambda domain. Note that R=A/B in lambda domain-(Richard Domain)
% A(lambda^2)=(a0)^2*(-1)^q*(lambda)^2q*[1-(lambda)^2]^k
% Here, in this function we count non zero terms both from bottom-
up with
% q=x; k=nA-y-q;
% Inputs:
%       a(lambda)
%       b(lambda
% Outputs:
%       k: Total number of cascaded UEs.
%       q: Total number of transmission zeros
%       (a0)^2:Leading coefficients of A(lambda)
%--------------------------------------------------------------------------
zero=1e-4;
% Norma=norm(a);a=a/Norma;b=b/Norma;
nb=length(b);B0=b(nb)*b(nb);
[A,B]=even_part(a,b);
Norm=norm(A);A=A/B0;B=B/B0;
%
nA=length(A);
% Determine q & k: start counting the unity roots (i.e. find non
zero A(i))
% Count from bottom-up
 [ q,nA ] = Richard_DCzeros( A );
% ----- end of bottom up counting ---------------------------------
j=1;s=0;
% Count from top-down
while abs(A(j))<zero;
    j=j+1;
    s=s+1;
end
    y=j;
% ----- end of top down  counting ---------------------------------
%
```

```
        k=nA-y-q;
% -----------------------------------------------------------------
a0=sqrt(abs(A(j)));
% Determine q
n=length(A)-1;
%
%[ q ] = Richard_DCzeros( A );
%
A=A/a0/a0;B=B/a0/a0;% Normalization of vectors A and B with respect
to a0*a*
% This means that non-zero leading coefficient of A(lambda^2) is 1.
end
```

PROGRAM LIST 4.37. FUNCTION RICHARD_NEWMINIMUMFUNCTION

```
function [a,b]=Richard_NewMinimumFunction(k,q,c,c0)
% Generate analytic form of Fmin(lambda)=a(lamda)/b(lambda)
% Generate B(-lambda^2)
% -----------------------------------------------------------------
% In this function first A(lambda^2)=(-1)^q*(1-lambda^2)^k is
generated
% with unit leading coefficient (a0=1).
% Then, B(Omega^2)=(1/2)[C(Omega)^2+C(-Omega)^2] is computed.
% where C(omega) is a full polynomial: C=[c c0].
% At the last step, a(lambda) and b(lambda) are normalized with
respect
% to norm of a(lambda).
% ----------------------------------------------------------------
% Inputs:
%            k: Total number of Cascaded UEs
%            q: Total number of Transmission zeros at DC (q=ndc)
%            c(Omega): Auxiliary polynomial of the denominator
B(Omega^2)
%            c0=last coefficient (n+1) of c(Omega)
% Outputs:
%            a(lambda): Numerator polynomial in lambda domain
%            b(lambda): Denominator polynomial in lambda domain
% ----------------------------------------------------------------
% It is important to note the following points:
% (a) Degree n=length(c) must be greater or equal to k+q
% (b) q=0 yields lowpass structure
% (c) if q=0 and if k=n then we only have n-cascaded UE
% (d) if q=0 and n>k then, we must have series short and shunt open
stubs
% ----------------------------------------------------------------
n=length(c);
if n<(q+k)
    Attention='n<(q+k) therefore you have unrealizable demand'
end
if n>=(q+k); % then start computations
        C=[c c0];
```

```
        BB=Poly_Positive(C);% This positive polynomial is in Omega-
domain
        B=polarity(BB);% Now, it is transferred to lamda-domain
% Generate A(-lamda^2) of R(-lamda^2)=A(-lamda^2)/B(-lamda^2)
nB=length(B);
A=Richard_kq(k,q); % A is specified in lambda-domain
nA=length(A);
if (abs(nB-nA)>0)
  A=fullvector(nB,A);% work with equal length vectors
end
% Generation of minimum immittance function using Bode or Parametric
method
ndc=q;
        [a,b]=RtoZ(A,B);% Here A and B are specified in p-domain
        na=length(a);
        if ndc>0;a(na)=0;end;
end
%
Norm=norm(a);a=a/Norm;b=b/Norm;
end
```

PROGRAM LIST 4.38. FUNCTION RICHARDCOMPLETE_IMPEDANCESYNTHESIS

```
function [Z_UE,a_new,b_new,CTF,CVF ] =
Richard_CompleteImpedanceSynthesis(a,b,k,q,R0,f0 )
% This function carries out complete synthesis in lambda domain
% Inputs:
%       k; Total number of cascaded UEs
%       q: Total number of high-pass elements as series open-stubs,
parallel shorted-stubs
%       Zin=a/b
%       R0: Normalization resistance
%       f0: Normalization frequency but it must be fixed at f0=1/2/pi
% Outputs:
%       Z_UE: Characteristic impedances of the Unit Elements
%       a_new,b_new: Z_new=a_new/b_new is the remaining impedance
after
%       extraction of k-UE.
%       CT,CV: Circuit codes and values of the ladder synthesis
%
ndc=q;
%
if k==0;a_new=a;b_new=b;
    Z_UE='k=0. There is no UE in the synthesis'
end
%
% ------ Definition of Algorithmic zero --------------------------
zero=1e-10;
zeroA=1e-10;
aux=[a b];
Kmax=max(aux);
```

```
a1=a/Kmax;b1=b/Kmax;na=length(a);
    if q>0;
    zeroA=a1(na);
    end
%
if zero<zeroA;zero=zeroA;end
% ------ End of Definition of Algorithmic zero --------------------
if abs(a(1))<zero;[a,b]=Richard_ImmittanceCorrection(a,b,k,q);end
if abs(b(1))<zero;[b,a]=Richard_ImmittanceCorrection(b,a,k,q);end
%------------------------------------------------------------------
n=length(a)-1;
if k<=n
    if k>0
%High precision-impedance based-Richard Extraction:
% [Z,a_new,b_new] = Richard_HighPrecisionUEExtraction( k,q,a,b );
% [Z,a_new,b_new] = Richard_RoughUEExtraction( k,q,a,b );
 [ Z,a_new,b_new ] = Richards_CameronExtractions( k,q,a,b );
  Z_UE=Z;
%------------------------------------------------------------------
    end
%------------------------------------------------------------------
a1=a_new;b1=b_new;
na=length(a_new);nb=length(b_new);
% a_new(na) might be zero and b_new(na) might be zero. In this case
degree
% reduction in "q" occurs.
if abs(a_new(na))<zero;
if abs(b_new(nb))<zero;
    clear a_new;clear b_new;
    for i=1:na-1
        a_new(i)=a1(i);
        b_new(i)=b1(i);

    end
    q=q-1;% Degree reduction in q
        Attention=('Degree reduction occurs in q=q -1')
        q1=q
end
end
% End of degree reduction loops
%------------------------------------------------------------------
end
%
nc=n;
if nc>k
%
if
abs(a_new(1))<zero;[a_new,b_new,q]=Check_immittance(a_new,b_new,q);end
if
abs(b_new(1))<zero;[b_new,a_new,q]=Check_immittance(b_new,a_new,q);end
    if abs(a_new(1))>zero;
```

```
        if abs(b_new(1))>zero;
            [a_new,b_new,q]=Check_immittance(a_new,b_new,q);
        end
    end
    %
[ CT CV ] = Synthesis_ImpedanceBased( a_new,b_new,q,R0,f0 );
end
if k==n
    CT=('k=n; Therefore, synthesis is completed only with cascade
connection of n-unit elements')
    CV=CT;


end
% Plot the remaining synthesized Circuit in lambda domain
% if n>k; Plot_Circuitv4(CT,CV);end
% ---------------------------------------------------------------------
% Plot of the complete commensurate transmission line circuit
nue=length(Z_UE);
for i=1:nue
    CT1(i)=20; CV1(i)=Z_UE(i);
end
if n>k
[ CTD,CVD ] = RichardsLadderTo_TrLineTopology( CT,CV );
CTF=[CT1 CTD];CVF=[CV1 CVD];
end
if n==k
  CTF=[CT1 9]; CVF=[CV1 a_new/b_new];
end
%Plot_Circuit4(CTF,CVF)
end
```

Chapter 5

Semi-analytic approaches to broadband matching problems: real frequency techniques

Binboga Siddik Yarman[1]

The analytic theory of broadband matching is essential to understand the gain–bandwidth limitations of the given impedances to be matched. However, its applicability is limited beyond simple problems. By simple we mean those problems of single or double matching in which the generator and load networks include at most one reactive element, either a capacitor or an inductor. For simple impedance terminations, lowpass equal ripple or flat gain prototype networks, which are obtained employing the analytic theory, may have practical usage. On the other hand, if the numbers of elements that increase in the impedance models is to be matched, the theory becomes inaccessible. If it is capable of handling the problem, the resulting gain performances turn out to be suboptimal. Equalizer structures become unnecessarily complicated, and even it may not be feasible to manufacture them. Therefore, in practice, computer-aided-design (CAD) techniques are preferred; commercially available programs are employed to solve the matching problems. Readily available tools are very good in analyzing and optimizing the given structures, but they do not include network synthesis procedures in literal sense. In other words, in designing a matching network or a microwave amplifier, a topology for the matching network with good initial element values should be supplied to a commercially available package. In this respect, many CAD packages work as fine trimming tools on the element values when the circuit is practically synthesized. Usually, a simple two-element, capacitor–inductor ladder network is a practical solution for narrow bandwidth matching problems. However, if the optimum topology of the equalizer is unknown, or if substantial bandwidth is requested, design task becomes difficult. In this case, modern CAD techniques which are known as the "real frequency techniques" are strongly suggested to design matching networks.

In all single and double-matching real-frequency algorithms, the goal is to optimize the transducer power gain (TPG), as high and flat as possible in the

[1]Department of Electrical and Electronics Engineering, College of Engineering, Istanbul University-Cerrahpasa, Istanbul, Turkey

band of operation. In the real frequency techniques, the matching network $[E]$, generator $[G]$ and load $[L]$ networks are considered as separate entities. TPG is expressed in terms of these entities. The lossless equalizer $[E]$ is either described in terms of its driving point "back-end impedance" Z_B (or equivalently admittance $Y_B = 1/Z_B$) or in terms of its unit normalized scattering parameters $[E_{ij}]$. Descriptive parameters of $[E]$ are chosen as the unknowns of the problem, and they are determined as the result of an optimization process. In this way, analytic extraction of generator and load networks is simply omitted.

In the following, first, the real frequency line-segment technique (RFLT) is introduced to handle single matching problems. Then, "Direct methods" via Bode or Gewertz procedures are discussed to solve double matching problems. Finally, simplified real frequency technique (SRFT) which is suitable to design single and multistage microwave amplifiers is detailed.

5.1 Real frequency line-segment technique

In 1977, a numerical approach known as the "RFLT" was introduced by H.J. Carlin for the solution of single matching problems [1]. RFLT utilizes measured load data, bypassing the analytic theory. Neither the equalizer topology nor the analytic form of a transfer function is assumed. They are the result of the design method. Measured data obtained from the devices to be matched is directly processed.

For single matching problems, Carlin describes the lossless equalizer in terms of its minimum driving point input immittance $F_B(j\omega) = R_B(\omega) + jX_B(\omega)$ at the back-end as shown in Figure 5.1.

In Carlin's approach $F_B(j\omega) = R_B(\omega) + jX_B(\omega)$ is assumed to be a minimum function [1]. This assumption does not disturb the general form of the driving point immittance since we can always add a Foster part to it.

The real part $R_B(\omega)$ is chosen as the unknown of the matching problem.

The imaginary part $X_B(\omega)$ is generated from $R_B(\omega)$ using the Hilbert transformation relation.

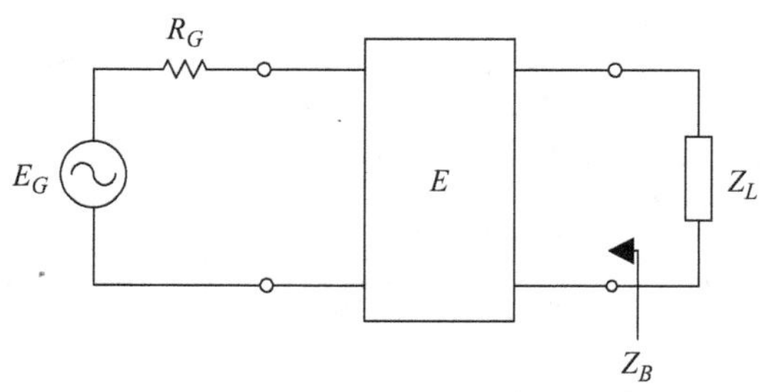

Figure 5.1 Single matching problem: power transfer from a resistive generator to a complex load

Furthermore, $R_B(\omega)$ is piecewise linearized as shown in Figure 5.2. In other words, $R_B(\omega)$ is expressed as the linear combination of line segments as follows:

$$R_B(\omega) = \left\{ \begin{matrix} a_j\omega + b_j & \text{for } \omega_j \le \omega \le \omega_{j+1}; \quad j = 1, 2, \ldots, (N-1) \\ 0 & \omega \ge \omega_N \end{matrix} \right\} \tag{5.1a}$$

where

$$a_j = \frac{R_j - R_{j+1}}{\omega_j - \omega_{j+1}} = \frac{\Delta R_j}{\omega_{j+1} - \omega_j} \tag{5.1b}$$

and

$$\Delta R_j = R_j - R_{j+1} \tag{5.1c}$$

In (5.1a)–(5.1c) ω_j is called *the break frequency* where R_B takes the value $R_j = R_B(\omega_j)$ Therefore, R_j is called *the break point*, ΔR_j is the *excursion* of the jth segment and N designates the number of break points.

By numerical Hilbert transformation relation, $X_B(\omega)$ is given as

$$X_B(\omega) = \sum_{j=1}^{N-1} B_j(\omega)\Delta R_j \tag{5.2}$$

where

$$B_j(\omega) = \frac{1}{\pi(\omega_j - \omega_{j+1})}\left[F_{j+1}(\omega) - F_j(\omega)\right] \tag{5.3}$$

and

$$F_j(\omega) = (\omega + \omega_j)\ln(|\omega + \omega_j|) + (\omega - \omega_j)\ln(|\omega - \omega_j|) \tag{5.4}$$

The immittance function $F_B(j\omega)$ can be selected as either minimum reactance or minimum susceptance depending on the variation of the load.

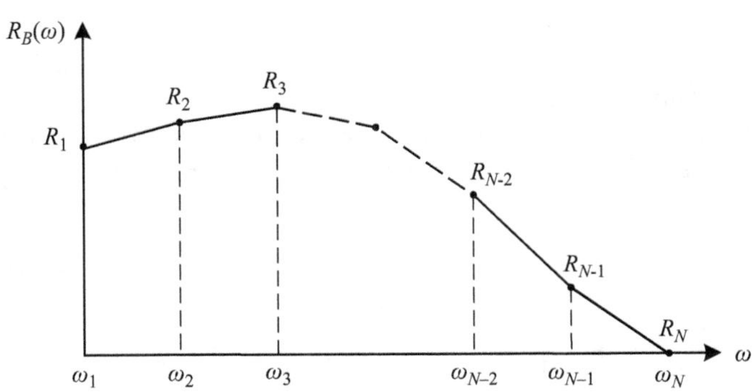

Figure 5.2 Piecewise linear approximation of $R_B(\omega)$

For example, if the load data is capacitive over the band of operation then, at the back end, the equalizer demands inductive driving point impedance so that possible reactance cancellation occurs to maximize the TPG. In this case, designer should prefer to start with a minimum susceptance admittance which avoids capacitive input. On the other hand, if the load is inductive, then the designer may prefer to start with minimum reactance impedance for the abovementioned reasons.

Let the measured load immittance be $F_L(j\omega) = R_L(\omega) + jX_L(\omega)$; then the transducer gain $T(\omega)$ is given by

$$T(\omega) = \frac{4R_B(\omega)R_L(\omega)}{[R_B(\omega) + R_L(\omega)]^2 + [X_B(\omega) + X_L(\omega)]^2} \tag{5.5a}$$

It should be emphasized that, in (5.5a) and (5.5b), the form of the TPG is generic in terms of F_B *and* F_L. In other words, the pair $\{F_B, F_L\}$ is either impedance or admittance preserving the form of (5.5a) and (5.5b).

The ultimate goal of the matching problem is to maximize the TPG as high and as flat as possible. Let physically attainable flat value of the gain be T_0. This level may be achievable by means of an optimization over the passband $B : \omega_{c1} \le \omega \le \omega_{c2}$.

Then, loosely speaking, we can state that[1]

$$T_0 \approx \frac{4R_B(\omega)R_L(\omega)}{[R_B(\omega) + R_L(\omega)]^2 + [X_B(\omega) + X_L(\omega)]^2} \ge 0; \quad \forall\omega \tag{5.5b}$$

or

$$T_0[R_B(\omega) + R_L(\omega)]^2 + [X_B(\omega) + X_L(\omega)]^2 - 4R_B(\omega)R_L(\omega) \approx 0 \tag{5.6}$$

By selecting a flat gain level T_0, and taking $R_B(\omega)$ as the unknown of the matching problem as specified by (5.1a)–(5.1c), and generating $X_B(\omega)$ as in (5.2), one can always come up with a reasonable spread for $R_B(\omega)$ by means of a nonlinear curve fitting algorithm. This is the crux of the Carlin's RFLT.

In short, the above is summarized in the following.

In RFLT, the unknowns of the single matching problem are the break points R_j *or equivalently the excursions* ΔR_j.

The imaginary part $X_B(\omega)$ is generated in terms of the same break points employing (5.2) and (5.3).

Thus, the single matching problem is converted to a nonlinear curve fitting problem with the following objective function which stems from (5.6):

$$error(\omega) = T_0[R_B + R_L(\omega)]^2 + \left[\sum_{j=1}^{N-1} B_j(\omega)\Delta R_j + X_L(\omega)\right]^2 - 4R_B(\omega)R_L(\omega) \tag{5.7a}$$

[1]The sign " $\approx$ " denotes the "approximately equal" concept.

or directly, we can state that

$$error(\omega) = T(\omega) - T_0 \tag{5.7b}$$

where $R_B(\omega)$ is also specified in terms of ΔR_j in linear manner as expressed by (5.10a)–(5.10c) and (5.11a)–(5.11d).

Once the break frequencies are selected by the designer, the objective function $error(\omega)$ is minimized to determine the unknown break points R_j *or equivalently* $\Delta R_j; j = 1, 2, 3, \ldots, N-1$.[2]

Clearly, the objective function $error(\omega)$ is quadratic in the unknown break points. Therefore, it is always convergent.

5.1.1 Solution to single matching problem with reactance cancellation: generation of initials for the nonlinear optimization

Close examination of (5.5a) and (5.5b) reveals that TPG is maximized if the perfect reactance cancellation occurs yielding $X_B = -X_L$.[3]

Under the perfect reactance cancellation, let T_0 be the physically attainable flat-gain level imposed by the theory of broadband matching. Then, the following equation must be satisfied:

$$T_0 = \frac{4R_B R_L}{(R_B + R_L)^2} \tag{5.8a}$$

or

$$T_0 R_B^2 - 2R_B R_L(1 - T_0) + T_0 R_L^2 = 0 \tag{5.8b}$$

In this case, idealized form of the real part $R_B(\omega)$ is obtained as

$$R_{B0}(\omega) = R_L(\omega)\left[\frac{(2 - T_0) + 2\text{sign}\sqrt{1 - T_0}}{T_0}\right] \geq 0; \quad \forall\omega \tag{5.9}$$

And the corresponding driving point immittance $F_{eq}(j\omega)$ of the equalizer is

$$F_{eq}(j\omega) = R_B + jX_{eq} \tag{5.10a}$$

$$F_{eq}(j\omega) = R_L(\omega)\left[\frac{(2 - T_0) + 2\text{sign}\sqrt{1 - T_0}}{T_0}\right] - jX_L(\omega) \tag{5.10b}$$

$$X_{eq} = -X_L \tag{5.10c}$$

[2]It is customary to distribute the break frequencies evenly within the passband and $R_B(\omega)$ is set to 0 at the stop. Therefore, for lowpass problems, total number of break points runs up to $N-1$ since the last break point $R_N = 0$ at ω_N for which $T(\omega_N) = 0$.

[3]Maximization of gain requires that $\frac{\partial T}{\partial X_B} = 0$ which in turn yields $X_B = -X_L$.

In general, over the entire real frequency axis, an arbitrary immittance $F_{eq}(j\omega)$ can be expressed as

$$F_{eq}(j\omega) = R_B(\omega) + jX_{eq}(\omega) = R_B(\omega) + jX_B(\omega) + jX_F(\omega) \tag{5.11a}$$

or

$$F_{eq}(j\omega) = F_B(j\omega) + jX_F(\omega) \tag{5.11b}$$

where

$$X_B(\omega) = \textit{Hilbert Transform of } \{R_B(\omega)\} \tag{5.11c}$$

$$X_F(\omega) = X_{eq}(\omega) - X_B(\omega) = -(X_L + X_B) \tag{5.11d}$$

In the above equation set $X_F(\omega)$ is the Foster portion of the driving point immittance and it must satisfy $dX_F(\omega)/d\omega = \geq 0 \; for \; \forall\omega$.

From equalizer design point of view, (5.11a) is quite important to understand if $F_{eq}(j\omega)$ is realizable with acceptable tolerance.

Let us investigate this situation by an example.

Example 5.1

Let the immittance $L + R//C$ with $L = 1.0, C = 3, R = 1$ be the load network to be matched to a resistive generator over the normalized angular frequency band $0 \leq \omega \leq 1$.[4]

1. Select $T_0 = 0.75$, generate the ideal driving point impedance $Z_{eq}(\omega) = R_B(\omega) + jX_B(\omega) + jX_F(\omega)$ over $N = 7$ break points by setting $\omega_N = 2$.
2. Note that, in (5.9), R_B has two values, namely, low $(sigma = -1)$ and $high(sigma = +1)$. Generate both high and low values of R_B and determine corresponding values of $X_B(\omega)$ *and* $X_F(\omega)$. Finally, plot $\{R_B(\omega), X_B(\omega)$ *and* $X_F(\omega)\}$ *vs* ω.
3. Comment on the realizability of $Z_{eq}(\omega)$ by examining the plot of the Foster part $X_F(\omega)$.
4. Plot the TPG of (5.5a) and (5.5b) with the minimum reactance function $F_B(j\omega) = Z_B(j\omega) = R_B(\omega) + jX_B(\omega)$.
5. Minimize the error function given by (5.7a) and (5.7b) which results in optimum $F_B(j\omega) = R_B(\omega) + jX_B(\omega)$.

Answer

1. This part of the example handled in two steps.

   In Step 1, we develop the following MATLAB® functions.

   $$\textit{Function WBR} = \textit{Break\_Frequencies}(N, ws1, ws2, wc1, wc2)$$

[4]Referring to Example (5.1), it is noted that this load has no monotone ladder Chebyshev solution since $(>L=1)>(L_{\max}<0.89)$

which generates the break frequencies for specified integer N (total number of break points), $ws1$ (lower edge of the stop band), $ws2$ (upper edge of the stop band frequency), $wc1$ (lower edge of the passband frequency) and $wc2$ (upper edge of the passband frequency). The first and the last break frequencies are placed at the frequencies $ws1$ and $ws2$, respectively. Then, the rests are evenly distributed over the passband ($wc2$–$wc1$). In *function WBR*, from the programming point of view, for lowpass problems ($wc1 = 0$ *case*), $ws1$ is chosen to be equal to $wc1$ to end up with $WBR(1) = ws1 = wc1 = 0$.

As specified, we select $N = 7$ break frequencies to sample the load impedance. Passband starts from $\omega_{c1} = 0$ and it ends at $\omega_{c2} = 1$. Therefore, we select $ws1 = wc1 = 0$. The first break point is placed at DC. Total number of $N - 1 = 6$ points are evenly distributed within the passband $0 \leq \omega \leq 1$. $WBR(7)$ is placed at $\omega = 2$. Thus, the break frequencies $WBR(or\ \omega_j)$ are given as

$$WBR = [0\ \ 0.2\ \ 0.4\ \ 0.6\ \ 0.8\ \ 1\ \ 2]$$

$$function[RL, XL] = RLC\_Load(w, R, C, L, KFlag);$$

generates the load data at a given angular frequency ω. Obviously, for the given $L + R//C$ load, the inputs are $L = 1.0, C = 3, R = 1$.

The input $KFlag$ controls the type of the immittance. If $KFlag = 1$, the user works with impedances. If $KFlag = 0$ then, admittances are employed throughout the program.

Then, over the break frequencies, the load impedance is generated as $ZLA = RLA + jXLA$ such that

$$RLA = [1\ \ 0.73529\ \ 0.40984\ \ 0.23585\ \ 0.14793\ \ 0.1\ \ 0.027027]$$

$$XLA = [0 - 0.44118 - 0.4918 - 0.42453 - 0.35503 - 0.3 - 0.16216]$$

In Step 2 of this part, the function called

$$[RBA, XBA, XFA] = Ideal\_SingleMatching(T0, \text{sign}, WBR, RLA, XLA)$$

generates the ideal driving point immittance as in (5.10a)–(5.12b).

In this case, function XB is determined employing the MATLAB function

$$XB = num\_hilbert(w, WBR, RBA)$$

All the above functions are gathered in the main program called Example5_1.m. Thus, the execution of Example5_1.m yields the following results.

In Table 5.1, low and high values of the driving point impedance and the corresponding Foster parts are listed.

Plots of the above quantities are depicted in Figures 5.3–5.5.

Thus, part (1) of the example is completed.

2. Close examination of $X_F(\omega)$vsω plot indicates that at higher frequencies ($\omega > 0.4$) $\partial X_F/\partial\omega < 0$. Therefore, the ideal solution does not exist.

Table 5.1 Generation of idealized driving point impedance using function $[RBA, XBA, XFA] = Ideal\_SingleMatching(T0, \text{sign}, WBR, RLA, XLA)$

WBR	R_{B-Low}	X_{B-Low}	R_{B-High}	X_{B-High}	X_{F-Low}	X_{F-high}
0	0.3333	0	3	0	0	0
0.2	0.2451	−0.13081	2.2059	−1.1773	0.37198	1.4184
0.4	0.13661	−0.16016	1.2295	−1.4414	0.25196	1.5332
0.6	0.078616	−0.13996	0.70755	−1.2597	−0.035507	1.0842
0.8	0.04931	−0.11594	0.44379	−1.0435	−0.32903	0.5985
1	0.033333	−0.09518	0.3	−0.85662	−0.60482	0.15662
2	0.009009	−0.051877	0.081081	−0.46689	−1.786	−1.3709

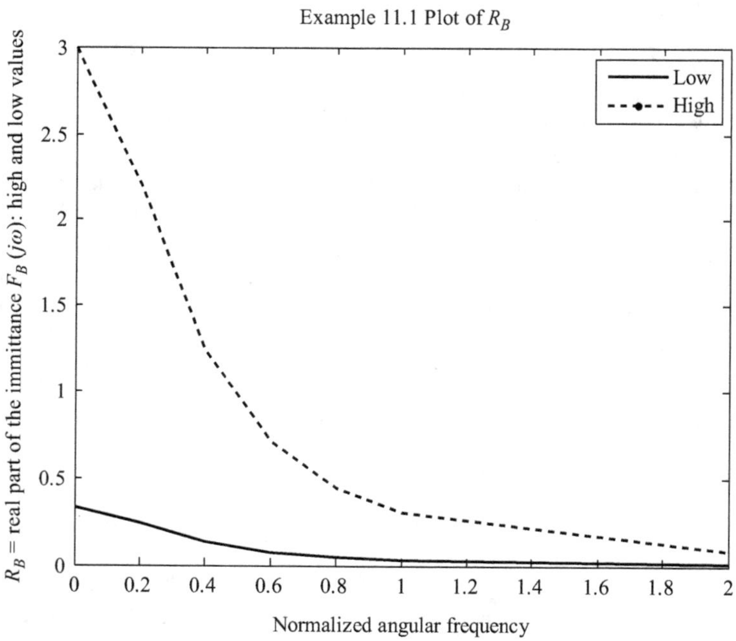

Figure 5.3 High and low values of the real part R_B of (5.9)

Nevertheless, at the lower frequencies $(\omega < 0.4), \partial X_F/\partial\omega > 0$. So, it may be possible to make a reasonable approximation for the Foster part at lower frequencies.

As we see from the above figure that, lower values of R_B results in smaller Foster part. In this case, we expect to end up with a reasonable initial guess for the minimization of the error function specified by (5.7a) and (5.7b).

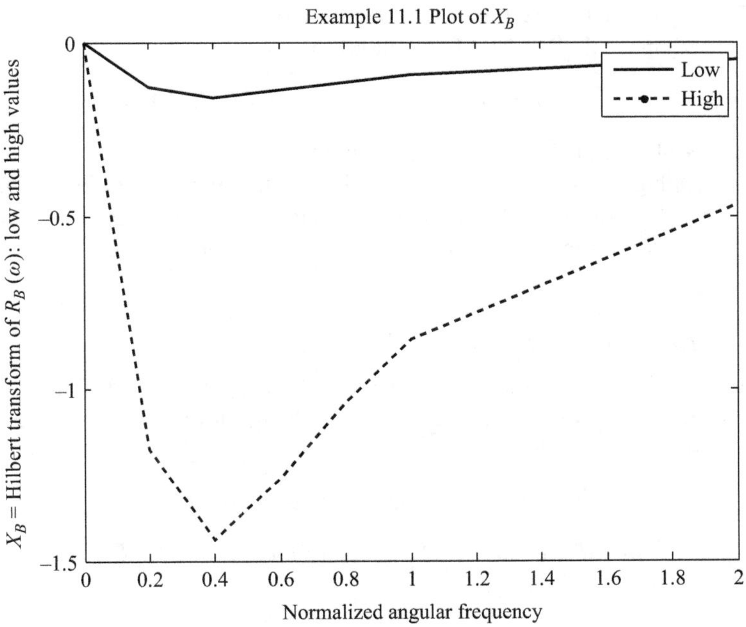

Figure 5.4 High and low values of X_B = Hilbert Transform$\{R_B\}$

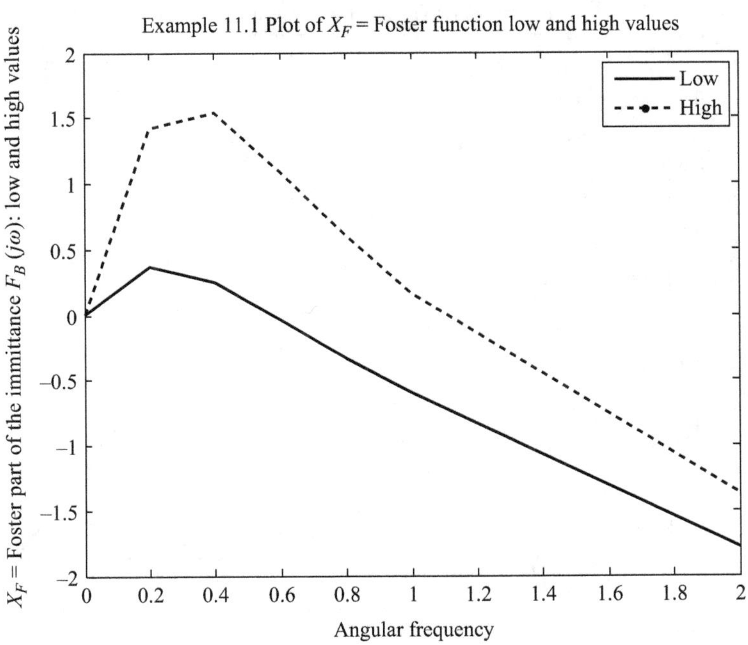

Figure 5.5 High and low values of the Foster part $X_F(\omega) = -(X_L + X_B)$

3. For this part of the example, the gain function of (5.5a) and (5.5b) is programmed under the MATLAB function

$$[Gain\_SM] = gain\_singleMatching\,(WBR, RLA, XLA, RBA, XBA)$$

and it is placed in the main program Example5_1.m so that we can plot the gain with high and low values of R_B. The result is depicted in Figure 5.6.

As pointed out in part (2), up to $\omega = 0.8$ lower values of R_B results in better gain approximation. However, for $0.8 \leq \omega \leq 1$ higher values of R_B yields higher gain.

5.1.2 Gain optimization for RFLT

4. In this part of the example, the error function given by (5.7a) and (5.7b) is minimized using the Levenberg–Marquardt algorithm which is called as a MATLAB function block as

$$OPTIONS = OPTIMSET(\text{'}MaxFunEvals\text{'}, 20{,}000, \text{'}MaxIter\text{'}, 50{,}000);$$

$$\begin{aligned} x = &\, lsqnonlin(\text{'}error\_RFLT\text{'}, x0, [\,], [\,], OPTIONS, WBR, wc1, wc2, \\ &\, N, T0, R, C, L, KFlag) \end{aligned}$$

error_RFLT is a user supplied function written by us which programs the error function for the RFLT.

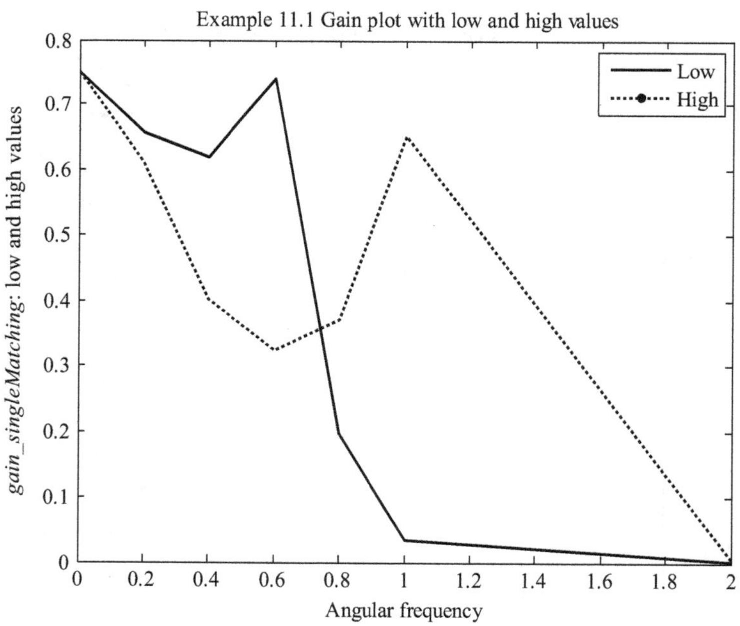

Figure 5.6 Initial gain plot for the single matching problem introduced by Example 5.1

At this point we should note that, in the main program, the optimization is nonlinear, and it requires an excellent initial guess. For the least square-nonlinear optimization (*lsqnonlin*), initial nonnegative break points are generated by a MATLAB function called

$$RBA = initials(R, C, L, KFlag, T0, \text{sign}, WBR)$$

as in (5.7a) and (5.7b). Furthermore, all the break points must be nonnegative. This simple constraint can easily be achieved by defining the unknown vector x as

$$RBA(j) = x^2(j) \geq 0 \tag{5.12a}$$

or the initials vector $x0$ is given by

$$x0(j) = \mp\sqrt{RBA(j)} \tag{5.12b}$$

where index j runs from 1 to $N-1$.

5. Execution of the program Example5_1.m optimizes the TPG which in turn results in the optimum minimum reactance driving point impedance $Z_B(j\omega) = R_B(\omega) + jX_B(\omega)$ in the form of break points.

   In Figure 5.7, optimized gain is depicted. In Figure 5.8, optimized break points are given as

$$RBA = [0.34077 \;\; 0.30775 \;\; 0.12076 \;\; 0.86689 \;\; 0.55783 \;\; 0.1635 \;\; 0]$$

and it is plotted in Figure 5.8.

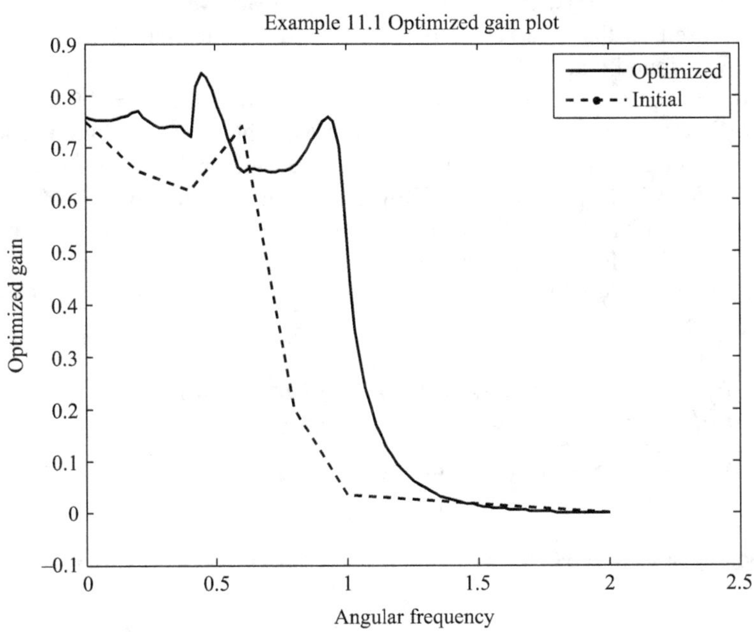

Figure 5.7 Optimized gain performance of Example 5.1

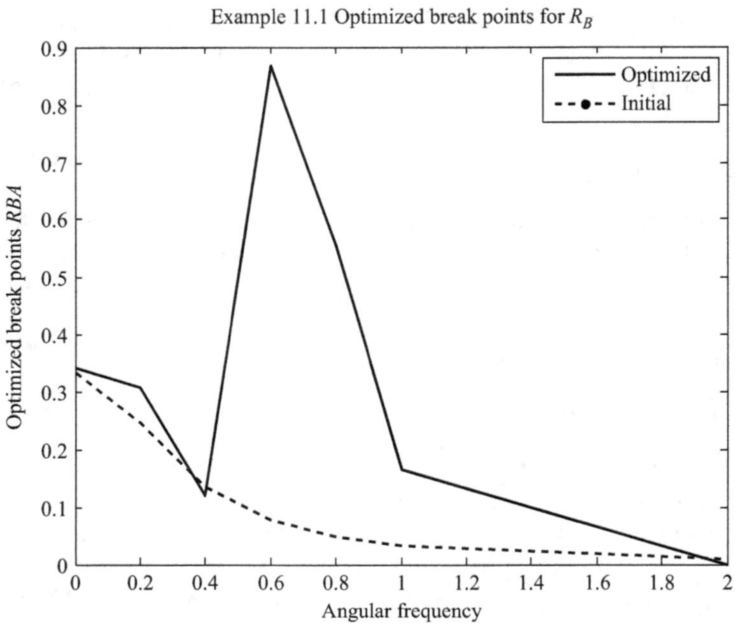

Figure 5.8 Optimized break points for Example 5.1

For the $R//C + L$ (with $R = 1$, $C = 3$) type load, we investigated the upper bound imposed on the value of inductor L to end with lowpass ladder equalizer. There, it was found that $L < 0.89$.

In other words, there is no simple analytic solution for $(R = 1)//(C = 3) + (L = 1)$. However, one may end up with a reasonable ladder solution when the minimum reactance driving impedance data is modeled analytically. In fact, this is the case which will be shown in the following section. Furthermore, RFLT solution can be improved by shifting the location of last break frequency toward *wc*2 and also by increasing the total number of break points.

Let us list the MATLAB main program and related functions developed for Example 5.1.

5.1.3 Effect of the last break point and total number of unknowns on the gain performance

Let us investigate the above issues in the following example.

Example 5.2

By choosing $ws2 = WBR(N) = 1.2$, solve the above problem for $N = 19$ break points.

Answer

In the new case, upper edge of the stop band frequency *ws*2 is shifted from $ws2 = 2$ to 1.2. In this case, power must be confined under the gain curve spread from $w = 0$

to 1.2 which in turns jacks the gain up at the cutoff frequency $wc2 = 1$. Furthermore, we have a better chance to smooth the wiggles of the gain function in the passband since the total number of break points has been drastically increased.

5.1.3.1 Performance measure of the matched system

Usually, the performance of the matched system is measured by means of the gain the fluctuation ΔT about the average level T_0. In this case, the gain function is expressed by

$$T(\omega) = T_0 + \mp\Delta T \tag{5.13}$$

In addition, the minimum and maximum of the passband gain are given as follows:

$$\begin{aligned} T_{\min} &= T_0 - \Delta T \\ T_{\max} &= T_0 + \Delta T \end{aligned}$$

or equivalently

$$T_0 = \frac{T_{\max} + T_{\min}}{2} \tag{5.14a}$$

and

$$\Delta T = T_0 - T_{\min} \tag{5.14b}$$

The objective of the broadband matching problem is to minimize fluctuation, while T_0 is maintained as high as possible. Therefore, in the optimization process, T_0 is provided by the user, and the mean square error is minimized employing the MATLAB optimization function *lsqnonlin*.

For the example under consideration, we run the main program Example5_1.m with the following inputs:

Enter total number of break points $N = 19$

$L + R//C$ *Load*: *Enter resistance of the load* $R = 1$

Enter Capacitor value of the load $C = 3$

Enter Inductor value of the load $L = 1$

Enter lower edge of the passband $wc1 = 0$

Enter upper edge of the passband $wc2 = 1$

Enter lower edge of the stopband; ($ws1 = 0$ *for Lowpass problems*) $ws1 = 0$

Enter upper edge of the stopband $ws2 = 1.2$

Enter ideal flat gain level $T0 = 0.75$

Enter KFlag, *for* $Z \geq 1$; *for* $Y \geq 0$; $KFlag = 1$

≥ 1 *high values*; ≥ -1 *low values*. *Enter sign* $= 1$

Notice that, for this example, we preferred to use initial break points with high values. Therefore, the "sign" of (5.9) is chosen as "sign = 1."

Results are given as follows:

The break points are shown as in Table 5.2 and it is plotted in Figure 5.9.

Optimized gain is depicted in Figure 5.10.

Close examination of Figure 5.10 reveals that by moving *WBR*(*N*) from 2 to 1.2 improves the passband gain performance of the system and by increasing the number of break points in the passband, fluctuations are drastically reduced. It is found that the average gain level is

$$T_{Average} = 0.75083 \mp \Delta T;$$

and

$$T_{\min} = 0.76453; \quad T_{\min} = 0.73712; \quad \Delta T = 0.013704$$

This is a superior result.

Form the theoretical point of view, we should mentioned that "any immittance function $F(j\omega) = R_B(\omega) + jX_B(\omega)$ generated employing the real frequency line technique is realizable since the real part $R_B(\omega)$ is nonnegative for all values of ω." However, from the practical point of view, one needs to end up with analytic form of the immittance. This can be done easily by immittance modeling techniques introduced in [2] employing Bode or Gewertz methods.

Table 5.2 Break frequencies and optimized beak points for Example 5.2

Break frequencies WBR	**Optimized break points RBA**
0	3.0991
0.058824	1.2522
0.11765	0.7985
0.17647	0.69104
0.23529	0.63639
0.29412	0.60763
0.35294	0.58588
0.41176	0.56766
0.47059	0.55006
0.52941	0.53192
0.58824	0.51264
0.64706	0.49138
0.70588	0.46774
0.76471	0.4411
0.82353	0.41028
0.88235	0.37439
0.94118	0.33228
1	0.27244
1.2	0

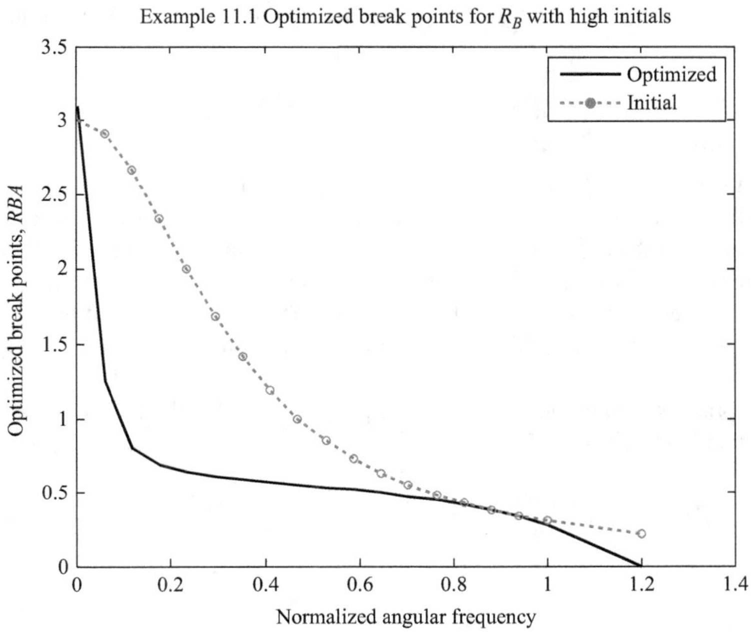

Figure 5.9 Optimized break points, RBA

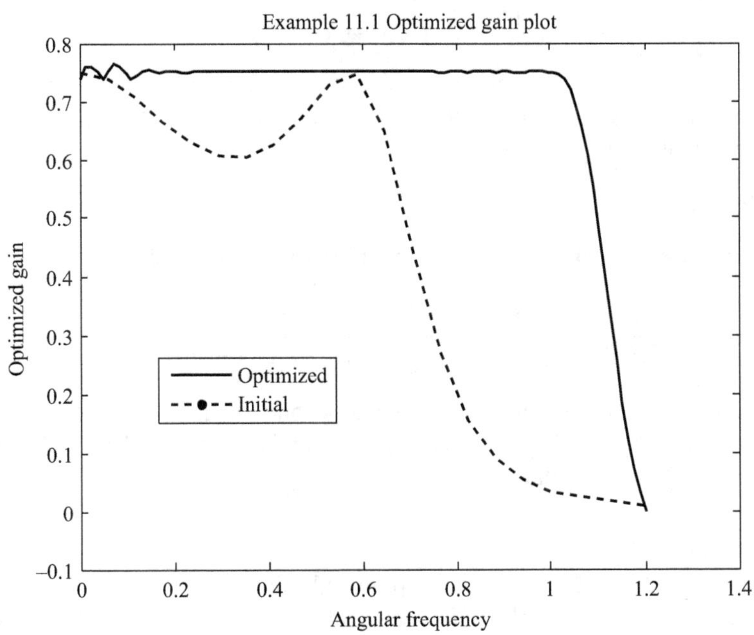

Figure 5.10 Gain variation for Example 5.2

In the following section, practical aspects of modeling techniques are discussed.

5.1.4 Practical models for RFLT generated minimum immittance data

The first step of the RFLT Single Matching approach is to determine the minimum immittance data which optimizes the TPG as discussed in the previous sections.

In the second step, generated immittance data is modeled as a practical positive real minimum function which can be realized as a lossless matching network.

At this step, it is sufficient to build a good analytic fit to the real part then, the analytic form of the immittance is generated employing either Bode or Gewertz procedure.

For practical matching problems, it is proper to choose the rational form of the real part as in (5.15a)–(5.15e):

$$R_B(\omega^2) = \frac{A_0\omega^{2ndc}\prod_{i=1}^{nz}\left(\omega_i^2 - \omega^2\right)^2}{B_1\omega^{2n} + B_2\omega^{2(n-1)} + \ldots + B_n\omega^2 + 1} = \frac{A(\omega^2)}{B(\omega^2)} \geq 0;\ \forall\omega \tag{5.15a}$$

where

$$B(\omega^2) = \frac{1}{2}\left[c^2(\omega) + c^2(-\omega)\right] > 0 \tag{5.15b}$$

such that

$$c(\omega) = c_1\omega^n + c_2\omega^{(n-1)} + \ldots + c_n\omega + 1 \tag{5.15c}$$

It should be noted that in (5.15a), the real constant A_0 must be positive. Therefore, it may be defined by means of an arbitrary real number a_0 which may take positive or negative real values. In this case, A_0 is defined as

$$A_0 = a_0^2 > 0 \tag{5.15d}$$

For simple cases, simplified version of (5.15a)–(5.15e) is given by

$$R_B(\omega^2) = \frac{a_0^2\omega^{2ndc}}{B_1\omega^{2n} + B_2\omega^{2(n-1)} + \ldots + B_n\omega^2 + 1} = \frac{A(\omega^2)}{B(\omega^2)} \geq 0;\quad \forall\omega \tag{5.15e}$$

That means, all the finite transmission zeros beyond DC is set to zero.

Hence, for specified break frequencies and break points, rational form of (5.15a)–(5.15e) can be obtained.

Example 5.3

Referring to Example 5.2, model the optimized minimum reactance driving point impedance data which matches $(R = 1)//(C = 3) + (L = 1)$ load to a resistive generator.

Answer

The new version of the modeling program is called Example5_3.m.

This program reads the generated data from the data files called

$WBR.txt$; *%the break frequencies*

$RBA.txt$; *%Break points for the real part of the immittance data*

$XBA.txt$; *%Imaginary part of the immittance data*

Inputs to the program are given as follows.

Enter transmission zeros at DC; $ndc = 0$(lowpass case)

Enter finite transmission zeros at$[W] = [\ldots] = 0$

Enter $a0 = 1$; $\left(\textit{Note that } A_0 = a_0^2\right)$

Enter initials for $[c0] = [\ldots] = [+1 + 1 + 1 - 1 - 1 + 1 - 1]$

$(ntr = 1, \textit{Yes, model with Transformer}, ntr = 0 \textit{ No Transformer})$; $ntr = 1$

Execution of the program yields the following results.
Coefficients of the polynomial $c(\omega)$ of (5.15c):

$$c = [156.55 \; -9.1399 \; -244.66 \; 34.512 \; 111.71 \; -21.195 \; -15.654]$$

which yields

$$R_B\left(\omega^2\right) = \frac{a_0^2}{B_B(\omega^2)}$$

with optimized values of

$$a_0 = 1.6299$$

$$BB = [24,507 \; -76,519 \; 94,206 \; -57,987 \; 18,659 \; -2,979.3 \; 202.66 \; 1]$$

Or equivalently in the complex p-plane,

$$R_B\left(p^2\right) = \frac{2.6565}{-24,507p^{14} - 76,519p^{12} - 94,206p^{10} - 57,987p^8 - 18,659p^6 - 2,979.3p^4 - 202.66p^2 + 1}$$

By Bode method, MATLAB function $[a, b] = RtoZ(A, B)$ yields the analytic form of the positive real minimum reactance impedance function $Z_B(p) = a(p)/b(p)$ as in Table 5.3.

5.1.5 Synthesis of the equalizer for RFLT

The third step of the RFLT is devoted to the synthesis of the driving point immittance.

Once the analytic form of the driving point impedance is generated, it is synthesized as a lossless two-port in resistive termination.

Table 5.3 *Analytic form of the driving point impedance* $Z_B(p) = a(p)/b(p)$

$a(p)$	$b(p)$
0	1
0.37787	0.6145
0.2322	1.75
0.5836	0.77759
0.24609	0.86865
0.23253	0.22864
0.050719	0.10579
0.016969	0.0063879

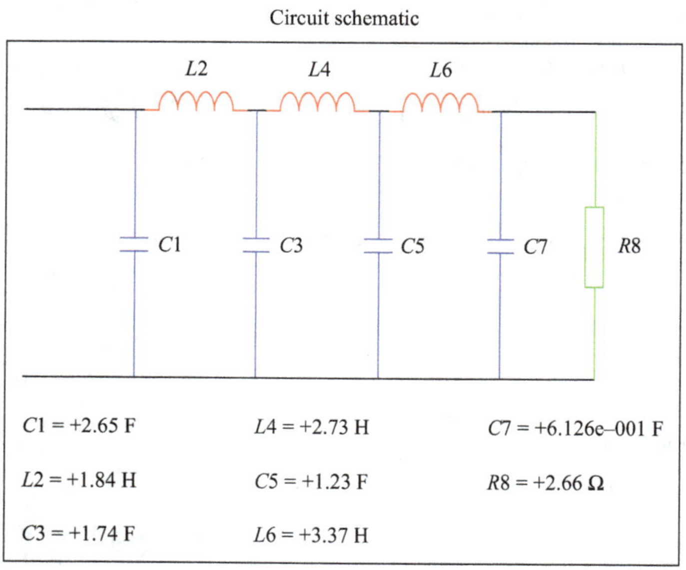

Figure 5.11 *Synthesis of the equalizer as an LC ladder in resistive termination* $R8 = 2.66$

For the problem under consideration, synthesis of the immittance can automatically be carried out using our MATLAB function

$$CVal = GeneralSynthesis\_Yarman(a, b)$$

For Example 5.3, equalizer is a typical LC ladder network as shown in Figure 5.11. In this figure, it is assumed that the load is connected to the left of the equalizer and the generator is on the right with termination resistance. $R8 = 2.66$. Obviously, $R8$ can be level to $R = 1\Omega$ *resistance* using an ideal transformer.

Corresponding data fit and gain plots are depicted in Figures 5.12, 5.13 and 5.14, respectively.

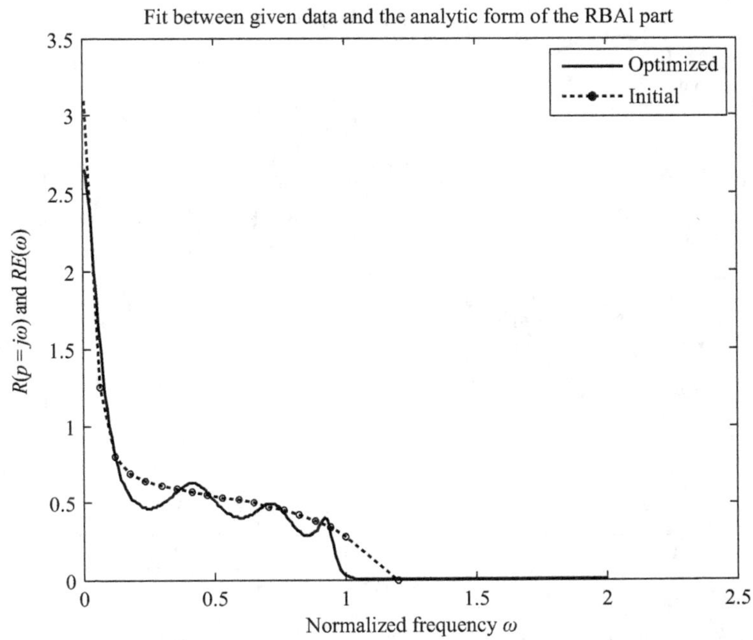

Figure 5.12 Fit between RFLT-generated real part data and the rational form of the real part $R_B(\omega)$ for Example 5.3

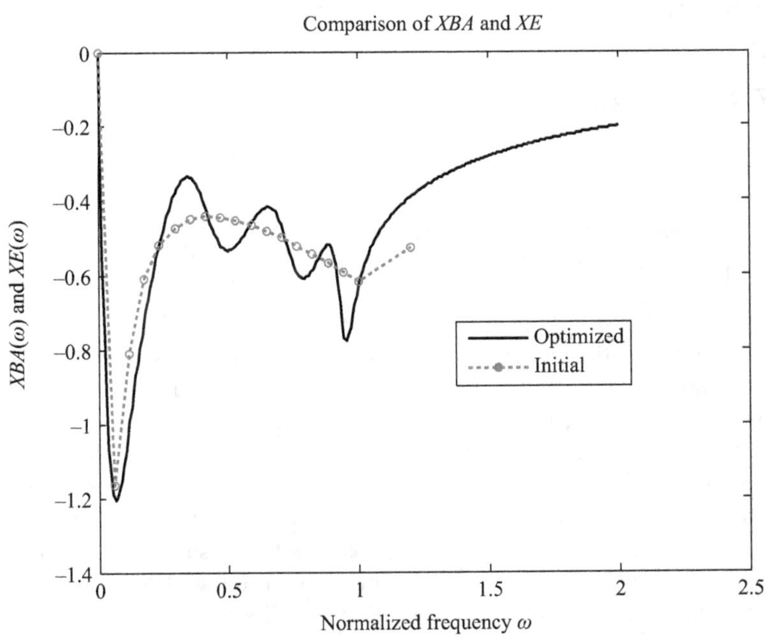

Figure 5.13 Fit between the imaginary parts of the driving point minimum reactance input impedance

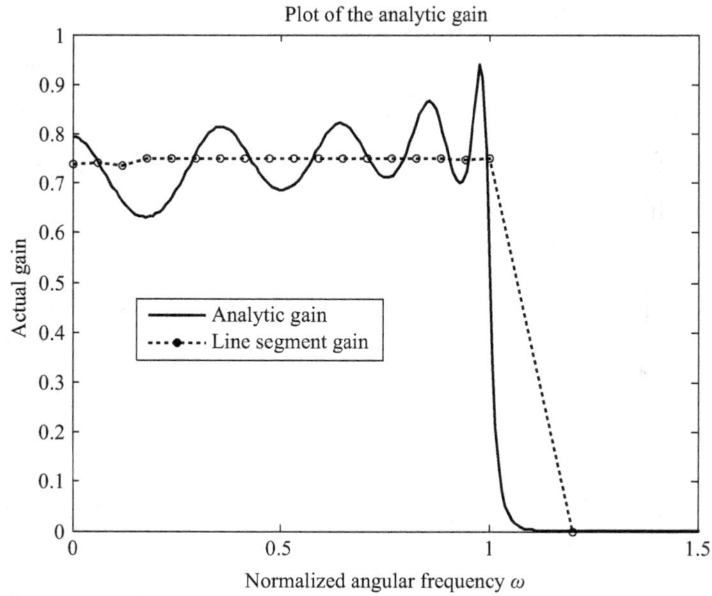

Figure 5.14 After nonlinear curve fitting, actual gain of the matched equalizer of Example 5.3

Performance parameters of the design is given as

$$\textit{Average passband gain is Tzero} = 0.78547 \mp \Delta T$$

$$T_{\min} = 0.63104 \; at \; \text{wmin} = 0.176;$$

$$T_{\max} = 0.9399 \; at \; \text{wmax} = 0.976$$

$$\Delta T = 0.154$$

5.1.6 Summary of RFLT algorithm

The RFLT consists of three major steps.

- *Step 1: Generation of optimum line segments*
  In the first step, the lossless equalizer is described in terms of its driving point input immittance $F_B(j\omega) = R_B(\omega) = JX_B(\omega)$ over the angular real frequency axis ω. Furthermore, it is assumed to be a minimum function.

  The real part of this immittance is approximated by means of lines in piecewise manner. These line segments are expressed in terms of break points $\{RBA = R_1, R_2, \ldots, R_{N-1}, R_N\}$ and user selected break frequencies $\{\omega_1, \omega_2, \ldots, \omega_{N-1}, \omega_N\}$. Usually, the real part is assumed to be practically zero for $\omega > WBR(N) = \omega_N$. Therefore, it is meaningful to produce the corresponding imaginary part XBA by means of Hilbert transformation.

Hence, in RFLT, the break points $\{RBA = R_1, R_2, \ldots, R_{N-1}, R_N\}$ are chosen as the unknowns of the single matching problem, and they are determined via optimization of the TPG function.

- *Step 2: Model for the line segments*
  In the second step, data obtained for the real part is modeled as a nonnegative rational function which in turn yields the minimum driving point immittance of the equalizer by means of Bode (or Parametric) approach or Gewertz procedure.
- *Step 3: Synthesis of the single matching Equalizer*
  Eventually, minimum driving point immittance is synthesized as a lossless network terminated in a resistance. Later this resistance is removed by an ideal transformer to end up with the original generator internal resistance.

Now, let us make the following remarks which are useful to implement the RFLT algorithm.

Remarks

- The first step of the RFLT algorithm can be regarded as the numerical or ideal design of the equalizer. It provides substantial inside to the problem. At this step, the design is virtual, but it theoretically yields a realizable driving point immittance for the matching network. Therefore, this step is excellent to understand the theoretical limits of the single matching problem by trial and error. Trials can be run for different flat gain levels T_0. In fact, one can generate a sweeping algorithm to end up with the highest level of T_0.
- For RFLT, the designer selects the break frequencies as desired. However, for lowpass designs, passband starts from DC ($\omega_{c1} = 0$) and runs up to ω_{c2}. In this case, for programming gimmick, we set $\omega_{s1} = \omega_{c1} = 0$. In general, choice for ω_{s1} and ω_{s2} determines the sharpness of the gain function as it rolls off at both lower and the upper edges of passband, which in turn yields the degree of the numerator and the denominator of the real part, or equivalently the number of reactive elements in the equalizer.
- For bandpass problems, at the band edges, the break frequencies and break points are fixed as follows:

  $$WBR(1) = \omega_{s1}; \quad R(1) = 0$$

  and

  $$WBR(N) = \omega_{s2}; \quad R(N) = 0$$

  and rest of the break frequencies are distributed over the passband as desired. However, we usually prefer even distribution, and they are paired with the unknown break points in the passband. Therefore, in the bandpass problems, if the number of break points is N, the total number of unknowns is specified as $N - 2$.

- In the second step, which is devoted to modeling, the designer should select a proper rational form for the real part $R_B(\omega)$. The following issues are quite important in this selection.
  - The selected rational form must provide a good fit to real part data.
  - The resulting equalizer must easily be built using the technology available for the designer. In this regard, one may wish to avoid finite transmission zeros on the real frequencies. If this is the case, one should set select the input vector $W = 0$, which includes user fixed transmission zeros in the main program.
  - For practical bandpass problems, it may be sufficient to select the numerator polynomial as $A(p^2) = (-1)^{ndc} A_0 p^{2ndc} = (-1)^{ndc} a_0^2 p^{2ndc}$.
  - Eventually, the degree of denominator polynomial n is chosen to end up with an acceptable fit. In the main program, the designer enters the initial coefficient for the nonlinear curve fitting which in turn yields the degree n.
- Selection of minimum reactance or minimum susceptance driving point immittance function does not violate the generalization of RFLT since one can always add a foster part to a minimum function which results in a general form of the driving point immittance. In this regard, when we start with a minimum function to design the equalizer, we can always extract a Foster part from it which is also included among the unknowns in a linear manner.

  For example, let the real part $R_B(\omega)$ be

$$R_B(\omega) = \sum_{i=1}^{N} a_i(\omega) R_i$$

  then, a practical form of a reactive part of a non-minimum immittance can be given as

$$X_{eq}(\omega) = \sum_{i=1}^{N} b_i(\omega) R_i + Foster(\omega)$$

  In above, the simple form of the frequency dependent Foster function is given by

$$f(\omega) R_{N+1} = \begin{Bmatrix} \omega \\ or \\ \dfrac{1}{\omega} \end{Bmatrix} R_{N+1}$$

  where R_{N+1} is included among the unknowns.
- The beauty of *RFLT* is the success of the gain optimization. No matter what the total numbers of break points are, one can always define an objective function which is quadratic in terms of the unknowns. Hence, nonlinear curve fitting is always convergent.[5]

[5]In optimization theory, it is well established that a quadratic or a convex objective function always leads to convergent optimization.

- In the course of optimization, one may employ either Bode or Gewertz procedure to generate the analytic form of the driving point immittance. However, Bode method provides robust numerical stability. Therefore, it may be preferred over the Gewertz method.
- Our final remark is that, once the analytic form of the driving point immittance is obtained via modeling, resulting gain performance of the matched system can be improved by re-optimization of the gain function directly, on the coefficients c_i of (5.15a)–(5.15e). In this case, RFLT is used as an initial solution for direct optimization. This approach is called the *Real Frequency Direct Method to matching problems*. It is detailed in the following section.

List of the MATLAB programs for Example 5.3 is given as the supplementary to this book.

5.2 Real frequency direct computational technique (RFDT) for double matching problems

Referring to Figures 5.15 and 5.16, in the real-frequency direct computational technique (RFDT), the TPG of the double-matched system is described in terms of the driving point immittances of the generator $[Z_G \text{ or } Y_G]$, equalizer $[Z_B \text{ or } Y_B]$ and the load $[Z_L \text{ or } Y_L]$.

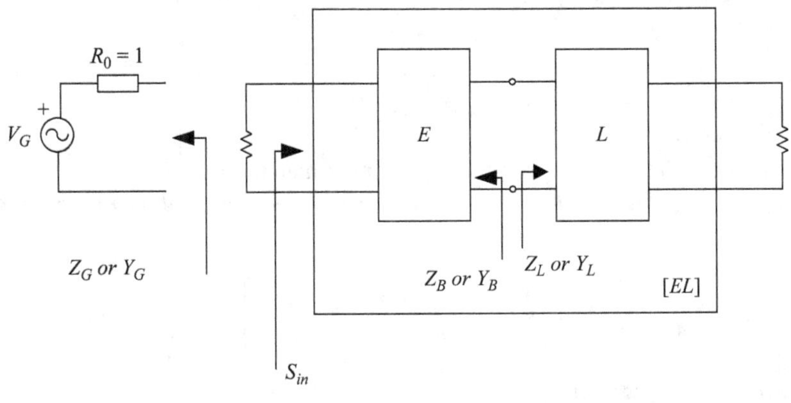

Program List 5.1 Cascaded connection of two lossless two-ports $[G]$ *and* $[EL]$

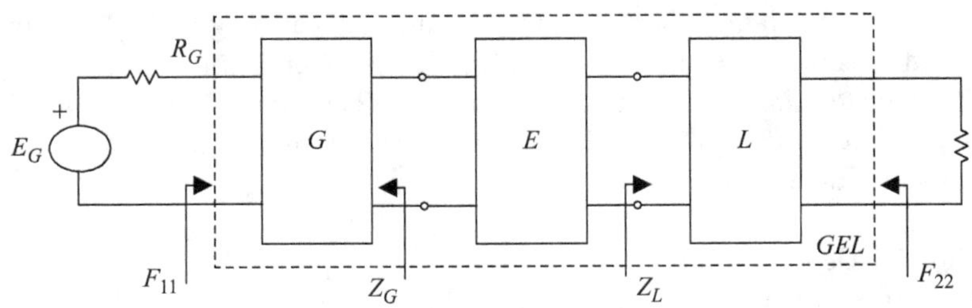

Figure 11.1 Double matching problem

In this case, TPG of Figure 5.11 is given by

$$T(\omega) = \frac{1 - |G_{22}|^2}{|1 - G_{22}S_{in}|^2} T_{[EL]} \tag{5.16}$$

where

$$G_{22} = \frac{Z_G - 1}{Z_G + 1} = \frac{1 - Y_G}{1 + Y_G} \tag{5.17}$$

is the unit normalized generator reflectance.

As proven by the main theorem of Yarman and Carlin [2,3] S_{in} is given by

$$S_{in} = \eta_B(j\omega)\left[\frac{Z_L - Z_B(-j\omega)}{Z_L(j\omega) + Z_B(j\omega)}\right] \tag{5.18}$$

which is the unit-normalized input reflectance of the lossless two-port $[EL]$.

In (5.18), the all-pass function $\eta_B(p)$ is given by

$$\eta_B(p) = \left[\frac{W_B(-p)}{W_B(p)}\right]$$

and its detailed description is postponed to (5.23).

It is interesting to observe that by (10.19), the TPG T_{EL} of the lossless two-port $[EL]$ is given by

$$T_{EL} = 1 - |S_{in}(j\omega)|^2 = \frac{4R_BR_L}{[R_B + R_L]^2 + [X_B + X_L]^2}$$

which is the immittance-based conventional single matching gain.

Assuming $F_B(j\omega)$ as a minimum function, (5.17) is expressed as a function of the real part R_B as

$$T_G(\omega) = \left[\frac{1 - |G_{22}|^2}{|1 - G_{22}S_{in}|^2}\right]\left\{\frac{4R_BR_L}{[R_B + R_L]^2 + [X_B + X_L]^2}\right\} \tag{5.19a}$$

where $X_B(\omega) = Hilbert\{R_B(\omega)\}$.

In the above formulation, the complex generator drives the lossless two-port $[EL]$. Therefore, $T_G(\omega)$ is referred to as the generator-based TPG of the double matching problem. On the other hand, we can turn the matching problem the other way around; feeding the equalizer $[E]$ with a complex generator of internal impedance Z_L while terminating it in Z_G at the front-end. This type of formulation of the TPG may be referred to as the load based. In this case, it may be useful to make the following definitions:

Definition 5.1 *Generator-based TPG $T_G(\omega)$*
T_G of (5.19a) is called the generator-based TPG of the double-matched system.

Definition 5.2 *Load-based TPG T_G*
Similarly, we can define the load-based TPG of the double-matched system as

$$T_L(\omega) = \left[\frac{1-|L_{11}|^2}{|1-L_{11}S_{in}|^2}\right]\left\{\frac{4R_BR_G}{[R_B+R_G]^2+[X_B+X_G]^2}\right\} \tag{5.19b}$$

where

$$L_{11} = \frac{Z_L-1}{Z_L+1} = \frac{1-Y_L}{1+Y_L} \tag{5.19c}$$

In this case, $F_B = R_B + jX_B$ is the driving point immittance of the resistively terminated equalizer at the front-end (or at the generator $[Z_G]$ end).

Obviously, generator and load-based defined TPGs must be identical. Thus,

$$T(\omega) = T_G(\omega) \equiv T_L(\omega) \tag{5.19d}$$

In (5.19a)–(5.19d), the fictitious function $W_B(j\omega)$ is defined on the explicit factorization of the even part $R_B(p^2)$ such that

$$F_B(p) = \frac{a(p)}{b(p)} \tag{5.20}$$

$$R_B(p^2) = Even\{F_B(p)\} = \frac{A(p^2)}{B(p^2)} = \frac{n_B(p)n_B(-p)}{b(p)b(-p)} = W_B(p)W_B(-p) \tag{5.21}$$

where

$$W_B(-p) = \frac{n_B(-p)}{b(p)} \tag{5.22}$$

It must be note that $n_B(-p)$ must include the proper RHP and $j\omega - axis$ zero of $R_B(p^2)$ as described in Chapter 10 of [4]. Thus, for the direct method of broadband matching, the unknown of the problem is the rational form of the real part R_B such that

$$0 \le R_B(\omega^2) = \frac{A_1\omega^{2m} + A_2\omega^{2(m-1)} + \ldots + A_m\omega^2 + A_{m+1}}{B_1\omega^{2n} + B_2\omega^{2(n-1)} + \ldots + B_n\omega^2 + B_{n+1}} \le \infty; \quad \forall\omega \tag{5.23}$$

In this case, the coefficients $\{A_i, B_j; i = 1, 2, \ldots, m, and\ j = 1, 2, \ldots, n\}$ must be determined in such a way that TPG of (5.19a) or, equivalently, (5.19b) is optimized as high and as flat as possible as in RFLT of the single matching problems.

However, we should note that the general from of R_B specified by (5.23) is not practical at all. It may result in complicated equalizer structures which cannot be built. Therefore, we prefer to work with simple form of R_B which has all its zeros on the $j\omega - axis$ as in (5.15a)–(5.16) such that

$$R_B(\omega^2) = \frac{A_0\omega^{2ndc}\prod_{i=1}^{nz}\left(\omega_i^2-\omega^2\right)^2}{B_1\omega^{2n} + B_2\omega^{2(n-1)} + \ldots + B_n\omega^2 + 1} = \frac{A(\omega^2)}{B(\omega^2)} \ge 0; \quad \forall\omega \tag{5.24a}$$

where

$$A_0 = a_0^2 > 0 \tag{5.24b}$$

$$B(\omega^2) = \frac{1}{2}\left[c^2(\omega) + c^2(-\omega)\right] > 0 \tag{5.24c}$$

$$c(\omega) = c_1\omega^n + c_2\omega^{(n-1)} + \ldots + c_n\omega + 1 \tag{5.24d}$$

By (5.24a)–(5.24d), it is straightforward to see that if all the zeros of $n_B(-p)$ are on the real frequency axis, the all-pass function $\eta_B(p) = W_B(-p)/W_B(p)$ is simplified as

$$\eta_B(p) = \frac{W_B(-p)}{W_B(p)} = (-1)^{ndc}\frac{b(-p)}{b(p)} \tag{5.25}$$

Once T_0 is selected and the coefficients $\{c_j; j = 1, 2, \ldots n\}$ and $A_0 = a_0^2 \geq 0$ of the real part $R_B(\omega)$ are initialized, we can generate the generator-based error function ε_G or equivalently the load-based error function ε_L as follows:

$$\varepsilon_G = \left[\frac{1 - |G_{22}|^2}{|1 - G_{22}S_{in}|^2}\right]\left\{\frac{4R_BR_L}{[R_B + R_L]^2 + [X_B + X_L]^2}\right\} - T_0 \tag{5.26a}$$

or

$$\varepsilon_L = \left[\frac{1 - |L_{11}|^2}{|1 - L_{11}S_{in}|^2}\right]\left\{\frac{4R_BR_G}{[R_B + R_G]^2 + [X_B + X_G]^2}\right\} - T_0 \tag{5.26b}$$

Then, the error function is minimized which in turn yields the realizable driving point input immittance $F_B(p) = a(p)/b(p)$ of the lossless equalizer. Eventually, $F_B(p) = a(p)/b(p)$ is synthesized yielding the desired lossless equalizer in resistive termination R. Finally, resistive termination is replaced by an ideal transformer with transformer ratio $[R = n^2 : 1]$ which completes the design.

As it is always the case, initial guess for the nonlinear optimizations are quite important. Therefore, in the following, we will investigate the degree of the nonlinearity of the double-matching gain function and introduce a method to generate the initials as in RFLT technique.

5.2.1 *Investigation on the nonlinearity of the double matching gain*

Close examination of (5.19a)–(5.19d) reveals that while Z_B converges to Z_L^*, if $S_{in}(j\omega)$ approaches to $G_{22}(-j\omega)$, gain of the double-matched system is maximized. However, instead of assuming perfect match, let us consider reactance cancellation at the load end. In this case, generator-based maximum flat gain level of the double-matched system may be approximated as

$$T_0 = T_{G0} \cong \left[1 - |G_{22}|^2\right]^{-1} \left\{ \frac{4R_B R_L}{[R_B + R_L]^2} \right\} \tag{5.27a}$$

when

$$X_B = -X_L \text{ and } S_{in} = G_{22}^* \tag{5.27b}$$

where

$$G_{22} = \frac{Z_G - 1}{Z_G + 1} = \frac{1 - Y_G}{1 + Y_G} \tag{5.27c}$$

Definition 5.3 *Generator-based attainable flat gain level T_{G0}*
T_{G0} is called the "generator based-attainable flat gain level" for the double-matched system $[G] - [E] - [L]$.

Definition 5.4 *Load-based attainable flat gain level T_{L0}*
Similarly, by (5.19a)–(5.19d), we can define the "load-based attainable flat gain level" as

$$T_0 = T_{L0} \cong \left[1 - |L_{11}|^2\right]^{-1} \left\{ \frac{4R_B R_G}{[R_B + R_G]^2} \right\} \tag{5.28a}$$

when

$$X_B = -X_G \text{ and } S_{in} = L_{11}^* \tag{5.28b}$$

where

$$L_{11} = \frac{Z_L - 1}{Z_L + 1} = \frac{1 - Y_L}{1 + Y_L} \tag{5.28c}$$

Definition 5.5 *Generator-based real part variation $R_{GB0}(\omega)$*
Employing (5.26a) and (5.26b), we can generate the generator-based initial variation for the real part function $R_{GB0}(\omega)$ as in RFLT as

$$R_{GB0}(\omega) = R_L(\omega) \left[\frac{(2 - W_{TG}) + 2\,\text{sign}\sqrt{1 - W_{TG}}}{W_{TG}} \right] \geq 0; \quad \forall \omega \tag{5.29}$$

provided that 1 $-W_{TG} \geq 0 \; \forall \; \omega$ in the passband.

Definition 5.6 *Frequency-dependent generator based weight factor W_{TG}*
The quantity W_{TG} is called the frequency dependent, generator-based weight factor and it is given by

$$W_{TG}(\omega) = \left[1 - |G_{22}(j\omega)|^2\right] T_0 \tag{5.30a}$$

for the best case and

$$W_{TG}(\omega) = \frac{T_0}{1 - |G_{22}(j\omega)|^2} = \frac{T_0}{|G_{21}|^2} \tag{5.30b}$$

for the worst case provided that $T_0 \leq 1 - |G_{22}(j\omega)|^2; \quad \forall\omega$ in the passband.

Definition 5.7 *Frequency-dependent load-based weight factor W_{TL}*
Similarly, frequency-dependent load-based weight factor can be defined as

$$W_{TL}(\omega) = \left[1 - |L_{11}(j\omega)|^2\right] T_0 \tag{5.31a}$$

for the best case and

$$W_{TL}(\omega) = \frac{T_0}{1 - |L_{11}(j\omega)|^2} = \frac{T_0}{|L_{11}|^2} \tag{5.31b}$$

for the worst case provided that $T_0 \leq 1 - |L_{11}(j\omega)|^2$; $\forall\omega$ in the passband.

Definition 5.8 *Load-based real part variation $R_{LB0}(\omega)$*
As in (5.29), one can generate the load-based real part variation $R_{LB}(\omega)$ as

$$R_{LB0}(\omega) = R_G(\omega)\left[\frac{(2 - W_{TL}) + 2\text{sign}\sqrt{1 - W_{TL}}}{W_{TL}}\right] \geq 0; \ \forall\omega \tag{5.32a}$$

provided that $1 - W_{TL} \geq 0 \ \forall\omega$ in the passband.

On the other hand, we see that the double matching gain is bounded by

$$\left[1 - |G_{22}|^2\right]\left\{\frac{4R_BR_L}{[R_B + R_L]^2}\right\} \leq T_G(\omega) \leq \frac{1}{1 - |G_{22}|^2}\left\{\frac{4R_BR_L}{[R_B + R_L]^2}\right\} \tag{5.32b}$$

In any case, the term $1 - |G_{22}|^2$ or its inverse can be considered as a weight factor on the single-matching gain. Therefore, we say, "the double matching gain function is quasi quadratic in terms of the real part function R_B."

The initial frequency variation of the unknown real part R_{B0} can be generated by means of either (5.24a)–(5.24d) or (5.32a) and (5.32b). Its rational approximation and corresponding analytic form for $F_{B0}(p)$ can be obtained employing Bode (parametric approach) or Gewertz methods as described in Chapter 4 or in [2], respectively.

In the light of the above explanations, we can develop a MATLAB code to solve the double matching problem by revising our previously developed single-matching programs.

Now, let us develop a MATLAB code for the solution of the double matching problem.

5.2.2 Algorithm for RFDT

Actually, this algorithm is straightforward. Main program and the objective function (or error function) developed for Example 5.3 can be revised to handle the double matching problems. Recall that, in Example 5.1, we generated the analytic form of the back-end driving point immittance $F_B(p) = a(p)/b(p)$ to fit the given real part data RBA. In the present case, all we have to do is to revise the error function to express it as the difference between double matching gain and the flat gain level as in (5.27). For this purpose, we use the rational form for the real part $R_B(\omega)$ as specified by (5.24a)–(5.24d). In this case, the designer must select the transmission zeros of the equalizer as in previous examples.

Now let us give the formal layout of the algorithm to generate the error function to be minimized.

Inputs

- $\boldsymbol{x_0}$: Initial values for the unknowns. Actually, the unknowns of the double matching problem are the coefficients a_0 and $\{c_i; i = 1, 2, \ldots, n\}$ of the rational form of $R_B(\omega)$ as specified by (5.24a)–(5.24d).
- $\boldsymbol{T_0}$: Expected flat gain level for the double matching problem. It is noted that T_0 may be estimated either by means of analytic theory if the generator and the load network is simplified to a single reactive element or by (5.33).
- $\boldsymbol{\omega_{c1}}$: Beginning of the optimization frequency (lower edge of the passband).
- $\boldsymbol{\omega_{c2}}$: End of optimization frequency (upper edge of the passband).
- ***KFlag***: Decide to work with impedance or admittance functions. If you wish to work with impedances set ***KFlag*** $= \mathbf{1}$, if you wish to work with admittances then set ***KFlag*** $= \mathbf{0}$.
- ***ntr***: If you wish to employ an ideal transformer in the equalizer set ***ntr*** $= \mathbf{1}$, otherwise ***ntr*** $= \mathbf{0}$.
- ***ndc***: Transmission zeros at DC. If there is no transmission zeros at DC set ***ntr*** $= \mathbf{0}$.

  At this point we should note that if ***ndc*** $\neq \mathbf{0}$, *ntr* must be ***ntr*** $= \mathbf{1}$ since the equalizer will be terminated in a non-unity resistance.
- ***W***: This vector includes the finite transmission zeros of the real part on the real frequency axis. Usually, we do not wish to employ finite transmission zeros on the real frequency axis. Therefore, for many problems, it is set to zero.
- $\boldsymbol{A_0}$: Coefficient of the numerator of (5.24a)–(5.24d). For transformerless design, it is fixed in advance; otherwise, it is included among the unknowns of the double matching problem.

- **Generator immittance** $\boldsymbol{F_{Gen}=R_G(\omega)+jX_G(\omega)}$ is supplied as the function of normalized angular frequencies.
- **Load immittance** $\boldsymbol{F_{Load}=R_L(\omega)+jX_L(\omega)}$ is supplied as the function of the normalized angular frequencies.

Computational steps

Step 1:
From the given inputs and the initial unknown $\boldsymbol{x_0}$, extract the coefficients c_i and $a_0 = \sqrt{A_0}$ to generate the realizable rational open-coefficient form of the real part $R_B(p) = A(p^2)/B(p^2)$. Actually, at the input of the main program, we enter the coefficients $a_0 and c_i$, then the ***initial vector*** $\boldsymbol{x_0}$ is formed accordingly.
Step 2:
From the generated $A(p^2)$ and $B(p^2)$, construct the minimum driving point immittance function $F_B(p)$ employing Bode or Gewertz procedure. At this step, we can use the previously developed MATLAB functions `[a,b] = RtoZ(A,B)` or `[a,b,x0,x,y0,y] = Gewertz(A,B)`

We should note that, as mentioned in Chapter 4, Bode procedure is numerically more stable than Gewertz method. Therefore, in the present MATLAB code, we prefer to use Bode approach.
Step 3:
Over the passband frequencies ω_j, generate the following quantities:
S_{in} as in (5.18)
Double matching gain as in (5.19a)–(5.19d)

Step 4:
Generate the error function as in (5.26a) and (5.26b) as

$$Error = Double\ Matching\ Gain - T0$$

Once the error function is developed, it is minimized using the MATLAB mean square function called ***lsqnonlin*** as in the previous examples.

Now let us run an example to utilize the above algorithm for the solution of the double matching problem.

Example 5.4

Solve the double matching problem for the following specification:

Generator network:
$R_G//C_G + L_G\ with\ R_G = 1, C_G = 0; L_G = 1.2$
Load network:
$R_L//C_L + L_L\ with\ R_L = 1, C_L = 3; L_L = 0.87$
Passband: $\omega_{c1} = 0; \omega_{c2} = 1$ (lowpass design which requires $ndc = 0$)
Total number of elements in the equalizer: $n = 3$
Transmission zeros at DC: $ndc = 0$ (no transmission zeros at DC)
Finite transmission zeros vector $\boldsymbol{W}$: $W = [0]$; (No finite $j\omega$ transmission zero)
Stop band frequency $\boldsymbol{ws1}$: $ws1 = 0$ (Generate gain plot starting from $ws1$)

Stop band frequency *ws2*: $ws2 = 1.5$ (Generate the gain function up to frequency $ws2$)
Flat gain level *T0*: $T0 = 0.85$

Answer
For this example, we developed an error function called

$$function\ error = direct\_RFDT$$

with the following inputs:

$$(x0, T0, wc1, wc2, KFlag, ntr, ndc, W, a0, RGen, CGen, LGen, RLoad, CLoad, LLoad)$$

as outlined above.

The error function is minimized by the main MATLAB program called Example5_4.

Inputs to the main program are given as follows:

Enter $T0 = 0.85$

Enter $KFlag = 1$ (We work with impedance function)

Enter $RGen = 1;$ *Enter* $CGen = 0;$ *Enter* $LGen = 1.2$

Enter $RLoad = 1;$ *Enter* $CLoad = 3;$ *Enter* $LLoad = 0.87$

Enter transmission zeros at DC $ndc = 0$ (No DC zero)

Enter finite transmission zeros at $[W] = [\ldots] = 0$ (No finite frequency zeros)

Enter $a0 = 0.5$ (First initial)

Enter initials for $[c0] = [\ldots] = [-1\ -1\ -1]$ (This is an ad hoc choice for the initials as it will be seen later it works fine)

$ntr = 1$, *Yes, work with Transformer*, $ntr = 0$ *No Transformer* $ntr = 1$ (Here, we employ an ideal transformer to provide an extra freedom for the gain optimization)

Enter $ws1 = 0;$ *Enter* $ws2 = 1.5$ (These values are used just plot the gain function at the end of the optimization. Therefore, they have no effect on the optimization)

Enter $wc1 = 0;$ *Enter* $wc2 = 1$ (Passband frequencies)

Results of the optimization are summarized as follows.

$a_0 = 0.64741;$ which yields $A_0 = a_0^2 = 0.41913$

Optimized coefficient c_i are given as

$$c = [-4.7604\ -0.59582\ 1.8461]$$

$$B(\omega^2) = \frac{1}{2}\left[c^2(\omega) + c^2(-\omega)\right]$$

which in turn yields the real part as in (5.24a)–(5.24d) with $p^2 = -\omega^2$ as.

$$R_B(p^2) = \frac{A_0}{B(p^2)} = \frac{0.41913}{-22.661p^6 - 17.221p^4 - 2.2164p^2 + 1}$$

By Bode method, the back-end minimum reactance driving point input impedance of the equalizer is given by

$$Z_B(p) = \frac{a(p)}{b(p)}$$

with

$$a = [0 \;\; 0.26835 \;\; 0.18861 \;\; 0.088047]$$

$$b = [1 \;\; 0.70287 \;\; 0.62698 \;\; 0.2100]$$

Gain performance of the double-matched system is depicted in Figure 5.15. Equalizer is constructed via synthesis of the driving point impedance.

$$Z_B(p) = \frac{a(p)}{b(p)} = \frac{0.26835p^2 + 0.18861p + 0.088047}{p^3 + 0.70287p^2 + 0.62698p + 0.21}$$

At the generator end, equalizer must include an ideal transformer with turn ratio

$$n^2 = \frac{0.088}{0.21} = 0.4191$$

Complete matched system is depicted in Figure 5.16.

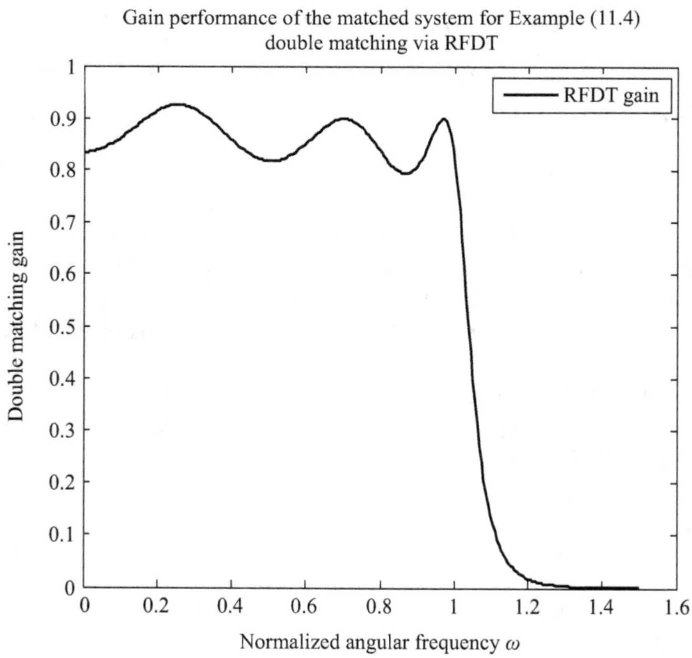

Figure 5.15 Gain plot for the matched generator and load of Example 5.4

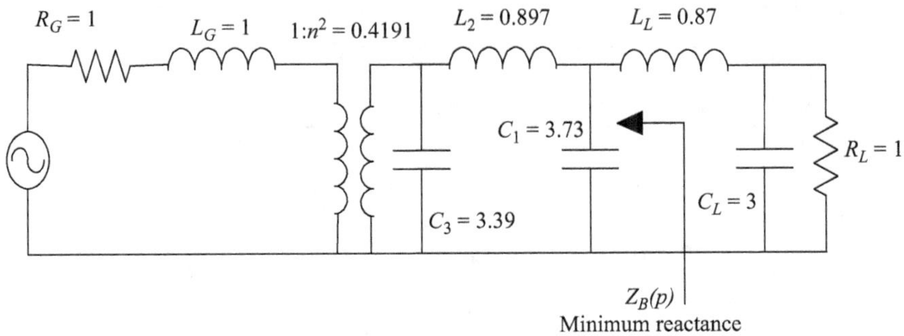

Figure 5.16 Matched generator and load networks for Example 5.4

Here is the summary of the gain performance:

$$wmax = 0.252, T\max = 0.92588$$
$$wmin = 0.87, T\min = 0.79332$$
$$T_{Average} = 0.8596 \mp 0.066284$$
$$T_0 = 0.85$$

Remarks

1. The above solution is pretty good for the optimization of the performance of the matched system.

   Notice that our initial target was $T_0 = 0.85$, and we ended up with

   $$T_{Average} = 0.8596 \mp 0.066284$$

   which is close enough.
2. For the sake of orientation of the reader, let us clarify the following:

   Our MATLAB function

   $$CVal = GeneralSynthesis\_Yarman(a, b)$$

generates the circuit shown in Figure 5.17 which corresponds to the actual equalizer design with transformer as shown in Figure 5.18.

In Figure 5.17, the load is considered on the left, and the termination resistance is replaced by an ideal transformer connected to the generator network as shown in Figure 5.16.

The "RFDT" developed for double-matching problems can be employed to improve the solutions obtained with RFLT. As an exercise, let us use it to improve the results obtained by Example 5.3.

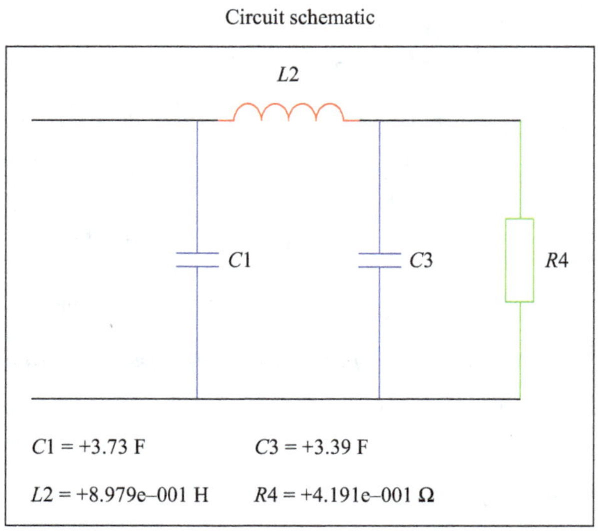

Figure 5.17 Direct synthesis of $Z_B(p) = a(p)/b(p)$

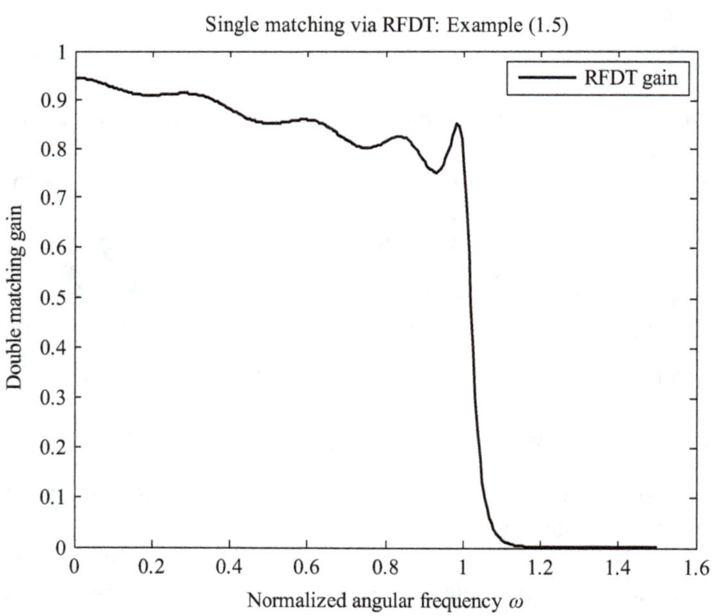

Figure 5.18 Gain performance of the single-matched system of Example 5.5 via RFDT

Example 5.5

Solve the single matching problem described in Example 5.3 employing the RFDT

Load network: *R//C+L* load with *R*=1, *CL*=3, *LL*=1

passband $0 \leq \omega \leq 1$.

Generator network: *RG*=1. *CG*=0. *LG*=0.

Answer

For this example, we run the Main Program Example5_4 with the following inputs:

Inputs

$$Enter\ T0 = 0.9;\ Enter\ KFlag = 1\ (Impedance\text{-}based\ design\ Choice)$$

$$Enter\ RGen = 1;\ Enter\ CGen = 0;\ Enter\ LGen = 0;$$

$$Enter\ RLoad = 1;\ Enter\ CLoad = 3;\ Enter\ LLoad = 1$$

$$Enter\ transmission\ zeros\ at\ DC\ ndc = 0$$

$$Enter\ finite\ transmission\ zeros\ at\ [W] = [\ldots] = 0$$

$$Enter\ a0 = 1.8;\ Enter\ initials\ for\ [c0] = [\ldots] = [111 - 1 - 11 - 1]$$

$$ntr = 1, Yes, model\ with\ Transformer, ntr = 0\ No\ Transformer\ ntr = 1$$

$$Enter\ ws1 = 0;\ Enter\ ws2 = 1.5;\ Enter\ wc1 = 0;\ Enter\ wc2 = 1$$

Results are given as follows (Table 5.4):

$$a0 = 0.78419$$

Gain plot is depicted in Figure 5.18.

Synthesis of $Z_B(p) = a(p)/b(p)$ is shown in Figure 5.19.

Finally, let us run an example to design a double matching equalizer employing RFDT which cannot be handled with the analytic theory of broadband matching.

Table 5.4 Solution to Example 5.5 $R_B = a_0^2/\left(\left(\frac{1}{2}\right)[c^2(\omega) + c^2(-\omega)]\right)$ *with* $a_0^2 = 0.78419$ $Z_B(p) = a(p)/b(p) T_{Average} = 0.84728 \mp 0.095872$

$c(\omega)$	$a(p)$	$b(p)$
31.851	0	1
−8.0279	0.30486	1.1742
−50.755	0.35797	2.2512
20.376	0.60295	1.7313
25.3	0.42994	1.4018
−11.388	0.28554	0.62125
−5.3639	0.098894	0.21194
	0.019307	0.031396

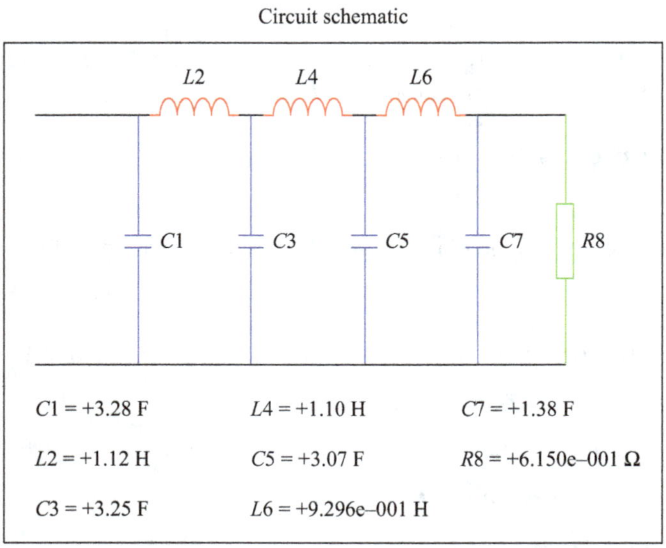

Figure 5.19 Synthesis of minimum reactance driving point impedance $Z_B(p)$ for Example 5.5

Example 5.6

Referring to Figure 5.21, construct an equalizer for the double matching problem which is described as follows.

Generator network: $\left\{\frac{1}{pC_G}+pL_G\right\} \parallel R_G$ with $C_G = 2, L_G = 0.3$

Load network: $\{R\|[pL_1]\} + [pL_2]$ with $R = 1, L_1 = 4, L_2 = 0.75$
Passband: $wc1 = 0.3$ $wc2 = 1$

Answer

This double matching problem can be solved employing the MATLAB main program developed for Example 5.4. However, for the present case, we should modify both the generator and the load functions to include the given specification. Hence, with the abovementioned modifications, we run the program with the name "Example5_6.m" with the following inputs:

Enter T0 = 0.85

In this problem, it is our objective to hit the flat gain level 0.85. If this does not work, it can be lowered in the next run.

KFlag = 0 > Admittance, KFlag = 1 > Impedance..Enter KFlag = 0

In the present design, we preferred to work with admittance functions. Therefore, *KFlag* is set to *KFlag*=0

Enter RGen = 1; Enter CGen = 2; Enter LGen = 0.3

Enter RLoad = 1; Enter L1 = 4; Enter L2 = 0.75

These are the specified parameters of the generator and the load networks.

Enter transmission zeros at DC ndc $= \mathbf{2}$

Since the matching problem is not lowpass, then, it may be proper to use DC transmission zeros in the equalizer. For the present example, we preferred to employ two DC transmission zeros in $[E]$. Eventually, these zeros will be realized as series capacitors and/or shunt inductors.

Enter finite transmission zeros at $[\boldsymbol{W}] = [\ldots] = \mathbf{0}$

Here, we do not wish to employ any finite transmission zero on the real frequency axis. Therefore, we set W=0.

***Enter a*0** $= \mathbf{1}$;

This is an ad hoc initialization for the unknown $a0$.

Enter initials for $[\mathbf{c0}] = [\ldots] = [\,\mathbf{1}\;\mathbf{1}\;\;-\mathbf{1}\;\;-\mathbf{1}\;\mathbf{1}]$

These are the ad hoc initialization of the unknown coefficients c_i of the real part R_B.

ntr $= \mathbf{1}$, ***Yes, model with Transformer, ntr*** $= \mathbf{0}$ ***No Transformer ntr*** $= \mathbf{1}$

Since we have chosen to work with $ndc = 2$ *case*, it is inevitable to employ a transformer in the final equalizer. Therefore, *ntr* must be set to $ntr = 1$.

***Enter ws*1** $= \mathbf{0}$; ***Enter ws*2** $= \mathbf{1.5}$

In this program, we use $ws1$ and $ws2$ as the edge frequencies to generate the gain function after optimization. In the present example, it is proper to plot the TPG function over the frequency range of $0 \le \omega \le 1.5$. Therefore, $ws1$=0 and $ws2$=1.5 are chosen.

***Enter wc*1** $= \mathbf{0.3}$; ***Enter wc*2** $= \mathbf{1.0}$

This is the specified frequency band of operation.
Results of optimization are given as follows:

$$a0 = 1.8332$$

$$c = [\,0.55556\;\;1.102\;\;-3.2435\;\;-6.7647\;\;0.0044049]$$

$$a = [\,0\;\;6.528\;\;14.241\;\;3.6597\;\;3.8283\;\;0]$$

$$b = [\,1\;\;2.1815\;\;6.2504\;\;12.999\;\;1.7207\;\;1.8]$$

Synthesis of $Z_B(p) = a(p)/b(p)$ is shown in Figure 5.20.

As mention earlier, $Z_B(p)$ is the driving point impedance of the equalizer, seen from the load end. Therefore, in Figure 5.21, load network must be connected to the left of Z_B. On the other hand, the generator network must be placed on the resistive termination site. Therefore, removing the termination resistor $R6$ by an ideal transformer of turn ratio $n^2 = R6$, we complete the equalizer design as shown in Figure 5.21.

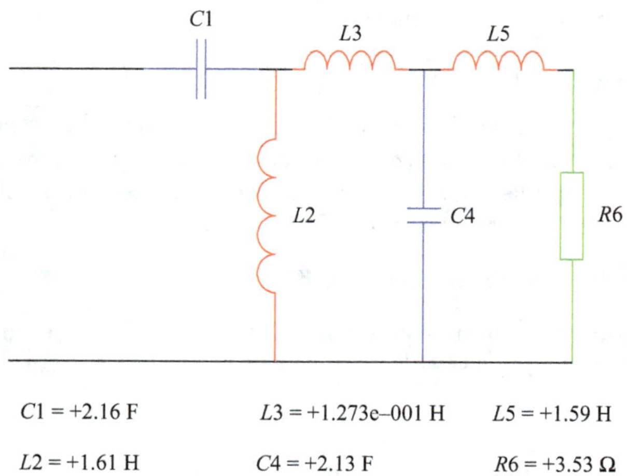

Figure 5.20 Double matching problem as described by Example 5.6

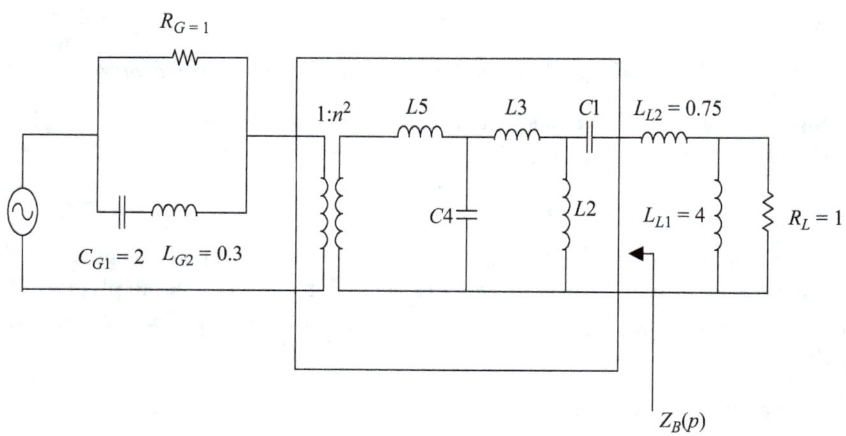

Figure 5.21 Construction of the double-matched equalizer with the ideal transformer of turn ration of 1 : n^2 = R6 = 3.53

Performance of the matched system is summarized as follows.

$$wmax = 0.336 Tmax = 0.87$$

$$wmin = 0.999 Tmin = 0.641$$

$$T_{Average} = 0.75739 \mp 0.11$$

$$T0 = 0.85$$

Gain plot of the matched system is depicted in Figure 5.22.

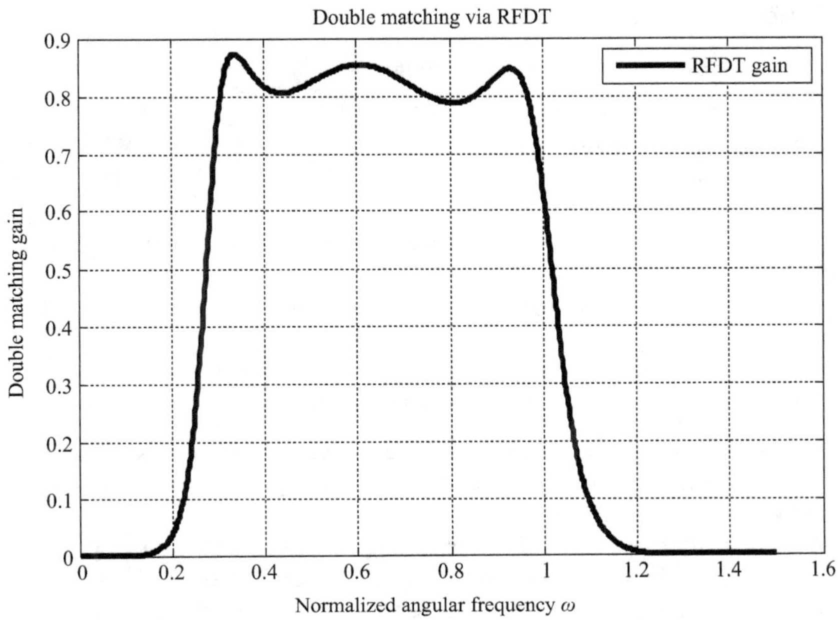

Figure 5.22 Gain performance of the double-matched system described by Figure 5.21

Obviously, performance of the matched system can be improved by reiterating the solution or by increasing the number of element in the equalizer.

For example, we run the program Example5_6 with 7 elements and obtained the following solution:

Ad hoc initials:

$$a0 = 1$$

$$c0 = [-1\ -1\ 1\ 1\ -1\ -1\ 1]$$

Results of the optimization

$$a0 = 1.2113$$

$$c = [-6.0379\ -11.425\ 12.706\ 25.361\ -5.3055\ -11.866\ 0.65382]$$

$$Z_B(p) = \frac{a(p)}{b(p)}$$

$$a = [0\ 3.0465\ 6.1713\ 3.424\ 5.4835\ 0.70472\ 0.73133\ 0]$$

$$b = [1\ 2.0257\ 3.9431\ 7.5108\ 2.8115\ 4.1226\ 0.36819\ 0.38209]$$

Performance summary of the equalizer is given as

$$\begin{aligned} \boldsymbol{wmax} &= \mathbf{0.3255}\ \boldsymbol{Tmax} = \mathbf{0.90468} \\ \boldsymbol{wmin} &= \mathbf{0.999}\ \boldsymbol{Tmin} = \mathbf{0.73466} \\ \boldsymbol{T\_Average} &= \mathbf{0.81967} \mp \mathbf{0.085012} \\ \boldsymbol{T0} &= \mathbf{0.85} \end{aligned}$$

Synthesis of the driving point impedance $Z_B(p) = a(p)/b(p)$ is shown in Figure 5.23.

Finally, the gain performance of the new matched system is depicted in Figure 5.24.

Clearly, gain performance of the equalizer has improved.

We should note the following important remarks.

- Example 5.6 cannot be handled with the analytic theory of broadband matching since there is no function that can embed the given generator and the load networks.
- Assume that we wish to solve this double matching problem by choosing an equalizer topology with unknown element values. Then, optimize the gain function to determine the unknown element values. First of all, how would we select the topology in advance? It is not clear. Second, even if we choose the topology for the equalizer, how would we initiate the element values to optimize gain? This is an ambiguous question.
- Based on our experience, we know by heart that it is not practically possible to determine the element values of the seven element equalizer by optimization of

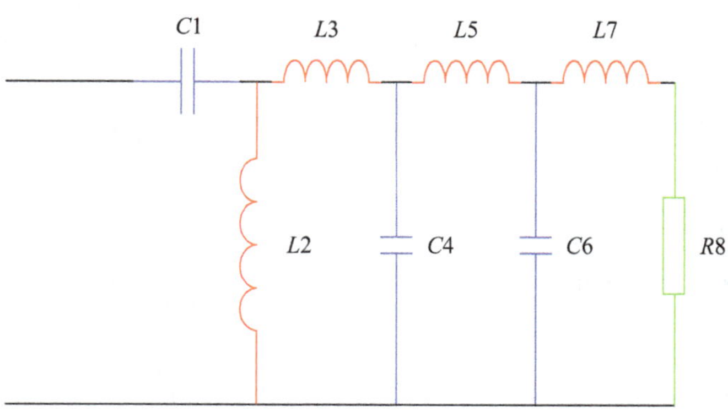

$C1$ = +1.91 F	$C4$ = +1.62 F	$L7$ = +5.64 H
$L2$ = +1.72 H	$L5$ = +3.24 H	R8 = +11.42 Ω
$L3$ = +4.057e–001 H	$C6$ = +7.533e–001 F	

Figure 5.23 Synthesis of $Z_B(p)$ with seven element

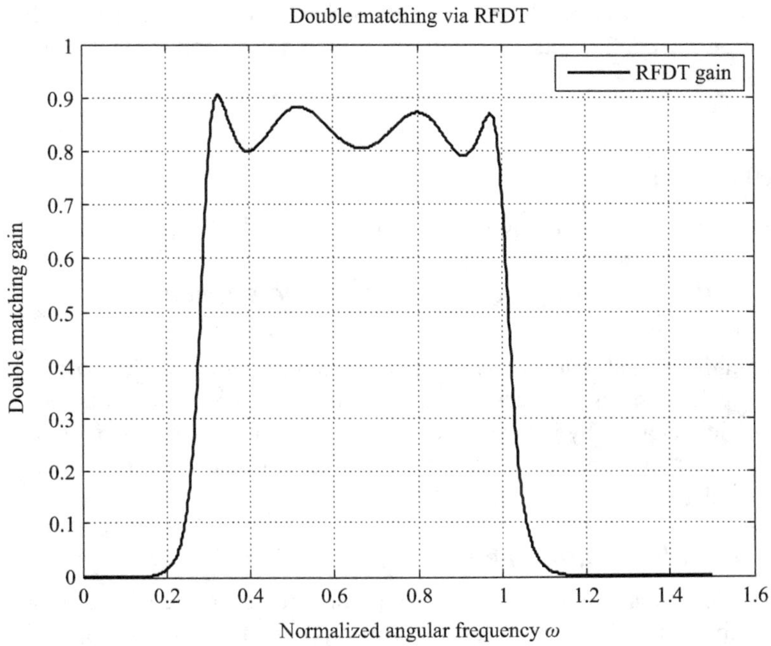

Figure 5.24 Gain performance of the matched system with seven-element equalizer

the gain function employing any commercially available design packages since the gain function of the double-matched system is highly nonlinear in terms of the unknown element values.

- However, the "RFDT" does not need to choose a transfer function in advance nor requires the selection of the equalizer topology. These are the natural consequences of RFDT.
- Therefore, we can state that the real frequency solutions to broadband matching problems are superior over the other available techniques if they provide solutions; otherwise, RFDT provides a unique solution to the matching problem under consideration.

In the following, we provide the MATLAB program lists and related functions.

5.3 Initialization of RFDT algorithm

5.3.1 Ad hoc initialization

As mentioned earlier, for nonlinear optimization problems, initial guess is quite important. As we have practiced so far, all the initials were selected on ad hoc basis for all the matching problems solved using RFDT.

In other words, it has been sufficient to initialize the double matching optimization on the coefficients by setting the initials for $a0$ and $c0$ as $+1$ or -1.

Such that

$$a0 = +1 \; or \; -1$$

$$c0(j) = \mp 1$$

The sign can be determined by trial and error.

5.3.2 *Initialization via real-frequency line-segment technique*

A meaningful initial guess may be generated using the single matching line-segment technique. In this case, the double matching problem can be considered like a single matching by ignoring the reactive elements of the generator. Then, the rational form of the real part is generated as in (5.25). Thus, the initial coefficients are obtained.

5.3.3 *Initialization on the best case solution*

Equation (5.20) may be utilized to generate the initials for the RFDT algorithm.

Assume that, in the best case, the phase variation of $G_{22}(j\omega)$ and $S_{in}(j\omega)$ occurs in such a way that the term:

$$|1 - G_{22}(j\omega_k)S_{in}(j\omega_k)|^2$$

becomes

$$\left(1 - |G_{22}(\omega_k)|^2\right)^2$$

In this case,

$$\frac{T_G}{1 - |G_{22}|^2}\left[1 - |G_{22}|^2\right]^2 = T_G\left[1 - |G_{22}|^2\right] = \frac{4R_BR_L}{\left[(R_B + R_L)^2 + (X_B + X_L)^2\right]}$$

Let

$$0 \le WG = T_G\left[1 - |G_{22}|^2\right] \le 1 \tag{5.33}$$

With the present assumption, we can try to hit the flat gain level T_0 such that perfect reactance cancellation occurs at the load end. In this case, the real part variation is given by as in (5.30a) and (5.30b):

$$R_{B0}(\omega) = R_L(\omega)\left[\frac{(2 - WG) + \text{sign}\sqrt{1 - WG}}{WG}\right] \tag{5.34}$$

Equation (5.36) can be modeled as an initial guess for the main optimization package as mentioned in RFLT.

We should note that, in working with RFDT, optimization of the TPG, both for single and double matching problems, is so well behaved in the coefficients of the real part $R_B(p^2)$ such that ad hoc choice for the initials provides even better numerical stability than that of forced initials. Therefore, we always prefer to work with ad hoc initial choices.

For example, if we wish to design an equalizer with five elements with an acceptable gain level, then we start RFDT optimization with

$$T0 = 0.85$$

$$a0 = 1$$

and

$$c0 = [1 - 11 - 11 - 1 \ldots 1]$$

Or we can start with

$$c0 = [+1 + 1 - 1 - 1 + 1]$$

In any case, a proper selection of the *sign* will make the optimization converge rapidly.

As far as the choice for $T0$ is concerned, the flat gain level can be found by trial and error. For example, when we start with 0.85, if the resulting gain performance hits 1 or higher values than 0.85, we can jack the gain, but if the passband gain makes huge variations down below 0.85 then, we must reduce the level of flat gain. Let us exercise the above-proposed approaches with the following actual antenna example.

5.4 Design of a matching equalizer for a short monopole antenna

Short monopole antennas over the frequency range $20\,\text{MHz} \leq f \leq 100\,\text{MHz}$ may have various commercial and military usages.

In this section, we exhibit the design of an actual matching network for a physical monopole antenna manufactured for a special purpose.

The measured impedance data for the antenna is given by Table 5.5, and it is depicted in Figure 5.25.

Before we start with the design, the data must be normalized with respect to both frequency and standard resistive termination $R0 = 50\,\Omega$.

Let us choose the frequency normalization with respect to $f0 = 100\,\text{MHz}$.

Close examination of Table 5.5 reveals that, at $f = 20\,\text{MHz}$, the real part is $RL(20\,\text{MHz}) = 0.6\,\Omega$. On the other hand, at $f = 70\,\text{MHz}$, $RL(70\,\text{MHz}) = 22.4\,\Omega$. Hence, the real part variation is $22.4/0.6 = 37.333$ which is quite high. Therefore, this is a very tough matching problem. At this point, we have no idea about the gain bandwidth limitation of the antenna. Actually, a model for the antenna can be

Table 5.5 Measured impedance data $Z_L = R_L + jX_L$ for the short monopole antenna

Actual frequency(MHz)	R_L (Ω)	X_L (Ω)
20	0.6	−6
30	0.8	−2.2
40	0.8	0
45	1	1.4
50	2	2.8
55	3.4	4.6
60	7	7.6
65	15	8.8
70	22.4	−5.4
75	11	−13
80	5	−10.8
90	1.6	−6.8
100	1	−4.4

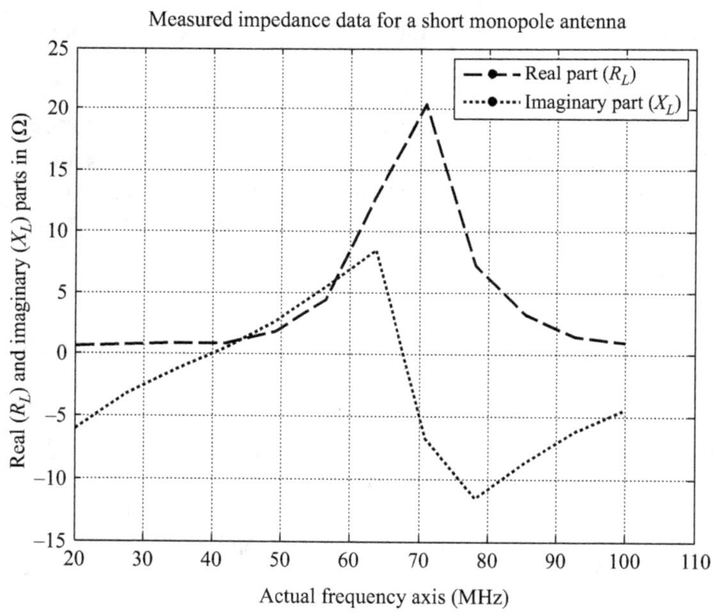

Figure 5.25 Measured impedance data of a short monopole antenna

obtained by means of the technique presented in the next chapter. We have done it, and it is shown in Figure 5.26.

As it is seen from Figure 5.26, into our knowledge, there is no known transfer function to embed the above-complicated load network to utilize the analytic theory of broadband matching. Therefore, the theory is not accessible to the present case.

From the practical point of view, we do not wish to deal with an antenna which delivers less than 50% of the available power of the generator. Therefore, we

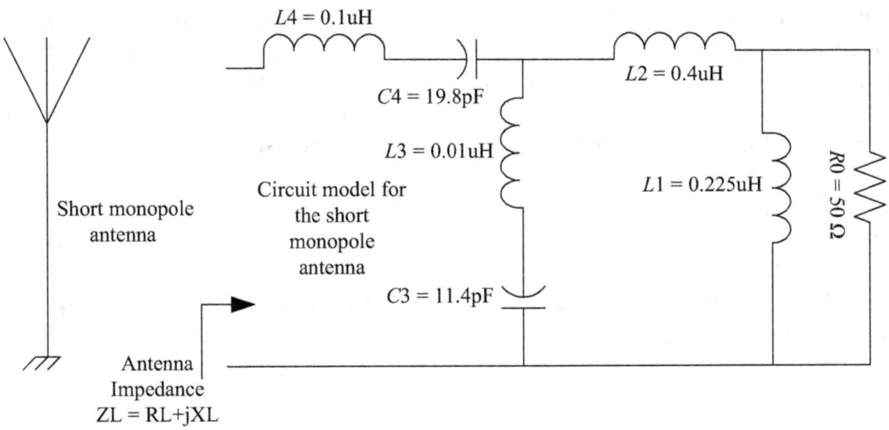

Figure 5.26 Model for the measured impedance data for the short monopole antenna

should seek for a useful bandwidth for the antenna such that it yields at least −3 dB or higher gain over the band of operation.

Our first inclination is to use the antenna over 20–100 MHz bandwidth. But, practically, the real part is almost zero (or about 0.6 Ω) at 20 MHz Therefore, we do not have much hope to get a gain at the lower edge of the measured frequencies. However, at 30 MHz, the real part slightly increases to 0.8 Ω.

Now, let us try to capture at least −3 dB gain over 30−100 MHz.

For this purpose, we developed a MATLAB code to construct an equalizer from the measured data. Name of the program is *Monopole_Antenna.m.*

In this program, RFDT is employed with nonlinear-optimization algorithm.

As in the previous examples, lossless equalizer is described in terms of the real part of the driving point input impedance *RB*. *RB* is described as in (11.25) with

$$a0, ndc, W, \text{ and the denominator coefficients } c_j; j = 1, 2, \ldots, n.$$

In other words, these are the unknowns of the matching problem.

The integer n refers to total number of elements in the equalizer. Minimum reactance driving point impedance $Z_B(p)$ is generated from *RB* using the Bode method.

For the RFDT algorithm and related MATLAB programs are provided for the reader as attached to this book.

The error function which is subject to minimization is generated under the MATLAB function called *direct_Monopole*. Eventually, we run the main program *monopole_Antenna.m* with the following inputs:

Inputs (Design Choices)

$$T0 = 0.625$$

This is the target values of the flat gain level

$$KFlag = 1$$

We should mention that, we can start the design either with a minimum reactance impedance ($KFlag = 1$ *case*) or a minimum susceptance admittance ($KFlag = 0$ *case*). Both choices are acceptable for the gain optimization.

Here, we have the liberty to select the minimum reactance driving point impedance for the equalizer which results in pretty good solution.

$$ndc = 0$$

In this design, we wish to work over a wide frequency of operation. Therefore, there is no need to introduce a zero of transmission at DC. Hence, we set $ndc = 0$. This choice will lead us to end up with a lowpass LC-ladder type equalizer structure which is quite easy to build.

$$W = 0$$

Here, we need no finite zero of transmission on the real frequency axis. Therefore, we set $W = 0$.

$$ntr = 1$$

We know by heart that the usage of an ideal transformer will help a lot to end up with a reasonable average gain level with smooth fluctuation. Furthermore, it is not difficult to realize transformers at the frequencies over 20–100 MHz. Therefore, we decided to employ an ideal transformer in the equalizer.

Initial coefficients for the optimization:

Here, we prefer an ad hoc choice for initial coefficients.

Therefore, we start with

$$a0 = 1$$

$$c0 = [-111 - 1 - 11]$$

which means that equalizer will include six elements.

Thus, we end with the following optimized solution:

$$a0 = 0.28103$$

$$A0 = a0 \times a0 = 0.0789$$

$$c = [-1.5256\ 1.8821\ 2.5705\ -4.3191\ -1.2094\ 2.1366]$$

$$B = [2.3274\ 4.3004\ -5.9611\ -17.429\ -11.853\ -2.1464\ 1]$$

$$a = [00.14475\ 0.35816\ 0.42631\ 0.40499\ 0.19034\ 0.051768]$$

$$b = [1\ 2.4742\ 3.9848\ 5.3703\ 4.0675\ 2.5009\ 0.65549]$$

$$wmax = 0.9596\ Tmax = 0.63906$$

$$wmin = 0.30505\ Tmin = 0.5212$$

$$T_{Average} = 0.58013 \mp 0.058932$$

$$dbmax = -1.9446;\ dbmin = -2.83$$

$$dB_{Average} = -2.3873 \mp 0.4427$$

Meaning that

$$R_B(p^2) = \frac{0.0789}{2.327p^{12} + 4.3p^{10} - 5.961p^8 - 17.429p^6 - 11.853p^4 - 2.146p^2 + 1}$$

and

$$Z_B(p) = \frac{0.144p^5 + 0.35p^4 + 0.42p^3 + 0.404p^2 + 0.19p + 0.0517}{p^6 + 2.47p^5 + 3.98p^4 + 5.73p^3 + 4.06p^2 + 2.5p + 0.655}$$

$Z_B(p)$ is synthesized as a lossless two-port in resistive termination which yields the equalizer as shown in Figure 5.27.

Actual element values of the equalizer can be computed as follows:

Actual capacitors are given by

$$C_A = \frac{C_{Normalized}}{2\pi f_0 R_0}$$

Actual Inductors are given by

$$L_A = \frac{L_{Normalized}}{2\pi f_0} R_0$$

Thus, we have

$$C1 = 2.1906e - 010\ Farad = 0.219\ nF;\ L2 = 1.1103e - 008 = 11.1\ nH$$
$$C3 = 7.6404e - 010\ Farad = 0.746\ nF;\ L4 = 9.5277e - 009 = 9.277\ nH$$
$$C5 = 5.3894e - 010\ Farad = 0.5389\ nF;\ L6 = 2.4538e - 009 = 2.4538\ nH$$
$$\frac{1}{n^2} = 12.6\ Turn\ ratio\ of\ the\ ideal\ transformer$$

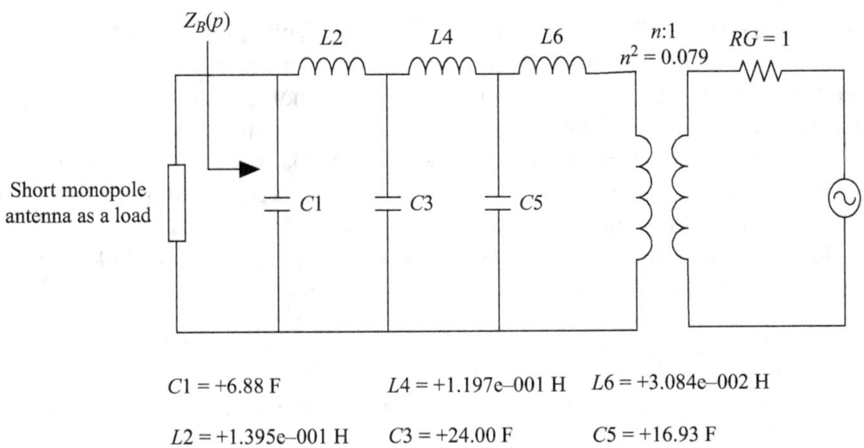

Figure 5.27 Single matching equalizer designed to match short a monopole antenna to a resistive generator

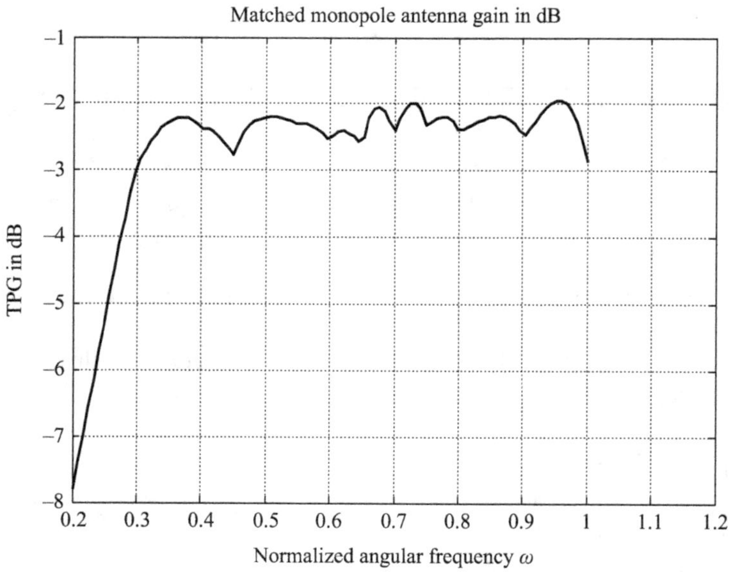

Figure 5.28 Gain performance of the matched monopole antenna

The gain performance of the matched antenna is depicted in Figure 5.28 in dB. It is clear that, more than 50% of the generator power is transferred to antenna over 30–100 MHz. This result is pretty satisfying.

We should mention that this is a unique solution which expands the band of the antenna more than three octaves.

Remarks

We tried to solve the same matching problem with commercially available packages by optimizing the gain function on the element values of the chosen circuit topology. Unfortunately, optimization was not successful. This clearly indicates that the solution obtained via RFDT is unique and superior.

Various bandpass solutions for the same antenna are given employing the SRFT. Therefore, we encourage the interested readers to go over those solutions as referred to in the footnote[6].

The authors of this book encourage the reader to run the monopole antenna programs on MATLAB as they are attached to this book.

[6]See Chapter 10 of the book by B.S. Yarman, "Design of Ultra Wideband Antenna Matching networks" Springer, 2008.

5.5 Design of a single matching equalizer for an ultrasonic transducer T1350

T1350 is a high-power piezoelectric ultrasonic transducer which resonates in the frequency range of 1,300—1,500 KHz. It is utilized for various commercial and military underwater applications.

For the ultrasonic piezoelectric T1350 transducer, the real and the imaginary parts of the measured impedance data are shown in Figures 5.29 and 5.30, respectively. Frequency range of the measurements is given by 600–2,100 KHz. We can normalize the data with respect to normalization frequency $f0 = 2{,}100$ KHz and the standard resistance $R0 = 50\,\Omega$. Normalized data is summarized in Table 5.6.

Close examination of Table 5.6 indicates that T1350 under consideration resonates at $f = 1{,}450$ KHz.

In this section, it is our purpose to match T1350 to a resistive generator over the frequency band of $f_{C1} = 1{,}100 \leq f \leq f_{c2} = 1{,}600$ KHz.

For this single matching problem, we modified the main program, error function for the optimization and the load function developed for the monopole

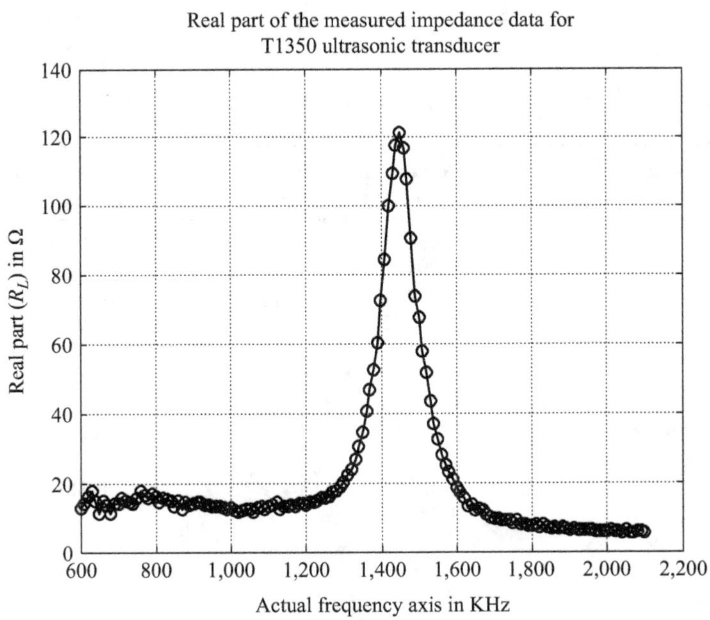

Figure 5.29 Real part variation of the measured impedance data of T1350

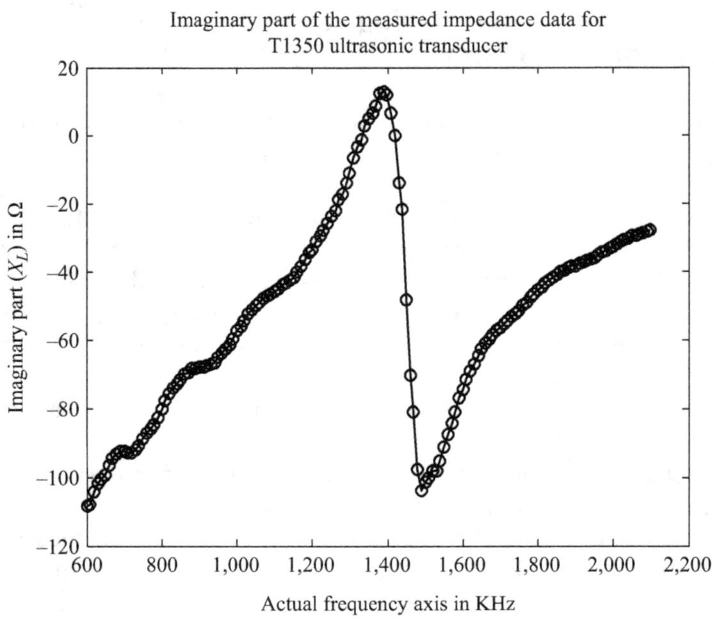

Figure 5.30 Imaginary part variation of the measured impedance for T1350

Table 5.6 Matching network design for the transducer T1350

Normalized angular frequency ω	Normalized real part data $RL(\omega)$	Normalized imaginary part data $RL(\omega)$
0.28571	0.25672	−2.169
0.34524	0.29164	−1.7846
0.40476	0.3032	−1.4787
0.46429	0.26372	−1.1999
0.52381	0.26356	−0.91012
0.58333	0.30754	−0.55386
0.64286	0.92886	0.008704
0.69048 resonance frequency	**2.4244**	**−0.96234**
0.69524	2.3375	−1.4045
0.70238	1.3801	−1.4592
0.7619	0.37162	−1.485
0.82143	0.18446	−1.0698
0.88095	0.13635	−0.8364
0.94048	0.11893	−0.67856
1	0.11039	−0.55752

Summary of the impedance measurements for the ultrasonic transducer T1350. Note that in this table all entrees are normalized with respect to $f0 = 2{,}100$ KHz, $R0 = 50\ \Omega$. Notice that resonance frequency of the transducer is at $f = 1{,}450$ KHz.

antenna to incorporate the measured load data of T1350. Thus, the new names are given as

$$Main\ Program : Ultrasonic\_Transducer.m$$

$$Error\ function\ for\ optimization$$

$$fun = direct\_Transducer(x0, KFlag, T0, wc1, wc2, ntr, ndc, W, a0)$$

$$function[RLN, XLN] = Load\_Transducer(w, KFlag)$$

The main program *Ultrasonic_Transducer.m* directly processes the normalized data with respect to $f0 = 2{,}100$ KHz *and* $R0 = 50\,\Omega$.

TPG of the matched system is optimized over the normalized frequency band of

$$fc1 = 1{,}100/2{,}100 = 0.52381\ to\ fc\,2 = 1{,}600/2{,}100 = 0.7619.$$

First, we will try to hit $T0 = 0.85$ flat gain level without using transformer in the equalizer. In this case, we must choose $ndc = 0, ntr = 1$. In order to see the final result over the complete measured frequency band, we choose

$$ws1 = \frac{600}{2{,}100} = 0.28571 \quad and \quad ws2 = \frac{2{,}100}{2{,}100} = 1$$

To make the equalizer as simple as possible, we set finite transmission zeros vector W=0.

First let us work with the impedance-based design. Then, *KFlag* is set to 1.

Hence, we run the main program Ultrasonic_Transducer with the following inputs:

INPUTS to Main Program Ultrasonic_Transducer :

$$T0 = 0.85$$

$$KFlag = 0 > Admittance, KFlag = 1 > Impedance..Enter\ KFlag$$

$$KFlag = 1(work\ with\ impedance\ functions)$$

$$Enter\ transmission\ zeros\ at\ DC\ ndc = 0$$

$$Enter\ finite\ transmission\ zeros\ at\ [W] = [\ldots] = 0$$

$$Initial\ guess\ on\ the\ coefficients\ of\ the\ real\ part\ RB$$

$$Enter\ a0 = 1; Enter\ initials\ for\ [c0] = [\ldots] = [11 - 1 - 11]$$

$$ntr = 1, Yes, work\ with\ Transformer, ntr = 0\ No\ Transformer\ ntr = 0$$

$$Enter\ ws1 = 600/2{,}100 = 0.28571$$

$$Enter\ ws2 = 2{,}100/2{,}100 = 1$$

$$Enter\ wc1 = 1{,}100/2{,}100 = 0.52381$$

$$Enter\ wc2 = 1{,}600/2{,}100 = 0.7619$$

Result of the optimization is summarized as follows:

$$R_B = \frac{a_0^2}{B(p^2)}; \quad Z_B(p) = \frac{a(p)}{b(p)} \text{ (minimum reactance driving point impedance)}$$

$$B(\omega^2) = \frac{1}{2}\left[c^2(\omega) + c^2(-\omega)\right]$$

$$a_0 = 1 \ (\textit{transformerless design})$$

$$c = [-33.448 \ \ -3.8855 \ 36.433 \ 1.7357 \ -9.0796 \ 1]$$

$$B(p) = [-1{,}118.8 \ \ -2{,}422.1 \ \ -1{,}921.2 \ \ -666.35 \ \ -85.911 \ \ 1]$$

$$a(p) = [0 \ \ 0.62956 \ \ 0.24185 \ \ 0.40182 \ \ 0.068798 \ \ 0.029897]$$

$$b(p) = [1 \ \ 0.38416 \ \ 1.1563 \ \ 0.30828 \ \ 0.30858 \ \ 0.029897]$$

$$wmax = 0.5619 \ Tmax = 0.85822$$

$$wmin = 0.7619 \ Tmin = 0.72742$$

$$T\_Average = 0.79282 \mp 0.065401$$

$$dbmax = -0.66401; \ dbmin = -1.3822$$

$$dB\_Average = -1.0231 \mp 0.35907$$

Gain performance of the matched piezoelectric ultrasonic transducer T1350 (in dB) is shown in Figure 5.31. The result is pretty good and confirms with the laboratory test.

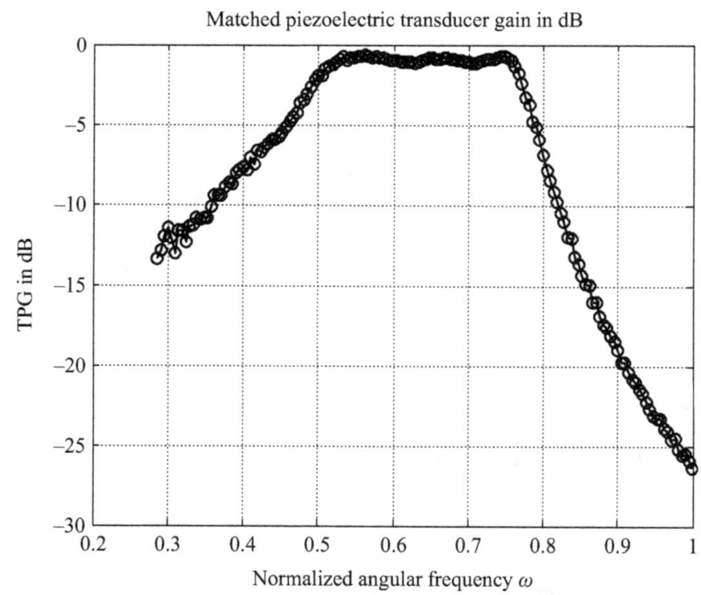

Figure 5.31 Transducer power gain of the matched transducer T1350

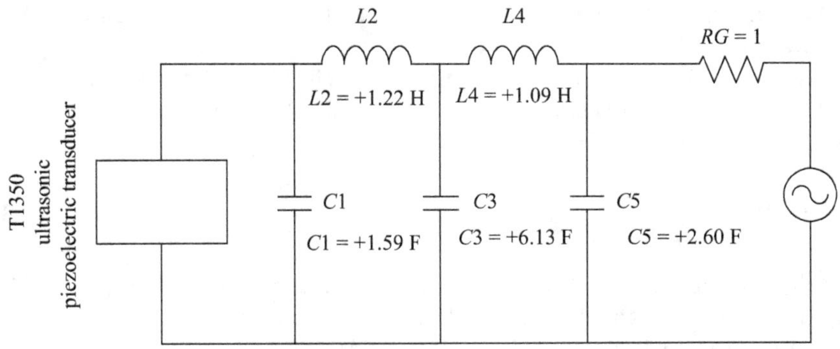

Figure 5.32 Design of single matching equalizer for T1350 piezoelectric transducer

We should mention that, by reiteration, gain performance of the matched transducer may be improved.

Synthesis of $Z_B(p)$ which in turn yields the desired equalizer is shown in Figure 5.32.

Finally, by de-normalization, actual element values are given as

$$C1 = 2.4077e - 009 = 2.4nF;$$

$$L2 = 4.6054e - 006 = 4.6\mu H;$$

$$C3 = 9.2917e - 009 = 9.29nF;$$

$$L4 = 4.1147e - 006 = 4.1\mu H;$$

$$C5 = 3.9457e - 009 = nF$$

5.6 Simplified real frequency technique (SRFT): "scattering approach"

In this chapter, we utilize SRFT to construct single and double matching equalizers as well as microwave amplifiers utilizing lumped circuit components.

Let us briefly summarized the SRFT as introduced [5,6].

In the SRFT, a lossless equalizer is simply described in terms of its unit normalized scattering parameters. Therefore, neither the impedance nor the admittance parameters are employed. All the computations are carried out with the scattering parameters of the blocks to be matched. Therefore, the gain optimization of the matched system is well-behaved numerically. It is faster than the other existing

CAD algorithms and easier to use. It is also naturally suited to design broadband microwave amplifiers.

Basis for SRFT or the scattering approach is to describe the lossless equalizer $[E]$ in terms of its unit normalized input reflection coefficient $E_{11}(p)$ in Belevitch form such that

$$E_{11}(p) = \frac{h(p)}{g(p)} \tag{5.35}$$

For selected transmission zeros, the complete scattering parameters of a lossless reciprocal equalizer can be generated from the numerator polynomial $h(p)$, using the para-unitary condition given by (6.37)–(6.54).

This idea constitutes the crux of the SRFT. In other words, TPG of the system to be matched is expressed as an implicit function of $h(p)$.

Replacing generator and load impedances by their Darlington equivalent lossless two-ports $[G]$ *and* $[L]$, and utilizing their unit normalized scattering descriptions given by (10.10) and (10.23), TPG of the doubly matched system can be generated in two steps as in (6.74)–(6.76) of Chapter 6.

In the first step, by multiplying the transfer scattering parameters of the lossless two-ports $[G]$ *and* $[E]$, gain of the cascaded duo $[GE]$ is generated as

$$T_1 = |G_{21}|^2 \frac{|E_{21}|^2}{|1 - E_{11}G_{22}|^2}$$

Then, in the second step, TPG of cascaded two-ports $[GE]$ *and* $[L]$ is obtained as

$$T = T_1 \frac{|L_{21}|^2}{\left|1 - \hat{E}_{22}L_{11}\right|^2}$$

Thus, using the open form of T_1 in the above equation, the double matching gain of the trio $\{[G] - [E] - [L]\}$ is expressed as follows:

$$T(\omega) = |G_{21}|^2 \frac{|E_{21}|^2}{|1 - E_{11}G_{22}|^2 \left|1 - \hat{E}_{22}L_{11}\right|^2} |L_{21}|^2 \tag{5.36}$$

where the scattering parameters of the generator $\{G_{ij}\}$ and the load $\{L_{ij}\}$ networks are specified either by measurements or they are provided as circuit models. More specifically, in terms of the generator $Z_G(j\omega)$ and the load $Z_L(j\omega)$ impedances, the generator reflectance is given by

$$G_{22} = \frac{Z_G - 1}{Z_G + 1}; \quad |G_{21}|^2 = 1 - |G_{22}|^2 \tag{5.37a}$$

the load reflectance is given by

$$L_{11} = \frac{Z_L - 1}{Z_L + 1}; \quad |L_{21}|^2 = 1 - |L_{11}|^2 \tag{5.37b}$$

and the back-end reflectance of the equalizer, while it is terminated in the complex generator is given by

$$\hat{E}_{22} = E_{22} + \frac{E_{21}^2 G_{22}}{1 - E_{11} G_{22}} \tag{5.37c}$$

Now, let us construct TPG, once $h(p)$ is initialized.

For simplicity, let us assume that $[E]$ is a minimum phase structure with transmission zeros only at $\omega = 0$ *and* $\omega = \infty$. This is a convenient assumption because it assures the realization of the equalizer without coupled coils, except possibly for an impedance level transformer. Then, the real normalized scattering parameters of a lossless-LC ladder network $[E]$ is given in Belevitch form as follows [2,5,7]:

$$E_{11}(p) = \frac{h(p)}{g(p)} = \frac{h_n p^n + h_{n-1} p^{n-1} + \ldots + h_1 p + h_0}{g_n p^n + g_{n-1} p^{n-1} + \ldots + g_1 p + g_0} \tag{5.38a}$$

$$E_{12}(p) = E_{21}(p) = \mp \frac{p^k}{g(p)} \tag{5.38b}$$

and

$$E_{22}(p) = (-1)^k \frac{h(-p)}{g(p)} \tag{5.38c}$$

where the integer n specifies the number of total reactive elements in $[E]$ and the integer k is the order of the transmission zeros at DC.[7]

Since the matching network is lossless, it follows that

$$G(p^2) = g(p)g(-p) = h(p)h(-p) + (-1)^k p^{2k} > 0 \textit{ for all } p \tag{5.39}$$

In the "SRFT algorithm," the goal is to optimize the TPG as high and as flat as possible over the band of operation as in the other real-frequency techniques.

The coefficients of the numerator polynomial $h(p)$ are selected as the unknowns of the matching problem. To construct the scattering parameters of $[E]$, it is sufficient to generate the Hurwitz denominator polynomial $g(p)$ from $h(p)$.

It can be readily shown that, once the coefficients of $h(p)$ are initialized, and the complexity of the equalizer $[E]$ is specified (i.e., n and k are fixed in advance), then $g(p)$ is generated as a Hurwitz polynomial by explicit factorization of (5.39). Thus, the bounded realness of the scattering parameters $\{E_{ij}; i,j = 1,2\}$ is already built into the design procedure.

It is noted that, in order to assure the realizability of the lossless equalizers, we cannot simultaneously allow $h(p)$ and $g(p)$ to become zero. Therefore, selection of

[7]Notice that, in (5.41), indexing used for polynomial coefficients $h(p)$*and* $g(p)$ are classical rather than MATLAB convention. However, replacing "i"*by*"$n - i + 1$" we can always end up with MATLAB indexing.

initial coefficients of $h(p)$ must be proper. For example, if $k > 0$, we cannot let $h_0 = 0$ since it makes $g(0) = 0$.

In generating the Hurwitz denominator polynomial $g(p)$ from the initialized coefficients of $h(p)$, we first construct $G(p^2)$ as in (5.41). That is,

$$G(p^2) = g(p)g(-p) = G_n p^{2n} + G_{n-1}p^{2(n-1)} + \ldots + G_1 p^2 + G_0 \tag{5.40}$$

where the coefficients G_i are given as the convolution of $h(p)$ *and* $h(-p)$ with the contributing term $(-1)^k p^{2k}$ as follows:

$$\begin{aligned} G_0 &= h_0^2 \\ G_1 &= -h_1^2 + 2h_2h_0 \\ &\ldots\ldots\ldots\ldots \\ G_i &= (-1)^i h_i^2 + 2\left(h_{2i}h_0 + \sum_{j=2}^{i}(-1)^{j-1}h_{j-1}h_{2j-i+1}\right) \\ &\ldots\ldots\ldots\ldots \\ G_k &= G_i\big|_{i=k} + (-1)^k \\ &\ldots\ldots\ldots\ldots \\ G_n &= (-1)^n h_n^2 \end{aligned} \tag{5.41}$$

Then, explicit factorization of (5.41) follows. At the end of the factorization process, polynomial $g(p)$ is formed on the left half plane zeros of $G(p^2)$.

Hence, the scattering parameters of $[E]$ are obtained as in (5.38a)–(5.38c) and the TPG $T(\omega)$ is generated employing (5.36). Then, as in the other real-frequency techniques, by selecting flat gain level T_0, the error function $error = T(\omega) - T_0$ is minimized. As the result of optimization, the unknown coefficients h_i are determined.

Details of the numerical work can be found in the reading list.

In brief, examination of (5.40) together with (5.38a)–(5.38c) indicates that TPG is almost inversely quadratic in the unknown coefficients h_i. Therefore, SRFT algorithm is always convergent. Furthermore, the numerical stability of the algorithm is excellent, since all the scattering parameters E_{ij} and reflection coefficients G_{22} and L_{11} are bounded real, i.e., $\{|E_{ij}|, |G_{22}|\ and\ |L_{11}|\} \leq 1$.

As is usually the case, an intelligent initial guess is important in efficiently running the program. It has been experienced that, for many practical problems, an ad hoc direct choice for the coefficients h_i $(e.g. h_i = \mp 1)$ provides satisfactory initialization to start the SRFT algorithm.

As indicated previously, SRFT is naturally suited to design microwave amplifiers since it employs scattering parameters for all the units to be matched. Over 30 years, matching networks and amplifiers designed using SRFT showed excellent agreement with laboratory performance measurements.

Now, let us develop the SRFT-matching algorithm to design a matching network for a helix antenna. The antenna can be utilized for general purpose global-positioning systems applications.

5.6.1 Antenna tuning using SRFT: design of a matching network for a helix antenna

For this section, a helix antenna was designed for a GPS receiver.

The antenna is supposed to operate in the neighborhood of the center frequency $f_0 = 1.575$GHz.

The real part *RL* and the imaginary part *XL* of the measured antenna impedance are depicted in Figures 5.33 and 5.34, respectively. Summary of the measurements is given in Table 5.7.

Here, the idea is to tune the antenna at 1.575 GHz employing SRFT.

For this purpose, we wish to design an equalizer to match the antenna to a 50 Ω generator over the passband B :1.5–1.6 GHz which also captures the tuning frequency $f_0 = 1.57$GHz. In this case, we must implement an SRFT algorithm for the given load to minimize the error function:

$$error = T(\omega) - T_0 \tag{5.42a}$$

or

$$error = |G_{21}|^2 \frac{|E_{21}|^2}{|1 - E_{11}G_{22}|^2 |1 - \hat{E}_{22}L_{11}|^2} |L_{22}|^2 - T_0 \tag{5.42b}$$

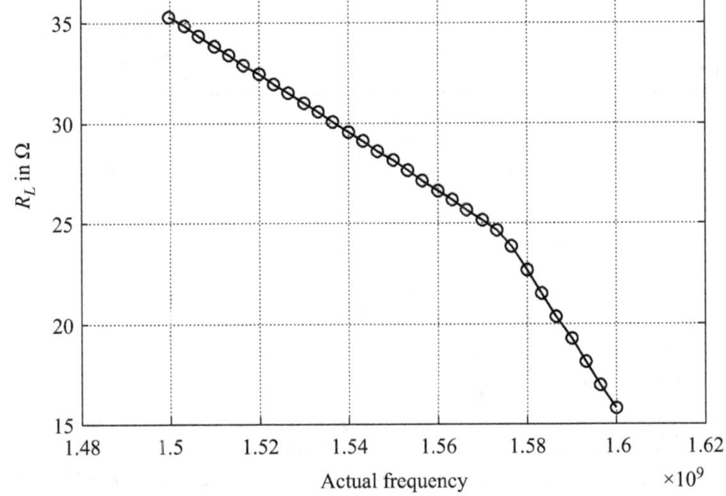

Figure 5.33 Measured real part data of the impedance of a helix antenna

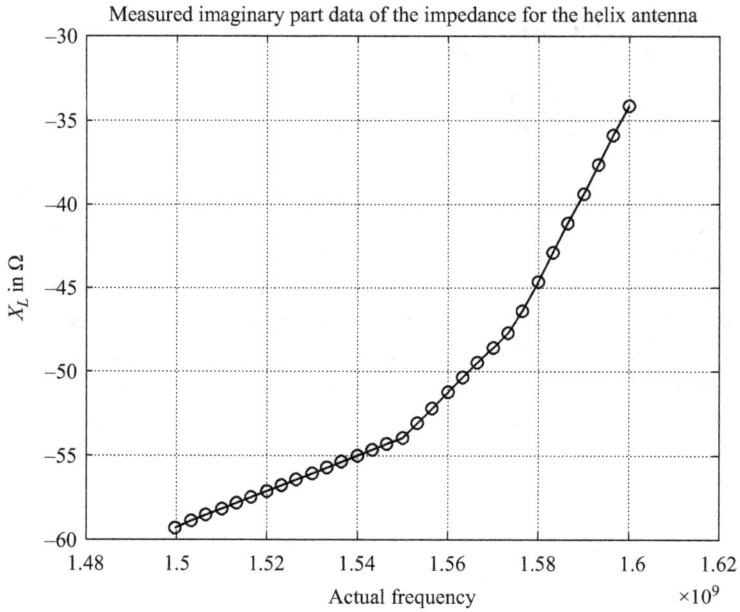

Figure 5.34 Imaginary part of the measured impedance data for helix antenna

Table 5.7 Impedance data for the helix antenna

Actual frequencies (GHz)	**Real part RL (Ω)**	**Imaginary part *XL*(Ω)**
1.5	35.31	59.3
1.55	28.14	−53.98
1.575	24.44	−47.32
1.6	15.77	−34.2

where $G_{21} = 1$ since the generator is resistive with $R_G = R_0 = 50\ \Omega$

$$L_{22} = \frac{Z_L - R_0}{Z_L + R_0}$$

and

$$\hat{E}_{22} = E_{22} = -(-1)^k \frac{h(-p)}{g(p)}$$

which is the back-end reflection coefficient of the equalizer. Obviously, here, it is presumed that all the transmission zeros of the equalizer are on the $j\omega - axis$. Furthermore, transmission zeros at DC is order of k.

In this regard, $f(p)$ is chosen as $f(p) = p^k$.

On the other hand, this is a narrow bandwidth problem. Therefore, during the optimization process, we may be able to hit the ideal flat gain level $T_0 = 1$.

It is our desire to make the matching network as simple as possible. So, we do not wish to employ an ideal transformer. Therefore, we set $ntr = 1$.

In this case, we can work with an LC ladder type structure with $k = 0$.

The MATLAB code developed for this problem is described in the following algorithm.

5.6.2 SRFT algorithm to design matching networks

This algorithm basically minimizes the error function specified by (5.41) which in turn determines the Belevitch form of the scattering parameters of the lossless equalizer $[E]$ as described above.

Within the algorithm first, the complexity of the matching network is selected by the designer. In other words, total number of elements n and the transmission zeros of the equalizer is set in advance.

For practical matching networks, we prefer to utilize transmission zeros only at DC and infinity. In this case, we set $f(p) = p^k$ so that

$$E_{21} = \frac{f(p)}{g(p)} = \frac{p^k}{g(p)}$$

This choice of equalizer is simple enough to be implemented employing the existing manufacturing technologies either with discrete components or with Silicon CMOS technology.

The unknown of the matching problem is the numerator polynomial

$$h(p) = h_n p^n + h_{n-1} p^{n-1} + \ldots + h_1 p + h_0.$$

Here, the coefficients $\{h_0, h_1, \ldots, h_n\}$ are determined to minimize the error function. For the optimization, we use standard MATLAB *Levenberg–Marquardt* function called *lsqnonlin* which minimizes the sum of mean square error.

The main program developed for this section is called *Helix_Antenna.m*, and it is provided with the reader as an attachment to this book.

Inputs to the main program are given as follows:

Inputs

- *FA: An array.Sampling Frequencies to measure the load data*
- *RLA: An array.Real part of the antenna impedance*
- *XLA: An array.Imaginary part of the antenna impedance*
- *T0: Targeted flat gain level*
- *ntr: Choice for transformer.*
  If you want to use an ideal transformer set $ntr = 1$,
  otherwise $ntr = 0$.
- *k: Order of the transmission zeros at DC*
- *Wlow: Low end of the optimization frequency*
  (*Normalized with respect to selected frequency which is* $f0 = 1.575$ GHz).

- *Whigh: High – end of the optimization frequency (Normalized as above)*
- *h0: initial value for the coefficient* $h0 = h(0)$
- *h: Initial values for the coefficients* $h(p)$

Computational steps

Step 1: Using the above initials, start optimization employing the error function is called

$$fun = sopt\_Helix(x, T0, ntr, k, nopt, wlow, whigh)$$

The input integer *nopt* is the total number of sampling points which is used to sample the error function over the normalized angular frequencies between *wlow and whigh*. For the present case, it may be sufficient to set $nopt = 30$.

The input x starts with the initials $x0$ on the coefficients $x0{=}\{h0, h1, h2, \ldots, hn\}$. Obviously, if there is no transformer used in the circuit, $h0 = 0$ and it is not part of the unknowns.

Step 2: Call optimization function
$x = lsqnonlin('sopt\_Helix', x0, [], [], OPTIONS, T0, ntr, k, nopt, wlow, whigh)$; which determines the optimum set for the numerator coefficients $x = \{h0, h1, \ldots, hn\}$ minimizing the error function $error = T(\omega) - T0$.

Step 3: Once the optimum $h(p)$ is determined, complete scattering parameters of the equalizer is constructed as specified by (5.38a)–(5.38c) and it is synthesized as a lossless two-port. At this step, it is useful to compare the performance of the antenna as it is matched and unmatched. The unmatched antenna performance can be analyzed from the input reflectance

$$L_{11} = \frac{Z_L - R_0}{Z_L + R_0} = \rho_L e^{j\varphi_L}$$

In this regard, the measure of mismatch is given by

$$L(\mathrm{dB}) = 10\log_{10}(\rho_L)$$

This is called the input mismatch. It is always nice to plot matched and unmatched reflection coefficients on the same sheet to understand the success of the optimization. The matched input reflectance in dB or the matched *input mismatch* is given by

$$\rho_M(\mathrm{dB}) = 10\log_{10}[1 - T(\omega)]$$

Mismatch is the measure of the power reflection from the load. As it groves in absolute values, power reflection becomes less. The ideal value of the mismatch is minus infinity dB which corresponds to full power transfer to the load. In this case, TPG becomes unity. The worst mismatch case is 0 dB which corresponds to full reflection from the load end. In this case, TPG becomes zero.

From the practical point of view, mismatch of −20 dB is pretty good. On the other hand, mismatch of −4 or −5 dB is considered quite poor.

Equalizer is synthesized by long division of its driving point impedance function

$$Z_{in}(p) = \frac{a(p)}{b(p)} = \frac{g(p) + h(p)}{g(p) - h(p)}$$

As in the other real frequency techniques, we use the function $CVal = General Synthesis\_Yarman(h, g)$ to synthesize the equalizer. In this function, $g(p)$ and $h(p)$ are defined as conventional MATLAB polynomials. Therefore, we have to shuffle the coefficients of the polynomials $h(p)$ and $g(p)$ as they are obtained from the optimizations.

Now, let us run the main program *Helix_Antenna.m* with the following inputs:

Inputs

$f0 = 1.575\ GHz(Normalization\ frequency)$

$R0 = 50(Normalization\ resistance)$

$FA = [1.5\ 1.55\ 1.575\ 1.6] Actual\ frequencies\ in\ GHz.$

$(Normalized\ input\ frequencies\ with\ respect\ to\ f0:$

$WBR = [0.95238\ 0.98413\ 1\ 1.0159]$

$RLA = [35.31\ 28.14\ 24.44\ 15.77]/R0(\text{real Part of the antenna impedance})$

$XLA = [-59.30\ \ -53.98\ \ -47.32\ \ -34.2]/R0(\text{imaginary part of the antenna impedance})$

$T0 = 1(we\ try\ to\ hit\ the\ best\ value\ over\ the\ band\ of\ operation);$

$k = 0;\ \left(Note\ that\ f(p) = p^k = 1\ which\ is\ fixed\ at\ the\ input\right).$

$(Here, we\ wish\ to\ work\ with\ lowpass\ LC\ ladder\ to\ ease\ the\ implementation$

$ntr = 0(in\ the\ equalizer\ design, we\ avoid\ using\ an\ ideal\ transformer)$

$(in\ the\ equalizer\ design, we\ avoid\ using\ an\ ideal\ transformer)$

$Wlow = 0.953\ which\ corresponds\ to\ 1.5\ GHz;$

$whigh = 1\ which\ corresponds\ to\ 1.6\ GHz$

Initial guess:

$h0 = 0, h = [1\ \ -1\ 1];\ ad\ hoc\ choice\ for\ the\ initials$

Result of optimization:

$h = [0.64211\ 0.28508\ 0](regular\ \text{MATLAB}\ polynomial\ form)$

$g = [0.64211\ 1.1685\ 1](\ regular\ \text{MATLAB}\ polynomial\ form)$

Optimized driving point input impedance of the equalizer is given as

$$Z_{in}(p) = \frac{a(p)}{b(p)} = \frac{g+h}{g-h} = \frac{1.2842p^2 + 1.4536p + 1}{0.88346p + 1}$$

$$= 1.4536p + \frac{1}{0.88346p + 1}$$

Here, the first component of the equalizer is a normalized inductor $L_{N1} = 1.4536$. The second component is a normalized capacitor $C_{N2} = 0.88346$ and the termination is a unit resistance as expected.

The actual element values of the matching network are obtained by de-normalization with respect to $f0 = 1.575$ GHz *and* $R0 = 50\ \Omega$. Hence, we end up with the following actual elements:

$$L_{A1} = \frac{L_{N1}}{2\pi f_0} R_0 = 7.3445e - 009H = 7.3445nH$$

Similarly, actual capacitor is found as

$$C_{A2} = \frac{C_{N2}}{2\pi f_{0R_0}} = 1.7855e - 012F = 1.7855pF$$

Performance of the matched system is summarized in Table 5.8.

Matched and unmatched input reflectance of the antenna is depicted in Figure 5.35.

Matched TPG is depicted in Figure 5.36.

Physical picture of the helix antenna is shown in Figure 5.37 and measured performance is depicted in Figure 5.38.

Finally, antenna with matching network is shown in Figure 5.39.

Table 5.8 Optimized performance of the tuned antenna

Actual frequency (GHz)	**Matched input reflectance (dB)**	**Unmatched input reflectance (dB)**	**Gain (dB)**
1.50	−12.394	−4.6122	−0.25777
1.51	−13.222	−4.5568	−0.21191
1.52	−14.175	−4.4931	−0.16932
1.53	−15.277	−4.4207	−0.13081
1.54	−16.545	−4.339	−0.097315
1.55	−17.966	−4.2478	−0.069927
1.56	−21.914	−4.2777	−0.028039
1.57	**−24.591**	**−4.2947**	**−0.015117**
1.58	−18.122	−4.2415	−0.067452
1.59	−11.659	−4.0438	−0.30704
1.6	−7.6666	−3.7053	−0.81516

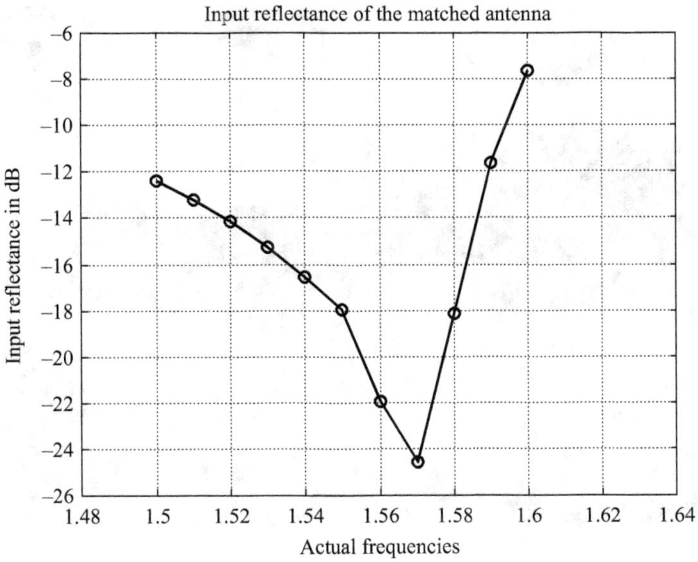

Figure 5.35 Optimized input reflectance of the matched helix antenna

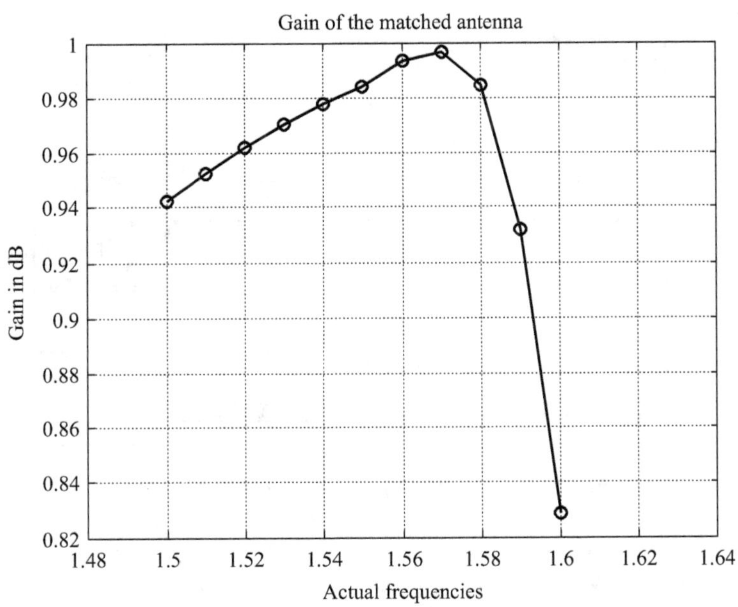

Figure 5.36 Optimized transducer power gain of the matched helix antenna

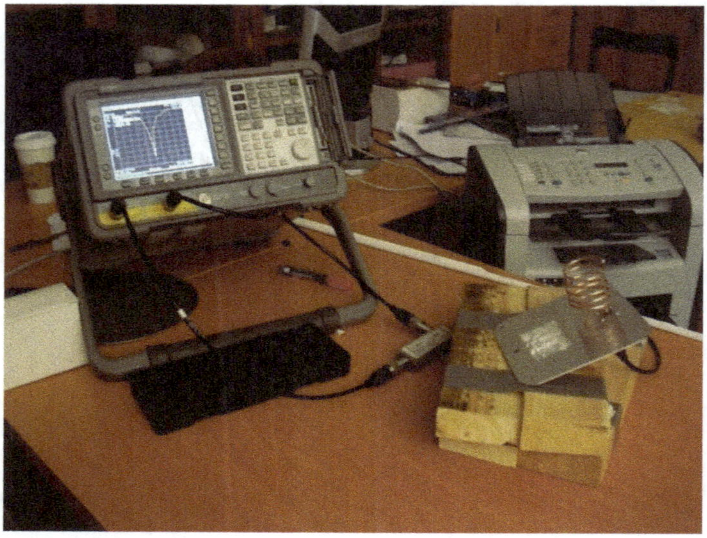

Figure 5.37 Picture of the helix antenna designed to operate at 1.575 GHz

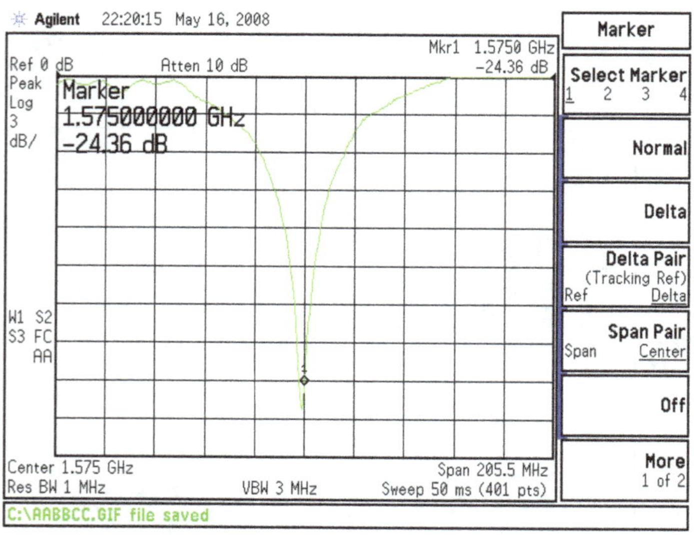

Figure 5.38 Measured performance of the helix antenna

Here, it is clearly shown that a simple LC ladder equalizer design using SRFT significantly improves the power transfer from the generator to the load. Especially, at $f0$=1.575 GHz, without equalizer, the reflection or mismatch is measured as **−4.3 dB**, whereas match input reflection is about −25 dB which is almost perfect match.

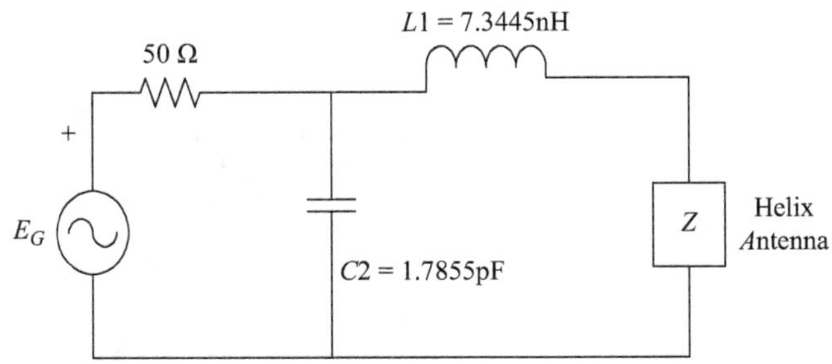

Figure 5.39 Tuned helix antenna employing SRFT

All the MATLAB design S/W for the Helix_Antenna.m is provided with the reader as the attachment to this book.

References

[1] H. Carlin, "A New Approach to Gain–Bandwidth Problems," *IEEE Trans CAS*, vol. 23, no. 4, pp. 170–175, 1977.

[2] B. Yarman, Design of Ultra Wide-Band Matching Networks via Real Frequency Techniques, London: Wiley, 2010.

[3] B. Yarman, Broadband matching from a complex source to a complex load, Ph.D. Thesis, Ithaca, NY: Cornell University, 1982.

[4] H. Carlin and B. Yarman, "The Double Matching Problem: Analytic and Real Frequency Solutions," *IEEE Trans CAS*, vol. 30, no. 1, pp. 15–28, 1983.

[5] B. S. Yarman, "A Simplified Real Frequency Technique for Broadband Matching Complex Generator to Complex Load," *RCA Rev*, vol. 43, no. 9, pp. 529–541, 1982.

[6] B. S. Yarman and H. J. Carlin, "A Simplified Real Frequency Technique is Applied to Broadband Multi-Stage Microwave Amplifiers," *IEEE MTT*, vol. 30, no. 12, pp. 2216–2222, 1982.

[7] V. Belevitch, Classical Network Theory, San Francisco, CA: Halden Day, 1968.

[8] D. C. Youla, "A New Theory of Cascade Synthesis," *IRE Trans Circuit Theory*, vol. 8, no. 3, pp. 244–260, 1961.

MATLAB® program lists

Program List 5.1: Main Program for Example 5.1

```
% Main Program Example5_1
% Solution to Single Matching Problem
% via Real Frequency Line Segment Technique
% Solution to Example 5.1
% For this problem load is given as R//C+L load.
clear
clc
% Inputs:
N=input('Enter total number of break points N=')%Total Number of Break
Points
R=input('L+R//C Load: Enter resistance of the load R=')
C=input('Enter Capacitor value of the load C=')
L=input('Enter Inductor value of the load L=')
wc1=input('Enter lower edge of the passband wc1=')
wc2=input('Enter upper edge of the passband wc2=')
ws1=input('Enter lower edge of the stop band;(ws1=0 for Lowpass problems)
ws1=')
ws2=input('Enter upper edge of the stop band ws2=')
T0=input('Enter ideal flat gain level T0=')
KFlag=input('Enter KFlag,for Z=>1; for Y=>0; KFlag=')
%
%--------------------------------------------------
%
% Step 1: Generate the load data
WBR=Break_Frequencies(N,ws1,ws2,wc1,wc2);
for i=1:N
    w=WBR(i);
    [RL,XL]=RLC_Load(w,R,C,L,KFlag);
    RLA(i)=RL;XLA(i)=XL;
end
% Step 2: Generate the low
%and the high values of RB,XB and XF
sign=-1;% Low Values
[RBA_Low,XBA_Low,XFA_Low]=Ideal_SingleMatching(T0,sign,WBR,RLA,XLA);
sign=+1;% High Values
[RBA_High,XBA_High,XFA_High]=Ideal_SingleMatching(T0,sign,WBR,RLA,XLA);
%
figure(1)
plot(WBR,RBA_Low, WBR,RBA_High)
xlabel('Normalized Angular Frequency')
ylabel('R_B=real part of the immittance F_B(jw): High and Low values')
title('Example 5.1 Plot of R_B')
legend('Low','High')
%
figure(2)
plot(WBR,XBA_Low, WBR, XBA_High)
xlabel('Normalized Angular Frequency')
ylabel('X_B=Hilbert Transform of R_B(w): Low & High values')
title('Example 5.1 Plot of X_B')
legend('Low','High')
%
figure(3)
plot(WBR,XFA_Low, WBR,XFA_High)
xlabel('Angular Frequency')
ylabel('XF=Foster Part of the immittance F_B(jw): Low & High values')
title('Example 5.1 Plot of XF=Foster Function Low & High values')
legend('Low','High')
```

```
%-----------------------------------------------
% Part (b) of Example 5.1
Gain_Low=gain_singleMatching(WBR,RLA,XLA,RBA_Low,XBA_Low)
Gain_High=gain_singleMatching(WBR,RLA,XLA,RBA_High,XBA_High)
%
figure(4)
plot(WBR,Gain_Low, WBR,Gain_High)
xlabel('Angular Frequency')
ylabel('Gain_Single Matching: Low & High values')
title('Example 5.1 Gain Plot with Low & High values')
legend('Low','High')
% Part (c) of Example (11.1):
%     Optimization of the Driving Point Immittance
%
%----------------------------------------------
% Optimization using Levenberg marquardt algorithm
% Generate Initials
sign=input('=>1 high values; =>-1 low values. Enter sign=')
RB0=initials(R,C,L,KFlag,T0,sign,WBR);
%
% Define unknowns for the optimization:
for j=1:(N-1)
       x0(j)=sqrt(RB0(j));%Initial
end
%
OPTIONS=OPTIMSET('MaxFunEvals',20000,'MaxIter',50000);
x=lsqnonlin('error_RFLT',x0,[],[],OPTIONS,WBR,wc1,wc2,N,T0,R,C,L,KFlag);
%--------END OF OPTIMIZATION---------------------%
% Generate optimized driving point impedance
 for j=1:N-1
    RBA(j)=x(j)^2;
 end
    RBA(N)=0.0;
%---------------------------------------------
N_Print=100;
DW=(ws2-ws1)/(N_Print-1);
w=wc1+1e-11;
NBR=length(WBR);
% Generate load and the driving point immittance
for i=1:N_Print
    WA(i)=w;
    [RL,XL]=RLC_Load(w,R,C,L,KFlag);
    RQ=line_seg(NBR,WBR,RBA,w);
    RQA(i)=RQ;
    XQ=num_hilbert(w,WBR,RBA);
    XQA(i)=XQ;
    RTSQ=(RQ+RL)*(RQ+RL);
    XTSQ=(XQ+XL)*(XQ+XL);
    TPG=4*RL*RQ/(RTSQ+XTSQ);
    TA(i)=TPG;
       w=w+DW;
end
%---------------------------------------------
%
%---------------------------------------------
figure(5)
plot(WA,TA,WBR,Gain_Low)
xlabel('Angular Frequency')
ylabel('Optimized Gain')
title('Example 5.1 Optimized Gain Plot')
legend('Optimized',' Initial')
```

```
figure(6)
plot(WBR,RBA,WBR,RBA_Low)
xlabel('Angular Frequency')
ylabel('Optimized Break Points RBA')
title('Example 5.1 Optimized Break Points for R_B')
legend('Optimized',' Initial')
```

Program List 5.2: Generation of Break Frequencies for specified inputs: *N*, *ws1*, *ws2*, *wc1*, *wc2*

```
function WBR=Break_Frequencies(N,ws1,ws2,wc1,wc2)
% Generate break frequencies
% Inputs:
%   N:   Total number of Break Points.
%   ws1: Lower edge of the stop band.
%   ws2: Upper edge of the stop band.
%   wc1: Left edge of the passband.
%   wc2: Upper edge of the passband.
% Outputs:
% WBR: Break Frequencies
%
% Generate Angular Break Frequencies
WBR(1)=ws1;
DW=(wc2-wc1)/(N-2);
W=WBR(1);
for j=2:(N-1)
    W=W+DW;
    WBR(j)=W;
end
WBR(N)=ws2;
```

Program List 5.3: Load function for Example (11.1)

```
function [RL,XL]=RLC_Load(w,R,C,L,KFlag)
% This function generates the impedance data for a given R//C load
% Inputs:
%       w,R,C,L,KFlag
% Generate the load impedance
    p=sqrt(-1)*w;
    YL=(1/R)+p*C;
    ZL=p*L+1/YL;
      if KFlag==1;% Work with impedances
        RL=real(ZL);
        XL=imag(ZL);
      end
        if KFlag==0;%Work with admittances
            RL=real(YL);
            XL=imag(YL);
        end
```

Program List 5.4: Ideal Solution to Single Matching Problems

```
function [RBA,XBA,XFA]=Ideal_SingleMatching(T0,sign,WBR,RLA,XLA)
% Computation of idealized immittance for single matching problems
% See Chapter 5, section 5.2 Eq.s (5.6,5.7,5.8,5.9,5.10,5.11 and
% 5.12)
% Inputs:
%       T0: Ideal flat gain level
%       WBR: Break frequencies of the load
%       RLA: Real Part of the immittance data
N=length(WBR);
for j=1:N
    RL=RLA(j);
   %RB(w)=RL(w)*((2-T0 )+2*sign*sqrt(1-T0 ))/T0
    RB=RL*((2-T0)+2*sign*sqrt(1-T0))/T0;
              %See Eq.(11.9)
    RBA(j)=RB;
end
% Computation of Hilbert Transform Loop
    for j=1:N
        w=1e-11+WBR(j);
   %XBA(w)=Hilbert Transform of {RB(w)}
    XB=num_hilbert(w,WBR,RBA);
    XBA(j)=XB;
    XA(j)= - XLA(j);
    XFA(j)=XA(j)-XBA(j);
         %see Eq. (11.12)
    end
```

Program List 5.5: User supplied objective function for the nonlinear least square optimization

```
function Func=error_RFLT(x,WBR,wc1,wc2,N,T0,R,C,L,KFlag)
% Objective function to be minimized
% It should be noted that
% number of sampling points N or N-1 must be much greater than the
% unknowns.
% RFLT for single matching problems
% Load: R//C+L
%-------------------------------------------
% Generate RBA from the given initials:
for j=1:N-1
    RBA(j)=x(j)^2;
end
    RBA(N)=0.0;
%-------------------------------------------
N_opt=30;
DW=(wc2-wc1)/(N_opt-1);
w=wc1+1e-11;
NBR=length(WBR);
% Generate load and the driving point immittance data
for i=1:N_opt
    [RL,XL]=RLC_Load(w,R,C,L,KFlag);
    RQ=line_seg(NBR,WBR,RBA,w);
    XQ=num_hilbert(w,WBR,RBA);
    RTSQ=(RQ+RL)*(RQ+RL);
    XTSQ=(XQ+XL)*(XQ+XL);
    TPG=4*RL*RQ/(RTSQ+XTSQ);
    Func(i)=TPG-T0;
    w=w+DW;
end
```

Program List 5.6: Initial Break Points for the optimization package

```
function RBA=initials(R,C,L,KFlag,T0,sign,WBR)
%function
%[RBA,XBA,XFA]=Ideal_SingleMatching(T0,sign,WBR,RLA,XLA)
% Computation of idealized immittance for single matching
% problems
% See Chapter 11, section 11.2 Eq.s (11.6-11.12)

%Inputs:
%       T0: Ideal flat gain level
%       WBR: Break frequencies of the load
%       RLA: Real Part of the immittance data
N=length(WBR);
for j=1:N
    w=WBR(j);
    [RL,XL]=RLC_Load(w,R,C,L,KFlag);
    RB=RL*((2-T0)+2*sign*sqrt(1-T0))/T0;
                %See Eq.(11.9)
    RBA(j)=RB;
end
```

Program List 5.7: Generation of Ideal solution for Single Matching Problems

```
function [RBA,XBA,XFA]=Ideal_SingleMatching(T0,sign,WBR,RLA,XLA)
% Computation of idealized immittance for single matching problems
% See Chapter 11, section 11.2 Eq.s (11.6.7.8.9.10,11 and 12)
% Inputs:
%       T0: Ideal flat gain level
%       WBR: Break frequencies of the load
%       RLA: Real Part of the immittance data
N=length(WBR);
for j=1:N
    RL=RLA(j);
   %RB(w)=RL(w)*((2-T0 )+2*sign*sqrt(1-T0 ))/T0
    RB=RL*((2-T0)+2*sign*sqrt(1-T0))/T0;
%(See Eq.(11.9))
    RBA(j)=RB;
end
% Computation of Hilbert Transform Loop
    for j=1:N
        w=1e-11+WBR(j);
   %XBA(w)=Hilbert Transform of {RB(w)}
    XB=num_hilbert(w,WBR,RBA);
    XBA(j)=XB;
    XA(j)= - XLA(j);
    XFA(j)=XA(j)-XBA(j);%see Eq. (11.12)
    end
```

Program List 5.8: Main Program for Example (5.3)

```
% Main Program Example5_3
clc
clear
% Nonlinear Immittance Modeling
%               via Positive Polynomial Approach:
%
%       Direct Modeling Method
%                  via Parametric Approach
%
% Immittance F(p)=a(p)/b(p)
% Real Part of F(p)is expressed as
%                        in direct computational
% techniques R(w^2)=AA(w^2)/BB(w^2)
% Where AA(w^2) if fixed by means of
%                      transmission zeros
%  BB(w^2)=(1/2)[C^2(w)+C^2(-w)]  where  C(w)=C(1)w^n+C(2)w^(n-
1)+...+C(n)w+1
% Note that AA(w^2) is computed by means of
%                function R_num which is
% specified in p-domain as
% A(-p^2)=(a0^2)*R_Num=(ndc,W)=
%           A(1)p^2m+A(2)p^2(m-1)+...+A(m)p^2+A(m+1)
% Inputs:
% Input Set #1:
% WBR.txt;   % Sampling Frequency Array to be loaded
% RBA.txt;  % Real Part of the Immittance data
%                                        to be loaded
% XBA.txt;  % Imaginary part of the immittance data
%                            to be loaded
%       ndc, W(i), B1: Transmission zeros of
%           the model immittance
%       W=[...] Finite transmission zeros of
%                                        even part
% Input Set #2:
%       a0: Constant multiplier of
%                   the denominator as a0^2
%       c0; Initial coefficients of BB(w^2)
%       ntr: Model with or without transformer/
%       ntr=1> Yes Transformer. ntr=0>No transformer
% Ouput:
%       Optimized values of a0,
%                  and coefficients of c(p)
% Direct form of Even Part R(p^2)=A(-p^2)/B(-p^2)
% Polynomials a(p) and b(p) of Minimum function F(p)
%-----------------
% Step 1: Load Data to be modeled
load WBR.txt;   % Sampling Frequency Array
load RBA.txt;  % RBAl Part of the Immittance data
load XBA.txt;  % Imaginary part of the Immittance data

% Step 2: Enter zeros of the RBAl part data
ndc=input('Enter transmission zeros at DC ndc=')
W=input('Enter finite transmission zeros at [W]=[....]=')
%
a0=input('Enter a0=')
[c0]=input('Enter  initials  for  [c0]=[....]=')%sigma(i):  RBAl
roots of b(p)
ntr=input('ntr=1,  Yes,  model  with  Transformer,  ntr=0  No
Transformer ntr=')
```

```
%
Nc=length(c0);
c0 =[172.53 -12.545 -266.47 39.425 119.77 -22.863 -16.164];
%
% Step 3: Generate inverse of the immittance data
NA=length(WBR);
cmplx=sqrt(-1);
for i=1:NA
    Z(i)=RBA(i)+cmplx*XBA(i);
    Y(i)=1/Z(i);
    GEA(i)=real(Y(i));
    BEA(i)=imag(Y(i));
end
% Step 4: Call nonlinear data fitting function
%%%%%%%%%% Preparation for the optimization
 OPTIONS=OPTIMSET('MaxFunEvals',20000,'MaxIter',50000);
%%%%%%%%%%%%%%%%%%%%%%%%%%%%%%%%%%%%%%%%%%%%%%%%%%%%%%
if ntr==1; x0=[c0 a0];Nx=length(x0);end;
                        %Yes transformer case
if ntr==0; x0=c0;Nx=length(x0);end;
                        %No transformer case
%
% Call optimization function lsqnonlin:
x=lsqnonlin('direct',x0,[],[],OPTIONS,ntr,ndc,W,a0,WBR,RBA,XBA);
%
% After optimization
%             separate x into sigma, alfa and beta
if ntr==1; %Yes transformer case
    for i=1:Nx-1;
    c(i)=x(i);
    end;
    a0=x(Nx);
end
if ntr==0;% No Transformer case
    for i=1:Nx;
        c(i)=x(i);
    end;
end
%
% Step 5: Generate analytic form of Fmin(p)=a(p)/b(p)
% Generate B(-p^2)
        C=[c 1];
        BB=Poly_Positive(C);
% This positive polynomial is in w-domain
        B=polarity(BB);
% Now, it is transferred to p-domain
% Generate A(-p^2) of R(-p^2)=A(-p^2)/B(-p^2)
nB=length(B);
        A=(a0*a0)*R_Num(ndc,W);
% A is specified in p-domain
nA=length(A);
if (abs(nB-nA)>0)
  A=fullvector(nB,A);
% work with equal length vectors
end

% Generation of minimum immittance function using
%Bode or Parametric method
        [a,b]=RtoZ(A,B);
% Here A and B are specified in p-domain
% Generate analytic form of Fmin(p)=a(p)/b(p)
```

```
        NA=length(WBR);
      cmplx=sqrt(-1);
% Step 6: Print and Plot results
Nprint=251;
ws1=0;ws2=2;
DW=(ws2-ws1)/(Nprint-1);
w=ws1;
Tmax=0.0;
Tmin=1.;
wc1=0;
wc2=1;
for j=1:Nprint
          WA(j)=w;
          p=cmplx*w;
          aval=polyval(a,p);
          bval=polyval(b,p);
          F=aval/bval;
          %
          RE(j)=real(F);
          XE(j)=imag(F);
          %
          RQ=RE(j);
          XQ=XE(j);
          %
          R=1;C=3;L=1;KFlag=1;
          [RL,XL]=RLC_Load(w,R,C,L,KFlag)
          %
          RTSQ=(RQ+RL)*(RQ+RL);
    XTSQ=(XQ+XL)*(XQ+XL);
    TPG=4*RL*RQ/(RTSQ+XTSQ);
    TA(j)=TPG;
    if max(TA(j))>Tmax
        wmax=WA(j);Tmax=TA(j);
    end
if w>=wc1
    if w<=wc2
    if min(TA(j))<Tmin
        wmin=WA(j);Tmin=TA(j);
    end
    end
end
    w=w+DW;
end
%
% Generation of optimized line segment gain
% for comparison purpose
N=length(WBR);
for i=1:N
w=WBR(i);
KFlag=1;
[RL,XL]=RLC_Load(w,R,C,L,KFlag)
RLA(i)=RL;XLA(i)=XL;
end
    Gain_Line=gain_singleMatching(WBR,RLA,XLA,RBA,XBA);

        figure(1)
        plot(WA,RE,WBR,RBA)
        title('Fit between given data and the analytic form of
the RBA1 part')
        xlabel('normalized angular frequency w')
        ylabel('R(p=jw) and RE(w)')
```

```
        legend('optimized' ,'initial')
  %
        figure(2)
        plot(WA,XE,WBR,XBA)
        title('Comparison of XBA and XE')
        xlabel('normalized frequency w')
        ylabel('XBA(w) and XE(w)')
        legend('optimized' ,'initial')
  %
        figure(3)
        plot(WA,TA, WBR,Gain_Line)
        title('Plot of the Analytic Gain')
        xlabel('normalized angular frequency w')
        ylabel('Actual Gain')
        legend('analytic Gain', 'Line Segment Gain')
  %
    CVal=general_synthesis(a,b)
```

Program List 5.9: Objective function for the main program Example5_3

```
   function fun=direct(x0,ntr,ndc,W,a0,WA,REA,XEA)
% Modeling via Direct - Parametric approach

if ntr==1; %Yes transformer case
    Nx=length(x0);
        for i=1:Nx-1;
            c(i)=x0(i);
        end;
        a0=x0(Nx);
end
if ntr==0;% No Transformer case
    Nx=length(x0);
    for i=1:Nx;
        c(i)=x0(i);
    end;
end
%
% Step 5: Generate analytic form of Fmin(p)=a(p)/b(p)
% Generate B(-p^2)
        C=[c 1];
        BB=Poly_Positive(C);
% This positive polynomial is in w-domain
        B=polarity(BB);
% Now, it is transferred to p-domain
% Generate A(-p^2) of R(-p^2)=A(-p^2)/B(-p^2)
nB=length(B);
        A=(a0*a0)*R_Num(ndc,W);
% A is specified in p-domain
nA=length(A);
if (abs(nB-nA)>0)
    A=fullvector(nB,A);
% Note that function RtoZ requires
%               same length vectors A and B
end
        [a,b]=RtoZ(A,B);
% Here, A and B are specified in p-domain
%
       NA=length(WA);
       cmplx=sqrt(-1);
```

```
%
        for j=1:NA
            w=WA(j);
            p=cmplx*w;
            aval=polyval(a,p);
            bval=polyval(b,p);
            Z=aval/bval;
            RE=real(Z);
            XE=imag(Z);
            error1=abs(REA(j)-RE);
            error2=abs(XEA(j)-XE);
            fun(j)=error1+error2;
        end
```

Program List 5.10: Main Program for Example 5.4 to design equalizers using RFDT

```
% Main Program Example5_4
clc
clear
%-----------------------------------------------------
% Direct Computational Technique for Double Matching Problems
%
% ----------------------------------------------------
%
% Optimize the back-end Immittance F(p)=a(p)/b(p)
% Real Part of F(p)is expressed as in direct
% computational techniques R(w^2)=AA(w^2)/BB(w^2)
% see Eq. (11. 25)
% Where AA(w^2) if fixed via jw transmission zeros
% BB(w^2)=(1/2)[C^2(w)+C^2(-w)]
% where C(w)=C(1)w^n+C(2)w^(n-1)+...+C(n)w+1
% Note that AA(w^2) is computed by means of function
%       R_num which is
% specified in p-domain as
% A(-p^2)=(a0^2)*R_Num=(ndc,W)=A(1)p^2m+A(2)p^2(m-
% 1)+...+A(m)p^2+A(m+1)
% Inputs:
% T0=input('Enter T0=')
%       Target value of the flat gain level
% Step 1: Generator and Load Data
%---------------------------------
% RGen=input('Enter RGen=')
% CGen=input('Enter CGen')
% Lgen=input('Enter LGen=')
%---------------------------------
% RLoad=input('Enter RLoad=')
% CLoad=input('Enter CLoad=')
% LLoad=input('Enter Load=')
%----------------------------------
% Step 2: Generic form of RB(w^2):
%       ndc, W(i), B1: Transmission zeros of the
%                      model immittance
%       W=[...] Finite transmission zeros of the
%                              even part
% Input Set #2:
% a0: Constant multiplier of the denominator as a0^2
%       c0; Initial coefficients of BB(w^2)
%       ntr: Model with or without transformer/
%       ntr=1> Yes Transformer. ntr=0>No transformer
% Output:
%  Optimized values of a0, and coefficients of c(p)
```

```
%  Direct form of Even Part R(p^2)=A(-p^2)/B(-p^2)
%  Polynomials a(p) and b(p) of Minimum function F(p)
%       -------------------------------------------
% Step 1: Enter T0, Generator and Load Networks
%--------------------------------
T0=input('Enter T0=')
KFlag=input('Enter KFlag=')
%
RGen=input('Enter RGen=')
CGen=input('Enter CGen=')
LGen=input('Enter LGen=')
%
%--------------------------------
RLoad=input('Enter RLoad=')
CLoad=input('Enter CLoad=')
LLoad=input('Enter LLoad=')
%
%---------------------------------
% Step 2: Enter zeros of the RB
ndc=input('Enter transmission zeros at DC ndc=')
W=input('Enter finite transmission zeros at [W]=[....]=')
a0=input('Enter a0=')
[c0]=input('Enter initials for [c0]=[....]=')%sigma(i): RBA1 roots of b(p)
ntr=input('ntr=1, Yes, model with Transformer, ntr=0 No Transformer ntr=')
%-----------------------------------
% Step 3 stop band and corner frequencies
ws1=input('Enter ws1=')
ws2=input('Enter ws2=')
wc1=input('Enter wc1=')
wc2=input('Enter wc2=')
%
%c0=[ 0.18045 0.12197 -5.183 -3.4992 3.1246]
%c0=[  -4.6046 -0.60396 1.8004];a0=0.64893;T0=0.9
Nc=length(c0);
%
% Step 4: Call nonlinear data fitting function
%%%%%%%%%%% Preparation for the optimization
%
OPTIONS=OPTIMSET('MaxFunEvals',20000,'MaxIter',50000);
%
%%%%%%%%%%%%%%%%%%%%%%%%%%%%%%%%%%%%%%%%%%%%%%%%%%%%

if ntr==1; x0=[c0 a0];Nx=length(x0);end;
%Yes transformer case
if ntr==0; x0=c0;Nx=length(x0);end;
%No transformer case
%
% Call optimization function lsqnonlin:
x=lsqnonlin('direct_RFDT',x0,[],[],OPTIONS,T0,wc1,wc2,KFlag,ntr,ndc,W,a0,RG
en,CGen,LGen,RLoad,CLoad,LLoad);
%
% After optimization separate x into coefficients
if ntr==1; %Yes transformer case
    for i=1:Nx-1;
    c(i)=x(i);
    end;
    a0=x(Nx);
end
if ntr==0;% No Transformer case
    for i=1:Nx;
        c(i)=x(i);
```

```
    end;
end
%
% Step 5: Generate analytic form of Fmin(p)=a(p)/b(p)
% Generate B(-p^2)
        C=[c 1];
        BB=Poly_Positive(C);
% This positive polynomial is in w-domain
        B=polarity(BB);
% Now, it is transferred to p-domain
% Generate A(-p^2) of R(-p^2)=A(-p^2)/B(-p^2)
nB=length(B);
        A=(a0*a0)*R_Num(ndc,W);
% A is specified in p-domain
nA=length(A);
if (abs(nB-nA)>0)
  A=fullvector(nB,A);
% work with equal length vectors
end

% Generation of minimum immittance function
%using Bode or Parametric method
        [a,b]=RtoZ(A,B);
% Here A and B are specified in p-domain
% Generate analytic form of Fmin(p)=a(p)/b(p)
        cmplx=sqrt(-1);
% Step 6: Print and Plot results
Nprint=251;
DW=(ws2-ws1)/(Nprint-1);
w=ws1;
Tmax=0.0;
Tmin=1.;
for j=1:Nprint
          WA(j)=w;
          p=cmplx*w;
          [RG,XG]=RLC_Gen(w,RGen,CGen,LGen,KFlag);
          [RL,XL]=RLC_Load(w,RLoad,CLoad,LLoad,KFlag);
        %
          aval=polyval(a,p);
          bval=polyval(b,p);
          bval_conj=conj(bval);
        %
          eta=(-1)^ndc*bval_conj/bval;
        %
          ZB=aval/bval;
          RB=real(ZB);
          XB=imag(ZB);
          %
          Sin=inputref_EL(KFlag,eta,RL,XL,RB,XB);
          %
           TEL=1-abs(Sin)*abs(Sin);
           ZG=complex(RG,XG);
           G22=(ZG-1)/(ZG+1);
           if KFlag==0
               G22=-G22;
           end
           Weight=(1-abs(G22)*abs(G22))/abs((1-G22*Sin))^2;
           T_Double=Weight*TEL;
           TA(j)=T_Double;
```

```
%
% Compute the performance parameters:
%Tmax,Tmin,Tave and detT
    if max(TA(j))>Tmax
        wmax=WA(j);Tmax=TA(j);
    end
if w>=wc1
    if w<=wc2
    if min(TA(j))<Tmin
        wmin=WA(j);Tmin=TA(j);
    end
    end
end
    w=w+DW;
end

  %

  %
        figure(1)
        plot(WA,TA)
        title('Double Matching Via RFDT')
        xlabel('normalized angular frequency w')
        ylabel('Double Matching Gain')
        legend('RFDT Gain')
  %
    CVal=general_synthesis(a,b)
    % Print Performance parameters:
wmax, Tmax,
wmin,Tmin,
Tzero=(Tmax+Tmin)/2
delT=Tzero-Tmin
%
```

Program List 5.11: error function for RFDT

```
   function
error=direct_RFDT(x0,T0,wc1,wc2,KFlag,ntr,ndc,W,a0,RGen,CGen,LGen,RLoad,CLo
ad,LLoad)
% Modeling via Direct - Parametric approach

if ntr==1; %Yes transformer case
    Nx=length(x0);
        for i=1:Nx-1;
            c(i)=x0(i);
        end;
        a0=x0(Nx);
end
if ntr==0;% No Transformer case
    Nx=length(x0);
    for i=1:Nx;
        c(i)=x0(i);
    end;
end
%
% Step 5: Generate analytic form of Fmin(p)=a(p)/b(p)
% Generate B(-p^2)
        C=[c 1];
```

```
        BB=Poly_Positive(C);
% This positive polynomial is in w-domain
        B=polarity(BB);
% Now, it is transferred to p-domain
% Generate A(-p^2) of R(-p^2)=A(-p^2)/B(-p^2)
nB=length(B);
        A=(a0*a0)*R_Num(ndc,W);
% A is specified in p-domain
nA=length(A);
if (abs(nB-nA)>0)
    A=fullvector(nB,A);
% Note that function RtoZ requires same length
% vectors A and B
end
        [a,b]=RtoZ(A,B);
% Here A and B are specified in p-domain
%
       N_opt=21;
       w=wc1;
       del=(wc2-wc1)/(N_opt-1);
       cmplx=sqrt(-1);
%
       for j=1:N_opt
           p=cmplx*w;
           [RG,XG]=RLC_Gen(w,RGen,CGen,LGen,KFlag);
           [RL,XL]=RLC_Load(w,RLoad,CLoad,LLoad,KFlag);
        %
           aval=polyval(a,p);
           bval=polyval(b,p);
           bval_conj=conj(bval);
        %
           eta=(-1)^ndc*bval_conj/bval;
        %
           ZB=aval/bval;
           RB=real(ZB);
           XB=imag(ZB);
         %
           Sin=inputref_EL(KFlag,eta,RL,XL,RB,XB);
         %
            TEL=1-abs(Sin)*abs(Sin);
            ZG=complex(RG,XG);
            G22=(ZG-1)/(ZG+1);
            if KFlag==0
                G22=-G22;
            end
            Weight=(1-abs(G22)*abs(G22))/abs((1-G22*Sin))^2;
            T_Double=Weight*TEL;
           error(j)=T_Double-T0;
           w=w+del;
       end
```

Program List 5.12: function for input reflectance Sin

```
function Sin=inputref_EL(KFlag,eta,RL,XL,RB,XB)
ZL=complex(RL,XL);
ZB=complex(RB,XB);
ZBC=conj(ZB);
S=(ZL-ZBC)/(ZL+ZB);
Sin=eta*S;
if KFlag==0
    Sin=-Sin;
end
```

Program List 5.13: function to generate Generator data

```
function [RGen,XGen]=RLC_Gen(w,RGen,CGen,LGen,KFlag)
% This function generates the impedance data
%    for a given R//C+L load
% Inputs:
%       w,RGen,CGen,LGen,KFlag
% Generate the load impedance
    p=sqrt(-1)*w;
    YGen=(1/RGen)+p*CGen;
    ZGen=p*LGen+1/YGen;
      if KFlag==1;% Work with impedances
        RGen=real(ZGen);
        XGen=imag(ZGen);
      end
        if KFlag==0;%Work with admittances
            RGen=real(YGen);
            XGen=imag(YGen);
        end
```

Program List 5.14: Main Program for Example 5.6

```
% Main Program Example5_6
clc
clear
input('Example (11.6)')
input('Welcome to RFDT Double Matching Program')
input('Generator Network is (1/pC+pL)//R ')
input(' Load Network is R//pL1+pL2')

%-----------------------------------------------------------
% Direct Computational Technique for Double Matching Problems
%
% ----------------------------------------------------------
%
% Optimize the back-end Immittance F(p)=a(p)/b(p)
% Real Part of F(p)is expressed as in direct
%                                 computational
% techniques R(w^2)=AA(w^2)/BB(w^2) see Eq. (11. 25)
% Where AA(w^2) if fixed by means of
%                              transmission zeros
% BB(w^2)=(1/2)[C^2(w)+C^2(-w)]
% where C(w)=C(1)w^n+C(2)w^(n-1)+...+C(n)w+1
% Note that AA(w^2) is computed by means of function
% R_num which is specified in p-domain as
% A(-p^2)=(a0^2)*R_Num=(ndc,W)=A(1)p^2m+A(2)p^2(m-
% 1)+...+A(m)p^2+A(m+1)
% Inputs:
% T0=input('Enter T0=')Target value of the flat gain
%                                                level
% Step 1: Generator and Load Data
%-------------------------------
% RGen=input('Enter RGen=')
% CGen=input('Enter CGen')
% Lgen=input('Enter LGen=')
%-------------------------------
% RLoad=input('Enter RLoad=')
% LL1=input('Enter CLoad=')
% LL2=input('Enter Load=')
%--------------------------------
% Step 2: Generic form of RB(w^2):
%  ndc, W(i), B1:
%   Transmission zeros of the model immittance
%  W=[...] Finite transmission zeros of even part
% Input Set #2:
%       a0: Constant multiplier of the denominator as a0^2
%       c0; Initial coefficients of BB(w^2)
%       ntr: Model with or without transformer/
%       ntr=1> Yes Transformer. ntr=0>No transformer
% Ouput:
%  Optimized values of a0, and coefficients of c(p)
%  Direct form of Even Part R(p^2)=A(-p^2)/B(-p^2)
%  Polynomials a(p) and b(p) of Minimum function F(p)
% ----------------------------------------------
% Step 1: Enter T0, Generator and Load Networks
%-------------------------------
T0=input('Enter T0=')
KFlag=input('KFlag=0>Admittance, KFlag=1>Impedance..Enter KFlag=')
%
```

```
RGen=input('Enter RGen=')
CGen=input('Enter CGen=')
LGen=input('Enter LGen=')
%
%--------------------------------
RLoad=input('Enter RLoad=')
LL1=input('Enter the parallel load inductor L1=')
LL2=input('Enter the series load inductor L2=')
%
%--------------------------------
% Step 2: Enter zeros of the RB
ndc=input('Enter transmission zeros at DC ndc=')
W=input('Enter finite transmission zeros at [W]=[....]=')
a0=input('Enter a0=')
[c0]=input('Enter initials for [c0]=[....]=')%sigma(i): RBAl roots of b(p)
ntr=input('ntr=1, Yes, model with Transformer, ntr=0 No Transformer ntr=')
%---------------------------------
% Step 3 stop band and corner frequencies
ws1=input('Enter ws1=')
ws2=input('Enter ws2=')
wc1=input('Enter wc1=')
wc2=input('Enter wc2=')
%
a0 =0.94437;
c0=[-3.5008 -5.8409  11.264 19.16  -5.8967  -10.624 0.9017];
    Nc=length(c0);
%
% Step 4: Call nonlinear data fitting function
%%%%%%%%%% Preparation for the optimization %%%%%

OPTIONS=OPTIMSET('MaxFunEvals',20000,'MaxIter',50000);
%
%%%%%%%%%%%%%%%%%%%%%%%%%%%%%%%%%%%%%%%%%%%%%%%%%%%%%

if ntr==1; x0=[c0 a0];Nx=length(x0);end;
%Yes transformer case
if ntr==0; x0=c0;Nx=length(x0);end;
%No transformer case
%
% Call optimization function lsqnonlin:
x=lsqnonlin('direct_RFDT',x0,[],[],OPTIONS,T0,wc1,wc2,KFlag,ntr,ndc,W,a0,RG
en,CGen,LGen,RLoad,LL1,LL2);
%
% After optimization separate x into coefficients
if ntr==1; %Yes transformer case
    for i=1:Nx-1;
    c(i)=x(i);
    end;
    a0=x(Nx);
end
if ntr==0;% No Transformer case
    for i=1:Nx;
        c(i)=x(i);
    end;
end
%
% Step 5: Generate analytic form of Fmin(p)=a(p)/b(p)
% Generate B(-p^2)
        C=[c 1];
        BB=Poly_Positive(C);
% This positive polynomial is in w-domain
```

```
        B=polarity(BB);
% Now, it is transferred to p-domain
% Generate A(-p^2) of R(-p^2)=A(-p^2)/B(-p^2)
nB=length(B);
        A=(a0*a0)*R_Num(ndc,W);
% A is specified in p-domain
nA=length(A);
if (abs(nB-nA)>0)
  A=fullvector(nB,A);
% work with equal length vectors
end

% Generation of minimum immittance function using Bode
%                 or Parametric method
        [a,b]=RtoZ(A,B);
% Here A and B are specified in p-domain
% Generate analytic form of Fmin(p)=a(p)/b(p)
        cmplx=sqrt(-1);
% Step 6: Print and Plot results
Nprint=1001;
DW=(ws2-ws1)/(Nprint-1);
w=ws1;
Tmax=0.0;
Tmin=1.;
for j=1:Nprint
           WA(j)=w;
           p=cmplx*w;
           [RG,XG]=RLC_Gen(w,RGen,CGen,LGen,KFlag);
           [RL,XL]=RLC_Load(w,RLoad,LL1,LL2,KFlag);
        %
           aval=polyval(a,p);
           bval=polyval(b,p);
           bval_conj=conj(bval);
        %
           eta=(-1)^ndc*bval_conj/bval;
        %
           ZB=aval/bval;
           RB=real(ZB);
           XB=imag(ZB);
         %
           Sin=inputref_EL(KFlag,eta,RL,XL,RB,XB);
         %
            TEL=1-abs(Sin)*abs(Sin);
            ZG=complex(RG,XG);
            G22=(ZG-1)/(ZG+1);
            if KFlag==0
                G22=-G22;
            end
            Weight=(1-abs(G22)*abs(G22))/abs((1-G22*Sin))^2;
            T_Double=Weight*TEL;
            TA(j)=T_Double;
%
    % Compute the performance
%     parameters: Tmax, Tmin, Taverage and delT
    if max(TA(j))>Tmax
        wmax=WA(j);Tmax=TA(j);
    end
if w>=wc1
    if w<=wc2
    if min(TA(j))<Tmin
        wmin=WA(j);Tmin=TA(j);
```

```
    end
    end
end
    w=w+DW;
end

  %


  %
        figure(1)
        plot(WA,TA)
        title('Double Matching Via RFDT')
        xlabel('normalized angular frequency w')
        ylabel('Double Matching Gain')
        legend('RFDT Gain')
  %
  if KFlag==1
    CVal=general_synthesis(a,b);
  end
  if KFlag==0
      CVal=general_synthesis(b,a);
  end
    % Print Performance parameters:
wmax, Tmax,
wmin,Tmin,
T_Average=(Tmax+Tmin)/2
delT=T_Average-Tmin
%
```

Program List 5.15: Generator function for Example (11.6)

```
function [RGen,XGen]=RLC_Gen(w,RGen,CGen,LGen,KFlag)
% This function generates the impedance data for a given R//C load
% Inputs:
%       w,RGen,CGen,LGen,KFlag
% Generate the load impedance
    p=sqrt(-1)*w;
    imp=1/p/CGen+p*LGen;
    adm=1/imp+1/RGen;
        YGen=adm;
        ZGen=1/YGen;
      if KFlag==1;% Work with impedances
        RGen=real(ZGen);
        XGen=imag(ZGen);
      end
        if KFlag==0;%Work with admittances
            RGen=real(YGen);
            XGen=imag(YGen);
        end
```

Program List 5.16: Load function for Example (11.6)

```
function [RL,XL]=RLC_Load(w,R,LL1,LL2,KFlag)
% This function generates the impedance data for a given R//C load
% Inputs:
%       w,R,C,L,KFlag
% Generate the load impedance
    p=sqrt(-1)*w;
    YL1=(1/R)+1/LL1/p;
    ZL=p*LL2+1/YL1;
    YL=1/ZL;
      if KFlag==1;% Work with impedances
        RL=real(ZL);
        XL=imag(ZL);
      end
        if KFlag==0;%Work with admittances
            RL=real(YL);
            XL=imag(YL);
        end
```

Program List 5.17:
Load function written for RFDT algorithm which process the measured data for the measured monopole antenna

```
function [RLN,XLN]=Load(w, KFlag)
Antenna_Data =[

          20          0.6          -6
          30          0.8        -2.2
          40          0.8           0
          45            1         1.4
          50            2         2.8
          55          3.4         4.6
          60            7         7.6
          65           15         8.8
          70         22.4        -5.4
          75           11         -13
          80            5       -10.8
          90          1.6        -6.8
         100            1        -4.4];
%      Normalization Numbers
R0=50; f0=100;
      N=length(Antenna_Data);
      for j=1:N
          WNA(j)=Antenna_Data(j,1)/f0;
          RLNA(j)=Antenna_Data(j,2)/R0;
          XLNA(j)=Antenna_Data(j,3)/R0;
      end
      RLN=line_seg(N,WNA,RLNA,w);
%Interpolation of the given real part data
      XLN=line_seg(N,WNA,XLNA,w);
%Interpolation of the given imaginary part data
% ya=line_seg(n,x,y,xa)is a line segment
%                                       approximation
      ZL=complex(RLN,XLN);
      YL=1/ZL;
      if KFlag==0;
          RLN=real(YL);
          XLN=imag(YL);
      End
```

Program List 5.18:
Error function written to design matching equalizer for the monopole antenna via real frequency direct computational method

```
   function fun=direct_Monopole(x0,KFlag,T0,wc1,wc2,ntr,ndc,W,a0)
% Modeling via Direct - Parametric approach

if ntr==1; %Yes transformer case
    Nx=length(x0);
        for i=1:Nx-1;
            c(i)=x0(i);
        end;
        a0=x0(Nx);
end
if ntr==0;% No Transformer case
    Nx=length(x0);
    for i=1:Nx;
        c(i)=x0(i);
    end;
end
%
% Generate analytic form of Fmin(p)=a(p)/b(p)
% Generate B(-p^2)
        C=[c 1];
        BB=Poly_Positive(C);
%           This positive polynomial is in w-domain
        B=polarity(BB);
% Now,             it is transferred to p-domain
% Generate A(-p^2) of R(-p^2)=A(-p^2)/B(-p^2)
nB=length(B);
        A=(a0*a0)*R_Num(ndc,W);
% A is specified in p-domain
nA=length(A);
if (abs(nB-nA)>0)
    A=fullvector(nB,A);
% Note that function RtoZ requires same length
%                                  vectors A and B
end
        [a,b]=RtoZ(A,B);
%        Here A and B are specified in p-domain
%
NA=100;
DW=(wc2-wc1)/(NA-1);
w=wc1;
       cmplx=sqrt(-1);
%
       for j=1:NA
           p=cmplx*w;
           aval=polyval(a,p);
           bval=polyval(b,p);
           Z=aval/bval;
           RB=real(Z);
           XB=imag(Z);
           [RL,XL]=Load(w,KFlag);
           T=4*RB*RL/((RL+RB)^2+(XB+XL)^2);
           error=T-T0;
           fun(j)=error;
      w=w+DW;
       end
```

Program List 5.19: Design of an equalizer fora monopole antenna

```
% Monopole Antenna Design
clear
clc
% User Information
input('Example: Monopole Antenna Matching Problem va RFDT Single
Matching')
input('The measured load Data is directly embedded into function
Load(w,KFlag)')
%
% Normalization numbers for the measured load
% Measured antenna data is normalized with respect to
%R0=50 ohm, f0=100 MHz
R0=50;
f0=100e6;
%
% Program inputs: Real Frequency Direct Computational
%                                Technique
ws1=0.2;ws2=1.0; %This is fixed by the measured data
% Step 2: Enter zeros of the RB
T0=input('Enter Flat Gain Level T0=')
KFlag=input('KFlag=0>Admittance, KFlag=1>Impedance..Enter KFlag=')
ndc=input('Enter transmission zeros at DC ndc=')
W=input('Enter finite transmission zeros at [W]=[....]=')
a0=input('Enter a0=')
[c0]=input('Enter initials for [c0]=[....]=')%sigma(i): Real roots of
b(p)
c0=[-4.1705 0.4066 9.2306 -1.887 -5.464 1.7575 ];
a0=0.19241;


ntr=input('ntr=1, Yes, model with Transformer, ntr=0 No Transformer
ntr=')
%
wc1=input('Enter wc1=')
wc2=input('Enter wc2=')
%
  Nc=length(c0);

% Optimization
% Step 4: Call nonlinear data fitting function
%%%%%%%%%% Preparation for the optimization

 OPTIONS=OPTIMSET('MaxFunEvals',20000,'MaxIter',50000);
%
%%%%%%%%%%%%%%%%%%%%%%%%%%%%%%%%%%%%%%%%%%%%%%%%%%%%%
%
if ntr==1; x0=[c0 a0];Nx=length(x0);end;
%Yes transformer case
if ntr==0; x0=c0;Nx=length(x0);end;
%No transformer case
%
% Call optimization function lsqnonlin:
x=lsqnonlin('direct_Monopole',x0,[],[],OPTIONS,KFlag,T0,wc1,wc2,ntr,ndc,W
,a0);
% fun= direct_Monopole (x0, KFlag, T0, wc1, wc2,
%                               ntr,ndc,W,a0)
%
```

```
% After optimization separate x into coefficients
%
if ntr==1; %Yes transformer case
    Nx=length(x);
        for i=1:Nx-1;
            c(i)=x(i);
        end;
        a0=x(Nx);
end
if ntr==0;% No Transformer case
    Nx=length(x);
    for i=1:Nx;
        c(i)=x(i);
    end;
end
%
% Step 5: Generate analytic form of Fmin(p)=a(p)/b(p)
% Generate B(-p^2)
        C=[c 1];
        BB=Poly_Positive(C);
% This positive polynomial is in w-domain
        B=polarity(BB);
        % Now, it is transferred to p-domain
% Generate A(-p^2) of R(-p^2)=A(-p^2)/B(-p^2)
nB=length(B);
        A=(a0*a0)*R_Num(ndc,W);
% A is specified in p-domain
nA=length(A);
if (abs(nB-nA)>0)
    A=fullvector(nB,A);
% Note that function RtoZ requires same length
%                       vectors A and B
end
        [a,b]=RtoZ(A,B);
%         Here A and B are specified in p-domain
%
% Initialize the performance measures.
Tmax=0;
Tmin=1;
NPrint=100;
DW=(ws2-ws1)/(NPrint-1);
w=ws1;
       cmplx=sqrt(-1);
%
       for j=1:NPrint
           WA(j)=w;
           p=cmplx*w;
           aval=polyval(a,p);
           bval=polyval(b,p);
           Z=aval/bval;
           RB=real(Z);
           XB=imag(Z);
           [RL,XL]=Load(w, KFlag)
           T=4*RB*RL/((RL+RB)^2+(XB+XL)^2);
           TA(j)=T;
           TdB(j)=10*log10(TA(j));
%
% Compute the performance parameters: Tmax,Tmin,Tave
%                                   and detT
    if max(TA(j))>Tmax
        wmax=WA(j);Tmax=TA(j);
```

```
    end
if w>=wc1
    if w<=wc2
    if min(TA(j))<Tmin
        wmin=WA(j);Tmin=TA(j);
    end
    end
end
% End of Performance measure loop
%
            w=w+DW;
        end
%

 figure(1)
 plot(WA,TA)
 xlabel('Normalized angular frequency w')
 ylabel('Transducer Power Gain')
 title('Matched Monopole Antenna Gain')
 %
 figure(2)
 plot(WA,TdB)
 xlabel('Normalized angular frequency w')
 ylabel('TPG in dB')
  title('Matched Monopole Antenna Gain in dB')
  %
  %Synthesis of the driving point immittance FB(p)
  if KFlag==1
    CVal=general_synthesis(a,b);
  end
  if KFlag==0
      CVal=general_synthesis(b,a);
  end
    % Print Performance parameters:
wmax, Tmax,
wmin,Tmin,
T_Average=(Tmax+Tmin)/2
delT=T_Average-Tmin
dbmax=10*log10(Tmax)
dbmin=10*log10(Tmin)
dB_Average=(dbmax+dbmin)/2
dbdel=dB_Average-dbmin
```

Program List 5.20: Main Program Helix_Antenna.m

```
% PROGRAM Helix_Antenna.m: Design of a Matching
% Network for a helix antenna
% employing Simplified Real Frequency Technique with simple optimization
% using MatlLab function lsqnonlin
display('Design of a matching network for a helix antenna using SRFT')
display('Measured antenna impedance data has been already put in the
arrays FA, RLA and XLA')
% Inputs:
% RLA: Real part of the measured impedance data
%XLA: Imaginary part of the measured impedance data
%FA : Frequency sampling points  for measurements
%
%T0: Desired flat gain level
%ntr: Control flag for equalizer design:
%ntr=1; equalizer design with a transformer.
%In this case k may be different than zero.
%ntr=0; equalizer is constructed without transformer.
%In this case k must be zero (k=0)
%and equalizer will be constructed as LC ladders
%
%initial values for h(p), and f(p)=p^k
%nopt: number of sampling points in optimization.
%wlow: Low frequency end of the optimization.
%whigh: High Frequency end of the optimization
%cccccccccccccccccccccccccccccccccccccccccc
% Outputs:
%Constructed antenna matching network
%Optimization using Levenberg-Marquardt Technique.
%Optimized Scattering Parameters of the Equalizer.
%Optimized Transducer Power Gain
%Realization of the Equalizer as an LC ladder
%           actual element values.
%           Plot of the Gain Performance
clear
clc
%  cccccccccccccccccccccccccccccccccccccccccc
% Input Data:
% Measured antenna impedance ZL=RL+jXL
% Normalization numbers:
f0=1.575; %Normalization frequency in GHZ
R0=50; % Normalization resistance
FA=[1.5 1.55 1.575 1.6]; %Actual frequencies in GHZ
RLA=[35.31 28.14 24.44 15.77]/R0;%
XLA=[-59.30 -53.98 -47.32 -34.2]/R0;
%
% Data Normalization
WBR=FA/f0;
% Normalized frequencies
display('Normalized Frequencies over the passband');
WBR'
T0=input('Enter desired flat gain level T0=')
ntr=input('Enter ntr; 0> for no transformer 1> with transformer ntr=')
k=input('Enter transmission zeros at DC k=')
wlow=input('Enter beginning of normalized optimization frequency wlow=')
whigh=input('Enter end of normalized optimization frequency whigh=')
h0=input('Enter initial value for h0=')
h=input('Enter initial values for h=[....] use brackets [.....]=')
```

```
nopt=30;
%Total number of sample points in optimization

           na=length(h);
 %
 %%%%%%%%%%%%%%%%%%%%%%%%%%%%%%%%%%%%%%%%%%%%%
%
 % Start Optimization using simplest version of the
%  LSQNONLIN
 %
 % Type design preferences:
 %ntr=1: design with transformer
 %ntr=0: design without transformer.
 % --------------------------------------
%
%%%%%%%%%%% Preparation for the optimization %%%%%%

OPTIONS=OPTIMSET('MaxFunEvals',20000,'MaxIter',50000);
%
%%%%%%%%%%%%%%%%%%%%%%%%%%%%%%%%%%%%%%%%%%
 %Design with transformer:
 if ntr==1,
            x0=h;
% Call optimization function lsqnonlin:

x=lsqnonlin('sopt_Helix',x0,[],[],OPTIONS,T0,ntr,k,nopt,wlow,whigh);
%
              h=x;
        end
%-----------------------------------------
%
% Design without transformer:
        if ntr==0,
                for r=1:na-1
                    x0(r)=h(r);
%Call optimization function lsqnonlin:
    x=lsqnonlin('sopt_Helix',x0,[],[],OPTIONS,T0,ntr,k,nopt,wlow,whigh);
                end
    h=x; %Regular MatLab polynomial Coefficients
    h(na)=0;
        end
 % Generate g(p) from optimized h(p):
    g=Hurwitzpoly_g(h,k);%General MatLab polynomial
% Compute the optimized transducer power gain
W1=wlow;
W2=whigh;
Nsample=100;
DW=(W2-W1)/Nsample;
W=wlow;
        for j=1:Nsample+1
            Wa(j)=f0*W;
            E22=Equalizer(W,h,g,k);
            % GENERATION OF THE LOAD DATA
             L11=reflection_Helix(W);
             LdB(j)=10*log10(abs(L11)*abs(L11));
             % (Load Mismatch factor)

%COMPUTATION OF GAIN
            T=gain(W,h,g,k,L11);
            Ta(j)=T;
```

```
            TdB(j)=10*log10(Ta(j));
            W=W+DW;
        end
%
        figure(1)
        plot(Wa,Ta)
        title('Gain of the Matched antenna')
        xlabel('Actual Frequencies in GHz')
        ylabel('Gain in dB')
%
Sin_dB=10*log10(1-Ta);%input mismatch factor
figure(2)
plot(Wa,Sin_dB, Wa,LdB)
title('Matched and unmatched Input Reflectance of the helix antenna')
xlabel('Actual Frequencies in GHz')
ylabel('Matched Reflectance in dB')
legend('matched',' unmatched')
%
figure (3)
plot(Wa,LdB)
title('Input Reflectance of the unmatched antenna')
xlabel('Actual Frequencies in GHz')
ylabel('unmatched Reflectance in dB')

CV=general_synthesis(h,g);
a=g+h;
b=g-h;
if a(1)<1e-7;kimm=0;
display('Start with a capacitor')
%Actual element values starts with an inductor
display('Actual Element Values AEV=')
fnorm=f0*1e9
AEV=Actual(kimm,R0,fnorm,CV)
end
if a(1)>1e-7; kimm=1;
display('start with an inductor')
%Actual element values starts with an inductor
display('Actual Element Values AEV=')
fnorm=f0*1e9
AEV=Actual(kimm,R0,fnorm,CV)
End
```

Program List 21: Error function subject to optimization for helix antenna problem

```
function fun=sopt_Helix(x,T0,ntr,k,nopt,wlow,whigh)
%Inputs:
% T0: Desired gain level
% ntr: Control flag for equalizer design:
% ntr=1; equalizer design with a transformer.
% In this case k may be different than zero.
% ntr=0; equalizer is constructed without
%transformer.
% In this case k must be zero (k=0)
% and equalizer will be constructed with Low Pass LC
%elements
%x=Row Vector includes the unknown coefficients
% of the polynomial h(p).
%for ntr=0 and k=0, dimension of x must be n.
%for ntr=1, dimension of x must be n+1.
%nopt: total number of sampling points within
% optimization.
```

```
% This is the dimension of the objective function
%invT subject to optimization.
% wlow: Low frequency end of the optimization.
% whigh: High Frequency end of the optimization
%%%%%%%%%%%%%%%%%%%%%%%%%%%%%%%%%%%%%%%%%%%%%%%%
% Output:
% fun: Objective function  Vector subject to
% optimization.
%       For the design of Antenna Matching Networks fun=T-T0
%%%%%%%%%%%%%%%%%%%%%%%%%%%%%%%%%%%%%%%%%%%%%%%%%%%%%
nx=length(x);
w=wlow;
DW=(whigh-wlow)/nopt;
% Start frequency loop to compute objective function
% within passband:
    for j=1:nopt
% Prepare to generate transducer power gain:
% Call load function to compute load reflection
%coefficient.
        L11=reflection_Helix(w);
        % Set h vector for the values of ntr:
        if ntr==1,
            h=x;
        end
        %
        if ntr==0,
                for r=1:nx
                    h(r)=x(r);
                end
            na=nx+1;
            h(na)=0;
        end
         %
         % Generate g polynomial from h:
            g=Hurwitzpoly_g(h,k);
            % Generate the Transducer power Gain
            %
                T=gain(w,h,g,k,L11);
            %
            % Generate objective function as
                fun(j)=T-T0;
                w=w+DW;
    end
            return
```

Program List 5.21: TPG of Helix Antenna

```
function T=gain(W,h,g,k,L11)
% Define complex variable p=jW; j=sqrt(-1)
p=sqrt(-1)*W;
%Compute the values of h and g using MatLab function
%polyval(p,X)

gval=polyval(g,p);
hval=polyval(h,p);
fval=p^k;
% Construct the input reflection coefficient of the equalizer from the
% back-end.
            E22=hval/gval;
%COMPUTATION OF GAIN
%
            Weight=fval*conj(fval)*(1-L11*conj(L11));
%Compute the denominator of the gain function as in
%Eq.(21).
            D=hval*conj(hval)*(1+L11*conj(L11))+fval*conj(fval)-
2*real(L11*hval*conj(gval));
            T=Weight/D;
            return
```

Program List 5.22: Reflectance of the Helix Antenna

```
function  SL=reflection_Helix(w)
% This function computes the Helix Antenna
%reflection coefficient provided by EDA KONAKYERI.
% SL is a complex quantity.
%%%%%%%%%%%%%%%%%%%%%%%%%%%%%%%%%%%%
RLA=[35.31 28.14 24.44 15.77]/50;
%Data provided by EDA of ITU
XLA=[-59.30 -53.98 -47.32 -34.2]/50;
%Data provided by EDA of ITU
WBR =[0.95238  0.98413 1 1.0159];
nL=length(WBR);
RL=line_helix(nL,WBR,RLA,w);
XL=line_helix(nL,WBR,XLA,w);
j=sqrt(-1);
ZL=RL+j*XL;
SL=(ZL-1)/(ZL+1);
```

Program List 5.23: Input reflectance of the equalizer

```
function SE=Equalizer(W,h,g,k)
p=sqrt(-1)*W;
            gval=polyval(g,p);
            hval=polyval(h,p);
            fval=p^k;
            SE=hval/gval;
            return
```

Program List 5.24: Generation of strictly Hurwitz polynomial from h(p) and $f(p) = p^k$

```
function g=Hurwitzpoly_g(h,k)
%Computation of g(p)=g1p^n+g2p^(n-
%                                1)+...+g(n+1)*p+g(n+1).
% Inputs: Row vector h which contains coefficients of h(p) in descending
order.
%              k which is the order of f(p)=p^k.
na=length(h);
%
% Step 1: Generate even polynomial H(-p^2)=h(p)h(-p);
%
% Computation of even polynomial with convolution:  W=conv(U,V)
                    U=h;
       % Generate row vector V
            sign=-1;
            for i=1:na
                sign=-sign;
                    V(na-i+1)=sign*U(na-i+1);
            end
                    W=conv(U,V);
% Select non-zero terms of W and form the vector H
            for i=1:na
                H(i)=W(2*i-1);
            end
%
%       Step 2: Generate G(-p^2)=H(-p^2)+(-1)^k
    for i=1:na
    G(i)=H(i);
    end
            ka=na-k; sign=(-1)^(k); G(ka)=H(ka)+sign;
%
%   Step 3:Generate the row vector X and Compute
%   the roots of G(-p^2)
% Arrange the sign of vector X.
sign=-1;
            for i=1:na
            sign=-sign;
                X(na-i+1)=sign*G(na-i+1);
            end
%
% Compute the RHP and LHP Roots of G(-p^2)
        Xr=roots(X);pr=sqrt(-Xr);prm=-pr;
%
%       Step 4: Generate the monic polynomial g'(p)
                    C=poly(prm);
%
%   Step 5: Generate Scattering Hurwitz polynomial g(p)
                    Cof=sqrt(G(1));
                for i=1:na
                    g(i)=Cof*C(i);
                end
```

Chapter 6

Broadband RF and microwave amplifier design employing real-frequency techniques

*Binboga Siddik Yarman**

There, the major idea was to match the input and the output of the amplifier to a given standard resistive source and load over a narrow frequency band. In this regard, first, the proper circuit topologies with two or three elements are selected for the input matching network (MN) (IMN) and the output MN (OMN). Then, the element values are determined by means of nonlinear optimization methods. On the other hand, when we deal with wideband matching problems, life becomes more difficult. First, we have no idea about the optimum choice for the circuit topology. Furthermore, even if we choose a practical circuit topology for the MNs, this topology would require more than two elements to achieve wideband designs. Moreover, the objective function to be optimized is highly nonlinear in terms of the component values and requires good initials. Eventually, we start the nonlinear optimization with ad hoc component values. At the end of the optimization, most probably, we will either hit a bad-local extremum or end up with non-divergent endless iterations.

Therefore, in this chapter, we introduce novel techniques to design wideband amplifiers employing "real-frequency techniques (RFTs)" [1–7]. In essence, RFTs are wideband semi-analytic design methods to realize lossless MNs and real frequency (RF) and microwave amplifiers with optimum circuit topologies.

6.1 Introduction

Referring to Figure 6.1, a typical RF or single-stage microwave amplifier consists of IMN and OMN [*F*] and [*B*], respectively.[1] The amplifier could be driven either with small or large signals (i.e., high-power input drives).

*Department of Electrical and Electronics Engineering, College of Engineering, Istanbul University-Cerrahpasa, Istanbul, Turkey

[1]The letter [*F*] designates the "front-end equalizer" or equivalently "input-matching network." Similarly, the letter [*B*] stands for the "back-end equalizer" or equivalently "output-matching network" as shown in Figure 6.1.

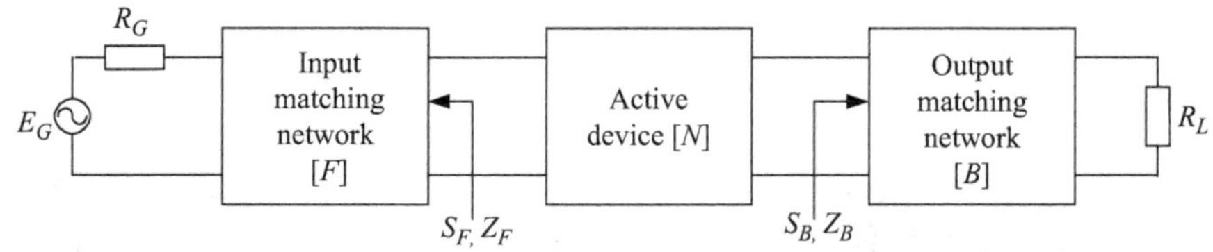

Figure 6.1 Schematic description of a single-stage amplifier

In RFTs, instead of choosing the circuit topologies for [F] and [B], they are fully described by means of their Darlington driving point network functions such as reflectance $\{S_F \text{ and } S_B\}$ or impedance $\{Z_F \text{ and } Z_B\}$ or equivalently admittance $\{Y_F = 1/Z_F, Y_B = 1/Z_B\}$ or in short immittance functions,[2] respectively.[3] In the impedance-based RFT, positive real (PR) functions $Z_F(p)$ and $Z_B(p)$ are determined to optimize the performance parameters of the amplifier such as "transducer power gain" (*TPG*), noise figures (*NF*), voltage standing wave ratios. In this representation, p designates the complex Laplace domain variable such that $p = \sigma + j\omega$, and the impedance functions $Z_F(p)$ and $Z_B(p)$ are specified as rational functions of complex variable p as follows[4]:

$$Z_F(p) = \frac{a_F(p)}{b_F(p)} = \frac{a_{1,F}p^n + a_{2,F}p^{n-1} + \ldots + a_{n,F}p + a_{(n+1),F}}{b_{1,F}p^n + b_{2,F}p^{n-1} + \ldots + b_{n,F}p + b_{(n+1),F}} \tag{6.1a}$$

$$Z_B(p) = \frac{a_B(p)}{b_B(p)} = \frac{a_{1,B}p^n + a_{2,B}p^{n-1} + \ldots + a_{n,B}p + a_{(n+1),B}}{b_{1,B}p^n + b_{2,B}p^{n-1} + \ldots + b_{n,B}p + b_{(n+1),B}} \tag{6.1b}$$

Similarly, in the reflectance-based RFT which is called "simplified RFT (or in short SRFT)," bounded real-reflectance functions are specified as[5]

$$S_F(p) = \frac{h_F(p)}{g_F(p)} = \frac{h_{1,F}p^n + h_{2,F}p^{n-1} + \ldots + h_{n,F}p + h_{(n+1),F}}{b_{1,F}p^n + b_{2,F}p^{n-1} + \ldots + b_{n,F}p + b_{(n+1),F}} \tag{6.2a}$$

$$S_B(p) = \frac{h_B(p)}{g_B(p)} = \frac{h_{1,B}p^n + h_{2,F}p^{n-1} + \ldots + h_{n,B}p + h_{(n+1),B}}{g_{1,B}p^n + g_{2,B}p^{n-1} + \ldots + g_{n,B}p + g_{(n+1),B}} \tag{6.2b}$$

In the course of MN design, RFTs determine the coefficients of (6.1) or (6.2) to optimize the performance parameters of the amplifier such as *TPG* over the

[2]In circuit theory, the expression "immittance" refers either to impedance or admittance.
[3]At this point we should emphasize that according to Darlington's theorem, a lossless two-port such as [F] or [B] can completely be described by means of its driving point reflectance or immittance.
[4]In classical circuit theory, complex Laplace variable is designated by the letter "$s = \sigma + j\omega$." This letter may be confused with those of scattering parameters of two ports $S = \{s_{ij}; i, j = 1, 2\}$ or reflectance parameters S_F and S_B. Therefore, we use the notation p for the Laplace domain variable.
[5]In this context, reflectance functions are assumed to be real normalized (unit normalized).

prescribed frequency band of operation. Then, network functions (either immittance or reflectance) are synthesized in Darlington sense as lossless two-ports in resistive terminations yielding the optimum circuit topologies for $[F]$ and $[B]$ with element values [8–12].

Generation of realizable network functions on the computer is an art. It is nice that RFTs provide robust and efficient algorithms to complete these tasks. Furthermore, our recently developed high-precision Darlington synthesis algorithms yield optimum circuit topologies for the MNs to construct RF and microwave amplifiers [6–8].

There are many variants of RFTs as in Chapter 5. In the following sections, we present the selected algorithms to design wideband practical power amplifiers (PAs) employing the RFTs.

6.2 Simplified real-frequency technique (SRFT) to design microwave amplifiers

It is well known that SRFT is well suited to design microwave amplifiers in two steps.

Referring to Figure 6.2, in the first step, front-end MN $[E_F]$ is designed, while the output port is terminated in 50 Ω (or in the normalization resistance $R_0 = 50\Omega$). At this step, TPG $\boldsymbol{T_1}$ which is specified by

$$\boldsymbol{T_1(\omega) = \frac{|S_{21F}|^2|S_{21}|^2}{|1-S_{22F}S_{11}|^2}} \tag{6.3}$$

is optimized to hit the ideal stable flat gain level.

$$\boldsymbol{T_{flat1} = minimum\ of\ T_{01} = \left\{\frac{|S_{21}(\omega)|^2}{1-|S_{11}(\omega)|^2}\right\} over\ B} \tag{6.4}$$

In (6.3) and (6.4), $\{S_{ijF}; i,j = 1,2\}$ and $\{S_{ij}; i,j = 1,2\}$ are the real normalized (50 Ω) scattering parameters of the front-end MN $[E_F]$ and the MOS transistor, respectively. B is the frequency band which runs from 11.23 to 22.39 GHz (more than 10 GHz bandwidth).

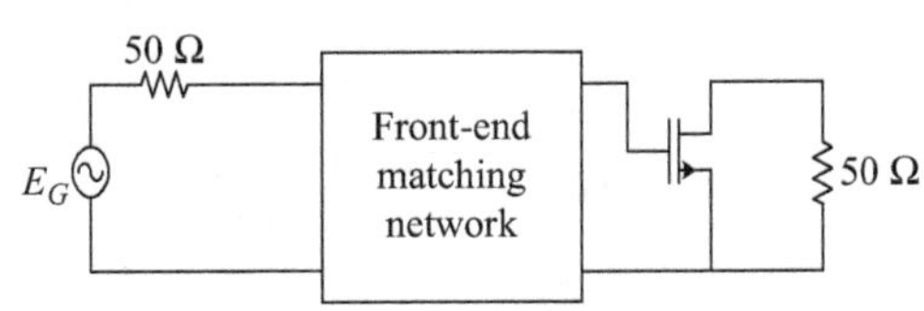

Figure 6.2 Step 1: Design of the front-end matching network

In the second step, the back-end equalizer $[E_B]$ is designed to optimize the overall TPG $T_2(\omega)$ as shown by Figure 6.3. In this step, $T_2(\omega)$ is specified by

$$T_2(\omega) = T_1(\omega).\frac{|S_{21B}|^2}{|1 - S_{2F}S_{11B}|^2} \tag{6.5}$$

At this step, the optimization algorithm targets the flat gain level of

$$T_{flat2} = minimum\ of\ T_{02}(\omega) = \frac{T_1(\omega)}{1 - |S_{22}(\omega)|^2} \tag{6.6}$$

In (6.5) the back-end reflection coefficient is given by

$$S_{2F} = S_{22} + \frac{S_{12}S_{21}}{1 - S_{11}S_{22F}} \tag{6.7}$$

In the SRFT, the front-end $[E_F]$ and the back-end $[E_B]$ MNs are described in terms of their unit normalized input reflection coefficients in Belevitch form as

$$S_{22F} = \frac{h_F}{g_F} = \frac{h_{0F} + h_{1F}p + \ldots + h_{n_F F}p^{n_F}}{g_{0F} + g_{1F}p + \ldots + g_{n_F F}p^{n_F}} \tag{6.8a}$$

$$S_{11B} = \frac{h_{0B} + h_{1B}p + \ldots + h_{n_B F}p^{n_B}}{g_{0B} + g_{1B}p + \ldots + g_{n_B B}p^{n_B}} \tag{6.8b}$$

In Step 1, selecting the total number of equalizer elements n_F in $[E_F]$ first, the coefficients $\{h_{0F}, h_{1F}, \ldots, h_{nF}\}$ are initialized. Then, the denominator polynomial $g_F(p)$ is generated as a strictly Hurwitz polynomial by the spectral factorization of

$$g_F(p)g_F(-p) = h_F(p)h_F(-p) + f_F(p)f_F(-p) \tag{6.8c}$$

In this step, S_{21F} is given as

$$S_{21F} = \frac{f_F(p)}{g_F(p)} \tag{6.8d}$$

and the monic polynomial $f_F(p)$ is selected by the designer to include transmission zeros of the front-end equalizer $[E_F]$. For example, if $[E_F]$ is a simple lowpass LC

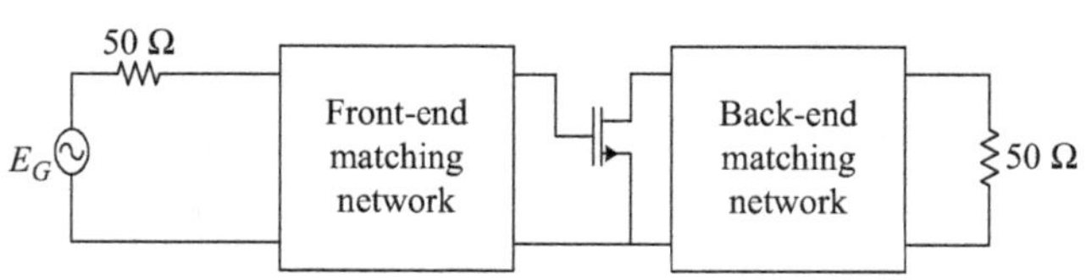

Figure 6.3 Step 2: Design of the back-end matching network

ladder, $f_F(p) = 1$ selected. If $[E_F]$ includes transmission zeros only at DC of degree n_{dc} and infinity, then $f_F(p)$ is selected as

$$f_F(p) = (-1)^{n_{Fdc}} p^{n_{Fdc}} \tag{6.8e}$$

where $n_{Fdc} \leq n_F$. It is noted that $n_{Fdc} = 0$ defines a lowpass ladder and $n_{Fdc} = n_F$ refers to a highpass ladder for the MN under consideration.

Eventually, coefficients $\{h_{0F}, h_{1F}, \ldots, h_{nF}\}$ are determined by minimizing the error function

$$\varepsilon_1(\omega) = T_1(\omega) - T_{flat1}$$

over the band of interest $B = f_2 - f_1$ where f_1 and f_2 designate the lower and the upper end of the frequency band, respectively.

Similarly, in Step 2, back-end equalizer $[E_B]$ is constructed by determining the unknown numerator coefficients $\{h_{0B}, h_{1B}, \ldots, h_{nB}\}$ of S_{11B} via optimization of the gain function. At this step, the losslessness condition of the back-end equalizer is specified in terms of the transfer scattering parameter S_{21B} such that

$$S_{21B} = \frac{f_B(p)}{g_B(p)} \tag{6.8f}$$

where

$$g_B(p)g_B(-p) = h_B(p)h_B(-p) + f_B(p)f_B(-p) \tag{6.8g}$$

For ladder structures, $f_B(p)$ is selected as

$$f_B(p) = \begin{cases} 1 & \textit{for lowpass ladders} \\ (-1)^{n_{Bdc}} p^{n_{Bdc}} & \textit{for bandpass ladders} \\ (-1)^{n_B} p^{n_B} & \textit{for highpass ladders} \end{cases}$$

with

$$n_{Bdc} \leq n_B \tag{6.8h}$$

In this case, the error function $\varepsilon_2(\omega) = T_2(\omega) - T_{flat2}$ is minimized over the same frequency band.

Once S_{22F} and S_{11B} are determined, they are synthesized as lossless two ports as in Darlington's procedure yielding the circuit topologies for the front-end and the back-end MNs with element values.

6.3 SRFT single-stage microwave amplifier design algorithm

In this section, the gain–bandwidth performance of a typical 0.18 μm NMOS-FET is assessed by designing an amplifier over 11.23–22.39 GHz.

The 50 Ω-based scattering parameters of the device are given in Table 6.1.

Table 6.1 Selected samples from the measured scattering parameters of 0.18 μm TSMC MOS transistor

Fr(*GH*)	*MS11*	*PS11*	*MS21*	*PS21*	*MS12*	*PS12*	*MS22*	*PS22*
10.0	0.90	−2.99	1.11	1.1	0.05	−0.34	0.77	−2.96
11.2	0.90	−3.01	0.97	1.1	0.05	−0.39	0.78	−2.96
12.6	0.91	−3.01	0.84	1.1	0.06	−0.44	0.79	−2.96
14.1	0.91	−3.01	0.73	1.0	0.05	−0.49	0.80	−2.96
15.8	0.92	−3.02	0.63	0.9	0.05	−0.54	0.82	−2.96
17.8	0.92	−3.02	0.55	0.8	0.05	−0.60	0.83	−2.96
20.0	0.93	−3.03	0.46	0.81	0.04	−0.66	0.85	−2.96
22.4	0.94	−3.04	0.39	0.74	0.04	−0.72	0.86	−2.96

Note: In the following, *MSij* are the amplitudes and *PSij* are the phases of the *S*-parameters. Phases are given in radian.

The biasing conditions and design parameters of the device is given as follows:

- VDD = 1.8 V; ID (drain DC current) = 200 mA.
- NMOS size is $W/L = 1{,}000/0.18$ μm.
- Simulation temperature is 50°C.
- Expected output power is approximately 100 mW up to 23 GHz when operated as a Class-A amplifier.

For this purpose, we developed a MATLAB® Main Program which is called *TSMC_amp.m.*

As explained above, this program constructs a microwave amplifier in two parts. In the first part, the front-end equalizer $[E_F]$ is constructed. Then, in the second part, the back-end MN of the amplifier is completed.

It is our desire to construct both equalizers employing LC ladder networks without transformer. Therefore, in the main program, we set

$$n_{Fdc} = n_{Bdc} = 0$$

and we fix

$$h_{0F} = h_{0B} = 0$$

Thus, in the first part, the objective function

$$function\ Func = levenberg\_TR1(x, k, w, f11, f12, f21, f22, TS1)$$

minimizes $\varepsilon_1(\omega) = T_1(\omega) - T_{flat1}$ which in turn generates the numerator polynomial $h_F(p) = h_{nF}p^n + \ldots + h_{1F}p + 0$.

In this case, the designer provides the following inputs to the main program TSMC_amp.m

Inputs to the main amplifier MATLAB design program *TSMC_amp.m* for Part I.

`TSMC_Spar.txt:` Scattering parameters of the CMOS transistor loaded from the text file.

In this file, *S*-parameters are read as amplitude and phase in degree in the following order:

$$MS11(i) = TMSC\_Spar(i,2); PS11(i) = TMSC\_Spar(i,3);$$

$$MS21(i) = TMSC\_Spar(i,4); PS21(i) = TMSC\_Spar(i,5);$$

$$MS12(i) = TMSC\_Spar(i,6); PS12(i) = TMSC\_Spar(i,7);$$

$$MS22(i) = TMSC\_Spar(i,8); PS22(i) = TMSC\_Spar(i,9);$$

where $i = 1$ to 48.

Program converts all the phases from degree to radian and then generates the maximum stable gain of the amplifier at the first step as

$$T_{01} = \left\{ \frac{|S_{21}(\omega)|^2}{1 - |S_{11}(\omega)|^2} \right\}$$

and plots it. For the CMOS transistor under consideration, $T_{01}(\omega)$ is depicted in Figure 6.4.

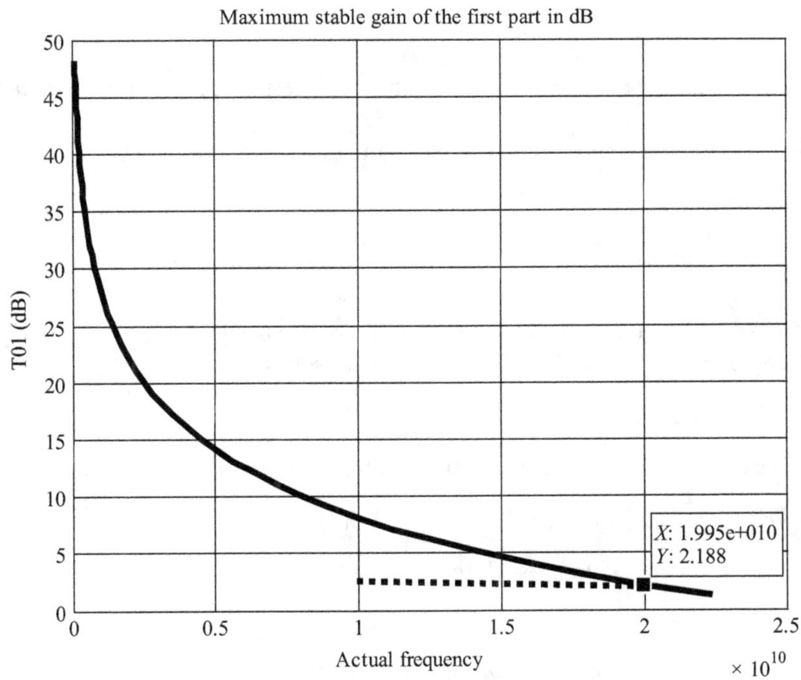

Figure 6.4 Maximum stable gain of Part I of the single-stage amplifier design

Close examination of the above figure reveals that one can obtain about 2.1 dB flat gain level even beyond 20 GHz.

At this point, the designer inputs the frequency band of operation.

In this case, the low (f_1) and the high (f_2) ends of the frequency band must be entered.

The low end of the frequency band is denoted by

$$f_1 = FR(nd1)$$

Similarly, the high end is denoted by

$$f_2 = FR(nd2).$$

However, the user only enters the index of the low and high ends of the frequency band.

The lower end of the optimization starts from the frequency $FR(nd1) =$ 11.23 GHz which is placed at the 42nd sampling point of the measured data.

Therefore, we set

$$nd1 = 42$$

The high end of the optimization frequency is $FR(nd2) = 22.39$ GHz which located at the 48th sampling point.

Hence, we set

$$nd2 = 48$$

Now, the program asks the designer to enter the desired flat gain level *Tflat*1 of Part I.

Here, it may be appropriate to select a the minimum or a little higher value of

$$T_{01} = \left\{ \frac{|S_{21}(\omega)|^2}{1 - |S_{11}(\omega)|^2} \right\}$$

in the passband to assure the absolute stability of the design.

Referring to Figure 6.4, it may be proper to select

$$TFlat1 = 2.1\text{dB}$$

As the last input, we should enter the initials for the unknown coefficients such that

$$hF(0) = 0, [xF(1) = hF(1), xF(2) = hF(2), \ldots, XF(n) = hF(n)]$$

Here, we use an ad hoc input such that

$$[xF(1) = hF(1), xF(2) = hF(2), \ldots, XF(n) = hF(n)] = [-1 - 1 - 1\ 1]$$

The above means that $nF = length(hF) = 4$. Therefore, we will employ only four elements in the LC ladder without ideal transformer, since we set $h0F = 0$ in advance.

6.3.1 Result of optimization

At this part of the main program, the optimization function is called as follows:

$$xF = lsqnonlin('levenberg\_TR1',\ xF0, [], [], OPTIONS, k, ww, f11, f12, f21, f22, Tflat1);$$

The function

function lsqnonlin

is a least mean square minimization algorithm which employs the error function *levenber_TR*1 such that

function Func = *levenberg_TR*1($x, k, w, f11, f12, f21, f22, Tflat1$)

This function basically generates the error term

$$\varepsilon_1 = T_1 - Tflat1$$

in decibel.

Minimization of $\varepsilon_1 = T_1 - Tflat1$ yields the optimum reflection coefficient

$$S_{22F}(p) = \frac{h_F(p)}{g_F(p)}$$

as shown in Table 6.2.

The above reflection coefficient leads to front-end driving point impedance

$$Z_F(p) = \frac{g_F + h_F}{g_F - h_F} = \frac{1.9293p^4 + 1.1928p^3 + 3.1183p^2 + 0.89332p + 1}{10.687p^3 + 6.6079p^2 + 10.888p + 1}$$

and it is synthesized by long division as

$$Z_F(p) = 0.18052p + \frac{1}{9.2703p + (1/(0.7128p + (1/(1.6174p + 1))))}$$

The normalized circuit elements of the front-end MN are given in Table 6.3.

The actual circuit element values are generated by de-normalization employing the *function AEVF* = *Actual*(*kimm*, *R*0, *fnorm*, *CVF*) as shown in the same table.

Thus, the first part of the amplifier design procedure is completed as shown in Figure 6.5.

Table 6.2 Front-end reflection coefficient $S_{22F} = h_F/g_F$

$h_{0F}=$	0	$g_{0F}=$	1
$h_{1F}=$	−4.9972	$g_{1F}=$	5.8905
$h_{2F}=$	−1.7448	$g_{2F}=$	4.8631
$h_{3F}=$	−4.7473	$g_{3F}=$	5.9401
$h_{4F}=$	0.96463	$g_{4F}=$	0.96463

Table 6.3 Normalized and actual element values of front-end equalizer for the microwave amplifier designed employing TSMC 180 μm Si CMOS transistor

Component type	Normalized element values (CVF)	Actual element values (AEVF)
$L1$(H)	0.18052	6.4167e − 011
$C2$(F)	9.2703	1.3181e − 012
$L3$(H)	0.7128	2.5337e − 010
$C4$(F)	1.6174	2.2996e − 013

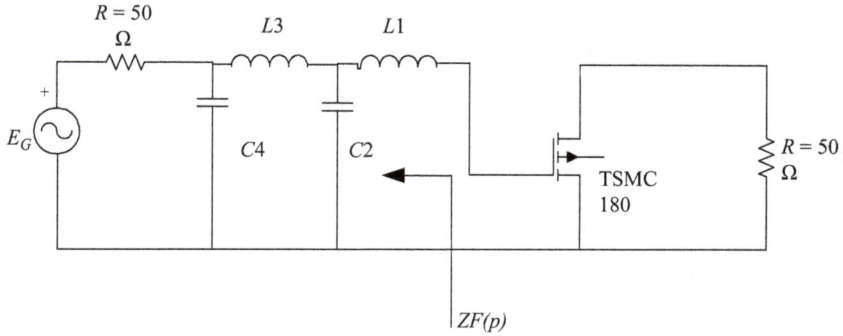

Figure 6.5 Completion of Part I of the single-stage amplifier design employing the Si CMOS transistor TSMC 180 μm

The optimized gain performance of the first part of the design problem is depicted in Figure 6.6.

In the second part of the design process, the back-end equalizer of the amplifier is constructed by minimizing the error

$$\varepsilon_2 = T_2 - Tflat2$$

At this point, the main program *TSMC_amp.m* calls the optimization routine *lsqnonlin* with error function

```
function Func = levenberg_TR2(x,k,w,T1,SF2,f11,f12,f21,f22,Tflat2)
```

The function *levenberg_TR2* generates $\varepsilon_2 = T_2 - Tflat2$ in decibel, and it is minimized by *lsqnonlin* which in turn yields the reflectance

$$S_{11B} = \frac{h_B(p)}{g_B(p)}$$

of the back-end equalizer of the amplifier.

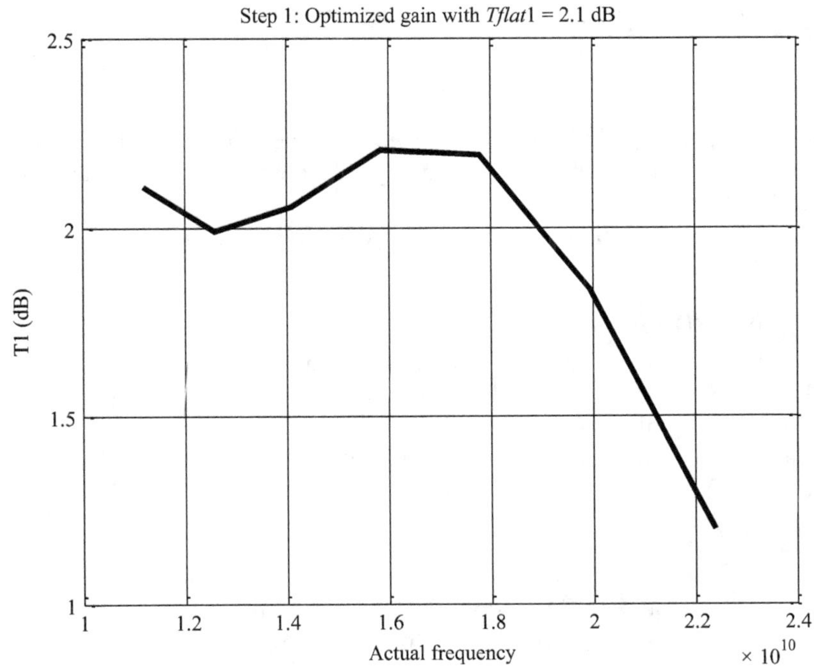

Figure 6.6 Optimized gain performance of Part I

Inputs to the main amplifier MATLAB design program "***TSMC_amp.m***" for Part II.

At this point, the main program generates the maximum stable gain at the high end of the passband which may be used as the possible target for the flat gain level for the amplifier.

For the present case, the main program returns to the designer with the following statements:

"*Suggested maximum flat gain level for the second part of the design*

$Tflat2 = 7.2059$

Enter the desired target gain value in dB for Step $2{:}Tflat2 =$ "

In this case, it may be appropriate to enter

$Tflat2 = 7.1$ dB

Then, the main program asks for the initial values for the back-end equalizer such that

"*enter back end equalizer*;

$[xB(1) = hB(1), xB(2) = hB(2), \ldots, XB(n) = hB(n)] =$ "

Here, we use an ad hoc initialization for the coefficient as

$$xB0 = [-1 \quad -1 \quad -1 \; 1 \quad -1 \; 1]$$

Meaning that, we will have $nB = length(hB) = 6$ elements in the back-end MN. We have already fixed the structure to LC ladder without ideal transformer by setting $k = 0$ and $hB0 = 0$ in advance.

Thus, the optimization returns with the following results.

6.3.2 Results of optimization

Back-end reflection coefficient is determined in Table 6.4.

Finally, $S_{11B} = h_B/g_B$ is synthesized as in Part I as shown in Figure 6.7.

The element values of the back-end equalizer are summarized in Table 6.5.

The gain performance of the amplifier is depicted in Figure 6.8 and listed in Table 6.6.

Table 6.4 Back-end reflection coefficient $S_{11B} = h_B/g_B$ (or in short S_B)

$h_{0B}=$	0	$g_{0B}=$	1
$h_{1B}=$	−5.9242	$g_{1B}=$	7.8643
$h_{2B}=$	−1.9747	$g_{2B}=$	13.375
$h_{3B}=$	−15.969	$g_{3B}=$	25.796
$h_{4B}=$	2.8824	$g_{4B}=$	20.767
$h_{5B}=$	−9.2939	$g_{5B}=$	17.764
$h_{6B}=$	6.4072	$g_{6B}=$	6.4072

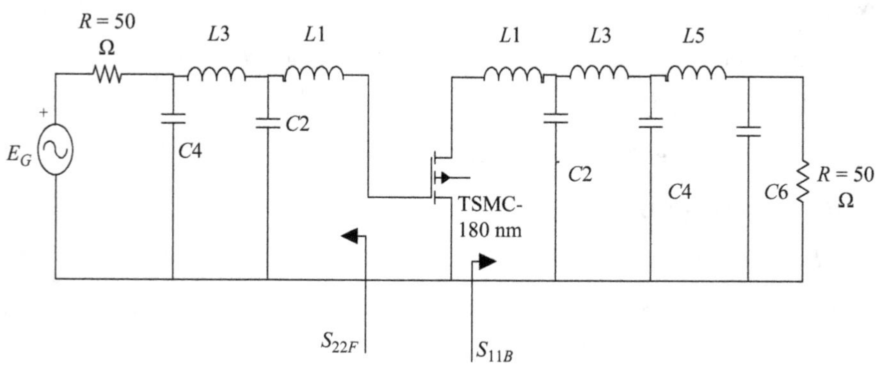

Figure 6.7 A single-stage amplifier designed using SRFT for a typical 0.18 μm NMOS transistor over 11.23–22.39 GHz

Table 6.5 Normalized and actual element values of back-end equalizer for the microwave amplifier designed employing TSMC 180 μm Si CMOS transistor

	Normalized element values (CVB)	**Actual element values (AEVB)**
$L1$(H)	0.4736	1.6834e − 010
$C2$(F)	6.9925	9.9422e − 013
$L3$(H)	0.50192	1.7841e − 010
$C4$(F)	5.2831	7.5117e − 013
$L5$(H)	0.96454	3.4285e − 010
$C6$(F)	1.5129	2.1511e − 013

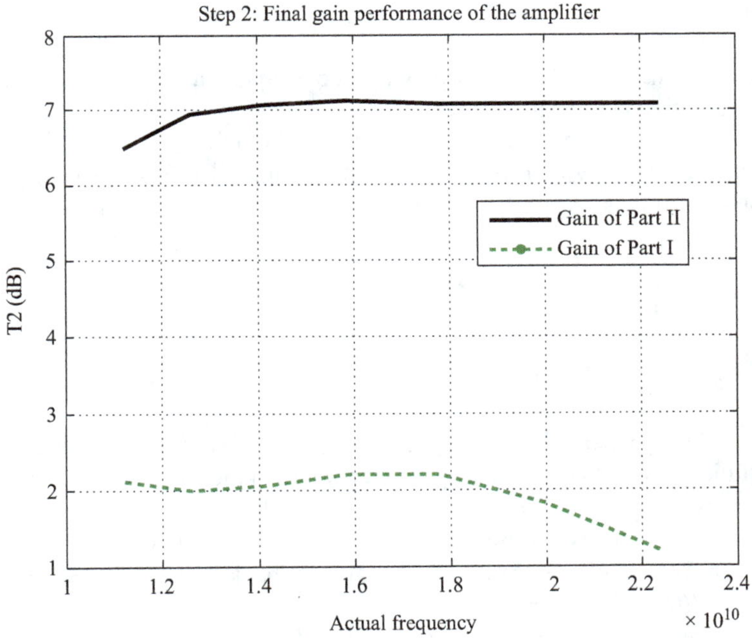

Figure 6.8 Ideal performance of the amplifier over the normalized frequencies from 0.55 (11.22 GHz) to 1 (22.39 GHz)

Table 6.6 Gain performance of the amplifier

Actual frequency	**Gain (dB)**
1.122e + 010	6.473
1.2589e + 010	6.9404
1.4125e + 010	7.0587
1.5849e + 010	7.1319
1.7783e + 010	7.0882
1.9953e + 010	7.098
2.2387e + 010	7.0987

6.4 Stability of the amplifier

In order to analyze the stability of an amplifier, the input and the output impedances of the matched amplifier must be computed.

Let $Z_{in}(j\omega) = R_{in}(\omega) + jX_{in}(\omega)$ and $Z_{out}(j\omega) = R_{out}(\omega) + jX_{out}(\omega)$ be the input and the output impedances of the matched amplifier.

Then, absolute stability of the amplifier requires that

$$R_{in}(\omega) \geq 0; \forall\omega \tag{6.9a}$$

and

$$R_{out}(\omega) \geq 0; \forall\omega \tag{6.9b}$$

Sometimes, it may be difficult to satisfy the above equation. In this case, we may be contended with conditional stability.

In general, let the amplifier be terminated in complex generator with internal impedance $Z_G = R_G(\omega) + jX_G(\omega)$ and a complex load $Z_L(j\omega) = R_L(\omega) + jX_L(\omega)$.

Then, conditional stability requires that

$$R_G(\omega) + R_{in}(\omega) \geq 0; \forall\omega \tag{6.10a}$$

and

$$R_L(\omega) + R_{out}(\omega) \geq 0; \forall\omega \tag{6.10b}$$

For the problem under consideration, $S_G = (Z_G - 1)/(Z_G + 1)$ is the input reflectance of the front-end equalizer which is generated in the first part of the amplifier design process as $SF = hF/gF$.

Similarly, complex termination $S_L = (Z_L - 1)/(Z_L + 1)$ of the active device is generated as the input reflectance of the back-end equalizer such that $S_L = S_B = hB/gB$.

In this regard, we need to generate input and the output impedances to investigate the stability situation of the matched amplifier.

The input reflectance S_{in} is given by

$$S_{in} = f_{11} + \frac{f_{12}f_{21}}{1 - f_{22}S_B}S_B \tag{6.11a}$$

where $\{f_{ij}; i,j = 1,2\}$ are the S-parameters of the active device and

$$S_B = \frac{h_B}{g_B} \tag{6.11b}$$

is the input reflectance of the back-end equalizer. Then, the input impedance is generated from S_{in}, and it is given by

$$Z_{in} = \frac{1 + S_{in}}{1 - S_{in}} \tag{6.11c}$$

such that

$$Z_{in}(j\omega) = R_{in}(\omega) + jX_{in}(\omega) \tag{6.11d}$$

Similarly, output reflectance is as

$$S_{out} = f_{22} + \frac{f_{12}f_{21}}{1 - f_{11}S_{11F}}S_F \tag{6.12a}$$

where

$$S_F = \frac{h_F}{g_F} \tag{6.12b}$$

is the input reflectance of the front-end equalizer. Then, the output impedance is generated from S_{out} as

$$Z_{out} = \frac{1 + S_{out}}{1 - S_{out}} \tag{6.12c}$$

such that

$$Z_{out}(j\omega) = R_{out}(\omega) + jX_{out}(\omega) \tag{6.12d}$$

All the above computations are carried out employing the following statements in the main program *TSMC_amp.m*.

Investigation on the stability of the amplifier: Stability check at the output port:

$$SB11 = hoverg(hB, k, ww);$$

$$(Input\ reflectance\ of\ the\ back-end\ equalizertance\ from\ hB)$$

$$SL = SB11;$$

$$ZL = stoz(SL);$$

$$RL = real(ZL);$$

$$(complex\ termination\ of\ theactive\ device)$$

$$Sout = SF2;$$

$$Zout = stoz(Sout);$$

$$Rout = real(Zout);$$

$$(output\ reflectance\ of\ the\ active\ devices\ when\ it\ is\ terminated\ in$$

$$front - end\ equalizer\ i.e\ in\ SG = SF)$$

$$StabilityCheck_{output} = RL + Rout \geq 0;$$

$$(it\ must\ be\ always\ non - negative)$$

Stability check at the input port:

$$SF = hoverg(hF, k, ww);$$

$$SG = SF;$$

$$(complex\ input\ termination\ of\ the\ active\ device)$$

$$ZG = stoz(SG);$$

$$RG = real(ZG);$$

$$[Sin] = INPUT\_REF(SB11, f11, f12, f21, f22);$$

$$Zin = stoz(Sin);$$

$$Rin = real(Zin);$$

$$StabilityCheck\_input = RG + Rin;$$

$$(it\ must\ be\ always\ nonnegative)$$

Now, let us investigate the stability of the amplifier as designed with the above front-end and back-end equalizers.

Table 6.7 indicates that the amplifier is absolutely stable from 12.6 to 22.39 GHz, but it is conditionally stable at 11.22 GHz.

Table 6.7 Stability check of the amplifier

Frequency	**Absolute stability check**		**Conditional stability check**	
(GHz)	***Rin***	***Rout***	***RG + Rin***	***RL + Rout***
11.22	−0.0067505	0.11592	0.050926	0.22373
12.589	0.0064048	0.092387	0.059	0.22254
14.125	0.017844	0.067221	0.067412	0.23336
15.849	0.023586	0.042898	0.071946	0.24681
17.783	0.025254	0.024308	0.073142	0.24775
19.953	0.023913	0.01286	0.068118	0.2474
22.387	0.016272	0.0033594	0.048029	0.078335

Note: All the impedances given in this table are normalized with respect to $R0 = 50\ \Omega$.

6.5 Practical aspects of the design

- It has been experienced that SRFT is naturally suited to design microwave amplifiers in a sequential manner. In fact, one can even design a multistage amplifier step by step as described by Yarman and Carlin [4]. Details are omitted here. Interested readers are referred to go through the reading list provided at the end of this chapter. In this regard, the MATLAB main program *TSMC_Amplifier.m* can easily be modified to design multistage amplifiers.
- Stability of equations (6.11a)–(6.12d) can be utilized as constraints in the course of TPG optimization.
- Upon the completion of the equalizer design with ideal elements, lumped circuit components can be realized on a selected substrate.
- Physical implementation of lumped circuit elements introduces material loss and some inductive and capacitive parasitic as expected. These additions degrade the final electrical performance of the power transfer networks designed via RF other techniques. Nevertheless, once we have an idealized solution, we can always re-optimize it with the commercially available software design packages employing the circuit models provided by the foundries. For example, an ideal inductor may be replaced by a practical model which is shown in Figure 6.9.
- Similarly, an ideal capacitor acts like a passive device which may be described by a model as shown in Figure 6.10.
- Parameters of the selected models are specified by geometric layouts and material properties of the components which are provided by the manufacturers.
- In order to end up with actual electrical performance of the MNs or microwave amplifiers designed employing the RFTs, we need to replace the ideal element values with those models which include loss and parasitic of the components, the

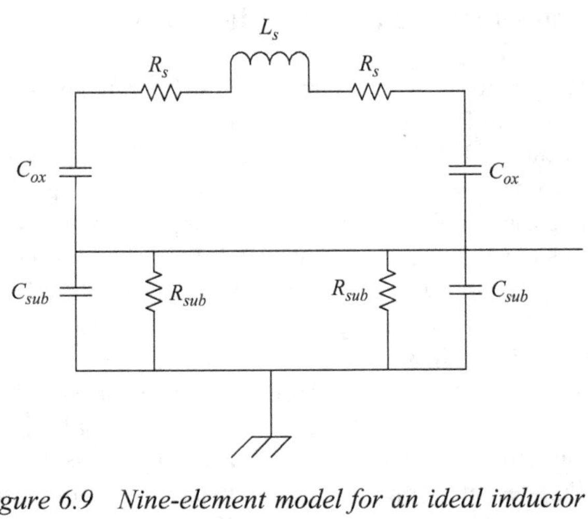

Figure 6.9 Nine-element model for an ideal inductor L_S

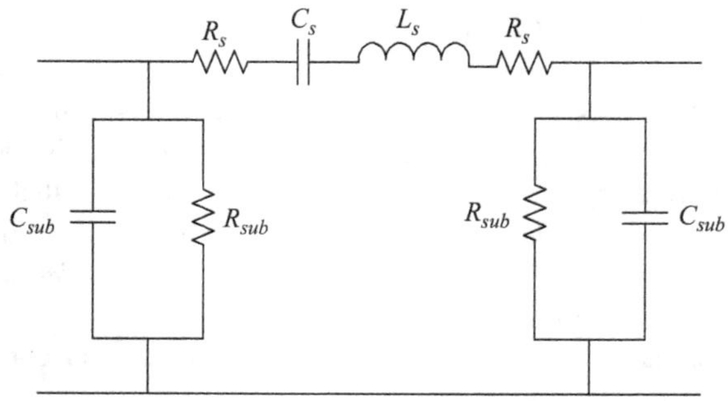

Figure 6.10 Eight-element model for an ideal capacitor C_S

way they are produced with the technology under consideration. On the other hand, degraded performance of the matched system can be improved by re-optimizing the gain of the systems on the element values of the preset models.

- It should also be noted that in practice, there exists neither an ideal lumped component nor an ideal distributed element. To be more realistic, physical passive layout may be modeled with both lumped and distributed elements which lead the engineer to design lossless two-ports with both lumped and distributed elements. From the theoretical point of view, designs consisting of mixed elements is highly complicated. However, in Chapter 13, we will present the extension of SRFT which handles mixed element designs.

This chapter is concluded with by listing all the SRFT MATLAB codes employed to run the SRFT examples presented in this section.

6.6 Design of an ultra-wideband microwave amplifier using commensurate transmission lines

Transmission lines are introduced in conjunction with the synthesis of driving point immittances described in Richards variable [5,13–15].

In the abovementioned chapter and references, it has been emphasized that, ideal transmission lines are very handy to introduce physical sizes and geometric layouts in microwave circuit design. In fact, lumped circuit components such as resistors, capacitors and inductors are realized in term of the geometric parameters of the layouts. A simple layout may be a transverse electromagnetic waves (TEM) line. Therefore, in this section, we will exhibit the design of a wideband microwave amplifier using commensurate transmission lines.

In the present case, the active device used in the microwave amplifier is a 180 nm complementary metal oxide semiconductors (CMOS) transistor [16] (Table 6.8).

The original amplifier chip consists of lumped components. In the present case, however, a similar design is made with commensurate transmission lines. The scattering parameters of the 180 nm silicon CMOS transistor is summarized in Table 6.7, and it is stored on a MATLAB text file called *newcmos.txt*

Table 6.8 $R_0 = 50\Omega$ *normalized scattering parameters of a 180 nm silicon CMOS transistor*

Frequency (GHz)	S_{11R}	S_{11X}	S_{21R}	S_{21X}	S_{12R}	S_{12X}	S_{22R}	S_{22X}
0.45	0.99	−0.05	−1.71	0.10	0.000	0.010	0.95	−0.04
0.7	1.00	−0.07	−1.71	0.15	0.001	0.015	0.94	−0.07
1	0.99	−0.10	−1.70	0.22	0.002	0.022	0.94	−0.09
2	0.97	−0.20	−1.64	0.43	0.009	0.042	0.92	−0.19
6	0.76	−0.51	−1.14	1.02	0.064	0.096	0.75	−0.49
10	0.51	−0.64	−0.56	1.19	0.127	0.101	0.54	−0.65

As in the previous example, here, we also design the microwave amplifier in two steps. In the first step, the front-end; and in the second step, the back-end equalizers are constructed. However, in this case, designs are made using the commensurate transmission lines. Therefore, the scattering parameters of the equalizers are described in Richard variable λ such that, on the RF axis $j\omega$:

$$\lambda = jtan(\omega\tau) \tag{6.13a}$$

where the constant delay τ is fixed as

$$\tau = \frac{1}{4(f_{end} \times kLength)} \tag{6.13b}$$

kLength is a user-defined parameter which refers to the fraction of the upper edge of the passband frequency f_{end}.

The amplifier design program developed for the previous section is revised to handle the design with commensurate transmission lines. The new program is called *Amplifier_Distributed.m*.

Let us summarize the design steps as follows:

First step of the amplifier design: description of the front-end MN.

The front-end equalizer is described in terms of its unit-normalized scattering parameters (normalization number is $R_0 = 50\Omega$).

At this step, the unknown of the design problem is the numerator polynomial in complex Richards variable:

$$\lambda = \Sigma + j\Omega \tag{6.14a}$$

where

$$\Omega = \tan(\omega\tau) \tag{6.14b}$$

when $\Sigma = 0$.

In this case, back-end input reflection coefficient of the front-equalizer is given in terms of Richard variable λ such that

$$S_{22F} = \frac{h_{F(\lambda)}}{g_F(\lambda)} \tag{6.15a}$$

where the numerator polynomial is defined in λ as

$$h_F(\lambda) = h_1\lambda^{nF} + h_2\lambda^{(nF-1)} + \ldots + h_{nF} + h_{nF+1} \tag{6.15b}$$

On the other hand, the transfer scattering parameter of the equalizer may be selected by the designer as

$$S_{21F}(\lambda) = \frac{\lambda^{qF}(1-\lambda^2)^{kF/2}}{g_F} \tag{6.16}$$

The above practical form of the transfer scattering parameter somehow restricts the lossless two-port topology such that it consists of k_F number of cascade connection of unit elements.[6] q_F specifies the total number of DC transmission zeros located at $\lambda = 0$ (or equivalently $\Omega = 0$ or $\omega = 0$). DC transmission zeros are realized either as open stubs in a series configuration (which is called Richards series capacitor) or short stubs in (which is called Richards inductor).

Once, the coefficients $h_F(\lambda)$ is initialized, strictly Hurwitz polynomial $g_F(\lambda)$ is generated by the losslessness condition as in (5.39).

Based on the above nomenclature, we set the following inputs to the MATLAB main program *Amplifier_Distributed.m*

Passband subject to optimization:

$$f_{Start} = f_{C1} = 1 \text{ GHz}, f_{end} = f_{C2} = 10 \text{ GHz}$$

where
f_{start} *is the beginning frequency of the optimization,*
f_{end} *is the high end of the passband where the optimization ends,*
$q_F = 0$ *is the count of transmission zeros at DC*
(it means that there is no transmission zero at DC),
$k_F = 6$ *is the total number of cascaded sections,*
$n_F = 6$ *is the total number of elements in the front–end equalizer.*

It should be noted that, by choosing $n_F = k_F$, we avoid using series short or open shunt stubs as described in Chapter 4. This fact makes the implementation easy. Thus, the front-end equalizer will consist of six cascade connections of unit elements.

$ntr = 0$ *means that no ideal transformer is used in the design*

$kLength = 0.37$; *this is our choice to fix the constant delay of UE*

$Tflat1 = 6.3\ dB$; *idealized flat gain level subject to optimization in the first part*

$xF = [-1\ -1\ -1\ \ 1\ -1\ \ 1]$;*initials for the optimization*

Here, it should be noted that the unknown of the front-end equalizer is the numerator polynomial $h_F(\lambda) = h_1\lambda^{nF} + h_2\lambda^{(nF-1)} + \ldots + h_{nF} + h_{nF+1}$. In other

[6] A unit element is a single ideal TEM transmission line with a fixed commensurate physical length such that $l = \tau v_p$. In this representation, v_p designates the propagation velocity of the TEM waves in the selected transmission medium.

words, by using MATLAB polynomial convention, at the beginning of the optimization, we set the numerator polynomial as

$$h_{F0} = [xF \;\; h_{F+1}] \;\; = [xF \;\; 0]$$

As you have noticed, in the above equation, we automatically set $h_{F+1} = 0$ since we desire no ideal transformer in the front-end equalizer. Actually, in order to simplify the main program, we set $h_{nF+1} = 0.0$ in advance. In this case, simplified version of the transfer scattering parameter is give as

$$S_{21F}(\lambda) = \frac{(1-\lambda^2)^{kF/2}}{g_F}$$

At this step, $error = T_1(\omega) - T_{flat1}$ which is generated by *function* ***Func = Distributed_TR1(x, kLength, qF, kF, w, f11, f12, f21, f22, Tfalt1)*** is minimized. List of this function together with the main program *Amplifier_Distributed.m* is given at the end of this chapter.

Second step of the amplifier design: description of the back-end MN.

In this step, back-end equalizer is described in terms of its input scattering parameters as

$$S_{11B} = \frac{h_{B(\lambda)}}{g_B(\lambda)}$$

where

$$h_B(\lambda) = h_1\lambda^{nB} + h_2\lambda^{(nB-1)} + \ldots + h_{nB} + h_{nB+1}$$

The transfer scattering parameter is specified by

$$S_{21B}(\lambda) = \frac{\lambda^{qB}(1-\lambda^2)^{kB/2}}{g_B}$$

Similarly, the denominator polynomial $g_B(\lambda)$ is generated via losslessness of the back-end equalizer as in (5.39).

At this step, error $error = T_2(\omega) - T_{flat2}$ which is generated by *function* ***Func = Distributed_TR2(x, kLength, qB, kB, w, T1, SF2, Tflat2)*** is minimized. List of this function together with the main program *Amplifier_Distributed.m* and other related MATLAB functions are given at the end of this chapter.

Now, let us use the following inputs for the back-end equalizer design.

Passband subject to optimization:

$$f_{Start} = f_{C1} = 1 \text{ GHz}, \; f_{end} = f_{C2} = 10 \text{ GHz}.$$

$q_B = 0$; *Total count of transmission zeros of at DC*
$k_B = 6$; *Total number of cascaded sections*
$n_B = 6$; *Total number of elements in the backend matching network*
$ntr = 0$; *No ideal transformer*

(Actually, the main program *Amplifier_Distributed.m* is organized in such a way that it makes designs without transformer automatically. Therefore, a user of the program does not need to bother with *ntr*. In this regard, h_{F+1} is also fixed at zero (i.e., $h_{F+1} = 0$).)

$kLength = 0.37$

$Tflat2 = 10.5\ dB$; *Idealized flat gain level subject to optimization in the first part*

$xB = [-1 - 1 - 1\ 1 - 1\ 1]$; *initials for the optimization*

As explained above, initials for the polynomial for $h_B(\lambda)$ is expressed in the main program as

$$h_{B0} = [xB\ \ 0]$$

6.6.1 Result of optimization

In the first step, optimization of the TPG $T_1(\omega)$ yields the results summarized in Table 6.9.

In the second step, optimization of the TPG $T_2(\omega)$ reveals the following results (Table 6.10).

Gain performance of Step 1 is depicted in Figure 6.11.

Overall gain performance is depicted in Figure 6.12.

The schematic of the amplifier is shown in Figure 6.13.

Table 6.9 Result of optimization of Step 1

Tflat1 = 6.3 dB **hF**	**kLength = 0.37** **gF**	**nd1 = 51** **freq(51) = 1 GHz** **zF(Normalized)**	**nd2 = 101** **freq(101) = 10 GHz** **ZFActual(Ω)**
−15.65	15.682	1.8171	Z1F = 90.854
42.125	63.873	2.6136	Z2F = 130.68
8.5322	64.792	0.5207	Z3F = 26.035
27.25	57.638	3.1297	Z4F = 156.49
4.1329	26.387	0.52992	Z5F = 26.496
2.4754	8.0561	1.9205	Z6F = 96.026
0	1	1	R0 = 50

Table 6.10 Result of optimization of Step 2

Tflat2 = 10.5 dB **hB**	**kLength = 0.37** **gB**	**nd1 = 51** **freq(51) = 1 GHz** **zB(Normalized)**	**nd2 = 101** **freq(101) = 10 GHz** **ZBActual**
5.0982	5.1954	5.7794	Z1B = 288.97
81.673	83.839	0.83948	Z2B = 41.974
23.821	57.315	4.0919	Z3B = 204.60
54.973	70.825	1.2459	Z4B = 62.293
7.3537	25.852	1.1117	Z5B = 55.587
5.8952	9.6154	2.4422	Z6B = 122.11
0	1	1	R0 = 50

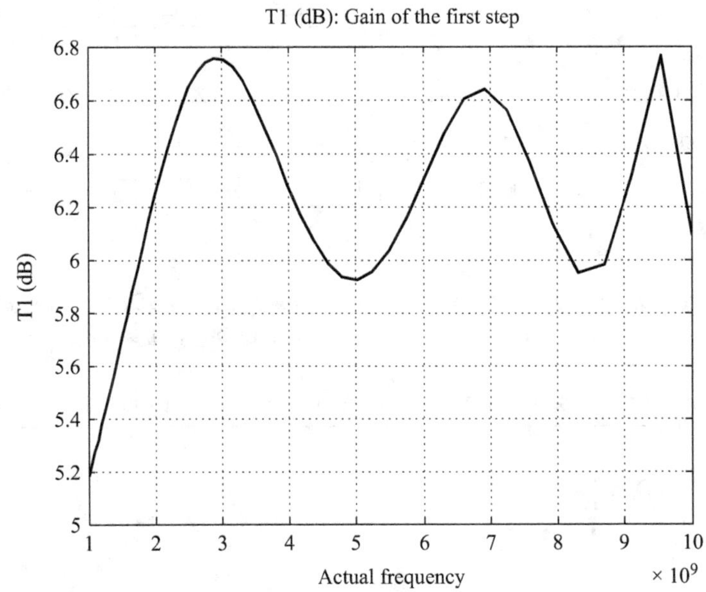

Figure 6.11 Gain performance of Step 1

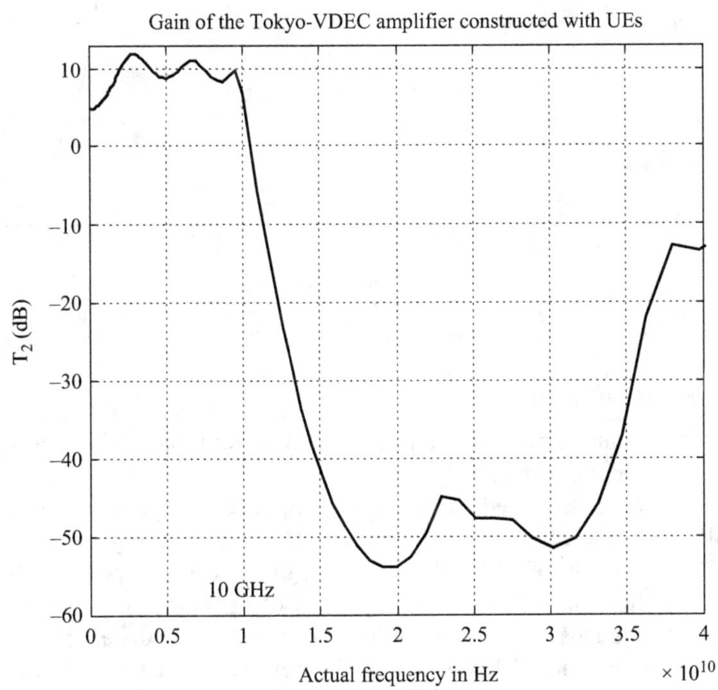

Figure 6.12 Gain performance of the overall amplifier constructed with commensurate transmission lines

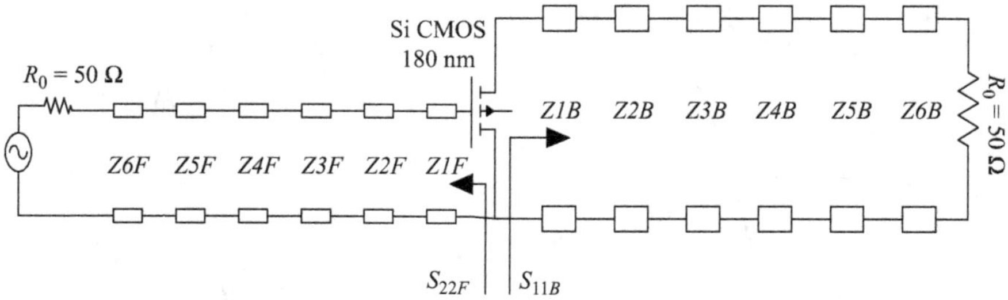

Figure 6.13 Design of a microwave amplifier using Si CMOS of 180 nm of VDEC of Tokyo University with TEM commensurate transmission lines

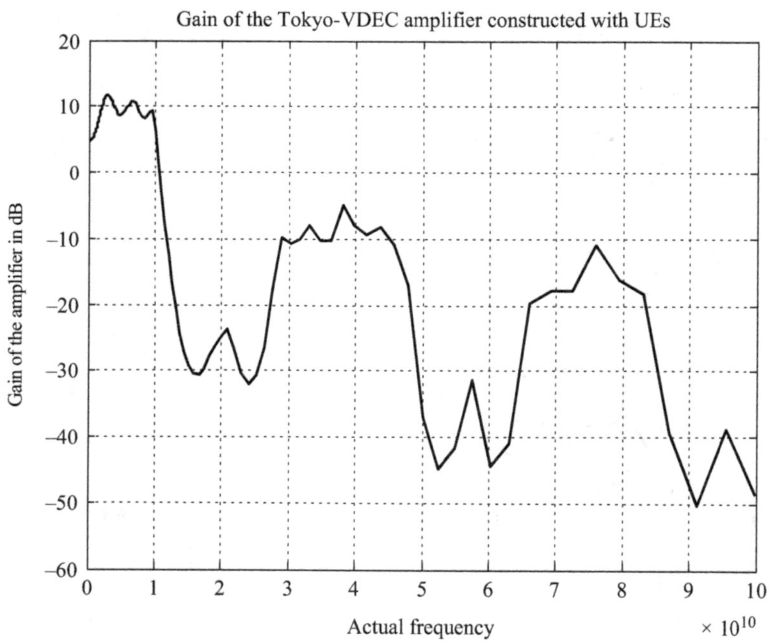

Figure 6.14 Gain performance of the amplifier designed for length = 0.3

6.6.2 Practical notes

Using the same transistor and the same passband, a similar amplifier was designed employing lumped elements [16].

It is observed that lumped element design offers better gain performance over the amplifier design with UEs as expected. The reason being lumped element design offers a smooth roll off in the passband. In other words, power delivered to the load is confined under the frequency band of operation, whereas the design made with UEs spreads the gain somewhat periodically over the entire frequencies with parasitic attenuation. This is due to the periodicity of the Richard variable $\lambda = j\tan(\omega\tau)$. This fact is clearly shown in Figure 6.14. After 10 GHz (high end of the passband), gain starts dropping drastically and it reaches its minimum value of -55 dB at 19 GHz. Then, it starts increasing up to -15 dB at 38 GHz. Of course, the shape is not exactly periodical as opposed to filter design with UE. This is due to the presence of active device parasitic components and losses. Remember that active devices are modeled with lumped elements which suppress the gain as the frequency increase maintaining the periodic nature of Richard frequency $\Omega = \tan(\omega\tau)$.

It is noted that a TEM line in cascade configuration resembles the operation of lumped elements. For example, if a UE has low characteristic impedance, it acts like a shunt capacitor. Similarly, if a UE has high characteristic impedance, it acts like a series inductor.

In this regard, we may wish to think that, the front-end and the back-end equalizers of the amplifier designed for the above example may resemble the operation of a lowpass CLC or LCL section. On the other hand, one should also keep in mind that this is a loose statement since we know exactly what the line impedance equations are.

It is crucial to point out that, in the process of gain optimization, fixed delay length τ may as well be included among the unknowns of the design problem. It also affects the characteristic impedance of the UEs.

For example, with the above inputs to the main program *Amplifier_Distributed.m*, if we choose *KLength* = 0.3 and end up with similar gain performance but in different, characteristic impedances such that

$$ZF\ Actual = [102.48\ \ 87.361\ \ 36.289\ \ 116.77\ \ 34.321\ \ 80.041]$$

$$ZB\ Actual = [229.98\ \ 54.205\ \ 157.95\ \ 65.844\ \ 69.276\ \ 90.29]$$

Corresponding gain performance of the amplifier is shown in Figure 6.14.

Clearly, new characteristic impedances are smaller and perhaps easier to realize as microstrip lines. The reader can run the main program *Amplifier_Distributed.m* with different values of *kLength* and investigate the effect of this parameter on the gain performance as well as on the characteristic impedances.

6.7 Physical realization of characteristic impedances

Physical delay length L_{actual} of the UEs can be determined after the selection of the substrate which is described by its thickness h, dielectric constant or permittivity $\varepsilon = \varepsilon_r\varepsilon_0$ and the permeability $\mu = \mu_r\mu_0$.

In this case,

$$L_{actual} = v\tau_{Actual} = \frac{\tau_{Actual}}{\sqrt{\mu\varepsilon}}$$

Based on our description

$$\tau = \frac{1}{4 \times f_{e-normalized} \times kLength} = \frac{1}{4 \times kLength}$$

Then, the actual delay is given by

$$\tau_{Actual} = \frac{\tau}{f_e}\,\mathrm{s}$$

For example, for a silicon substrate, $\varepsilon_r = 3.8, \mu_r = 1$. Then, for *kLength* = 0.3

$$\tau = \frac{1}{4kLength} = \frac{1}{4 \times 0.3} = 0.83333$$

And

$$\tau_{Actual} = \frac{\tau}{f_e} = \frac{0.8333}{10 \times 10^9} = 0.0833 \text{ ns}$$

Thus, the actual physical length is,

$$L_{actual} = \frac{0.0833 \times 10^{-9}}{\sqrt{3.8}} \times 3 \times 10^8 = 0.0042732 \text{ m} = 4.2732 \text{ mm}$$

If UEs are realized as microstrip TEM lines, width of the lines can be determined by means of readily available formulas given for the characteristic impedance calculation as in [17–19].

For example, the main program

Microstrip_CharateristicImpedance.m

calculates the characteristic impedance Z of a given microstrip line for specified ε_r, width w and substrate thickness h. List of the program is given as follows:

Program List for *Microstrip_CharateristicImpedance.m*

```
% Main Program Microstrip_CharateristicImpedance.m
clear
% Inputs:
% er : relative dielectric constant
% W : width of track
% h : thickness of the substrate
%
er=input('Enter relative permittivity: ');
W=input('Enter width of the substrate (mm): ');
h=input('Enter thickness of the substrate (mm): ');

if (W/h<1)
    e_eff=(er+1)/2+((er-1)/2)*((1/(sqrt(1+(12*h/W))))+0.04*(1-W/h)^2);
else
  e_eff=((er+1)/2)+((er-1)/2)*(1/sqrt(1+12*h/W));
end
%
if (W/h<1)
  Z_0=60/sqrt(e_eff)*log(8*h/W+W/4*h);
else
  Z_0=(120*pi/sqrt(e_eff))*(1/(W/h+1.393+0.677*log((W/h)+1.444)));
end
%
[s,errmsg]=sprintf('Z_0 = %d',Z_0);
disp(s);
%
```

Just to orient the reader, let us run the example given next.

6.8 A hypothetical example of the calculation of characteristic impedance

Let $\varepsilon_r = 3.8, w = 0.02$ mm $= 20$ μm *and* $h = 1$ mm, the characteristic impedance Z is found as

$$Z = 226.87\Omega$$

which is a high value of characteristic impedance acts like a series inductor.

On the other hand, if $w = 1$ mm, the characteristic impedance becomes $Z = 75.3\ \Omega$.

If $w = 4$ mm, $Z = 32.7\ \Omega$.

Similarly, if $w = 10$ mm, $Z = 15.8\ \Omega$ as shown in Figure 6.15.

All the above calculations are symbolic and may not necessarily be feasible to print on a silicon chip.

We should mention that, this book is intended to provide the reader design capability of PAs with ideal circuit elements. However, just to give some insights, we have also included elementary design concepts of circuit components stems from the electromagnetic field theory. However, details and the related technologies to manufacture circuit components are out of scope of this book. Therefore, interested may search related references such as [17–19] and perhaps more.

6.9 Design of broadband multistage microwave amplifiers via SRFT

Employing SRFT, a multistage amplifier can be designed step by step as described in [4]. Referring to Figure 6.16, for the first k cascaded amplifier stages, the TPG $T_k(\omega)$ is given by (see the Appendix):

$$T_k(\omega) = T_{(k-1)}\left[\frac{|e_{21_k}|^2|l_{21_k}|^2}{|1 - e_{11_k}S_{G_k}|^2|1 - \hat{e}_{22_k}S_{l_k}|^2}\right] = T_{(k-1)} \cdot E_K(\omega^2), \quad k \geq 1 \quad (6.17)$$

where $T_{(k-1)}$, is the TPG of the first $(k-1)$ stages with resistive terminations.

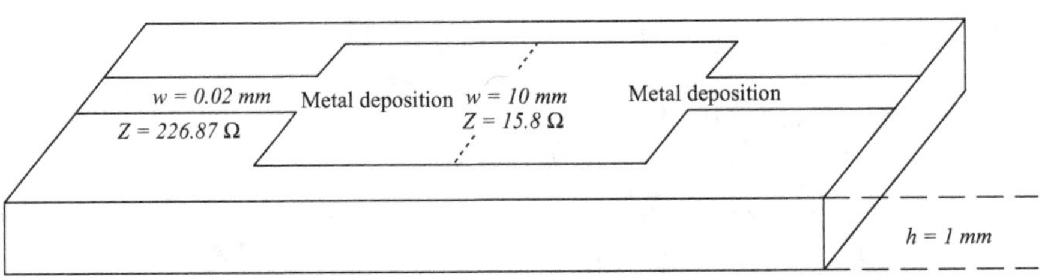

Figure 6.15 Realization of UE microstrip lines on a selected substrate with fixed actual length L_{actual} = 4,273 μm

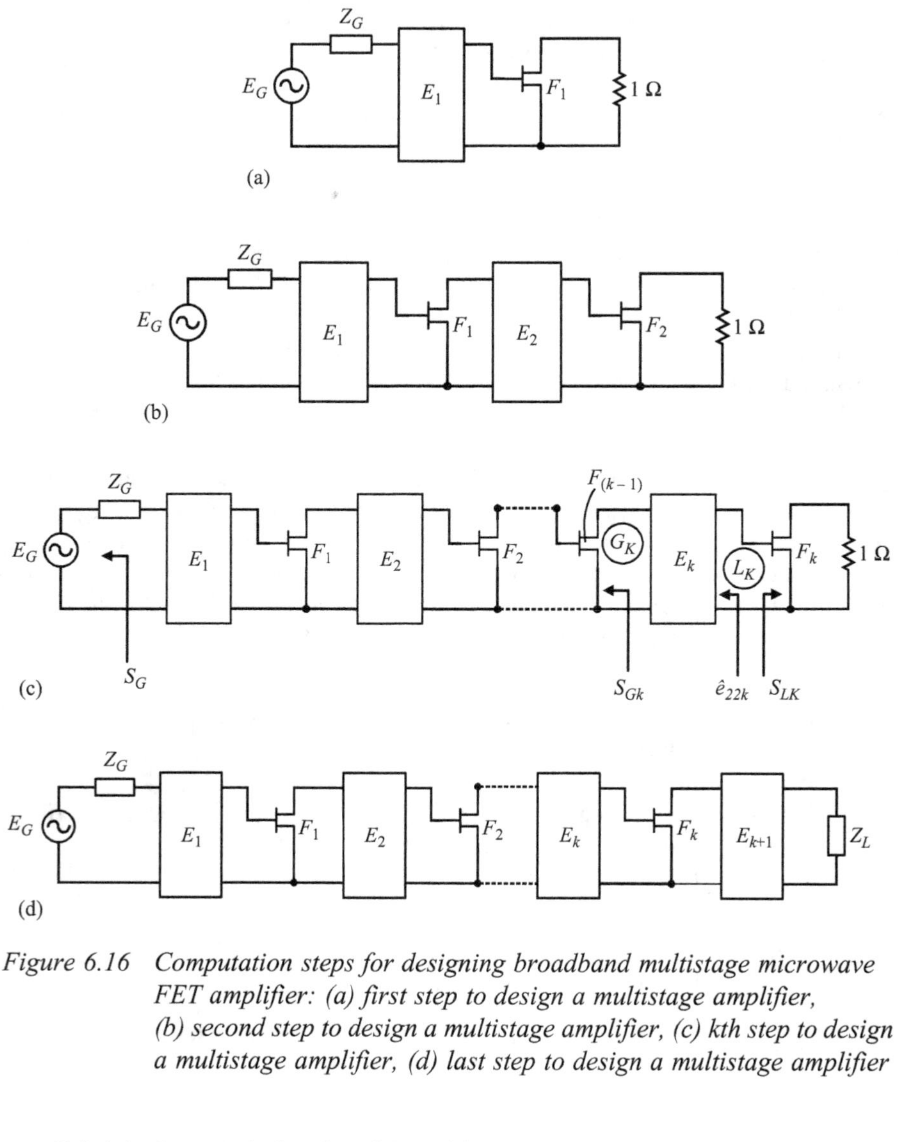

Figure 6.16 Computation steps for designing broadband multistage microwave FET amplifier: (a) first step to design a multistage amplifier, (b) second step to design a multistage amplifier, (c) kth step to design a multistage amplifier, (d) last step to design a multistage amplifier

$E_k(\omega)$ is the term in brackets [ ], and it is given by

$$E_k(\omega) = \frac{|e_{21_k}|^2 |l_{21_k}|^2}{|1 - e_{11_k} S_{G_k}|^2 |1 - \hat{e}_{22_k} S_{l_k}|^2} \tag{6.18}$$

e_{ij_k} are the unit normalized scattering parameters of the kth equalizer E_k, S_{G_k} is the unit-normalized reflection coefficient measured at port G_K to the left.

$\hat{e}_{22_k}$ and S_{l_k} are the unit normalized reflection coefficients measured at port L_K to the left and right, respectively. It is straightforward to generate $\hat{e}_{22_k}$ and S_{l_k} using the information obtained from the previous stages and the S parameter data for the

FET's. Let $\{f_{lJ_k}\}$ be the unit-normalized scattering parameters of the kth FET (F_k). For the system shown in Figure 6.16(c), above-introduced qualities are specified by

$$l_{21_k} = f_{21_k}, \qquad S_{l_k} = f_{11_k} \tag{6.19}$$

and

$$\hat{e}_{22_k} = e_{22_k} + \frac{e_{21_k}^2 S_{G_k}}{1 - S_{G_k} e_{11_k}} \tag{6.20}$$

where

$$S_{G_k} = f_{22_{(k-1)}} + \frac{f_{12_{k-1}} \cdot f_{21_{k-1}} \hat{e}_{22_{(k-1)}}}{1 - f_{11_{(k-1)}} \cdot \hat{e}_{22_{(k-1)}}}, \quad k \geq 2 \tag{6.21}$$

with

$$S_{G_1} = S_G = \frac{Z_G - 1}{Z_G + 1} \tag{6.22}$$

the known reflection coefficient of the source network.

Utilizing the SRFT described above, a multistage microwave FET amplifier may be designed as follows.

6.10 Algorithm: step-by-step multistage amplifier design

- **Step 1:** Construct the front-end equalizer $E_1 = \{e_{ij_1}\}$ for the first stage such that $T_1(\omega) = T_G \cdot E_1(\omega)$ is optimized (Figure 6.16(a)), where $T_G = 1 - |S_{G_1}|^2$.
- **Step 2:** The second equalizer-FET stage is now cascaded with the first one. The interstage equalizer $E_2 = \{e_{ij_2}\}$ is computed so that $T_2(\omega^2) = T_1 \cdot E_2(\omega)$ is optimized. Notice that in $T_2(\omega)$, $T_1(\omega)$ is regarded as a weighting function (Figure 6.16(b)).
- **Step 3:** Repeat Step 2 for designing the remaining cascaded stages (Figure 6.16(c)) up to k.
- **Step 4:** Finally, the back-end equalizer $E_{(k+1)} = \left\{e_{ij_{(k+1)}}\right\}$ determined for a given complex load $S_L(j\omega)$ so that the overall TPG $T(\omega)$ is optimized:

$$T(\omega) = (T_1 \cdot T_2 \cdots T_k) E_{(k+1)}(\omega) \tag{6.23}$$

In (6.23), the term $(T_1 \cdot T_2 \cdots T_k)$ is known from the previous stages and acts as a weighting term on $E_{(k+1)}(\omega)$, which is a function of the back-end equalizer (Figure 6.16(d)) and obtained by replacing k with "$k+1$" in $E_k(\omega)$ of (6.18). In this case,

$$S_{l_{(k+1)}} = S_L = \frac{Z_L - 1}{Z_L + 1} \tag{6.24}$$

and

$$\left| l_{21_{(k+1)}} \right|^2 = 1 - |S_L|^2 \tag{6.25}$$

In the course of the above design process, the gain taper of each FET is compensated at the port connected to the preceding equalizer in the cascade. The term $E_{(k+1)}(\omega)$ of (6.23) only provides impedance matching. It should further be noted that (6.23) exactly specifies the gain with all equalizers and FET's in place, and the *non-unilateral behavior of the FET's taken into account*. However, the design procedure can be iterated one or more times in order to improve the maximum flat gain level.

The design technique that has been discussed is applicable to optimizing a variety of objective functions as in [20]. Thus, it can be used for maximizing the minimum passband gain or for minimizing maximum NF, etc.

6.11 Examples

In the present section, we will exhibit an application of the above procedure. The following example was solved using the computer program called Main_Yarman_MultiStageDesign2018.

In the present instance, we used available data for the Mitsubishi transistor MGF 2124 FET (Table 6.11) for illustrating a three-stage design over the band 11.7–12.2 GHz.

The actual design starts by constructing a single-stage amplifier. Then a two-stage and finally three-stage FET amplifier design is completed. Throughout the examples, $R_0 = 50\Omega$ was chosen as the normalization number, and no ideal transformers are employed in the MNs. Therefore, each equalizer is selected as a lowpass ladder, with $h(0) = 0$, for each of the $e_{11}(s)$ function.

At stage k (Figure 6.16(c)), the target flat gain level, T_{0k}, is estimated as in (6.26) over the passband:

$$T_{0k} = \text{minimum of} \left\{ T_{k-1} \frac{|f_{21_k}|^2}{1 - |f_{11_k}|^2} \right\} \tag{6.26}$$

Table 6.11 Measured scattering parameters $[f_{ij}]$ for the X-band Mitsubishi transistor MGF 2124

	f_{11}		f_{21}		f_{12}		f_{22}	
Freq. MHz	**MAG**	**DEG**	**MAG**	**DEG**	**MAG**	**DEG**	**MAG**	**DEG**
11,700	0.741	71.00	0.865	−91.8	0.059	−74.8	0.532	151.8
11,800	0.748	68.20	0.883	−94.8	0.061	−76.5	0.530	149.3
11,900	0.751	65.40	0.886	−97.4	0.061	−79.6	0.528	146.9
12,000	0.755	62.30	0.895	−99.6	0.061	−82.5	0.524	144.7
12,100	0.755	58.60	0.910	−103.	0.061	−84.7	0.522	142.2
12,200	0.758	55.00	0.905	−106.	0.061	−86.1	0.520	139.4

At the last step (Figure 6.16(d)), the goal is to reach the flat gain level T_0 as

$$T_0 = \text{minimum of } T_k\left\{\frac{1}{1-|f_{22_k}|^2}\right\} \tag{6.27}$$

In the above step, the unknown coefficients h_i are simply initialized as +1 or −1.
Let us design a single-stage amplifier.

6.12 Design of a microwave power amplifier with mixed lumped and distributed elements: comparative results

In the past decade, advances in *GaN* technology have made it possible to design RF PAs with wide bandwidths and high efficiencies because of *GaN* FET's higher port impedances and lower capacitance [21]. But as the operational frequency increases and higher output power levels are required, device periphery increases, consequently port impedances decrease. Therefore, most engineers in the industry prefer to use microwave monolithic integrated circuit (MMIC) devices rather than matching the discrete FET and start designing from scratch [22–26].

Vendors of discrete FETs, like CREE, offer dies of the discrete FETs that can be used in the designs. These FET dies are usually broadband, but the problem is, as the FET's output power increases so does the number of cells, and due to this, toward the higher edge of the band device's, impedance takes a nose downfall from several ohms to few milliohms, i.e., way off from 50 Ω and more close to short on Smith chart. Matching such impedances require large microstrip structures that are not realistic at 10 GHz. Such a problem is the bases of this chapter, and CREE's CGHV1J070D, a 70 W die, has been chosen for this design. It is a wideband discrete FET that can operate from 10 MHz to 18 GHz but suffers from the same phenomena, i.e., it has load impedance of $11+j8.57$ at 1 GHz and $0.279+j1.18$ at 18 GHz. This means, matching at 1 GHz is far easier than matching at 18 GHz, which is very close to short on the Smith chart [21].

In the following sections, for CREE CGHV1J070D HEMT FET, IMN and OMN is designed over 8–10 GHz using the RF direct computational technique via parametric representation of Brune functions employing lumped, distributed and mixed elements as described in [6,7,15,27].

It is shown that MNs constructed using the lumped elements yields high power added efficiency (PAE) over the frequency band of interest. They can be realized as MMIC or with discrete lumped components. On the other hand, mixed-lumped and distributed element MNs are practical to be built with discrete components on a selected substrate and they also yield high-power efficiency over the selected band of operations.

Conventionally, it is known that, MNs designed using commensurate transmission lines are very handy to be printed as microstrip lines on commercially available substrates such as alumina, droid, silicon. However, for the present case, characteristic impedances of the lines become either too high in the order of several hundred ohms or to low such as several tenths of ohms. Therefore, they are not realizable.

In the following sections, operation class and the biasing scheme of the active device is introduced. Then, RF design of MNs and their electrical performance is summarized. Finally, it is concluded that broadband PAs in X-band are built utilizing mixed lumped and distributed element results in technically and economically optimum solutions for production.

6.12.1 Operation class of 50 W power amplifier

In designing microwave/RF PAs, operation class is crucial. In Figure 6.17, generic block diagram of the PA is shown as it is placed on the test bench.

In Figure 6.17, active device CGHV1J070D is DC biased over the resistor R1 via BIASEDTEEs X3 and X2; gate and drain terminals, respectively. IMN is placed between gate and source. Source is grounded. OMN is placed between drain and ground.

In the present case, Class AB operation is selected where the drain voltage is $V_D = -40$V and the quiescent drain current is $I_{DQ} = 535$ mA. For CGHV1J070D, drain current (I) and drain–source voltage (V) characteristics, so-called DC IV-plot is depicted in Figure 6.18.

For the above DC biasing, load-pull measured source Z_S and load Z_L impedances of the FET CGHV1J070D is given in Table 6.12.

Complex conjugate of the source and load impedances must be the input Z_{in} and the output Z_{out} impedances of the FET under consideration, respectively.

6.12.2 Design of matching networks for the power amplifier

Simplified design schematic (bias free) of the PA is depicted in Figure 6.19.

Considering the maximum power transfer over the IMN and the OMN, PA design can be completed in three steps as follows:

In the first step, IMN is designed to match $R_G = 50\ \Omega$ to the load-pull input impedance $Z_{in}(j\omega) = Z_S^*(j\omega)$. This is a single matching problem over the band of interest. Similarly, in the second step, OMN is generated to match the load-pull measured output impedance $Z_{out}(j\omega) = Z_L^*(j\omega)$ of the active device to $R_L = 50\ \Omega$. This is also a standalone single matching problem. As the result of the above steps, optimum MN topologies with component values are obtained for both IMN and OMN via RFT.

In the third step, overall electrical performance of the amplifier is re-optimized using a commercially available S/W packages such as AWR yielding the best solution.

For the problem under consideration, lumped and distributed element prototype IMN and OMN are designed employing the RFT via parametric approach [5–7,15].

In RFTs, a lossless MN is described in terms of its PR Darlington's driving point input impedance Z as shown in Figure 6.20.

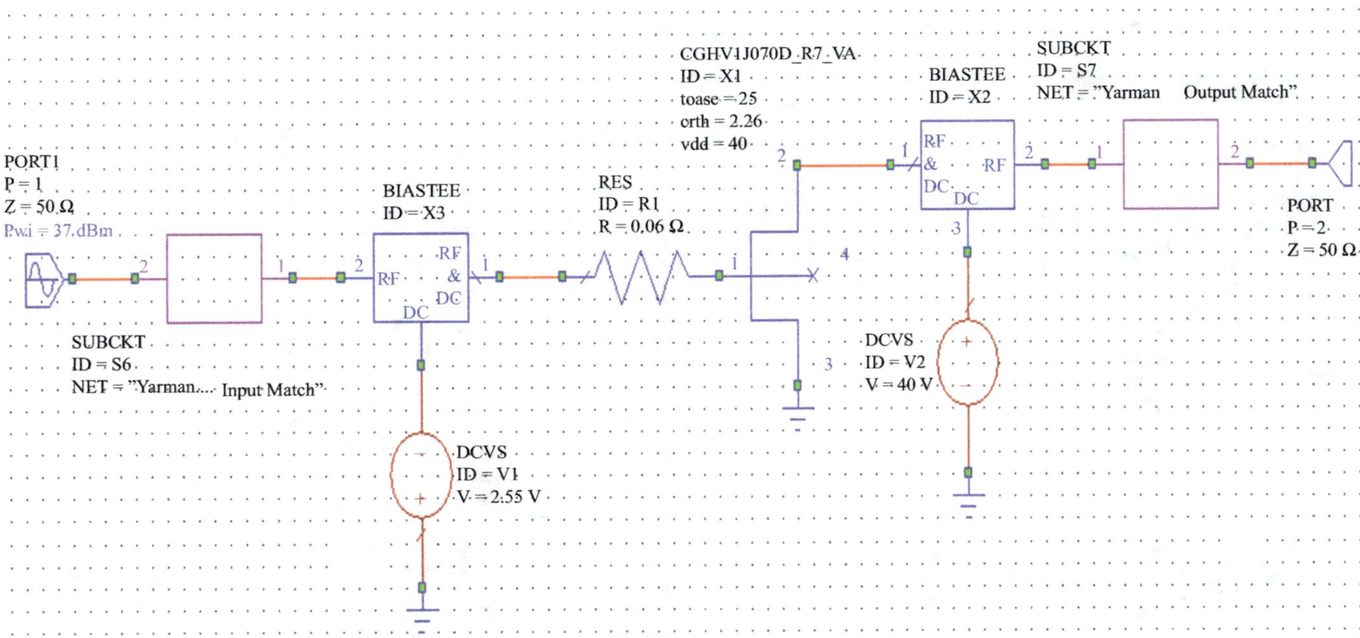

Figure 6.17 Generic block diagram of the PA as it is on the test bench

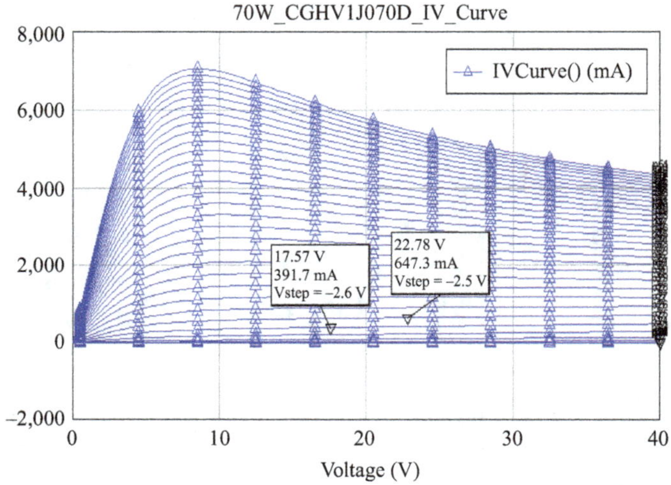

Figure 6.18 DC-IV characteristics for CGHV1J070D HEMT FET

Table 6.12 Load-pull measured Z_s and Z_L for CGHV1J070D

Fr **(GHz)**	Z_S (Ω)	Z_L (Ω)
8	$0.453+j0.190$	$0.785+j3.20$
9	$0.399+j0.096$	$0.71+j2.86$
10	$0.328+j0.010$	$0.63+j2.56$

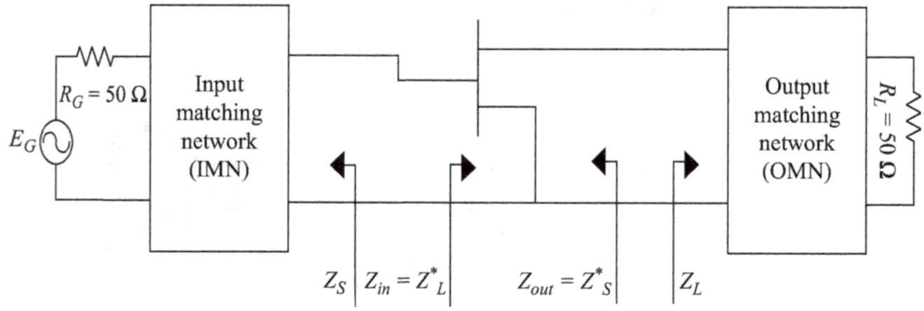

Figure 6.19 Bias free-simplified PA schematic

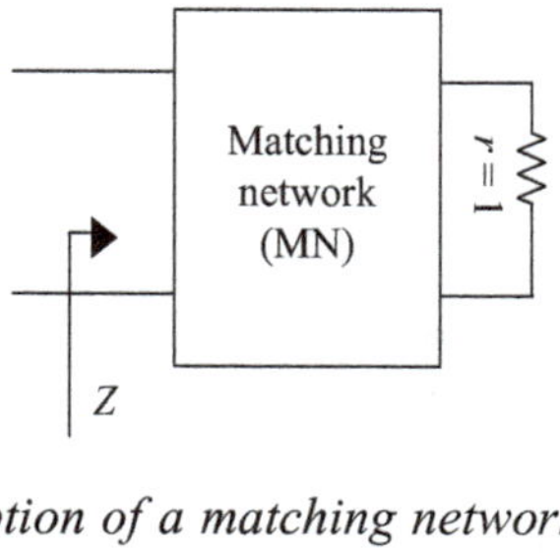

Figure 6.20 Description of a matching network in Darlington sense

In RFT-single matching problems, lumped element MN prototype is constructed by optimizing the TPG, which in turn determines the rational form of the driving point impedance $Z(p) = N(p)/D(p)$, where $p = \sigma + j\omega$ is the classical Laplace variable. Eventually, optimum MN topology is obtained as the result of the driving point impedance synthesis as in [10–12,28].

6.12.3 Design of lumped element power amplifier

In short, employing the RFT-parametric approach in Laplace domain, we end up with the broadband PA consisting of lumped elements, and the solution is re-optimized for the best performance as shown in Figure 6.21.

Element values of Figure 6.21 is given in Table 6.13.

In Figure 6.22, under the large signal operations, electrical performance of the amplifier is shown. Close examination of this figure reveals that, at 8 GHz, output power of the PA is 47.6 dBm which is about 54W and it is almost constant up to 9.5 GHz. At 10 GHz, output power drops to 46.32 dBm. On the other hand, over the band of 8.2 – 9.75 GHz, PAE is more than 50% which is acceptable for many practical applications. The large signal TPG of the PA is over 10 dB over 8–10 GHz.

Above results are generated using the nonlinear device model provided by the manufacturer.

As far as physical realization of the IMN and OMN is concerned, lumped-element PA can be constructed using either discrete components or it may be realized as MMIC. In IMN, value of $L3$ (231pH) and in OMN value of $L2$ (217pH) is considered as high-value components; therefore, they require large chip area which in turn increases the overall loss of the amplifier.

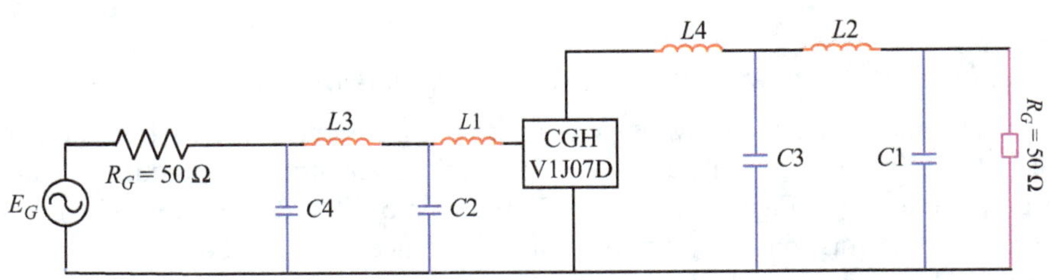

Figure 6.21 Lumped element PA designed via RFT-parametric approach

Table 6.13 Element values of the lumped element PA

Input matching network				Output matching network			
*L*1	*C*2	*L*3	*C*4	*C*1	*L*2	*C*3	*L*4
pH	pF	pH	pF	pF	pH	pF	pH
24.17	13.91	231.56	1.4	1.564	217	10.45	79.88

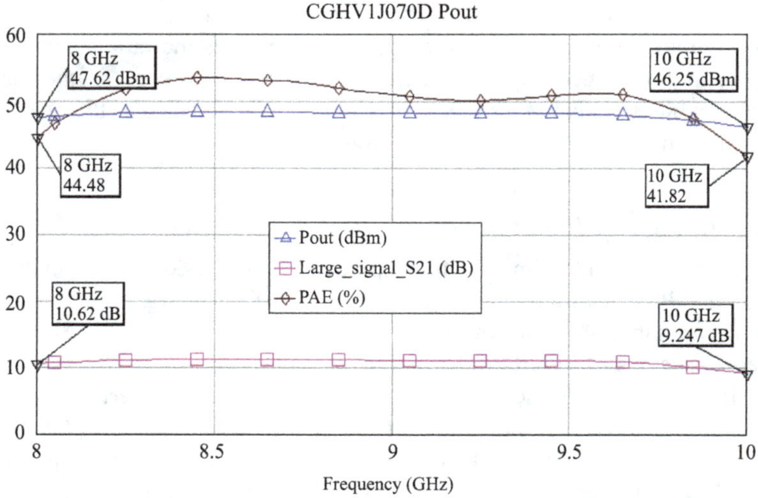

Figure 6.22 Performance of the PA constructed with lumped elements

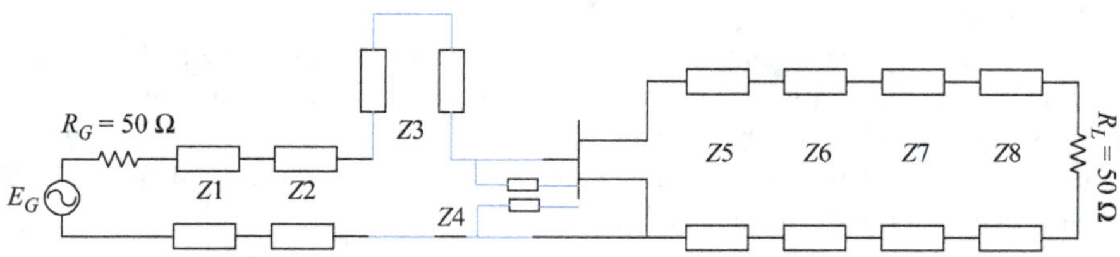

Figure 6.23 Design of PA with commensurate transmission lines

6.12.4 Design with commensurate transmission lines

For many practical situations, MNs are designed using the commensurate transmission lines. Ideal lines can be realized by employing the microstrip line-production technology. Referring to Figure 6.23, when RFT is employed to design MNs with commensurate transmission lines, driving point impedance Z of an MN is expressed in terms of Richards variable λ such that $Z(\lambda) = N(\lambda)/D(\lambda)$ where $\lambda = \tanh(p) = \Sigma + j\Omega$. As in the lumped-element design case, $Z(\lambda)$ is determined by optimizing the TPG of the matched system over the band of interest. Eventually, $Z(\lambda)$ is synthesized as a lossless two-port which is the desired MN with optimum topology. Using the large signal impedance data given by Table 6.14, PA design is completed in two steps as shown in Figure 6.23.

In the above design, electrical lengths of the commensurate transmission lines are fixed as $\theta = 1.1$ rad $= 63°$ at $f = 10$ GHz. Characteristic impedances of the lines are given in Table 6.14.

Under large signal operations, as standalone single matching, both IMN and OMN yields TPG above 0.02 dB (i.e., more than 99% power transfer at the ports)

Table 6.14 Actual characteristic impedances of the lines given by Figure 6.23

Input matching network (Ω)				Output matching network (Ω)			
Z1	Z2	Z3	Z4	Z5	Z6	Z7	Z8
1.8	3.9	15.3	45.9	4.046	61.9	63.3	80.95

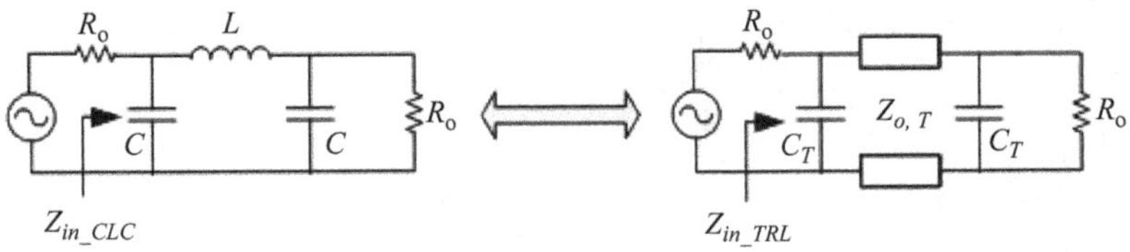

Figure 6.24 Replacement of a lowpass CLC section with a $C_T - TRL - C_T$ section

over the frequency band of 8–10 GHz. This means that electrical performance of the amplifier is supposed to be as good as the lumped element PA. Unfortunately, it is not possible to realize the low-value characteristic impedances of $Z1 = 1.8\ \Omega$, $Z2 = 3.9\ \Omega$ and $Z5 = 4\ \Omega$ with commercially available substrates. Therefore, we conclude that the design of broadband PAs in X-band using microstrip lines is not technically feasible.

6.12.5 Design with mixed lumped and distributed elements

Lumped element PA design can be transformed to a mixed element design as described above. In this approach, a lumped element lowpass symmetric *CLC* section replaced with it is almost equivalent $C_T - TRL - C_T$ section as shown in Figure 6.24.

Element values of $C_T - TRL - C_T$ is given in terms of the component values of symmetric lowpass section *CLC* such that

$$\tau = \left(\frac{1}{2\pi f_0}\right)\sin^{-1}\left(\frac{2\pi f_0 L}{Z_0}\right) \tag{6.28a}$$

$$C_T = \frac{\cos(2\pi f_0 \tau) + (2\pi f_0)^2 LC - 1}{(2\pi f_0)^2 L} \tag{6.28b}$$

In the above formulation, once a reasonable value for the line characteristic impedance is selected, delay length τ of the transmission line is calculated as in (6.28a). Then, line loading capacitors C_T are determined by (6.28b). In the above equations, f_0 is selected as the high end of the frequency band. For our case, it is 10 GHz.

Regarding the PA design problem under consideration, our lumped element prototype of Figure 6.24 can easily be converted to mixed element design. Furthermore, resulting amplifier performance can be re-optimized on a commercially available S/W package such as AWR, yielding the final mixed element PA to be realized.

In Figure 6.25, we present the realizable mixed element PA design ready to build.

Resulting large signal-electrical performance of the PA is depicted in Figure 6.26.

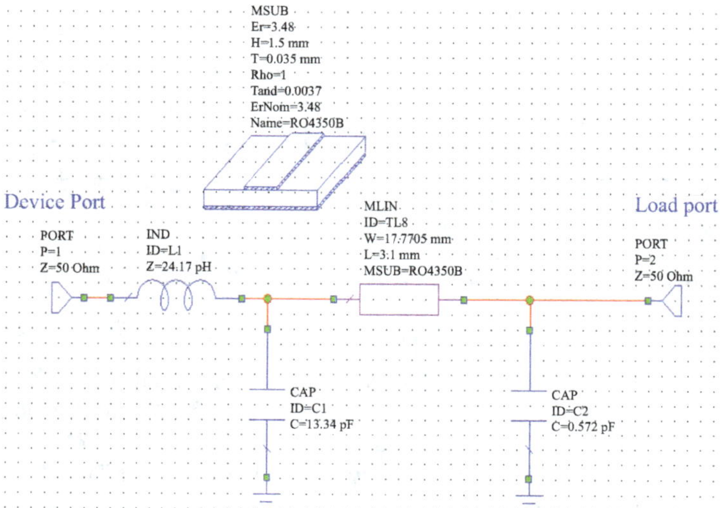

(a)

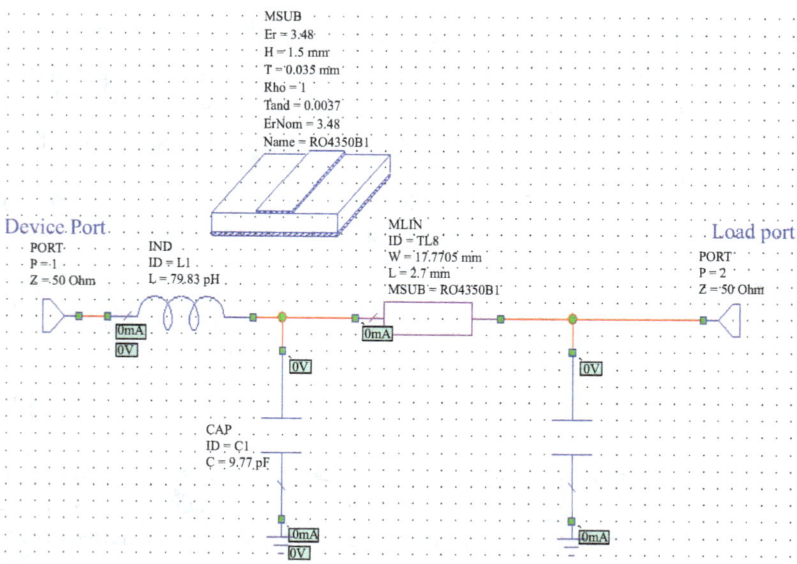

(b)

Figure 6.25 (a) Mixed element solution for the input matching network, (b) mixed element solution for the output matching network

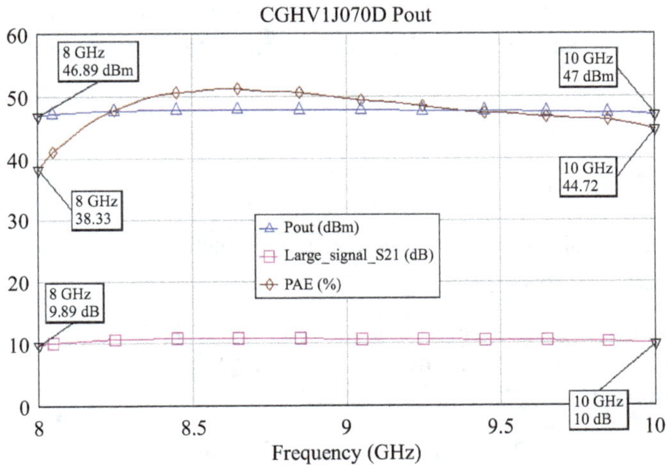

Figure 6.26 Large signal electric performance of the mixed element PA

Examination of Figure 6.26 indicates that all the component values of the mixed element design are technically feasible to be realized using discrete components. Furthermore, electrical performance of the amplifier is pretty good. Large signal output power is approximately 47 dBm over the entire frequency band which is about 50 W as desired. PAE is over 50% between 8.25 and 9.5 GHz. Large signal TPG is flat up to 10 dB.

APPENDIX

In the appendix, we show the derivation of the TPG for cascaded, multistage FET amplifiers.

First assume that the generator is resistive. This restriction will later be removed without the loss of generality.

Referring to Figure 6.16(c), let $S_{(k-1)} = \left[s_{ij_{(k-1)}}\right]$, $E_k = \left[e_{ij_k}\right]$ and $F_k = \left[f_{ij_k}\right]$ be the scattering parameters of the $(k-1)$ cascaded stages, the kth equalizer and the FET, respectively. Let us also denote the corresponding transmission matrices as $T_{k-1} = \left[t_{ij_{k-1}}\right]$, $T_{E_k} = \left[t_{ij_{E_k}}\right]$ and $T_{F_k} = \left[t_{ij_{F_k}}\right]$, respectively. All normalization numbers are the same and real.

For any k-cascaded stages, the transmission matrix $T_k = \left[t_{ij_k}\right]$ is given as

$$T_k = \hat{T} \cdot T_{F_k} \tag{A1}$$

where the transmission matrix $\hat{T}$ is

$$\begin{aligned}\hat{T} &= T_{(k-1)} T_{E_k} \\ &= \left[\hat{t}_{ij}\right].\end{aligned} \tag{A2}$$

The relation between the scattering and the transmission matrices is given in [29].

Let $T = [t_{ij}]$ be the transmission matrix of a two-port and $S = [s_{ij}]$ be its corresponding scattering matrix. Then, the entries of matrix T are given as follows:

$$\begin{aligned} t_{11} &= s_{12} + \frac{s_{11}s_{22}}{s_{21}} & t_{12} &= \frac{s_{11}}{s_{21}} \\ t_{21} &= -\frac{s_{22}}{s_{21}} & t_{22} &= -\frac{1}{s_{21}}. \end{aligned} \tag{A3}$$

It should be emphasized that the inverse of t_{22} is directly related to the TPG of the two-port $\left(|s_{21}|^2 = 1/|t_{22}|^2\right)$.

Thus, employing (A3) and evaluating (A2) and (A1) sequentially, one calculates the transmission matrix of the k-cascaded stages.

The term t_{22} in (A2) is found as

$$\begin{aligned} t_{22} &= \frac{1 - s_{22(k-1)}e_{11_k}}{s_{21_{(k-1)}}e_{21_k}} \\ &= \frac{1}{\hat{s}_{21}} \end{aligned} \tag{A4a}$$

or

$$\hat{s}_{21} = s_{21_{(k-1)}} \cdot \frac{e_{21_k}}{1 - s_{22(k-1)} \cdot e_{11_k}} \tag{A4b}$$

where $\hat{s}_{21}$ is the transfer scattering parameter of the cascaded connection of $(k-1)$ stages with the kth MN E_k.

Similarly, s_{21_k} is evaluated:

$$s_{21_k} = \hat{s}_{21}\frac{F_{21_k}}{1 - \hat{s}_{22} \cdot F_{11_k}}. \tag{A5}$$

Employing (A4b) in (A5), one obtains the expression for the TPG of the k-cascaded stages:

$$T_k = |s_{21_k}(j\omega)|^2 = |s_{21_{k-1}}(j\omega)|^2 \frac{|e_{21_k}|^2|F_{21_k}|^2}{\left|1 - s_{22_{(k-1)}}e_{11}\right|^2|1 - \hat{s}_{22_k}F_{11_k}|^2}. \tag{A6}$$

Note that throughout the text, the following notation has been used:

$$\begin{aligned} s_{22_{(k-1)}} &= S_{G_{(k-1)}} \\ \hat{s}_{22_k} &= \hat{e}_{22_k} \\ \left|s_{21_{(k-1)}}\right|^2 &= T_{(k-1)} \\ |s_{21_k}|^2 &= T_k \end{aligned} \tag{A7}$$

Assuming that the complex generator and load networks are passive one-ports, one can represent generator and load networks as lossless-Darlington two-ports with resistive termination. Then, lossless generator and load two-ports may also be included in the cascaded structure. This new structure is also considered as resistively terminated, including the complex generator and load. It should be noted that with this modification, the TPG of the overall amplifier (including complex generator and load) remains unchanged [3,30].

Let

$$G = [g_{ij}]$$

and

$$L = \left[l_{ij(k+1)}\right]$$

be the unit normalized scattering parameters associated with the lossless two-ports in the Darlington representation of the generator and load networks, respectively.

Employing (A6), at the first step of the design procedure, TPG T_1 is calculated

$$T_1 = |g_{21}|^2 \frac{|e_{21_1}|^2 |F_{21_1}|^2}{|1 - g_{22}e_{11_1}|^2 |1 - \hat{e}_{22_1}F_{11_1}|^2}$$

where $g_{22} = S_{G_1}$ is the generator reflection coefficient, and by losslessness

$$|g_{21}|^2 = 1 - |S_{G_1}|^2$$

which was defined as T_g.

Similarly, at the last step, TPG of the overall amplifier is given as

$$T = T_k \frac{\left|e_{21_{(k+1)}}\right|^2 \left|l_{21_{(k+1)}}\right|^2}{\left|1 - e_{11_{(k+1)}} S_{G(k+1)}\right|^2 \left|1 - \hat{e}_{22_{(k+1)}} F_{11_{(k+1)}}\right|^2}$$

where $\left[e_{ij(k+1)}\right]$ is the scattering parameter of the back-end MN, $l_{11_{(k+1)}} = S_L$ is the reflection coefficient of the load and by losslessness,

$$\left|l_{21_{(k+1)}}\right|^2 = 1 - |S_L|^2$$

$e_{22_{(k+1)}}$ and $S_{G(k+1)}$ are then evaluated using (A9) and (A10).

Thus, we have completed the derivations of the TPGs for all the $(k+1)$ steps of the design procedure as given in Section 6.4.

References

[1] H. Carlin, "A New Approach to Gain–Bandwidth Problems," *IEEE Trans CAS*, vol. 23, no. 4, pp. 170–175, 1977.

[2] B. S. Yarman, "A Simplified Real Frequency Technique for Broadband Matching Complex Generator to Complex Load," *RCA Rev*, vol. 43, no. 9, pp. 529–541, 1982.

[3] H. Carlin and B. Yarman, "The Double Matching Problem: Analytic and Real Frequency Solutions," *IEEE Trans Circuits Syst*, vol. 30, no. 1, pp. 15–28, 1983.

[4] B. S. Yarman and H. J. Carlin, "A Simplified Real Frequency Technique is Applied to Broadband Multi-Stage Microwave Amplifiers," *IEEE MTT*, vol. 30, no. 12, pp. 2216–2222, 1982.

[5] B. Yarman, Design of Ultra Wide-Band Matching Networks via Real Frequency Techniques, London: Wiley, 2010.

[6] B. S. Yarman, A. Aksen and A. Fettweis, "An Integrated Design Tool to Construct Lossless Two-Ports with Mixed Lumped and Distributed Elements for Matching Problems," in *Symposium on Recent Advances in Microwave Tech.*, August 1991.

[7] B. S. Yarman and A. Aksen, "An Integrated Design Tool to Construct Lossless Matching Networks with Mixed Lumped and Distributed Elements," *IEEE Trans CAS*, vol. 39, no. 9, pp. 713–723, 1992.

[8] S. Darlington, "Synthesis of Reactance 4-Poles Which Produce Prescribed Insertion Loss Characteristics," *J Math Phys*, vol. 18, no. 9, pp. 257–353, 1939.

[9] H. Carlin, "Darlington Synthesis Revisited," *IEEE Trans Circuits Syst I, Fundam Theory Appl*, vol. 46, no. 1, pp. 14–21, 1999.

[10] A. Kilinc and B. Yarman, "High Precision LC Ladder Synthesis Part I: Lowpass Ladder Synthesis via Parametric Approach," *IEEE Trans Circuits Syst I, Regul Pap*, vol. 60, no. 8, pp. 2074–2083, 2013.

[11] B. S. Yarman and A. Kilinc, "High Precision LC Ladder Synthesis Part II: Immittance Synthesis with Transmission Zeros at DC and Infinity," *IEEE Trans Circuits Syst I, Regul Pap*, vol. 60, no. 10, pp. 2719–2729, 2013.

[12] B. S. Yarman, A. Aksen, R. Kopru, C. Aydın, and C. Atilla, "Computer Aided High Precision Darlington Synthesis for Real Frequency Matching," in *IEEE Benjamin Franklin Symposium on Microwave and Antenna Sub-Systems*, Philadelphia, 2014.

[13] P. I. Richards, "Resistor-Transmission Line Circuits," *Proc IRE*, vol. 36, no. 2, pp. 217–220, 1948.

[14] H. J. Carlin, "Distributed Circuit Design with Transmission Lines," *Proc IEEE*, vol. 59, no. 7, pp. 1059–1081, 1971.

[15] A. Grebennikov, N. Kumar and B. S. Yarman, "Chapter 5, Section 5.15, Design of Ultrawideband Microwave Amplifiers Using Commensurate Transmission Lines," in *Broadband RF and Microwave Amplifiers*, Boca Raton, 33487-2742 FL, USA: CRC Press, Taylor & Francis Group, 2016, pp. 219–488.

[16] B. S. Yarman, N. Retdian, N. Fujii and S. Takagi, "Gain Bandwidth Limitations of 0.18 μm Si-CMOS RF Transistor," in *ECCTD*, Sevilla, Spain, August 26–30, 2007.

[17] I. Bahl, Lumped Elements for RF Microwave Circuits, London: Artech House, 2003.

[18] I. Bahl and P. Bhartia, Microwave Solid State Circuit Design, Hoboken, NJ, USA: Wiley Interscience, 2003.

[19] K. Chang, Handbook of RF Microwave Components and Engineering, New York, NY, USA, Chichester, UK: Wiley, 2003.

[20] H. J. Carlin and J. J. Komiak, "A New Method of Broadband Equalization Applied to Microwave Amplifiers," *IEEE Trans Microw Theory Tech*, vol. 27, no. 2, pp. 93–99, 1979.

[21] R. S. Pengelley, S. M. Wood, S. T. Milligan, W. L. Sheppard and W. L. Pribble, "A review of GaN on SiC High Electron Mobility Power Transistors and MMICs," *IEEE Trans Microw Theory Tech*, vol. 60, no. 6, pp. 1764–1783, 2008.

[22] S. Guolin and R. H. Jansen, "Broadband Doherty Power Amplifier via Real Frequency Technique," *IEEE Trans Microw Theory Tech*, vol. 60, no. 1, pp. 99–11, 2012.

[23] A. Khalil, M. Liero, R. Rudolph, R. Lossy and W. Heinrich, "GaN HEMT Potential for Low Noise Highly Linear RF Applications," *IEEE Microwave Wireless Compon Lett*, vol. 18, no. 9, pp. 606–607, 2008.

[24] T. Martin, M. Reveyrand, R. Compovecchio, *et al.*, "Design of GaN Based Balanced Cascode Cells for Wide-band Distributed Power Amplifier," in *Microwave Integrated Circuit Conference (EU MIC-2007)*, 2007.

[25] V. Bridier, D. Ducatteau, M. Oliver, H. J. Simon, F. Graux and G. Dambrine, "Nonlinear Measurement Dedicated to Non Periodic Pulse Train for Radar Power Amplifier Characterization," in *IEEE International Microwave Symposium*, 2014.

[26] Cree Inc, "Research Triangle Park," wwww.wolfspeed.com, NC 27709, 2018.

[27] B. S. Yarman and A. Fettweis, "Computer Aided Double Matching via Parametric Representation of Brune Functions," *IEEE Trans Circuits Syst*, vol. 37, no. 2, pp. 212–222, 1990.

[28] B. S. Yarman, R. Kopru, N. Kumar and C. Prakash, "High Precision Synthesis of a Richards Immittance via Parametric Approach," *IEEE Trans Circuits Syst I, Regul Pap*, vol. 61, no. 4, pp. 1055–1067, 2014.

[29] B. Yarman, Design of Ultra Wideband Matching Networks via Simplified Real Frequency Technique, Netherlands: Springer, 2008.

[30] B. Yarman, Broadband matching from a complex source to a complex load, Ph.D. Thesis, Ithaca, NY: Cornell University, 1982.

Index

www.ingramcontent.com/pod-product-compliance
Lightning Source LLC
LaVergne TN
LVHW011542200926
848484LV00001B/286